Strüngmann Forum Reports

Series Editor

Julia R. Lupp, Ernst Strüngmann Forum, Frankfurt Institute for Advanced Studies, Frankfurt, Germany

This series publishes the work of the Ernst Strüngmann Forum, an independent academic initiative dedicated to advancing knowledge in basic science.

The Ernst Strüngmann Forum provides a collaborative environment for invited experts to address contemporary research challenges. Guided by an international advisory board, topics chosen are of high-priority interest to various areas of research, requiring multidisciplinary input to expand understanding of the issues. Past topics include digital ethology, emergent brain dynamics, deliberate ignorance, psychiatric genetics, agrobiodiversity, and computational psychiatry. The insights and future research strategies that emerge from a topic are disseminated through the Strüngmann Forum Report series.

Established in 2006, the Ernst Strüngmann Forum has gained a reputation as a place where intellectual "dead ends" are overcome, where novel ways of conceptualizing issues are articulated, and where new collaborations are established. For further information, see https://esforum.de

This is an Open Access book series.

Nora S. Newcombe · Ken Cheng
Editors

Challenges in Navigation Research

Mapping New Directions

Springer

Editors
Nora S. Newcombe
Department of Psychology
and Neuroscience
Temple University
Philadelphia, PA, USA

Ken Cheng
School of Natural Sciences
Macquarie University
Sydney, NSW, Australia

ISSN 3059-3786 ISSN 3059-3794 (electronic)
Strüngmann Forum Reports
ISBN 978-3-032-20562-9 ISBN 978-3-032-20563-6 (eBook)
https://doi.org/10.1007/978-3-032-20563-6

This work was supported by the Ernst Strüngmann Foundation.

This Springer imprint is published by the registered company Springer Nature Switzerland AG
The registered company address is: Gewerbestrasse 11, 6330 Cham, Switzerland

Preface

Science is a highly specialized enterprise—one that enables areas of inquiry to be minutely pursued, establishes working paradigms and normative standards, and supports rigor in experimental research. All too often, however, researchers encounter problems that fall outside the scope of their immediate area, and to progress, new insights and conceptual frameworks are needed.

The Ernst Strüngmann Forum was established in 2006 to address such complex, cross-disciplinary issues. Founded on the principles of scientific independence and open discourse, we provide a platform for experts to scrutinize topics that require input from multiple fields.

At the core of our philosophy lies the following generative approach: A proposal, submitted by leading members of the scientific community, outlines the overarching topic and defines aspects of the problem to be explored. Following a review process, and once the proposal is accepted, we convene a steering committee to transform it into a framework that will support a multifaceted discourse among researchers from diverse disciplines. The committee delineates thematic areas for each of the four working groups and identifies potential participants for the central gathering, or *Forum*. The committee also determines what background information will be needed to enable effective dialogue. This material takes the form of invited *background papers*, which replace traditional keynote addresses and initiate discussion in advance.

A Forum is best envisioned as a week-long brain-storming activity. Existing perspectives are questioned, knowledge gaps are exposed, and strategies to address those gaps are proposed. Throughout, contentious issues and ideas for future research are highlighted. To ensure that the essence of each working group's discussions is not lost, *draft reports* are written during the week.

After a Forum, efforts shift toward refinement and synthesis. The background papers are revised to incorporate feedback from the discussions and to address peer-review comments. The draft reports are finetuned to ensure a coherent synthesis of these multifaceted discussions. The results of the entire project are then disseminated through the *Strüngmann Forum Report* series.

This volume is the culmination of a proposal put forth by Nora Newcombe and Ken Cheng "to assist the study of navigation across taxa, stages of development, and levels of description—in robotic as well as biological agents—and to integrate multidisciplinary concepts into a usable toolbox." After its approval, Kate Jeffery and Thomas Wolbers joined us on the steering committee to construct a framework for discussion. To bridge existing gaps across taxa, levels, and methods of study, four primary areas were defined for the working groups:

- Integration across levels (from cells and circuits to brains and behavior)
- Integration across model systems (e.g., cross-species, multiple tasks)
- Brain systems beyond the hippocampus (emotion, planning, memory storage)
- From fundamental science to the clinic

Then, from December 1–6, 2024, participants gathered in Frankfurt for lively debate. As delineated in Chapter 1, initial questions were posed for the groups to consider, and eleven invited background papers opened up the discussion. This volume synthesizes the ideas and perspectives that emerged from the entire process.

Engaging in this type of discourse requires the ability to question one's own perspectives even as one probes those of others. It is an intense experience, and I would like to thank each person who participated for their time, effort, and willingness to contribute. A special word of gratitude goes to the members of the organizing committee as well as to the authors of the background papers. In addition, I would like to highlight the efforts of the group moderators—Nora Newcombe, Nachum Ulanovsky, Sepiedeh Keshavarzi, and Thomas Wolbers—as well as the rapporteurs—Freyja Ólafsdóttir, Sylvia Wirth, Zita Patai, and Coco Newton. To enable and guide a lively debate and transform it into a coherent, multiauthor report is never a simple matter. Finally, I wish to extend my appreciation to Nora Newcombe and Ken Cheng for their efforts in proposing and guiding this topic.

Good science requires collaboration and steadfast support, and I wish to gratefully acknowledge the essential partnerships that sustain our work. The generous support of the Ernst Strüngmann Foundation, established by Dr. Andreas and Dr. Thomas Strüngmann in honor of their father, enables us to conduct our work in the service of science and society. The expertise of our scientific advisory board is invaluable in reviewing proposals and guiding our scientific program. I also wish to thank the Ernst Strüngmann Institute for sharing its vibrant setting with us during the Forum.

It is never easy to expand the boundaries of knowledge, as long-held views can be difficult to set aside. Yet once we recognize our limitations, the process of developing strategies to move forward can be invigorating. Such was the spirit of this 38th Ernst Strüngmann Forum.

The ability to navigate is fundamental to many species throughout the life course. In humans, this ability is often compromised with age, creating challenges and imposing significant financial costs on individuals and society alike. Sound science is needed to address these challenges. On behalf of all involved in this Forum, I hope this volume will inspire further action toward this goal.

Julia R. Lupp, Director, Ernst Strüngmann Forum
Frankfurt Institute for Advanced Studies
Ruth-Moufang-Str. 1, 60438 Frankfurt am Main, Germany
https://esforum.de/

Contents

Preface v
Julia R. Lupp

List of Contributors ix

List of Acronyms and Abbreviations xiii

1 **Addressing the Challenges of Navigational Research** 1
Nora S. Newcombe and Ken Cheng

Universal Navigational Principles

2 **Taking a Functional Look at Bird Navigation** 11
David J. Pritchard

3 **How Roboticists Think about Navigation and Major Unsolved Challenges** 33
Michael Milford

4 **Universal Principles of Navigation across Species and Spatial Scales** 63
Sylvia Wirth, Ken Cheng, Pauline Fleischman, Kenneth J. Lohmann, Michael Milford, Henrik Mouritsen, David J. Pritchard, Benjamin Robira, Nachum Ulanovsky, and Kirsty Wan

Linking Cells and Circuits to Brains and Behavior

5 **What Are the Contributions of Modeling to Bridging the Gaps between Methods and across Species?** 139
Andrej Bicanski

6 **Do Data from Intracranial Techniques Align (or Not) with What We Have Learned about Human Navigation from fMRI?** 163
Philipp K. Büchel and Lukas Kunz

7 **What Do We Know about Navigation Development from the Perspective of Cognitive Neuroscience in Humans?** 189
Daniel D. Dilks, Rebecca J. Rennert, and Yaelan Jung

8 **How Can Development Inform Us about the Neural Mechanisms of Navigation in Rats?** 201
Thomas J. Wills and Francesca Cacucci

9 **Spatial Navigation in Humans and Rodents: Bridging Levels of Analyses** 221
H. Freyja Ólafsdóttir, Andrej Bicanski, Sarah H. Creem-Regehr, Daniel D. Dilks, Flavio Donato, Russell A. Epstein, Emily K. Farran, Joshua Jacobs, Lukas Kunz, Thomas J. Wills, and Nora S. Newcombe

Navigation in a Wider Perspective

10 **Planning and Imagination in the Navigational Domain:The Role of the Hippocampus and Prefrontal Cortex** 259
Shantanu P. Jadhav

11 The Role of Motivation and Emotion in the Hippocampal Navigation System **273**
Thomas Forro and Stéphane Ciocchi

12 The Brain's Extended Navigation Network: Insights from Single-Neuron Studies **291**
E. Zita Patai, Thackery I. Brown, Stéphane Ciocchi, Michael Goard, Shantanu P. Jadhav, Sepiedeh Keshavarzi, Julija Krupic, Jean Laurens, Koen Vervaeke, and Kate Jeffery

Translating Knowledge into Practice

13 How Does Aging Impact the Structure and Function of Parallel Navigation Circuits across Mammals? **321**
Carol A. Barnes, Sahana Srivathsa, Paul F. Hill, Sabina Srokova, and Arne D. Ekstrom

14 How Do Aging and Neurodegenerative Diseases Affect Spatial Navigation at the Behavioral Level? **379**
Jan Wiener and Michael Hornberger

15 Designing Effective Clinical Tools for Measuring Navigation Deficits Associated with Cognitive Impairment: Insights from Human-Computer Interaction Research **399**
Victor R. Schinazi, Giorgio Colombo, and Karolina Minta

16 From Fundamental Science to the Clinic (and Back Again): A Translational Framework for Navigation and Aging **431**
Coco Newton, Carol A. Barnes, Virginia L. Flanagin, Anna K. Gillespie, Michael Hornberger, Scott Moffat, Victor R. Schinazi, Vladislava Segen, Jan Wiener, Steven M. Weisberg, and Thomas Wolbers

Subject Index **469**

Strüngmann Forum Reports **479**

List of Contributors

Barnes, Carol A. Department of Psychology, Evelyn F. McKnight Brain Institute, University of Arizona, Tucson, AZ 85724, USA

Bicanski, Andrej Neural Computation Group, Max-Planck Institute for Human Cognitive and Brain Sciences, Department of Psychology, 04103 Leipzig, Germany

Brown, Thackery I. School of Psychology, Georgia Institute of Technology, Atlanta, GA, 30332, USA

Büchel, Philipp K. Department of Epileptology, University Hospital Bonn, 53127 Bonn, Germany

Cacucci, Francesca Department of Neuroscience, Physiology and Pharmacology, University College London, London WC1E 6BT, UK

Cheng, Ken School of Natural Sciences, Macquarie University, Macquarie Park, NSW 2109 Australia

Ciocchi, Stéphane Laboratory of Systems Neuroscience, Department of Physiology, University of Bern, CH-3012 Bern, Switzerland

Colombo, Giorgio Future Health Technologies, Singapore-ETH Centre, Campus for Research Excellence and Technological Enterprise (CREATE), Singapore 138602, Singapore; Chair of Cognitive Science, ETH Zürich, 8092 Zürich, Switzerland

Creem-Regehr, Sarah H. Department of Psychology, University of Utah, Salt Lake City, UT 84112, USA

Dilks, Daniel D. Department of Psychology, Emory University, Atlanta, GA 30322, USA

Donato, Flavio Biozentrum, University of Basel, 4056 Basel, Switzerland

Ekstrom, Arne D. Department of Psychology, Evelyn F. McKnight Brain Institute, University of Arizona, Tucson, AZ 85724, USA

Epstein, Russell A. Department of Psychology, University of Pennsylvania, Philadelphia, PA 19104, USA

Farran, Emily K. School of Psychology, University of Surrey, Guildford, Surrey GU2 7XH, UK

Flanagin, Virginia L. German Centre for Vertigo and Balance Disorders (DSGZ), LMU Klinikum, 81377 Munich, Germany

Fleischman, Pauline N. Institute of Biology and Environmental Sciences, Carl von Ossietzky Universität Oldenburg, 26129 Oldenburg, Germany; Research Center Neurosensory Science, Carl von Ossietzky Universität Oldenburg, 26111 Oldenburg, Germany

Forro, Thomas Laboratory of Systems Neuroscience, Department of Physiology, University of Bern, CH-3012 Bern, Switzerland

Gillespie, Anna K. Depts. of Neurobiology and Biophysics, Lab Medicine and Pathology, University of Washington, Seattle, WA 98195, USA

Goard, Michael Department of Molecular Cellular and Developmental Biology, Department of Psychological and Brain Sciences, University of California, Santa Barbara, Santa Barbara, CA, 93106, USA

Hill, Paul F. Department of Psychology, University of Arizona, Tucson, AZ 85724, USA

Hornberger, Michael Clinical Neurosciences, Department for Clinical and Experimental Sciences, Faculty of Medicine, University of Southampton, SO14 3DT, UK

Jacobs, Joshua Department of Neurology, University of Chicago, Chicago, IL 60637, USA

Jadhav, Shantanu P. Neuroscience Program, Department of Psychology, and Volen National Center for Complex Systems, Brandeis University, Waltham, MA 02453, USA

Jeffery, Kate School of Psychology and Neuroscience, College of Medical, Veterinary and Life Sciences, University of Glasgow, Glasgow G12 8QB, UK

Jung, Yaelan Department of Psychology, Emory University, Atlanta, GA 30322, USA

Keshavarzi, Sepiedeh Department of Physiology, Development and Neuroscience, University of Cambridge, Cambridge CB2 3EG, UK

Krupic, Julija UCL Dementia Research Institute, University College London, London WC1E 6BT, UK

Kunz, Lukas Department of Epileptology, University Hospital Bonn, 53127 Bonn, Germany

Laurens, Jean Ernst Strüngmann Institute (ESI) for Neuroscience in Cooperation with Max Planck Society, 60528 Frankfurt am Main, Germany

Lohmann, Kenneth J. Department of Biology; University of North Carolina; Chapel Hill, NC 27599; USA

Milford, Michael QUT Centre for Robotics and School of Electrical Engineering and Robotics, Queensland University of Technology, Brisbane QLD 4000, Australia

Minta, Karolina Future Health Technologies, Singapore-ETH Centre, Campus for Research Excellence and Technological Enterprise (CREATE), Singapore 138602, Singapore

Moffat, Scott Department of Psychology, Georgia Institute of Technology, Atlanta, GA 30332, USA

Mouritsen, Henrik Carl von Ossietzky Universität Oldenburg, Institute of Biology and Environmental Sciences, 26129 Oldenburg, Germany

Newcombe, Nora S. Department of Psychology, Temple University, Philadelphia, PA 19122, USA

Newton, Coco Institute of Cognitive Neuroscience, University College London, London WC1E 6BT, UK; Engineering Design Centre, University of Cambridge, Cambridge CB2 1PZ, UK

Ólafsdóttir, H. Freyja Donders Institute for Brain, Cognition and Behaviour, Radboud University, 6525 EN Nijmegen, The Netherlands

Patai, E. Zita Department of Neuropsychology, Institute of Cognitive Neuroscience, Ruhr-Universität Bochum, IB 6/185 Bochum, Germany

Pritchard, David J. School of Psychology, Newcastle University, Newcastle upon Tyne NE1 7RU, UK

Rennert, Rebecca J. Department of Psychology, Emory University, Atlanta, GA 30322, USA

Robira, Benjamin Department of Evolutionary Biology and Environmental Studies, University of Zurich, 8057 Zurich, Switzerland

Schinazi, Victor R. Department of Psychology, Bond University, Robina, 4226, Queensland, Australia; Future Health Technologies, Singapore-ETH Centre, Campus for Research Excellence and Technological Enterprise (CREATE), Singapore 138602, Singapore

Segen, Vladislava Aging, Cognition and Technology Group, German Center for Neurodegenerative Diseases (DZNE), 39120 Magdeburg, Germany

Srivathsa, Sahana Evelyn F. McKnight Brain Institute, University of Arizona, Tucson, AZ 85724, USA

Srokova, Sabina Department of Psychology, Evelyn F. McKnight Brain Institute, University of Arizona, Tucson, AZ 85724, USA

Ulanovsky, Nachum Department of Brain Sciences, Weizmann Institute of Science, Rehovot 7610001, Israel

Vervaeke, Koen Department of Physiology, University of Oslo, 0372 Oslo, Norway

Wan, Kirsty Y. Living Systems Institute and Department of Mathematics and Statistics, University of Exeter, Exeter EX4 4QD, UK

Weisberg, Steven M. Department of Psychology, University of Texas at Arlington, Arlington, TX 76019 USA

Wiener, Jan School of Psychology, University of Bournemouth, Bournemouth BH12 5BB, UK

Wills, Thomas J. Department of Cell and Development Biology, University College London, London WC1E 6BT, UK

Wirth, Sylvia Institut des Sciences Cognitives Marc Jeannerod, CNRS 5229, Bron, France

Wolbers, Thomas Aging, Cognition and Technology Group, German Center for Neurodegenerative Diseases (DZNE), 39120 Magdeburg, Germany

List of Acronyms and Abbreviations

ACC	anterior cingulate cortex
ADN	anterior dorsal thalamic nucleus
AI	artificial intelligence
AMN	anterior medial nucleus
AND	anterior dorsal nucleus
AVN	anterior ventral nucleus
BB model	Bicanski–Burgess model
BOLD	blood-oxygen-level-dependent
BPI	boundary-supported path integration
BVC	boundary vector cell
CAN	continuous attractor networks
CX	central complex
DARPA	U.S. Defense Advanced Research Projects Agency
DG	dentate gyrus
dlPFC	dorsolateral
dmPFC	dorsomedial
DNN	deep neural network
DTN	dorsal tegmental nucleus
EC	entorhinal cortex
EEG	electroencephalogram
fMRI	functional magnetic resonance imaging
GPS	global positioning system
HC	hippocampus
HD	head direction
HiLAM	hierarchical look-ahead model
HUD	heads-up display
IL	infralimbic cortex
IMU	inertial measurement unit
JSON	JavaScript Object Notation
LC	locus coeruleus
LEC	lateral entorhinal cortex
LFP	local field potential
LiDAR	light detection and ranging
LMN	lateral mammillary nucleus
LNT	Leiden Navigation Test
LPI	landmark-supported path integration
LTD	long-term depression
LTP	long-term potentiation
MEC	medial entorhinal cortex
MEG	magnetoencephalography
mHealth	mobile health
MMSE	Mini-Mental State Examination
MoCA	Montreal Cognitive Assessment
mPFC	medial prefrontal cortex
MRI	magnetic resonance imaging
MTL	medial temporal lobe
MWM	Morris water maze

NASA-TLX	NASA-Task Load Index
NCL	nidopallium caudolaterale
NeuroSLAM	neuro-inspired SLAM system
NHP	nonhuman primate
NMB	net monetary benefit
NOR	novel object recognition
OFC	orbital frontal cortex
OPA	occipital place area
PCA	principal component analysis
PCC	posterior cingulate cortex
PET	positron emission tomography
PFC	prefrontal cortex
PHC	parahippocampal cortex
PKC	protein kinase C
PoS	postsubicular cortex
PP2B	protein phosphatase 2B
PPA	parahippocampal place area
PPI	pure path integration
PPN	pedunculopontine tegmental nucleus
Prb neuron	perirhinal neuron that codes for boundaries
PRo neuron	perirhinal neuron that codes for objects
PWb	parietal window boundary coding cells
PWo	object-coding cells in the parietal window
PrL	prelimbic cortex
RatSLAM	robotic simultaneous localization and mapping
Re	Reynolds number
RRT	rapidly exploring random tree
RSC	retrosplenial cortex
RSComp	retrosplenial complex
SatNav	satellite navigation
SHQ	Sea Hero Quest
SLAM	simultaneous localization and mapping
SPACE	Spatial Performance Assessment for Cognitive Evaluation
SPL	superior parietal lobule
SR	successor representation
SUS	System Usability Scale
UAVs	unmanned aerial vehicle
UEQ	User Experience Questionnaire
UI	user interface
UMAP	uniform manifold approximation and projection
UV	ultraviolet
UX	user experience
VE	virtual environment
VR	virtual reality
VTA	ventral tegmental area
VTE	vicarious trial and error
VTN	ventral tegmental nuclei

1

Addressing the Challenges of Navigational Research

Nora S. Newcombe and Ken Cheng

Navigation occurs across many modes and scales of travel, ranging from tiny microbes that move micrometers to globe-trotting sea turtles that cross the Earth's oceans. How animals traverse across complex terrain or featureless oceans or deserts, or find their way home after a foraging trip, has long fascinated and puzzled observers in equal measure. Because successful navigation is essential to survival, understanding its mechanisms has been a major goal for researchers in biology, neuroscience, psychology, and robotics. Easing the task of wayfinding has also been a focus for practical inventors of human navigational tools, which have ranged over time from traditional systems that rely on a complex mix of terrestrial and celestial cues to modern global positioning systems.

The wealth of behavioral and neural data generated by these investigations, combined with formal computational specificity, offers an excellent opportunity to formulate pluralistic explanations linking brain and behavior and relating behavioral patterns and neural firing patterns. The field of navigation, however, has been fragmented, with little communication across disciplines and levels of analysis. For example, neuroscientists recording cells may not think about, or find it difficult to tackle experimentally, the ecological validity of the behaviors their animals exhibit, while biologists seeking to understand natural behavior may not consider the neural substrates that support it, thereby limiting their hypotheses.

An important reason for this fragmentation is that the variety of findings and the sheer volume of data generated by these endeavors can be overwhelming. One way to cope with the vast mass of information, so that findings can be synthesized across these disparate fields, is to formulate an organizing rubric. The *navigation toolbox*, proposed by Wiener et al. (2011) to identify common underlying principles

Nora S. Newcombe (✉)
Dept. of Psychology, Temple University, Philadelphia, PA 19122, USA
Email: newcombe@temple.edu

Ken Cheng
School of Natural Sciences, Macquarie University, Macquarie Park, NSW 2109 Australia

N. S. Newcombe and K. Cheng (eds.), *Challenges in Navigation Research*,
Strüngmann Forum Reports, https://doi.org/10.1007/978-3-032-20563-6_1

that may operate across many different taxa as well as differentiate them, provides one such example (see also Jeffery et al. 2024). However, many vital questions remain unanswered by the toolbox, including issues concerning methodology, the processes of development and aging, and the application of this knowledge to real-world problems.

To overcome this impasse, we approached the Ernst Strüngmann Forum to ask for their help in convening a multidisciplinary discourse that would aim at

- integrating research on biological navigation across species and agents and across different levels of analysis (behavioral data with cellular, circuit, and systems neuroscience),
- addressing variations in navigation across individuals, ages, environments, and cultures, and
- identifying major questions for future research to address.

As described in the Preface, after the proposal was approved, Kate Jeffery and Thomas Wolbers joined us to develop a discussion framework that would bridge existing gaps across taxa, levels, and methods of study. This discourse unfolded within four working groups, the foci of which are described below.

1 Universal Navigational Principles

Evolutionary theory provides one kind of organizing rubric (and in fact, Darwin wrote a bit about it). Since navigation is a well-defined problem, at least within ecological niches, it lends itself readily to solutions at the computational, algorithmic, and neurobiological levels—levels spelled out by the vision scientist David Marr (1982). To identify some universal or at least ubiquitous principles of navigation across taxa and scales, this working group (Sylvia Wirth et al., Chapter 4) looked for commonalities in navigational strategies and mechanisms across species that inhabit different ecotypes (e.g., water, air, land, vertical travel, horizontal travel) as well as across scales of movement. It also investigated how differences in these strategies and mechanisms have been shaped throughout evolution by the ecological contexts in which organisms move. Two background chapters prepared the discussion by exploring different aspects of navigation.

In Chapter 2, David Pritchard discusses functional aspects of avian navigation. The examples presented here describe how birds are well equipped to respond to emergent issues encountered during navigation: variable environmental information, navigation at different scales, conflicting information, ecological barriers, and the impact of wind. These five areas provide an ecological framework by which to compare bird species and other taxa. The challenge for future research, however, is to understand how such ecological factors give rise to navigational solutions.

In Chapter 3, Michael Milford reviews highly simplified robot navigation systems (e.g., micro- and nano-robot systems), examining both the challenges and points of intersection with biological systems. He argues that artificial systems must emulate the adaptability, introspective awareness, and efficiency of biological systems. As work continues to overcome these challenges, Milford posits that insights from artificial systems can be used, in turn, to inform the biological systems from which inspiration has been drawn.

In the report of the working group, Sylvia Wirth et al. (Chapter 4) present two key themes that apply to a wide array of organisms. The first centers on different scales of travel:

- A long-distance phase, dominated by the challenge of keeping a direction of travel
- A middle-distance phase, within a familiar area that permits a range of strategies to be used
- A fine-scale phase that pinpoints the goal of the navigational challenge and might rely on a distinct set of cues from those used in other phases

The second theme centers on the sensory, cognitive, representational, decisional, and output (motor) components needed for navigation. Further, Wirth et al. identify and discuss components involved in processing sensory information, determining direction and position, and generating movement. Between information processing and movement generation lies a central component responsible for error correction, decision making, and action planning. By examining these themes across a wide range of taxa in which the authors have particular expertise, they consider the ecological context in which navigation takes place, including abiotic, environmental, and biotic factors (e.g., predators and prey) that affect navigational decisions and movement. They conclude by outlining areas for future research directions. Suggested trajectories for further study include the mechanisms underlying transitions between phases of navigation; processes of introspection, error monitoring, and correction; the sensing and processing of magnetic information; and the neurobiological foundations of large-scale navigation under natural conditions—scales that far exceed those typically examined in laboratory studies of spatial cognition.

2 Linking Cells and Circuits to Brains and Behavior

An exciting development in specifying the neural bases of navigation emerged from research, conducted largely in rats, that identified cells that process place and heading direction, provide a grid system, code for boundaries, and more. At this level of analysis, however, research on humans is technically not feasible. Instead, most human research uses functional magnetic resonance imaging (fMRI) to identify brain areas that form a navigation circuit. This body of evidence can be cross-checked against individuals with damage to navigation-relevant brain areas and, most recently, by studying epilepsy patients who, during presurgical preparation, are implanted with intracranial electrodes that target navigation-relevant areas. How do these lines of research, conducted with different species using different techniques, inform each other?

To facilitate comparisons and productive exchange across labs working with different species and techniques, this working group asked: Is it possible to understand behavioral variation among individuals using neural tools? Can we describe how environmental input affects navigation at both the neural and the behavioral levels?

Are there emerging technologies or analytic techniques to aid translation across levels of analysis? To support this group's discussions, four background papers examined different ways of linking cells and circuits to brains and behavior.

In Chapter 5, Andrej Bicanski explores how modeling informs our understanding of biological navigation across diverse model organisms, neural subsystems, and levels of abstraction. Using two case studies, direction coding and vector navigation, he demonstrates how Marr's (1982) three levels of analysis can be applied to address increasing complexity and argues that large-scale, systems-level models are essential for investigating interactions among multiple neural subsystems, in both successful and failed navigational outcomes.

In Chapter 6, Philipp K. Büchel and Lukas Kunz examine how neural correlates of human navigation, observed through noninvasive fMRI, align with those obtained from invasive recordings. Focusing on five key components of spatial navigation within a defined set of brain regions, they argue that grid and scene representations in the entorhinal and parahippocampal cortices, respectively, exhibit the strongest correspondence across methodologies. To advance understanding of how spatial neural representations derived from different neuroscientific methods relate to each other, they advocate for research that integrates simultaneous noninvasive and invasive recordings in animal models, sequential noninvasive and invasive recordings in humans, and computational modeling.

In Chapter 7, Daniel D. Dilks et al. explore the development of navigational abilities in humans. After reviewing behavioral and fMRI evidence implicating the medial temporal lobe and retrosplenial complex in the map-based navigation system, and the occipital place area in visually guided navigation in adulthood, they note that surprisingly little is known about how these abilities emerge. The authors propose, perhaps counterintuitively, that map-based navigation develops at around age five years whereas visually guided navigation emerges late, around age eight. They emphasize that although human fMRI has clarified the functional roles of key brain regions in navigation, these systems remain poorly understood in animal models, thus setting up a central challenge for future research.

In Chapter 8, Thomas J. Wills and Francesca Cacucci hold that development provides a unique opportunity to identify the cognitive building blocks underlying navigation and to delineate the neural circuits that support them. Using the rat as a model organism, they review current knowledge on the development of navigational behavior and the emergence of spatially modulated neural activity, presenting a concise timeline of functional development of navigation at both behavioral and neural levels. They emphasize the importance of designing behavioral protocols to enable direct comparisons; for instance, between the emergence of map-based and route navigation strategies and the neural circuits that underpin them.

In the report of this working group (Chapter 9), Freyja Ólafsdóttir et al. stress the importance of connecting work across species, at both methodological and behavioral levels, as well as integrating studies of development and mature functioning. They conclude that human research involving epilepsy patients aligns reasonably well with findings on healthy adults studied using fMRI, although gaps are inevitable given the methodologies. Ólafsdóttir et al. argue that formal modeling should serve as a useful bridge across species and methodologies. In their discussion of neural coding of information essential for navigation (e.g., orientation, distance, and boundaries), they compare humans and rodents, along with methods for obtaining neural data, and conclude that mapping neuronal activity to behavior remains a significant gap to be addressed. The authors view the study of development as a promising avenue for bridging divides.

3 Navigation in a Wider Perspective

Traveling to a destination requires more than map and compass systems. To begin, an agent must decide where to go—a process that engages motivational and emotional systems as well as decision making. Next, a route must be selected and planned, and knowledge of the spatial environment must be translated into action, which likely involves a transformation of spatial reference systems. To date, neurobiological work on mammalian navigation has focused on the hippocampal place-cell system, but many parts of the brain, distributed across wide regions, both cortical and subcortical, may be involved. Here, research has been challenging because clear neural signals (e.g., activity patterns of place, head-direction, and grid cells) are rarely found in other brain areas.

To approach the overall goals of the Forum, this working group began by asking: How do we get traction on these more complex, integrated and distributed signals? What new technology or experimental paradigms could help? Two chapters helped to initiate the discussions.

In Chapter 10, Shantanu P. Jadhav outlines hippocampal-prefrontal spatial representations and their interactions in rodents and their roles in navigational planning. Addressing the importance of planning and imagination, which involve both conscious effort and internal computations, Jadhav reviews evidence of specific brain systems involved in navigational planning and argues that a holistic approach is required to understand the underlying mechanisms. Further, he emphasizes the need to consider multiple timescales of operation. Fast timescale sequences that occur during "offline" states should be incorporated into models of navigational planning, with hippocampal-prefrontal interactions serving as a good starting point. To achieve a complete understanding of navigational systems, future research will need to include broader distributed networks across sensory, cognitive, and motor regions.

In Chapter 11, Thomas Forro and Stéphane Ciocchi focus on the role of motivation and emotion in navigation. Whereas emotions guide an organism toward the most efficient and least risky path to a reward, the hippocampus integrates sensory information to form cognitive maps of the environment. Here, the authors examine how the intertwining of emotions and goals within the hippocampus can shape spatial maps and memories. The links between emotional systems and hippocampal systems have broad implications, ranging from the treatment of mental disorders to advancing our understanding of fundamental principles of brain function.

Building on these background chapters, this working group views navigation as a complex set of interacting cognitive processes, supported by a brain-wide network of structures whose individual functions are not well understood. In their summary report (Chapter 12), Zita Patai et al. review studies of single neurons in freely exploring rodents and note that although these studies have revealed an array of "canonical neurons" (e.g., place, head-direction, and grid cells), much more of the brain is involved in navigation, and many navigation-related neurons lack canonical space-coding properties. Understanding how these other areas function, and the nature of the less obvious coding schemes, remains a major challenge and should be prioritized for future inquiry. Throughout, Patai et al. identify key open questions and suggest that sustained interactions between neurobiological, behavioral, and ethological studies are required to achieve a comprehensive understanding of the mechanistic basis of navigation.

4 Translating Knowledge into Practice

The ability to navigate is fundamental to our everyday life, and when this ability becomes impaired, the consequences can be devastating. Although our understanding of the neural and behavioral mechanisms underlying spatial navigation has grown substantially, this knowledge has not been effectively translated into the clinical context. We still lack a comprehensive understanding of how neurological conditions affect navigation, as well as a standardized, reliable method for assessing navigational ability in clinical settings. Moreover, predicting how navigational deficits influence real-world mobility remains a major challenge. The goal of this working group, therefore, was to relate findings from basic research to the reality of clinical practice. Three background chapters provided impetus for their discussions.

In Chapter 13, Carol A. Barnes et al. stress the need to consider how parallel navigation circuits contribute to age-related differences in navigational behavior. To date, most studies that examine aging in a navigational context have not accounted for changes in synaptic connections and plasticity within these circuits, yet doing so might reveal differences in strategies across the lifespan. To address the strategies that older individuals may use interchangeably during navigation, the authors recommend moving beyond known hierarchical levels of navigation. They emphasize that although the dynamic interplay between parallel circuits can lead to flexible behaviors, it can also create competition between strategies that impair navigation performance.

The focus of Chapter 14, by Jan Wiener and Michael Hornberger, is on the effects of cognitive aging and Alzheimer disease on spatial navigation at the behavioral level. Although research offers a detailed understanding of the aging navigational system, they emphasize that our knowledge of how age-related navigational deficits impact everyday navigation and people's experience of living with declining navigation abilities is surprisingly limited. In addition, despite what we know about spatial disorientation in Alzheimer disease, our understanding of how these deficits progress is limited, as is our understanding of their real-world implications. Wiener and Hornberger argue that future efforts must focus on everyday navigation so that targeted, effective interventions and diagnostics can be developed.

In Chapter 15, Victor Schinazi et al. review the development of technologies and argue that this process must ensure patient and clinician usability. In addition, because navigational deficits are early indicators of cognitive impairment in dementia and neurodegenerative diseases, we must develop an accurate way to measure deficits so that early diagnosis can be achieved. Using the Spatial Performance Assessment for Cognitive Evaluation (SPACE), Schinazi et al. illustrate how insights from research into human-computer interactions can guide and support the design of effective clinical tools. Schinazi et al. emphasize that there is no such thing as a universal design, since the needs and abilities of individuals differ greatly across age groups. In addition, when transitioning from the lab to the clinic, modularity must be built in to enable rapid assessment and feedback collection.

Working to identify gaps between basic research, clinical science, and clinical practice, this discussion group sought ways to guide and stimulate translational navigation research in aging. As reported in Chapter 16, the discussions by Coco Newton et al. yielded the "Triadic Navigation Framework." The intent of this comprehensive, holistic platform is to guide researchers as they consider the full spectrum of factors relevant to understanding the brain–behavior relationship in navigation, in both healthy and unhealthy aging individuals. The authors also suggest ways to expand the BB model of navigation (Bicanski and Burgess 2018; see also Chapter 5). The navigation system operates in a context, and the system shows considerable individual differences, even in typically functioning young adults. Newton et al. offer suggestions for applying the Triadic Navigation Framework, outline directions for future research, and recommend ways to advance both basic and applied understanding of navigation in the context of healthy and unhealthy aging.

5 The Way Forward

The ability to navigate is ubiquitous across the animal kingdom. For humans, it is fundamental to managing day-to-day activities, maintaining a quality of life, and ensuring our very survival. The loss of navigational ability can compromise independent living, yet it may also serve as an early marker of neurodegenerative disorders. Even in normative aging, navigational behavior changes over the lifespan, posing significant yet diverse challenges to individuals, their families, communities, and society as a whole. The economic consequences of this are immense.

The toolbox that science currently possesses is not nearly sufficient to address all these challenges, and many more will inevitably come to light as our knowledge base continues to expand. By unpacking and assessing the existing arsenal, and imagining future trajectories, we believe that this Forum has taken an important first step. We wish to thank all who participated for their contributions, their openness to debate, and willingness to think outside the box. On behalf of everyone involved, we hope this volume will provide fresh impetus for the field of navigation research to advance further in a spirit of collaboration and commitment to the challenges ahead.

References

Bicanski A, and Burgess N (2018) A Neural-Level Model of Spatial Memory and Imagery. eLife 7 (7052):1–3. https://doi.org/10.7554/elife.33752

Jeffery KJ, Cheng K, Newcombe NS, Bingman VP, and Menzel R (2024) Unpacking the Navigation Toolbox: Insights from Comparative Cognition. Proc R Soc Lond B Biol Sci 291 (2016):20231304. https://doi.org/10.1098/rspb.2023.1304

Marr D (1982) Vision: A Computational Investigation into the Human Representation and Processing of Visual Information. MIT Press, Cambridge, MA

Wiener JM, Shettleworth S, Bingman VP, et al. (2011) Animal Navigation: A Synthesis. In: Animal Thinking: Contemporary Issues in Comparative Cognition, ed. R Menzel and J Fischer. Strüngmann Forum Reports, vol. 8, JR Lupp series editor. MIT Press, Cambridge, MA

Universal Navigational Principles

2

Taking a Functional Look at Bird Navigation

David J. Pritchard

Abstract Birds are some of the most wide-ranging species on the planet, with many navigating with unerring accuracy. They are also ecologically diverse, making birds a promising model for functional studies of navigation. Despite this, most research on avian navigation has focused on the mechanistic basis of navigation, particularly which sensory cues birds use. In this chapter, avian navigation is placed in a functional context. Issues that navigating birds face and resolve are outlined, including navigation at different scales, variation in environmental information, how conflicting information is managed, detouring around ecological barriers, and navigation through wind. These five areas highlight the potential of a systematic cognitive ecology of avian navigation, placing navigation by birds and other taxa in a comparative, ecological framework. To guide future work, open questions regarding the problems that birds face, and the solutions they have evolved, are presented to facilitate such an ecological approach.

Keywords Avian navigation, cognitive ecology, environmental variability, path integration, scales of navigation, spatial cognition

1 Introduction

Birds are some of nature's premiere navigators, with some operating on a global scale. For example, Razorbills (*Alco torda*) forage across the North Atlantic Ocean and yet reliably find their way home over kilometers of open ocean to bring fish for their chicks (Shoji et al. 2016). The annual spectacle of billions of birds undertaking seasonal migrations includes species that cross deserts, mountains, and even entire continents with remarkable accuracy (Alerstam 2006). How birds manage these navigational feats is one of the classic questions in experimental biology, one marked by over 120 years of research (Wiltschko and Wiltschko 2003). As it has

David J. Pritchard (✉)
School of Psychology, Newcastle University, Newcastle upon Tyne NE1 7RU, UK
Email: David.Pritchard2@newcastle.ac.uk

N. S. Newcombe and K. Cheng (eds.), *Challenges in Navigation Research*,
Strüngmann Forum Reports, https://doi.org/10.1007/978-3-032-20563-6_2

developed, the study of avian navigation has increasingly focused on the mechanisms underlying navigation, particularly the sensory cues birds use and how these are sensed. This line of research has led to some incredible discoveries, such as a magnetic compass in the retina of songbirds (Ritz et al. 2004) or the role of smell in long-distance navigation (Gagliardo et al. 2013), but sensory mechanisms alone cannot explain how birds navigate.

Some 13 years after Gustav Kramer kicked off the modern study of bird navigation with his description of the sun compass in 1950 (Kramer 1950), Niko Tinbergen set out his famous four questions (Tinbergen 1963): any behavior we observe in an animal can be described in terms of the underlying mechanisms, the developmental history, the survival function, and the evolutionary history. A holistic explanation of a behavior, Tinbergen argued, requires understanding of all these different aspects. Even navigation, despite arguably being a mechanism rather than a behavior, can be examined in this way. The way in which animals acquire and use information during navigation can be understood in terms of its mechanistic basis, its ontogeny, its evolutionary history, and even its function (Wynn and Liedvogel 2023). Fields such as sensory ecology have demonstrated how a functional perspective can complement mechanistic understanding (Stevens 2013). While the cells in the retinas of different bird species all work in largely the same way, the distribution of cells on the retina, and even the placement of the eyes themselves, can vary wildly between species (Martin 2017). This variation can be understood if we consider the function of vision in these species (i.e., what they use vision for) and the environments they inhabit. Although less established than sensory ecology, cognitive ecologists have made similar arguments for the study of cognitive processes (Dukas 1998; Real 1993). Understanding animal cognition is not simply predicated on the neural basis of cognitive processes, but also in how these processes are used in the ecology of the species being studied.

In this chapter, I will discuss various strategies that birds use to navigate through the environment. There are already many excellent reviews on this topic, usually structured around the different mechanisms birds possess, the sensory modalities birds use, or the different navigational scale birds travel (Mouritsen 2018). To approach a well-worn topic from a different perspective, I will focus on what navigating birds *need to do*. To achieve this, I discuss five common issues that birds face during navigation and how birds solve these issues. These particular issues were not chosen because I believe they are necessarily the most important issues birds face, but because these are the issues for which we have some data. The vast majority of research in avian navigation does not aim to answer functional questions (Wynn and Liedvogel 2023). Thus, a systematic study of the relationship between the task birds face and the solutions they use is not yet possible. Toward the end of this review, I highlight open questions which could help in building a more systematic approach and facilitate a true cognitive ecology of avian navigation.

2 What Issues Do Birds Face during Navigation?

In its simplest form, navigation is about getting from point A to point B, yet in the natural world, a number of different factors can complicate such a journey. Below I discuss five possible issues that a navigating bird might need to address as they navigate: different scales, environmental variation, conflicting information, the need to detour around barriers, and travel through wind (Figure 2.1).

Male rufous hummingbirds (*Selasphorus rufus*) navigate from their wintering grounds in Mexico to the Canadian Rockies, where they form a territory and navigate between learned patches of flowers (Healy and Hurly 2013). Hummingbirds, and indeed many bird species, need to navigate over a range of scales, from migratory journeys over hundreds of kilometers to foraging decisions over meters or even centimeters. To solve this challenge, birds appear to use a toolkit of different mechanisms, each appropriate for a different scale.

In functional terms, "long-distance" navigation usually refers to situations in which birds are traveling without direct access to any cues from their goal or familiar surroundings (Dyer 1998). In these situations, birds appear to rely on "compasses" to provide orientation and follow a consistent course. The compass cues that birds

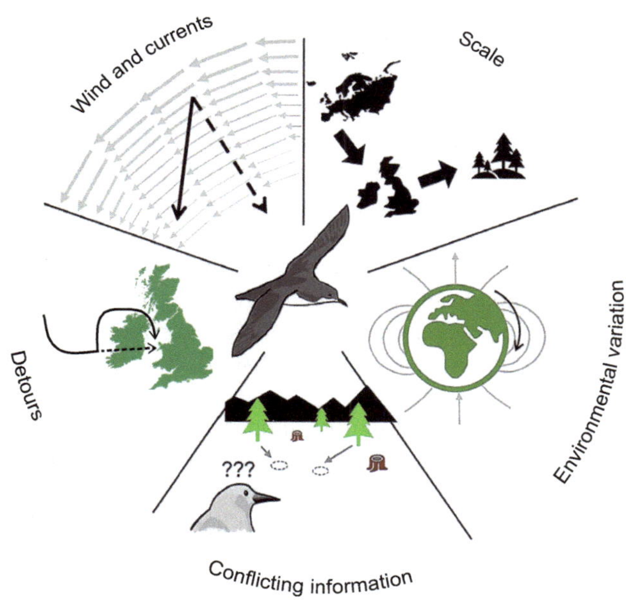

Figure 2.1 Possible issues that birds face during navigation. Clockwise from top right: Many birds have to find their way across a range of scales, from intercontinental migrations to pinpointing specific resources. The information available in the environment can vary across both space and time; for instance, the angle of magnetic inclination guiding birds south becomes shallower as birds get closer to the equator. Different sources of information can disagree regarding where a bird should go. Thus mechanisms are required to decide between options or integrate them into a single course of action. The direct route to a goal might not be feasible due to geographic barriers, requiring birds to detour around these barriers to reach their goal. Birds flying in the air or swimming in water have to contend different scales. Map of the British Isles from Wikipedia, https://creativecommons.org/licenses/by-sa/4.0/deed.en

can use include magnetic inclination (poleward vs. equatorward) (Wiltschko and Wiltschko 1972), constellations around the center of celestial rotation (Emlen 1970), and the position of the Sun in the sky (Kramer 1952). Using a sun compass requires birds to compensate for the time of day (Schmidt-Koenig 1958). Shifting the circadian clock of birds artificially by a number of hours ("clock-shifting") produces predictable shifts in the flight direction when birds rely on a sun compass, providing a useful tool for investigating navigation mechanisms (e.g., Padget et al. 2018).

Compasses only help birds stay on course. Setting a course over long distances either involves birds following a vector or determining their position with a "navigational map" (Berthold 1991). Vector navigation involves following a compass bearing for either a set period of time (a "clock-and-compass" strategy; Mouritsen 1998) or until they reach an environmental signal or "waypoint" telling them to stop or change to a different bearing (Wynn et al. 2022). This strategy allows birds to travel to locations they have not visited before, but it means that they are unable to correct for any unexpected displacement if they are blown off course (or moved by biologists). Utilizing a navigational map, in contrast, allows birds to set a course toward their destination even after being displaced hundreds of kilometers, an ability known as "true navigation" (Able 2001). The classic model of true navigation involves a two-step process: the map stage in which birds ascertain their position relative to the goal ("I am southwest of home, I need to go northeast"), and the compass stage in which birds follow a compass trajectory toward their goal ("northeast is that way") (Kramer 1957). These steps appear to be independent of one another, as shifting the compass does not seem to affect where birds think they are—just the direction in which they then travel. The nature of the navigational map underlying true navigation has been one of the great questions in bird navigation. One of the best supported explanations involves birds acquiring a "grid map," by learning how multiple gradients change in different directions within the familiar environment and then extrapolating these gradients into unfamiliar space (Able 2001). As long as the same gradients are perceptible, birds can fix their position relative to home by determining their position in these gradients. Over very large distances, magnetic fields might provide the gradients necessary for a grid map (Packmor et al. 2024). Over shorter distances, it has been suggested that olfactory gradients might be preferred (Wallraff 2004).

While birds traveling shorter distances through familiar environments can continue to use true navigation strategies, the repeated exposure to a landscape allows birds to build a "familiar area map" to guide their journey (Holland 2003). This map is based on visual landmarks, and research with homing pigeons (*Columba livia*) suggests that it can take several forms. Birds may associate a release location with a compass bearing home, build a sparse representation of landmarks connected by compass bearings, or not rely on their compass at all and follow a route of familiar landmarks (Guilford and Biro 2014). The strategy birds use depends on the environment and experience. After repeated flights from the same location, experienced homing pigeons are more likely to form routes than inexperienced pigeons (Griffiths et al. 2021). In addition, route fidelity appears to be higher in rural landscapes than in urban landscapes (Mann et al. 2011).

The smallest scale of navigation involves pinpointing a location, such as the nest, a perch, or a resource. In some cases, this pinpointing needs to be very precise.

Birds that cache food, for example, need to remember the location of their cache within centimeters or they might never recover it. To do this, birds can use a range of cues, including large-scale global cues (Gould-Beierle and Kamil 1996), landmarks at different distances to the goal (Vander Wall 1982), and visual or olfactory beacons (Bonadonna and Bretagnolle 2002). Birds can learn the position of a goal relative to a landmark in terms of its distance (Goodyear and Kamil 2004), its absolute or relative direction (Jones and Kamil 2001), or view of the landmark from the goal's location (Pritchard et al. 2018).

Together, the strategies above make up the majority of what we consider to be the avian navigation toolkit; each reflects the different challenges that come with navigating at a particular scale. The information available to birds, for example, depends on the scale. Local variation in magnetic fields makes these cues too noisy to be useful over small scales, but they are reliable over larger scales and so a possible basis for the navigational map. Routes of landmarks provide a precise means of traveling home over smaller scales, but limited detection ranges and the cognitive costs of keeping so many landmarks in memory could make them less useful over longer distances. These different challenges could also explain some patterns in spatial behavior. For example, food-storing Clark's nutcrackers (*Nucifraga columbiana*) travel widely when harvesting seeds in the autumn but cache the vast majority of the seeds they discover within their home range, even when this means transporting seeds over 20 km to do so. This centralized depositing might seem less efficient than caching close to where seeds are harvested, but by caching in their familiar territory (median size 4.13 km^2), the nutcrackers might be better able to use familiar landmarks to recover these seeds in the winter rather than rely on larger-scale navigation strategies (Lorenz et al. 2011).

2.1 Environmental Variability

Another key challenge for navigating birds involves changes in the environment. The cues that birds use to navigate (e.g., landmarks, magnetic fields, or celestial cues) can change over time, and the way in which birds cope with this change depends on how fast it occurs. Very slow changes, with cues remaining stable across multiple generations, can still allow for natural selection to favor a set strategy, such as a particular preferred compass direction during migration (Delmore et al. 2020). If cues are not stable enough across generations for natural selection to act on specific preferences, then birds can still adapt to changes via learning. Manx shearwaters (*Puffinus puffinus*), for example, seem to imprint on the magnetic inclination they experienced at their natal colony prior to fledging and use this information when first returning to the colony (Wynn et al. 2020b). Imprinting provides a possible solution for acquiring information too unstable to be captured by natural selection, but stable enough within a single bird's lifespan. When cues change within an animal's lifetime but are predictable, then birds can learn to compensate for these changes (Dyer 1998). The time-compensated sun compass studied in homing pigeons is a prime example of how birds learn to deal with a predictably changing cue. Time compensation is learned when pigeons are young; it requires birds to have experience of the Sun's position at different times of day, possibly associating these positions not only with their internal sense of time but also with directions from a magnetic compass (Wiltschko et al. 1998).

In addition to changes over time, the sun compass might also have to account for changes between locations. As birds move across longitudes (east or west), the rhythm of the Sun's movements can become uncoupled from the bird's own internal circadian rhythm. Moving east, for example, would result in the Sun rising earlier than expected and setting sooner. Such changes in the Sun's movements and their relationship with circadian time are unlikely to affect short-ranging birds, but they will affect long-distance migrants or seabirds on long foraging trips (Åkesson and Bianco 2017). To adapt, birds must adjust their compass timing, which may mean stopping over at a location and spending a few days. Some migrants do not have that luxury and instead use another way to adapt. Arctic waders, such as bar-tailed godwits (*Limosa lapponica*), commit to long endurance flights over the ocean, with flights showing a characteristic curved path (Gill et al. 2005). While these curves sometimes approximate a great circle or orthodromic route, following the shortest distance between two points on the globe, a curved path is also consistent with birds continuing to use a sun compass without compensating for the change in time zone across longitudes. Rather than adapting their compass, these birds may instead be adapting their initial flight direction, taking the resulting curvature into account, such that the uncompensated compass leads the birds to their intended direction (Åkesson and Bianco 2017).

As birds move across latitudes (north or south), multiple navigational cues change (for a review, see Åkesson and Bianco 2017). Toward the poles, the Sun's arc varies more across the seasons, resulting in permanent darkness or sunlight at the extremes. Either condition can make keeping track of time difficult: birds in continuous daylight show apparent drifting in their circadian rhythm and possibly in their sun compass. Continuous daylight also restricts the ability of young birds to see the night sky and acquire a star compass, whereas the steep angles of magnetic inclination can make using the magnetic compass challenging (Åkesson et al. 2001). The equator presents similar problems. Around the equator, magnetic inclination becomes horizontal, providing no clear poleward direction. The Sun also rises and sets faster at the equator, presenting problems for compasses calibrated for higher latitudes; it also passes directly overhead at midday, making the angle to travel relative to the Sun ambiguous. Crossing the equator presents another set of problems. Birds using an inclination compass can only orient poleward versus equatorward, meaning that birds crossing the equator need to switch from traveling to the equator to traveling to the pole (Wiltschko and Wiltschko 1992). The position of the Sun also changes as birds cross the equator, with the Sun changing from a clockwise azimuthal progression in the northern hemisphere to an anticlockwise progression in the southern hemisphere. Finally, the star compass, which is based on the center of rotation in the night sky, can be compromised at the equator as potentially neither center of rotation is visible (Wehner 1998). As birds cross the equator from north to south, the northern center of rotation and any learned constellations give way to the southern center of rotation, complete with a different set of constellations. There are two main solutions to the problems of navigating at the poles and equator. The first requires some means for birds to change the nature of the mechanisms they are using. For instance, when garden warblers (*Sylvia borin*) cross the magnetic equator, they use the horizontal inclination lines as a cue to flip from equatorward navigation to poleward (Wiltschko and Wiltschko 1992). Birds using a star compass at the equator for the first time could also use the opportunity to learn new constellations around the opposite center of rotation and rely on these cues in the future. The second solution is to use a combination of compasses. In the Arctic, for example, when the Sun is not available, birds could use the stars to orient. Both celestial compasses could be used when the inclination angle is too steep or too shallow to use (Åkesson 1993), as could magnetic declination (the difference between magnetic and geographic north) (Åkesson and Bianco 2017). This kind of redundancy is not just seen in compass orientation. Homing pigeons traveling over familiar areas are able to use both visual landmarks and an olfactory map to guide their way home and may rely on each to different degrees, depending on context (Gagliardo et al. 2020). Both homing pigeons and Clark's nutcrackers are able to switch between using a sun compass or the configuration of landmarks to guide them to their goal, adjusting their reliance on each cue based on the environment and their own experience (Gagliardo et al. 2009; Kelly et al. 2019). For both finding direction and determining position, by using multiple redundant cues, birds gain a versatile toolkit for navigating across a range of environments.

2.2 Conflicting Information

Not only does redundancy allow birds to navigate a range of different environments, it also equips birds to manage future changes to the environment. If a prominent landmark changes appearance or disappears (e.g., following a fire), then birds would still be able to use other information to navigate. But learning redundant cues raises another set of challenges. What should a bird do if the cues it has learned conflict with each other?

One solution is to form a hierarchy of different cues, with some preferentially used while others are only called upon if the preferential cues are unavailable. Black-capped chickadees (*Poecile atricapillus*), for example, will preferentially relocate a rewarding feeder using its absolute location in space (Brodbeck 1994). If this is unsuccessful, the chickadee will search whichever feeder is at the correct position relative to other feeders. If neither of these spatial cues are available, the chickadee will then use the color or pattern on the feeder. The precise ordering of different cues is not consistent for all species. The prioritization of spatial information by chickadees may reflect the important role of space for food-storing birds. For the dark-eyed junco (*Junco hyemalis*), a nonstoring species, no evidence of any hierarchy has been found; birds favor all cues equally. Even when birds do use a hierarchy to choose between different cues, the order of this hierarchy can depend on the current task or environment. Rufous hummingbirds, for example, will prioritize the relative position of a flower in a patch over its absolute location, as long as the surrounding flowers are within 40 cm of the target flower (Healy and Hurly 1998). If the surrounding flowers are 80 cm away or further, the preference reverses and hummingbirds prefer the absolute location over the relative position in the patch. Pigeons also show context-dependent hierarchies of relative and absolute information when searching for a rewarded location on a touchscreen. In all cases, pigeons learned the reward in terms of both relative position (local cues) and absolute location (global cues). However, when the possible locations were presented in a horizontal line, pigeons prioritized relative position, and when locations were presented in a vertical or diagonal line, pigeons prioritized absolute location (Legge et al. 2009). Exactly why these hierarchies are structured this way, and why they respond to these specific changes in task or context, is not known. What is clear, however, is that using a hierarchy of cues does not imply that this hierarchy is fixed across species, or even within species.

A particularly contentious example of hierarchical processing regards the interactions between compass cues. Multiple studies have demonstrated that birds will use one compass or orientation cue to recalibrate another (Muheim et al. 2006). Savannah sparrows (*Passerculus sandwichensis*) and white-throated sparrows (*Zonotrichia albicollis*), for example, can both use the pattern of polarized light near the horizon at sunrise and sunset to recalibrate their magnetic compass (Muheim et al. 2009), which can then be used to recalibrate the star compass. This pattern suggests a simple hierarchy of cues in nocturnal migrants, but as with hierarchies of global and local cues, the picture is a bit more complicated. Different patterns of recalibration are found between different species, including different patterns observed in Europe compared to North America (Muheim et al. 2006). These differences could be due to actual species differences, regional differences in methodology, or slight variations in the types of additional cues birds might access. A view of landmarks along the horizon appears to be important for any recalibration using polarized light, and the degree to which the magnetic compass is recalibrated seems to depend on observing polarization patterns at both sunrise and sunset (Muheim et al. 2006; cf. Schmaljohann et al. 2012). Even taking these factors into account, some birds still show different kinds of interactions between compasses, suggesting that a single hierarchy across species, or even within species, might be less likely than a flexible system in which compasses can be weighted to different degrees based on multiple factors.

The idea that multiple cues might be weighted more or less favorably leads us to the second solution to conflicting information: integrating different cues into a single output. Some cues tend to be used in combination with others to specify a location, such as when a Clark's nutcracker determines the location of a buried seed using multiple surrounding landmarks. If one of these landmarks were to move, creating conflict between the information it provides and that provided by the other landmarks, where should the nutcracker search? Bayes theorem suggests that the answer depends on the weight given to the displaced landmark (Cheng et al. 2007). If all landmarks were weighted equally, the nutcracker might ignore the movement of any single landmark, as most remaining landmarks indicate the same location. This pattern has been observed in pigeons using a configuration of landmarks to find a reward. Moving any one landmark in the configuration was insufficient to change where the birds searched (Sutton 2002). If, however, the displaced landmark had been particularly preferred because, for example, it had been the closest landmark to the buried seed, then the nutcracker might be expected to shift its search in the direction of the displaced landmark. This weighting of landmarks by distance appears to be the case for Clark's nutcrackers, which respond to the movement of landmarks in a manner proportional to the distance of the landmark to the goal (Vander Wall 1982). Still, integration is not inevitable. If a highly weighted landmark moves too far, the birds might end up searching in two locations: the new location indicated by the displaced landmark as well as the original location indicated by the remaining landmarks. Pigeons, using a wall and a cylinder, integrated information from both cues when the cylinder was shifted 35 cm in a direction parallel to the wall or 47 cm in an orthogonal direction, but searched in two discrete locations when the cylinder was shifted 83 cm in the parallel direction or 98 cm in the orthogonal direction (Legge et al. 2016).

Together, hierarchical cue use and cue integration demonstrate the importance of flexibility for navigating birds. Being able to learn multiple kinds of information about a location and then adjust how this information is prioritized or combined enables birds to adapt to a wide variety of environments and considerable changes in those environments.

2.3 Detours

A navigation system which reliably indicates the direction of your destination is all well and good, but what happens if a straight-line trip home is neither desirable nor possible? Navigating birds take detours for a wide range of reasons. In some cases, direct travel is prevented by environmental barriers, such as oceans, deserts, or mountain ranges. In others, birds might have been blown off course by strong winds or a storm and need to get back on track. How birds deal with these detours depends on several of the issues discussed (e.g., the scale of journey and the stability of the environment). How well birds respond to detours, or fail to do so, provides insights into the flexibility of avian navigation.

When migratory species encounter environmental barriers, they are faced with two options: (a) Some species, like the bar-tailed godwit, simply fuel up and fly nonstop across the barriers. This decision could be due to the potential costs of alternative paths, including the risks of predation or competition over more habitable land. (b) Other birds go around these barriers. If barriers are stable across generations, then the detours might be reflected in an inherited navigation program, with preferred directions changing with time or in response to environmental waypoints. Pied flycatchers (*Ficedula hypoleuca*) appear to use both an endogenous time program and magnetic waypoints; Siberian populations spontaneously modify their migratory direction over time to avoid the deserts of Central Asia (Chernetsov et al. 2008), whereas European populations use the local magnetic conditions to switch direction and avoid the Mediterranean Sea and Alps (Beck and Wiltschko 1982). These inherited detours do not require birds to know where they are relative to their goal and can evolve as natural selection responds to variation in the time or conditions that trigger the change in direction.

For wide-ranging seabirds, inherited responses are less useful. Although homing seabirds may need to detour around inhospitable regions, such as large landmasses, foraging trips themselves can show little similarity across journeys, with birds ending up homing from a large area of possible starting locations. This unpredictability makes detouring more of a cognitive challenge. For Manx shearwaters nesting off the British coast, the island of Ireland provides one such barrier, as birds need to travel past Ireland to forage in the Atlantic Ocean and to return home (Figure 2.2). Despite this, shearwaters heading home from foraging in the Atlantic do not seem to recognize the need to detour around Ireland; their leaving times reflect an expected straight-line trip home (Padget et al. 2019). The navigational map of these birds is likely based on olfactory gradients while at sea, as in other petrels (Padget et al. 2017), and does not appear to contain any information about the presence of Ireland, which must be taken into account to minimize homing distance around the

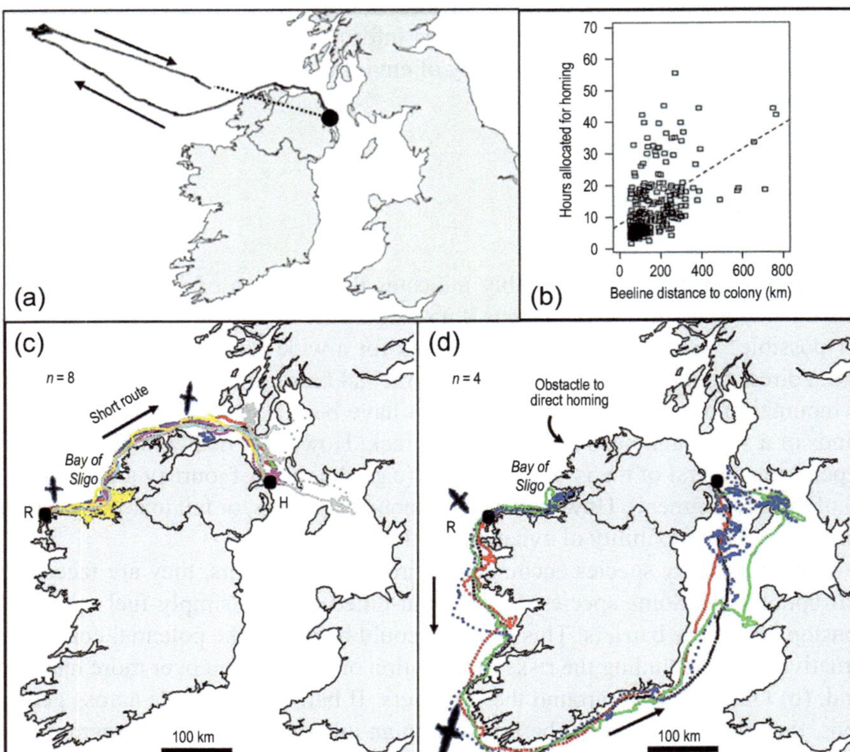

Figure 2.2 Manx shearwaters appear to know the distance and direction to home but do not encode the intervening island of Ireland. (a) Example track of a shearwater headed out to forage and is returning home to its burrow (black circle). The return journey closely follows the beeline home over land (dashed line), even though these birds do not travel over land and will thus have to detour around the northern Irish coast to get home (Padget et al. 2019; http://creativecommons.org/licenses/by/4.0/). (b) Shearwaters foraging further away from home (x-axis) allocated more time for homing (y-axis), suggesting they can represent their beeline distance to their burrow (Padget et al. 2019; http://creativecommons.org/licenses/by/4.0/). Despite being capable of taking the shorter detour around the island of Ireland during normal foraging, when caught at the burrow and released on the western Irish coast (R), all birds initially attempted to take a beeline direction home: (c) eight birds detoured the shorter distance around the northern coast http://creativecommons.org/licenses/by/4.0/; (d) four birds backtracked to detour along the much longer southern route (Padget et al. 2022; http://creativecommons.org/licenses/by/4.0/). This suggests that birds might use outbound information to select the shorter detour during usual homing.

island. Instead, Manx shearwaters might require outbound information to successfully detour around Ireland. When birds were displaced to the northwestern Irish coast, many birds did not take the shortest route home, with some flying the long way south around Ireland to reach their colony (Padget et al. 2022).

The need for outbound information when making detour decisions could imply a role for path integration in avian navigation. Despite being well studied in mammals and insects, path integration has been surprisingly understudied in birds. Explicit evidence for path integration in birds is restricted to a single study in which geese of various species (e.g., *Anser anser, Branta leucopsis, A. indicus*) were passively transported via an indirect route to a release location (Von Saint Paul 1982). On release, the birds headed directly homeward, but only if they had visual access to the top third of the enclosure upon release. It is unclear what cues they used, but this study suggests that path integration is available to birds and does not rely on actively walking or flying. Studies of flight control suggest that birds can use optic flow to control their speed and distance to a wall, and it is possible that this information could also be used for visual odometry (Bhagavatula et al. 2011). Indeed, the route taken during translocation experiments also appears to affect how birds respond when released. When pigeons were taken to a release site via a long detour to either the right or left, they showed a compensatory direction upon release, flying left of the direct line home if detoured left and right of the line if detoured right. This effect may reflect accumulated errors from a path integrator (Tögell and Wiltschko 1992). It was also observed only when pigeons were released outside their familiar area, suggesting that when familiar landmarks are available, pigeons are able to either reset or ignore the effects of the detour.

While path integration and the navigational map provide a means of estimating a direct vector home, sometimes a direct vector is not useful. In such cases, outbound information and spatial memory might allow birds to better plan their behavior on detouring routes. Balearic shearwaters (*P. mauretanicus*), for example, nest on colonies in the Mediterranean but travel past Gibraltar to forage in the Atlantic and English Channel. Similar to Manx shearwaters, Balearic shearwaters can estimate the distance they need to travel home and accordingly time their departure from feeding grounds. Unlike Manx shearwaters, however, the Balearic shearwaters do not appear to use a straight-line distance when making this estimation, which would involve travel over Western Europe, but instead use the distance of their actual route. These seabirds are therefore able to ignore the information that is presumably provided by their navigational map in favor of their own memory of their outbound journey (Lewin et al. 2024).

2.4 Wind

For flying birds, getting from A to B is complicated by the fact that they are moving through a flowing, fluid medium. While an onboard compass might indicate the direction in which a bird needs to travel, crosswinds, vortices, and storms can make it difficult to translate that direction into successful navigation. Wind itself may even act as a selection pressure on migratory routes; the balance of costs and benefits for any particular route may depend on the patterns of wind that birds experience (Alerstam 2011). It is thus unsurprising that birds have evolved strategies to cope with wind during navigation.

When strong winds are encountered, one strategy for navigating birds is to use the wind rather than work against it. This may involve route flexibility, such as in Eleonora's falcons (*Falco eleonorae*), which show considerable variation in their crossing points from mainland Africa to Madagascar, selecting crossing locations based on favorable wind conditions. Taking advantage of wind can also occur at different stages of a journey, with the optimal use of wind requiring a trade-off between different stages. Honey buzzards (*Pernis apivorus*), migrating north from West Africa, fly initially against headwinds to gain an advantage of favorable winds later in their journey, when they cross the Sahara (Shamoun-Baranes et al. 2017). How these strategies interact with the navigational mechanisms used by birds remains unclear. For example, the flexible routes of falcons crossing to Madagascar might require nothing more than responding to local wind cues and updating their relative heading using the navigational map. Alternatively, birds could be learning the spatiotemporal distribution of favorable winds and plan routes accordingly. Desertas petrels (*Pterodroma deserta*) foraging across the Atlantic Ocean appear to not only exploit predictable clockwise wind patterns, but do so in a way that suggests that they know where and when those winds are going to occur (Ventura et al. 2020). Wind regimes have been suggested as a driving force for some migratory detours; however, whether these detours indicate that birds possess knowledge of the wind regimes they encounter is currently unknown.

Another solution is to compensate for wind when setting a travel direction. Without compensation, even birds that regularly reorient via the navigational map will fail to reach their target (McLaren et al. 2014). The degree to which birds compensate for wind appears to be highly variable: some compensate almost completely whereas others under- or overcompensate. This variation in wind compensation may reflect a strategic choice. Partial compensation, for example, could represent a trade-off between staying on course and minimizing the energetic cost of fully compensating against sidewinds. Later in the journey, these birds may then overcompensate against winds to correct against their previous drift (Chapman et al. 2011).

Compensating for wind requires some means to sense how the wind is flowing. This can be direct, through sensing turbulence on feathers, but the best evidence suggests that birds sense wind indirectly. Although wind itself is invisible, when birds are flying over land, the visual effects of wind drift can be used to compensate. Hovering hummingbirds, for example, maintain stability by compensating for any apparent movement in the world around them (Goller and Altshuler 2014). If the world appears to be moving toward the bird, the hummingbird compensates by moving backward. Birds following a compass bearing could also be using optic flow to estimate and compensate for wind drift. Wind compensation appears to be most robust in diurnal migrants flying at lower altitudes, consistent with a need to see the ground below (Chapman et al. 2011).

Wind compensation in birds appears to be learned. In both ospreys (*Pandion haliaetus*) and honey buzzards, for example, migrating adults show significantly less wind drift than juveniles (Thorup et al. 2003). The development of this compensation has been observed in frigatebirds (*Fregata minor*) which fly within sight of land at the beginning and end of foraging but spend the majority of their foraging over the open ocean. As fledglings increased their number of foraging trips, the effect of wind drift decreased (Wynn et al. 2020a). The difference between more and less experienced birds was significantly reduced, however, when the birds were in sight of land, with all birds compensating to a similar degree. This pattern suggests that learning to compensate for wind in the presence of visual landmarks might scaffold the ability to compensate when visual landmarks are not available.

3 What Is Needed for a Cognitive Ecology of Navigation?

The promise of a cognitive ecology of navigation is not new. In 1998, Fred Dyer set out an excellent road map for such an enterprise (Dyer 1998). Noting a lack of interest in functional explanations for navigation, accompanied by a familiar focus on sensory mechanisms, he suggested an approach inspired by optimal foraging theory (Stephens and Krebs 1986). At the heart of his proposal was the need to define (a) the navigational task an animal is facing, (b) the possible currencies animals are optimizing, and (c) the constraints and trade-offs affecting this optimization. Despite similar approaches being used to study "optimal migration" (Alerstam 2011), Dyer's proposal did not make much headway. Part of the reason, I suspect, is that the questions motivating most navigation researchers do not result in the kind of data amenable to the approach Dyer advocated. A cognitive ecology of navigation should aim to explain the variation observed in the natural world; however, there has been little apparent interest in species-level differences in navigation, let alone the underlying reasons for their existence. In some cases, bird navigation is well suited to exploring species differences in navigation, as the known avian navigation toolkit includes well-understood mechanisms that can be compared across different species (e.g., differences in reliance or precision of different compasses). However, in other areas, we simply do not know enough to enable meaningful interspecies comparisons. This may reflect the traditional approaches used to study avian navigation, which tend to focus disproportionately on certain behaviors and species, thereby limiting our scope of the wider problems birds face. In addition, emphasis on sensory mechanisms may have constrained insights into the underlying cognition, limiting our ability to discern how evolved solutions differ across species. Specific open questions on each of these areas are detailed below.

3.1 Problem-Focused Questions: Putting Navigational Tasks in Context

To understand the problems navigating birds are facing, we need to put navigation in its ecological context and analyze how birds use navigation in their natural lives. This involves more than just thinking about scale. The task a species faces depends on the habitats in which it is navigating, the goals toward which it is heading, and any constraints on its freedom to move. The navigation challenges a small songbird faces when returning to its nest in a dense woodland are likely different from those faced by a large raptor as it migrates over a stretch of ocean. Open questions in this area include:

- *How does habitat affect navigation?* Most studies of navigation focus on birds traveling over landscapes rather than moving within them. Although migrating birds spend time foraging during stopovers, homing pigeons fly at altitudes of 50 m or more over relatively open landscapes, making navigation in this situation difficult to compare to movements closer to the ground or through dense vegetation. Does the physical structure of an environment and its effects on available information influence navigation strategy?
- *Are there differences in navigation between homing and foraging?* Homing birds travel from one (potentially unpredictable) location to another. In foraging, birds might need to travel directly between multiple sites (Rose et al. 2006; Tello-Ramos et al. 2021), search areas as they track where they have been (Healy and Hurly 1995), and incorporate other information such as time (Tello-Ramos et al. 2015). Does foraging require a different strategy than homing, or is foraging just a series of discrete homing tasks?
- *How does predation affect navigation strategy?* Many small birds might avoid open areas or travel too quickly to minimize the risk of attracting predators (Gaynor et al. 2019). How does this impact the way in which these birds navigate? Are routes prioritized between sheltered locations rather than direct flights to a goal?
- *What effect does mode of locomotion have on navigation?* Most studies focus on flying birds, yet multiple species cannot or only rarely fly. Despite this, some nonflying species, including penguins, show impressive homing abilities and appear to use spatial information as they forage (Shiomi et al. 2020). Does swimming or walking affect the way in which birds navigate?
- *What role does vertical dimension play in navigation?* While movements over large distances may be best modeled as navigating across a two-dimensional plane, over smaller distances many important locations are defined by their three-dimensional position (Flores-Abreu et al. 2013). Caches, nests, roosts, and foraging sites might all require birds to remember their precise locations. Soaring birds might also need to attend to their own altitude when deciding where to travel after leaving a thermal (Harel et al. 2016). Does paying attention to height influence how birds navigate, and does this differ between birds that recognize their own altitude and those that seek a specific three-dimensional location?
- *What makes a "good" navigation strategy?* "Better" navigation can be defined in many ways, including increased accuracy, higher speed, less energetically expensive, and increased efficiency. Not all of these will be as important to all species, and some (e.g., speed and accuracy) may be mutually exclusive (Chittka et al. 2009). Do birds appear to prioritize a particular currency? If so, does this vary between species in different ecological settings?

3.2 Solution-Led Questions: Making Spatial Cognition Comparable

Just as behavioral ecologists seek to explain variation in behavior and sensory ecologists aim to explain variation in sensory systems, a cognitive ecology of navigation is interested in explaining variation in spatial cognition, broadly defined to encompass all information processing relevant to navigation. This poses two problems. First, unlike behavior and sensory systems, cognition is not directly observable and thus difficult to measure. Robustly identifying variation, therefore, is going to be a challenge. Second, spatial cognition is arguably not a single trait but rather a web of different mechanisms, including attention, learning, memory, decision making, and many others. Differences in navigation performance may arise when there is variance in any of these components. Despite this, we do see signs of species differences in spatial cognition. In some migrating species, only experienced individuals can correctly reorient after being displaced (Perdeck 1958; Thorup et al. 2007). In other species, naïve birds can successfully reorient to their original destination (Kishkinev et al. 2015; Thorup et al. 2020). In yet others, even experienced adults cannot reorient (Kishkinev et al. 2016). To understand these apparent differences in ability, we need a firmer understanding of how navigational information is learned and organized, which may require questioning some central tenets of avian navigation. Open questions in this area include:

- *What role do psychological "general processes" play in avian navigation?* In most studies of long-distance navigation, compasses and maps are treated as isolated adaptations, with restricted inputs and outputs that are processed differently to other information. This modular perspective can be contrasted with smaller-scale spatial learning in birds, which follows the rules of general associative learning and generalization. Is this difference due to actual differences in biology or to differences in the fields that study each topic? Can behavior like map navigation be explained using known learning mechanisms (Guilford and de Perera 2017)?
- *Do experimental paradigms capture mechanisms that birds actually use?* Large-scale translocations, sensory disorientation or deprivation, and clock-shifting all introduce artificial scenarios designed to disentangle the cues birds use for navigation. Does the artificiality of these paradigms make it harder to apply resultant findings to birds navigating under natural conditions? Do any experienced birds actually use true navigation, regularly updating a constant vector to their goal, or is it an artifact of an extreme experimental manipulation (Wynn and Liedvogel 2023)?
- *Can the navigational map or familiar area map store multiple concurrent locations?* Navigational and familiar maps are built to describe how birds reach a single location. They can be contrasted with the "cognitive map" described by (Tolman 1948) and O'Keefe and Nadel (1979), which supposedly stores the spatial relationships between multiple locations and enables animals to make spatial computations. While homing pigeons can navigate from an unfamiliar location to different locations based on current needs (Blaser et al. 2013), it is unclear whether they "know" where these locations are in relation to one another. Can birds use either map to form an efficient,

novel route between known locations?

- *What nonpositional information can avian maps contain?* Some birds that travel over large distances appear to know their position with respect to a goal and are able to learn where and when to find favorable winds to reach the goal. Other birds, however, seem not to recognize what would appear to be important features, such as large landmasses (Padget et al. 2019). Are large-scale representations incorporated into a single navigational map, or do they exist separately?
- *How do birds encode outbound information?* Studies with shearwaters suggest that outbound information is important for estimating travel distance and planning detours, but there are multiple ways to acquire this information. How do birds with and without visual landmarks estimate the distance they travel? What role does path integration play in avian navigation?
- *How does landmark use at small scales relate to landmark use at larger scales?* Landmark use over familiar areas is often discussed separately to landmark use over small scales, even employing different terminology, such as "pilotage" (Able 2001). To what extent can strategies seen in birds navigating over small scales—such as panorama-matching (Pritchard et al. 2018; Gee et al. 2024), triangulating bearings (Jones et al. 2002; Kamil et al. 2001), and Bayesian integration (Legge et al. 2016; Sutton 2002)—explain landmark use over larger distances? How important are map-and-compass-style strategies over small scales?

4 Conclusion

Reviews of navigation in birds often sweepingly refer to "bird navigation" as a whole, as though there is no difference between the thousands of species of birds. This seems unlikely. Even though the ability to navigate is incredibly useful, not all bird species need to navigate to the same extent, use the same cues, or traverse the same kinds of terrain. To adapt a metaphor from Alan Kamil (1987), avian navigation could be like feathers: ubiquitous among birds but still shaped by the demands each species faces. As seen in the examples presented in this chapter, birds are well equipped to address the issues that emerge during navigation. Our next step, and the real challenge, is to understand how ecological challenges give rise to navigation solutions. By looking at patterns and variation in the navigational challenges faced by birds and other taxa, as well as the solutions they have evolved, we can build upon the advances made over the past 120 years and deepen our understanding of how birds have become the spectacular navigators they are today.

Acknowledgments I would like to thank Ken Cheng, Shoko Sugasawa, and Oliver Padget for comments on earlier drafts of this review, and for the useful comments and inspiring conversations with attendees at this Ernst Strüngmann Forum.

References

Able KP (2001) The Concepts and Terminology of Bird Navigation. J Avian Biol 32 (2):174–183. https://doi.org/10.1034/j.1600-048X.2001.320211.x

Åkesson S (1993) Effect of Geomagnetic Field on Orientation of the Marsh Warbler, *Acrocephalus palustris*, in Sweden and Kenya. Anim Behav 16 (6):1157–1167. https://doi.org/10.1006/anbe.1993.1305

Åkesson S, and Bianco G (2017) Route Simulations, Compass Mechanisms and Long-Distance Migration Flights in Birds. J Comp Physiol A Neuroethol Sens Neural Behav Physiol 203 (6):475–490. https://doi.org/10.1007/s00359-017-1171-y

Åkesson S, Morin J, Muheim R, and Ottosson U (2001) Avian Orientation at Steep Angles of Inclination: Experiments with Migratory White-Crowned Sparrows at the Magnetic North Pole. Proc R Soc Lond B Biol Sci 268 (1479):1907–1913. https://doi.org/10.1098/rspb.2001.1736

Alerstam T (2006) Conflicting Evidence About Long-Distance Animal Navigation. Science 313 (5788):791–794. https://doi.org/10.1126/science.1129048

——— (2011) Optimal Bird Migration Revisited. J Ornithol 152 (S1):5–23. https://doi.org/10.1007/s10336-011-0694-1

Beck W, and Wiltschko W (1982) The Magnetic Field as a Reference System for Genetically Encoded Migratory Direction in Pied Flycatchers (*Ficedula hypoleuca* Pallas). Z Tierpsychol 60 (1):41–46. https://doi.org/10.1111/j.1439-0310.1982.tb01075.x

Berthold P (1991) Spatiotemporal Programmes and Genetics of Orientation. In: Berthold P (ed) Orientation in Birds, vol 60. Birkhäuser, Basel, pp 86–105

Bhagavatula PS, Claudianos C, Ibbotson MR, and Srinivasan MV (2011) Optic Flow Cues Guide Flight in Birds. Curr Biol 21 (21):1794–1799. https://doi.org/10.1016/j.cub.2011.09.009

Blaser N, Dell'Omo G, Dell'Ariccia G, Wolfer DP, and Lipp HP (2013) Testing Cognitive Navigation in Unknown Territories: Homing Pigeons Choose Different Targets. J Exp Biol 216 (16):3123–3131. https://doi.org/10.1242/jeb.083246

Bonadonna F, and Bretagnolle V (2002) Smelling Home: A Good Solution for Burrow-Finding in Nocturnal Petrels? J Exp Biol 205 (16):2519–2523. https://doi.org/10.1242/jeb.205.16.2519

Brodbeck DR (1994) Memory for Spatial and Local Cues: A Comparison of a Storing and a Nonstoring Species. Anim Learn Behav 22 (2):119–133. http://link.springer.com/10.3758/BF03199912

Chapman J, Klaassen R, Drake VA, et al. (2011) Animal Orientation Strategies for Movement in Flows. Curr Biol 21 (20):R861–R870. https://doi.org/10.1016/j.cub.2011.08.014

Cheng K, Shettleworth SJ, Huttenlocher J, and Rieser JJ (2007) Bayesian Integration of Spatial Information. Psychol Bull 133 (4):625–637. https://doi.org/10.1037/0033-2909.133.4.625

Chernetsov N, Kishkinev D, Gashkov S, Kosarev V, and Bolshakov CV (2008) Migratory Programme of Juvenile Pied Flycatchers, *Ficedula hypoleuca*, from Siberia Implies a Detour around Central Asia. Anim Behav 75 (2):539–545. https://doi.org/10.1016/j.anbehav.2007.05.019

Chittka L, Skorupski P, and Raine NE (2009) Speed–Accuracy Tradeoffs in Animal Decision Making. Trends Ecol Evol 24 (7):400–407. https://doi.org/10.1016/j.tree.2009.02.010

Delmore KE, Van Doren BM, Conway GJ, et al. (2020) Individual Variability and Versatility in an Eco-Evolutionary Model of Avian Migration. Proc R Soc Lond B Biol Sci 287 (1938):20201339. https://doi.org/10.1098/rspb.2020.1339

Dukas R (1998) Cognitive Ecology: The Evolutionary Ecology of Information Processing and Decision Making. Univ Chicago Press, Chicago

Dyer FC (1998) Cognitive Ecology of Navigation. In: Dukas R (ed) Cognitive Ecology: The Evolutionary Ecology of Information Processing and Decision Making. Univ Chicago Press, Chicago, pp 201–260

Emlen ST (1970) Celestial Rotation: Its Importance in the Development of Migratory Orientation. Science 170 (3963):1198–1201. https://www.science.org/doi/10.1126/science.170.3963.1198

Flores-Abreu IN, Hurly TA, and Healy SD (2013) Three-Dimensional Spatial Learning in Hummingbirds. Anim Behav 85 (3):579–584. https://doi.org/10.1016/j.anbehav.2012.12.019

Gagliardo A, Bried J, Lambardi P, et al. (2013) Oceanic Navigation in Cory's Shearwaters: Evidence for a Crucial Role of Olfactory Cues for Homing after Displacement. J Exp Biol 216 (15):2798–2805. https://doi.org/10.1242/jeb.085738

Gagliardo A, Ioalè P, Savini M, Dell'Omo G, and Bingman VP (2009) Hippocampal-Dependent Familiar Area Map Supports Corrective Re-Orientation Following Navigational Error during Pigeon Homing: A GPS-Tracking Study. Eur J Neurosci 29 (12):2389–2400. https://doi.org/10.1111/j.1460-9568.2009.06793.x

Gagliardo A, Pollonara E, and Wikelski M (2020) Pigeons Remember Visual Landmarks after One Release and Rely Upon Them More If They Are Anosmic. Anim Behav 166:85–94. https://doi.org/10.1016/j.anbehav.2020.05.009

Gaynor KM, Brown JS, Middleton AD, Power ME, and Brashares JS (2019) Landscapes of Fear: Spatial Patterns of Risk Perception and Response. Trends Ecol Evol 34 (4):355–368. https://doi.org/10.1016/j.tree.2019.01.004

Gee M, Gonzalez-Granda N, Joshi S, et al. (2024) Navigating the Landscape of Fear. bioRxiv Preprint:608007. https://doi.org/10.1101/2024.08.15.608007

Gill RE, Jr., Piersma T, Hufford G, Servranckx R, and Riegen A (2005) Crossing the Ultimate Ecological Barrier: Evidence for an 11 000-Km-Long Nonstop Flight from Alaska to New Zealand and Eastern Australia by Bar-Tailed Godwits. Condor 107 (1):1–20. https://doi.org/10.1093/condor/107.1.1

Goller B, and Altshuler DL (2014) Hummingbirds Control Hovering Flight by Stabilizing Visual Motion. PNAS 111 (51):18375–18380. https://doi.org/10.1073/pnas.1415975111

Goodyear AJ, and Kamil AC (2004) Clark's Nutcrackers (*Nucifraga columbiana*) and the Effects of Goal—Landmark Distance on Overshadowing. J Comp Psychol 118 (3):258–264. https://doi.org/10.1037/0735-7036.118.3.258

Gould-Beierle KL, and Kamil AC (1996) The Use of Local and Global Cues by Clark's Nutcrackers, *Nucifraga columbiana*. Anim Behav 52 (3):519–528. https://doi.org/10.1006/anbe.1996.0194

Griffiths C, Schiffner I, Price E, et al. (2021) Repeated Training of Homing Pigeons Reveals Age-Dependent Idiosyncrasy and Visual Landmark Use. Anim Behav 177:159–170. https://doi.org/10.1016/j.anbehav.2021.05.004

Guilford T, and Biro D (2014) Route Following and the Pigeon's Familiar Area Map. J Exp Biol 217 (2):169–179. https://doi.org/10.1242/jeb.092908

Guilford T, and de Perera TB (2017) An Associative Account of Avian Navigation. J Avian Biol 48 (1):191–195. https://doi.org/10.1111/jav.01355

Harel R, Duriez O, Spiegel O, et al. (2016) Decision-Making by a Soaring Bird: Time, Energy and Risk Considerations at Different Spatio-Temporal Scales. Philos Trans R Soc Lond B Biol Sci 371 (1704):20150397. https://doi.org/10.1098/rstb.2015.0397

Healy SD, and Hurly TA (1995) Spatial Memory in Rufous Hummingbirds (*Selasphorus rufus*): A Field Test. Anim Learn Behav 23 (1):63–68. https://doi.org/10.3758/BF03198016

——— (1998) Rufous Hummingbirds' (*Selasphorus rufus*) Memory for Flowers: Patterns or Actual Spatial Locations? J Exp Psychol Anim Learn Cogn 24 (4):396–404. https://doi.org/10.1037/0097-7403.24.4.396

——— (2013) What Hummingbirds Can Tell Us about Cognition in the Wild. Comp Cogn Behav Rev 8:13–28. https://doi.org/10.3819/ccbr.2013.80002

Holland RA (2003) The Role of Visual Landmarks in the Avian Familiar Area Map. J Exp Biol 206 (11):1773–1778. https://doi.org/10.1242/jeb.00365

Jones JE, Antoniadis E, Shettleworth SJ, and Kamil AC (2002) A Comparative Study of Geometric Rule Learning by Nutcrackers (*Nucifraga columbiana*), Pigeons (*Columba livia*) and Jackdaws (*Corvus monedula*). J Comp Psychol 116 (4):350–356. https://doi.org/10.1037/0735-7036.116.4.350

Jones JE, and Kamil AC (2001) The Use of Relative and Absolute Bearings by Clark's Nutcrackers, *Nucifraga columbiana*. Anim Learn Behav 29 (2):120–132. https://doi.org/10.3758/BF03192821

Kamil AC (1987) A Synthetic Approach to the Study of Animal Intelligence. Nebr Symp Motiv 35:258–308. https://pubmed.ncbi.nlm.nih.gov/3332032/

Kamil AC, Goodyear AJ, and Cheng K (2001) The Use of Landmarks by Clark's Nutcrackers: First Tests of a New Model. J Navig 54 (3):429–435. https://doi.org/10.1017/S0373463301001436

Kelly DM, Cheng K, Balda R, and Kamil AC (2019) Effects of Sun Compass Error on Spatial Search by Clark's Nutcrackers. Integr Zool 14 (2):172–181. https://doi.org/10.1111/1749-4877.12302

Kishkinev D, Chernetsov N, Pakhomov A, Heyers D, and Mouritsen H (2015) Eurasian Reed Warblers Compensate for Virtual Magnetic Displacement. Curr Biol 25 (19):R822–R824. https://doi.org/10.1016/j.cub.2015.08.012

Kishkinev D, Heyers D, Woodworth BK, et al. (2016) Experienced Migratory Songbirds Do Not Display Goal-Ward Orientation after Release Following a Cross-Continental Displacement: An Automated Telemetry Study. Sci Rep 6 (1):37326. https://doi.org/10.1038/srep37326

Kramer G (1950) Weitere Analyse der Faktoren, Welche die Zugaktivität des Gekäfigten Vogels Orientieren. Naturwiss 37 (16):377–378. https://doi.org/10.1007/BF00626007

——— (1952) Experiments on Bird Orientation. Ibis 94 (2):265–285. https://doi.org/10.1111/j.1474-919X.1952.tb01817.x

——— (1957) Experiments on Bird Orientation and Their Interpretation. Ibis 99 (2):196–227. https://doi.org/10.1111/j.1474-919X.1957.tb01947.x

Legge ELG, Madan CR, Spetch ML, and Ludvig EA (2016) Multiple Cue Use and Integration in Pigeons (*Columba livia*). Anim Cogn 19 (3):581–591. https://doi.org/10.1007/s10071-016-0963-8

Legge ELG, Spetch ML, and Batty ER (2009) Pigeons' (*Columba livia*) Hierarchical Organization of Local and Global Cues in Touch Screen Tasks. Behav Processes 80 (2):128–139. https://doi.org/10.1016/j.beproc.2008.10.011

Lewin PJ, Wynn J, Arcos JM, et al. (2024) Climate Change Drives Migratory Range Shift via Individual Plasticity in Shearwaters. PNAS 121 (6):e2312438121. https://doi.org/10.1073/pnas.2312438121

Lorenz TJ, Sullivan KA, Bakian AV, and Aubry CA (2011) Cache-Site Selection in Clark's Nutcracker (*Nucifraga columbiana*). The Auk 128 (2):237–247. https://doi.org/10.1525/auk.2011.10101

Mann R, Freeman R, Osborne M, et al. (2011) Objectively Identifying Landmark Use and Predicting Flight Trajectories of the Homing Pigeon Using Gaussian Processes. J R Soc Interface 8 (55):210–219. https://doi.org/10.1098/rsif.2010.0301

Martin GR (2017) The Sensory Ecology of Birds. Oxford Univ Press, Oxford

McLaren JD, Shamoun-Baranes J, Dokter AM, Klaassen RHG, and Bouten W (2014) Optimal Orientation in Flows: Providing a Benchmark for Animal Movement Strategies. J R Soc Interface 11 (99):20140588. https://doi.org/10.1098/rsif.2014.0588

Mouritsen H (1998) Modelling Migration: The Clock-and-Compass Model Can Explain the Distribution of Ringing Recoveries. Anim Behav 56 (4):899–907. https://doi.org/10.1006/anbe.1998.0826

——— (2018) Long-Distance Navigation and Magnetoreception in Migratory Animals. Nature 558 (7708):50–59. https://doi.org/10.1038/s41586-018-0176-1

Muheim R, Moore FR, and Phillips JB (2006) Calibration of Magnetic and Celestial Compass Cues in Migratory birds—A Review of Cue-Conflict Experiments. J Exp Biol 209 (1):2–17. https://doi.org/10.1242/jeb.01960

Muheim R, Phillips JB, and Deutschlander ME (2009) White-Throated Sparrows Calibrate Their Magnetic Compass by Polarized Light Cues during Both Autumn and Spring Migration. J Exp Biol 212 (21):3466–3472. https://doi.org/10.1242/jeb.032771

O'Keefe J, and Nadel L (1979) Précis of O'Keefe & Nadel's the Hippocampus as a Cognitive Map. Behav Brain Sci 2 (4):487–494. https://doi.org/10.1017/S0140525X00063949

Packmor F, Kishkinev D, Zechmeister T, Mouritsen H, and Holland RA (2024) Migratory Birds Can Extract Positional Information from Magnetic Inclination and Magnetic Declination Alone. Proc R Soc Lond B Biol Sci 291 (2034):rspb20241363. https://doi.org/10.1098/rspb.2024.1363

Padget O, Bond SL, Kavelaars MM, et al. (2018) *In Situ* Clock Shift Reveals That the Sun Compass Contributes to Orientation in a Pelagic Seabird. Curr Biol 28 (2):275–279.e272. https://doi.org/10.1016/j.cub.2017.11.062

Padget O, Dell'Ariccia G, Gagliardo A, González-Solís J, and Guilford T (2017) Anosmia Impairs Homing Orientation but Not Foraging Behaviour in Free-Ranging Shearwaters. Sci Rep 7 (1):9668. https://doi.org/10.1038/s41598-017-09738-5

Padget O, Gillies N, Syposz M, Lockley E, and Guilford T (2022) Shearwaters Sometimes Take Long Homing Detours When Denied Natural Outward Journey Information. Biol Lett 18 (2):20210503. https://doi.org/10.1098/rsbl.2021.0503

Padget O, Stanley G, Willis JK, et al. (2019) Shearwaters Know the Direction and Distance Home but Fail to Encode Intervening Obstacles after Free-Ranging Foraging Trips. PNAS 116 (43):21629–21633. https://doi.org/10.1073/pnas.1903829116

Perdeck AC (1958) Two Types of Orientation in Migrating Starlings, *Sturnus vulgaris* L., and Chaffinches, *Fringilla coelebs* L., as Revealed by Displacement Experiments. Ardea 46:1–36. https://doi.org/10.5253/arde.v1i2.p1

Pritchard DJ, Hurly TA, and Healy SD (2018) Wild Hummingbirds Require a Consistent View of Landmarks to Pinpoint a Goal Location. Anim Behav 137:83–94. https://doi.org/10.1016/j.anbehav.2018.01.014

Real LA (1993) Toward a Cognitive Ecology. Trends Ecol Evol 8 (11):413–417. https://doi.org/10.1016/0169-5347(93)90044-P

Ritz T, Thalau P, Phillips JB, Wiltschko R, and Wiltschko W (2004) Resonance Effects Indicate a Radical-Pair Mechanism for Avian Magnetic Compass. Nature 429 (6988):177–180. https://doi.org/10.1038/nature02534

Rose E, Haag-Wackernagel D, and Nagel P (2006) Practical Use of GPS-Localization of Feral Pigeons *Columba livia* in the Urban Environment. Ibis 148 (2):231–239. https://doi.org/10.1111/j.1474-919X.2006.00499.x

Schmaljohann H, Rautenberg T, Muheim R, Naef-Daenzer B, and Bairlein F (2012) Response of a Free-Flying Songbird to an Experimental Shift of the Light Polarization Pattern around Sunset. J Exp Biol 216 (8):1381–1387. https://doi.org/10.1242/jeb.080580

Schmidt-Koenig K (1958) Experimentelle Einflußnahme auf die 24-Stunden-Periodik bei Brieftauben und Deren Auswirkungen unter besonderer Berücksichtigung des Heimfindevermögens. Z Tierpsychol 15 (3):301–331. https://doi.org/10.1111/j.1439-0310.1958.tb00568.x

Shamoun-Baranes J, Liechti F, and Vansteelant WMG (2017) Atmospheric Conditions Create Freeways, Detours and Tailbacks for Migrating Birds. J Comp Physiol A Neuroethol Sens Neural Behav Physiol 203 (6):509–529. https://doi.org/10.1007/s00359-017-1181-9

Shiomi K, Kokubun N, Shimabukuro U, and Takahashi A (2020) Homing Ability of Adélie Penguins Investigated with Displacement Experiments and Bio-Logging. Ardea 107 (3):333–339. https://doi.org/10.5253/arde.v107i3.a12

Shoji A, Aris-Brosou S, Owen E, et al. (2016) Foraging Flexibility and Search Patterns Are Unlinked during Breeding in a Free-Ranging Seabird. Mar Biol 163 (4):72. https://doi.org/10.1007/s00227-016-2826-x

Stephens DW, and Krebs JR (1986) Foraging Theory. Princeton Univ Press, Princeton
Stevens M (2013) Sensory Ecology, Behaviour, and Evolution. Oxford Univ Press, Oxford
Sutton JE (2002) Multiple-Landmark Piloting in Pigeons (*Columba livia*): Landmark Configuration as a Discriminative Cue. J Comp Psychol 116 (4):391–403. https://doi.org/10.1037/0735-7036.116.4.391
Tello-Ramos MC, Hurly TA, Barclay M, and Healy SD (2021) Hummingbirds Modify Their Routes to Avoid a Poor Location. Learn Behav 50:89–98. https://doi.org/10.3758/s13420-021-00476-3
Tello-Ramos MC, Hurly TA, Higgott C, and Healy SD (2015) Time–Place Learning in Wild, Free-Living Hummingbirds. Anim Behav 104:123–129. https://doi.org/10.1016/j.anbehav.2015.03.015
Thorup K, Alerstam T, Hake M, and Kjellén N (2003) Bird Orientation: Compensation for Wind Drift in Migrating Raptors Is Age Dependent. Proc R Soc Lond B Biol Sci 270 (suppl_1):S8–S11. https://doi.org/10.1098/rsbl.2003.0014
Thorup K, Bisson I-A, Bowlin MS, et al. (2007) Evidence for a Navigational Map Stretching across the Continental U.S. In a Migratory Songbird. PNAS 104 (46):18115–18119. https://doi.org/10.1073/pnas.0704734104
Thorup K, Vega ML, Snell KRS, et al. (2020) Flying on Their Own Wings: Young and Adult Cuckoos Respond Similarly to Long-Distance Displacement during Migration. Sci Rep 10 (1):7698. https://doi.org/10.1038/s41598-020-64230-x
Tinbergen N (1963) On Aims and Methods of Ethology. Z Tierpsychol 20 (4):410–433. https://doi.org/10.1111/j.1439-0310.1963.tb01161.x
Tögell A, and Wiltschko R (1992) Detour Experiments with Homing Pigeons: Information Obtained during the Outward Journey Is Included in the Navigational Process. Behav Ecol Sociobiol 31 (2):73–79. https://doi.org/10.1007/BF00166339
Tolman EC (1948) Cognitive Maps in Rats and Men. Psychol Rev 55 (4):189–208. https://doi.org/10.1037/h0061626
Vander Wall SB (1982) An Experimental Analysis of Cache Recovery in Clark's Nutcracker. Anim Behav 30 (1):84–94. https://doi.org/10.1016/S0003-3472(82)80240-6
Ventura F, Granadeiro JP, Padget O, and Catry P (2020) Gadfly Petrels Use Knowledge of the Windscape, Not Memorized Foraging Patches, to Optimize Foraging Trips on Ocean-Wide Scales. Proc R Soc Lond B Biol Sci 287 (1918):20191775. https://doi.org/10.1098/rspb.2019.1775
Von Saint Paul U (1982) Do Geese Use Path Integration for Walking Home? In: Papi F, Wallraff HG (eds) Avian Navigation. Springer, Berlin, Heidelberg, pp 298–307
Wallraff HG (2004) Avian Olfactory Navigation: Its Empirical Foundation and Conceptual State. Anim Behav 67 (2):189–204. https://doi.org/10.1016/j.anbehav.2003.06.007
Wehner R (1998) Navigation in Context: Grand Theories and Basic Mechanisms. J Avian Biol 29 (4):370–386. https://doi.org/10.2307/3677156
Wiltschko R, and Wiltschko W (2003) Avian Navigation: From Historical to Modern Concepts. Anim Behav 65 (2):257–272. https://doi.org/10.1006/anbe.2003.2054
Wiltschko W, Weindler P, and Wiltschko R (1998) Interaction of Magnetic and Celestial Cues in the Migratory Orientation of Passerines. J Avian Biol 29 (4):606–617. https://doi.org/10.2307/3677181
Wiltschko W, and Wiltschko R (1972) Magnetic Compass of European Robins. Science 176 (4030):62–64. https://doi.org/10.1126/science.176.4030.62
——— (1992) Migratory Orientation: Magnetic Compass Orientation of Garden Warblers (*Sylvia borin*) after a Simulated Crossing of the Magnetic Equator. Ethology 91 (1):70–74. https://doi.org/10.1111/j.1439-0310.1992.tb00851.x
Wynn J, Collet J, Prudor A, et al. (2020a) Young Frigatebirds Learn How to Compensate for Wind Drift. Proc R Soc Lond B Biol Sci 287 (1937):20201970. https://doi.org/10.1098/rspb.2020.1970
Wynn J, and Liedvogel M (2023) Lost: on What Level Should We Aim to Understand Animal Navigation? J Exp Biol 226 (10):jeb245441. https://doi.org/10.1242/jeb.245441

Wynn J, Padget O, Mouritsen H, et al. (2022) Magnetic Stop Signs Signal a European Songbird's Arrival at the Breeding Site after Migration. Science 375 (6579):446–449. https://doi.org/10.1126/science.abj4210

Wynn J, Padget O, Mouritsen H, Perrins C, and Guilford T (2020b) Natal Imprinting to the Earth's Magnetic Field in a Pelagic Seabird. Curr Biol 30 (14):2869–2873.e2862. https://doi.org/10.1016/j.cub.2020.05.039

3

How Roboticists Think about Navigation and Major Unsolved Challenges

Michael Milford

Abstract To expand interdisciplinary consideration of the mechanisms involved in navigation, this chapter provides an overview of research into robot navigation. Discussion is intended for a multidisciplinary audience, including non-roboticists, and focuses on modular, probabilistic, and learning-based systems. Ultra-simplistic robot navigation systems such as those found in micro- and nano-robot systems are discussed, and recent developments driven by advances in machine learning, including large language models and vision language models, are outlined. Key components of robot navigation as well as challenges in the field are presented, and points of intersection between artificial and biological systems highlighted. To overcome current challenges, artificial systems must be able to emulate the adaptability, introspective awareness, and efficiency of biological systems. Solving these challenges in robots may, in turn, generate new insights into the biological systems from which inspiration has been drawn.

Keywords Robotics, navigation, mapping, localization, sensing

1 Introduction

Robot navigation systems range in complexity from simple to sophisticated, varying from explicitly hard-coded and rigid "classical" navigation approaches to end-to-end learning systems. In addition, they are characterized by numerous sensors, computational hardware types, and platforms (navigation is performed by robot arms, wheeled robots and autonomous vehicles, flying drones or fixed-wing aircraft, submersible robots, walking robots, and even robots inside the human body), which vary in form nearly as much as organisms do in the natural kingdom.

Michael Milford (✉)
Centre for Robotics and School of Electrical Engineering and Robotics, Queensland University of Technology, Brisbane QLD 4000, Australia
Email: michael.milford@qut.edu.au

N. S. Newcombe and K. Cheng (eds.), *Challenges in Navigation Research*,
Strüngmann Forum Reports, https://doi.org/10.1007/978-3-032-20563-6_3

The field of robotics encompasses the design, construction, and operation of semi- or fully automated machines capable of performing tasks that typically require human intelligence or physical labor. Robotics has its roots in automation, which first appeared primarily in industrial settings where repetitive, precise but highly constrained tasks were being performed, and navigation was not required. The field of robotics has been marked by significant advances in both hardware and software, including of course artificial intelligence (AI). Today, robots are utilized to various extents across a range of domains, including manufacturing, health care, environmental monitoring and service industries, each requiring different levels of autonomy and specialization. But despite their high profile and specific success stories, widespread deployment—a robot in every home, as Bill Gates (2008) once coined—has yet to occur. There are many challenges that limit their proliferation: one of the major ones is navigating effectively and reliably in their intended operational environments.

1.1 Purpose and Scope of this Article

Within a single article of limited length intended for a multidisciplinary audience including non-roboticists, it is not feasible to provide a comprehensive and detailed treatment of all aspects of robot navigation. There are several dimensions that one could theoretically cover, such as the complete history of robot and related navigation systems, the spectrum of robot navigation system complexity (from simple to sophisticated), and the variation from explicitly hard-coded and rigid "classical" navigation approaches to end-to-end learning systems. In addition, there are numerous sensors, computational hardware types, and platforms, which vary in form nearly as much as organisms do in the natural kingdom.

Given the context of this chapter, written both before and after a highly interdisciplinary symposium on navigation, my aim is to provide some broad context on robot navigation systems, with a particular focus on robot navigation research and development over the past 25 years, and the modular, probabilistic, and learning-based systems that have dominated the field. I provide only a brief treatment of ultra-simplistic robot navigation systems—such as those found in micro- and nano-robot systems—and only highlight recent developments driven by advances in machine learning, including large language models and vision language models. These fields advance rapidly and hence any detailed treatment of them is likely to be soon out of date. Thus, they are mentioned for awareness but not discussed in detail.

1.2 Evolution of Robotics: Industrial Automation to Autonomous Systems

Some of the earliest industrial robots, like the Unimate developed in the 1960s, revolutionized manufacturing by automating repetitive tasks such as welding and assembly (Gasparetto and Scalera 2018). These robots were confined to structured environments, performing specific, preprogrammed tasks without the need for external sensing or adaptability.

As technology advanced, robots began to transition from simple automation to autonomy. This shift was largely driven by commercial opportunity as well as developments in AI, machine learning, and sensor technologies, which enabled robots to operate better in more unstructured and dynamic environments. Modern autonomous systems are now capable of perceiving their surroundings, making decisions in real time, and adapting to changing conditions. This transformation expanded the use of robots beyond industrial applications, enabling them to perform tasks such as exploring lunar and planetary bodies, facilitating surgery, and autonomous driving. Still, the balance between full automation and human operation remains varied.

1.3 Recent Developments: Drones, Autonomous Vehicles, and the Rebirth of Humanoids

In recent years, the field of robotics has been marked by significant advancements in three major areas, all with significant navigational components: drones, autonomous vehicles, and (the resurgence of) walking robots, including quadrupeds and humanoids (Figure 3.1). All three of these platform types have significant navigation requirements and challenges. Walking robots and drones, in particular, embody aspects of different natural navigators, ranging from four-legged animals to flying birds and insects.

Drones, also known as unmanned aerial vehicles (UAVs), have rapidly gained traction across various industries, including agriculture, logistics, defense, and surveillance. The evolution of drone technology, particularly in terms of navigation, sensing, and battery efficiency, has enabled these systems to autonomously perform complex missions, such as monitoring crop health, delivering packages, and surveying remote or dangerous areas with minimal human input (Floreano and Wood 2015).

Autonomous vehicles represent another transformative leap in robotics. Companies such as Waymo and Tesla have pioneered self-driving technologies that rely on sophisticated sensor arrays and computing hardware. LiDAR (light detection and ranging), which is not used by Tesla, cameras, and radar are coupled with advanced machine-learning algorithms to navigate roads, enable autonomous vehicles to avoid obstacles (especially pedestrians, cyclists, and other vehicles), and make real-time driving decisions. These vehicles are poised to potentially transform aspects of transportation, logistics, and public transit on a global scale. For example, Waymo recently passed the milestone of delivering 150,000 paid autonomous rides involving 1 million miles of driving per week (Ohnsman 2024). While debate rages over their exact level of driving competence, financial viability, the ethics of testing around the general population, and the impact on transportation jobs, it is clear that the technology, including their navigation systems, is at least approaching technical viability in specific environments.

Simultaneously, there has been renewed interest in the development of walking robots, initially as a four-legged quadruped. After two decades of little activity, recent advances in AI, robotics, and material sciences have brought humanoids

(a)
(b)
(c)

Figure 3.1 Three classes of robots currently under intense development: (a) bipedal and quadrupedal walking robots, (b) drones, and (c) autonomous vehicles. Photos by Michael Milford: (a) taken at the IROS2024 conference, (b) at Queensland University of Technology Centre for Robotics, and (c) in Los Angeles.

back into the spotlight. Companies like Boston Dynamics, Honda, Agility Robotics, Unitree, and Figure AI have introduced robots such as Atlas and Optimus that are capable of performing sophisticated tasks involving manipulation, mobility, and human-robot interaction (Ackerman 2024). These humanoid robots are designed to operate in human-centric environments, where their physical configuration matches a world designed and built for humans, and are proposed to revolutionize both industrial (e.g., manufacturing) and consumer (e.g., home helper robots) applications. To be viable, these humanoids will need to effectively navigate these human-centric environments.

2 Robot Navigation: An Overview

2.1 The Importance of Navigation in Robotics

Navigation is one of the most fundamental challenges in robotics, enabling autonomous systems to move through and interact with their environment as they pursue specific goals. Whether it is a robot delivering a package, an autonomous vehicle navigating through a city, or a drone mapping a disaster zone, accurate and reliable navigation is an essential enabler of the robot's mission. Navigation systems provide the critical information needed for robots to determine their location, map their surroundings, and plan a path to their destination (although simpler systems may employ more reactive, behavioral strategies). Without effective navigation, even the most advanced robots are reduced to static, noninteractive machines, or must be remote-controlled by people. Importantly, navigation involves more than getting from point A to point B: it requires real-time decision making, obstacle avoidance, and adaptation to dynamic, often unpredictable environments.

2.2 Historical Milestones

Early robot navigation systems in the 1960s and 1970s relied heavily on fixed paths and environmental cues and had a limited capacity for real-time decision making. Consequently, these early robots were restricted to structured environments such as offices or factories, where predictable conditions permitted the use of simple navigation systems based on physical markers or preprogrammed routes.

In the 1980s and 1990s, sensors like sonar and LiDAR—both of which provide a range estimate to the environment around the sensor—as well as advancements in computational power enabled robots to begin perceiving and mapping their environments in more sophisticated ways. Interestingly, most sensors used by robots were not initially designed specifically for robot navigation: instead, they were co-opted from consumer camera technology, factory safety monitoring laser sensors, and gaming console depth sensor technology—a point of difference from natural systems where sensors have coevolved with the creatures using them. Landmark developments included the introduction of SLAM (simultaneous localization and mapping), which enabled robots to build maps (*mapping*) of unknown environments while *simultaneously* keeping track of their location (*localizing*) within those maps. SLAM became a cornerstone of autonomous navigation and remains widely used in robotics today (Cadena et al. 2016), although it is a capability that is not required for all navigational contexts.

Further advancements in the 2000s, including ongoing work on SLAM, were driven by breakthroughs in probabilistic robotics and the introduction of machine-learning techniques, especially following the rapid growth of deep learning in 2012. Probabilistic methods allowed robots to address uncertainty in sensor data, enhancing their ability to operate in dynamic environments. Autonomous vehicles like those developed for the DARPA Grand Challenge[1] in the mid-2000s, which used assistive navigational tools like GPS waypoints, pushed the boundaries of what robots could achieve in outdoor, unstructured environments. It has been argued that DARPA kicked off the "modern" autonomous-vehicle technology race, even though autonomous-vehicle prototypes were being developed as early as the 1980s (Pomerleau 1988).

[1] A competition for autonomous vehicles, funded by the U.S. Defense Advanced Research Projects Agency.

2.3 Current State of the Art

Today, the state of the art in robot navigation continues to evolve rapidly. Advanced sensor technologies, such as LiDAR, depth cameras, and high-precision GPS, provide robots with rich, navigationally relevant information about their surroundings. This information is processed using sophisticated algorithms that integrate perception, mapping, and motion planning, both at local and global levels.

One of the most significant advancements in recent years is the integration of deep learning techniques with traditional navigation algorithms. These hybrid approaches allow robots to recognize objects, understand their context, and make more informed decisions while navigating at a higher level of capability than previously possible, at the cost of extra system complexity. In addition, multiagent systems are becoming more common; robots can coordinate their movements with other robots, enhancing their effectiveness in large-scale operations such as warehouse automation or search-and-rescue missions (Ebadi et al. 2023). Even more recently, large language models and vision language models have provided navigating robots with an unprecedented level of common sense about the world (or at least the ability to act as if they have some degree of common sense understanding), as well as the ability to communicate naturally and iteratively with humans in that world (Rana et al. 2023). Other advances (including but not limited to transformers, neural radiance fields, and Gaussian splatting) have revolutionized parts of the robot navigation stack, such as the ways that these systems encode maps and representations of the world.

2.4 How Roboticists Think about Navigation

It is impossible to fully represent the spectrum of attitudes and interests in the robotics field toward navigation. The commentary here is meant to be a broad overview, noting there will be numerous exceptions.

Typically, roboticists focus on engineering solutions for navigation using sensors, algorithms, and data processing. Navigation is typically viewed as a problem of perception, mapping, localization or position estimation, and motion planning, where sensors provide data that must be processed to create maps and generate paths. The goal is to enable robots to operate autonomously, efficiently, and safely in human environments or remote areas.

A small subset of roboticists have attempted to draw inspiration from nature, both in terms of behavioral and neural discoveries about navigation, in systems such as RatSLAM (Milford et al. 2004), NeuroSLAM (Wu et al. 2019), and insect-inspired robot navigation systems (Dijk et al. 2024; Dupeyroux et al. 2019; Mangan et al. 2023). Many computational neuroscientists and mathematicians have modeled these systems in simulation (Banino et al. 2018) or mathematical models, abstracted heavily from real-world robotic requirements but with appealing properties such as noise rejection (Fiete et al. 2008). Finally, some approaches, especially those targeted at small or cheap robot platforms, eschew the conventional navigation pipeline and employ instead reactive behavioral approaches, sometimes implemented using prioritization architectures like subsumption (Brooks 1986). The subsumption architecture directly links sensory information with action selection over multiple levels in a hierarchy, where higher level abilities (like exploring) utilize lower-level abilities (like obstacle avoidance).

3 Key Components of Robot Navigation

"Standard" robotic navigation systems rely on several core components, each of which plays a crucial role in enabling robots to move autonomously through their environment. These components are typically divided into three main categories:

1. Perception and sensing
2. Representation (mapping and localization)
3. Path planning and execution

All areas should generally be tightly integrated, as the robot must accomplish all three tasks: perceive its surroundings, interpret and represent this data in a meaningful way, and then plan and execute movements based on that information. There are, however, an endless array of different mechanisms and techniques by which to implement each of these components.

3.1 Perception and Sensing

Perception and sensing are fundamental to any robotic navigation system. The ability to perceive the environment in which a robot operates provides the necessary data for all subsequent (and higher-level) navigation processes, from localization to path planning. Sensing technologies capture various forms of environmental data, such as distances to obstacles and visual cues, which are then processed to build a coherent understanding of the robot's surroundings and its location within them. The quality and type of sensors used can significantly impact a robot's navigation performance, as poor or inaccurate perception leads to errors in navigation, such as collisions or misdirected movements. The evolution of sensor technology has been a major driver of advancements in robotic navigation, enabling robots to operate in increasingly complex and unstructured environments.

Notable robot navigation sensors include cameras of many types. For example, they employ LiDAR—a range sensor providing precise ranges to objects up to several hundred meters away from the robot in all directions—and global navigation satellite systems, of which GPS is one (Figure 3.2). Robots can also make use of radar and sonar, both of which provide ranges to objects around the robot but typically not with the acuity that a LiDAR sensor achieves. These sensors, however, have other benefits: radar, for example, is particularly good at "seeing through" fog, where a conventional camera would be blinded. The process of "dead reckoning" or "path integration" that animals perform can be achieved on robots with inertial measurement units (IMUs), which provide information on the robot's orientation

Figure 3.2 An example of (a) a stereo (two-) camera sensor configuration and a 360° LiDAR sensor from an autonomous-vehicle research platform; (b) a panoramic (360°) grayscale image from an indoor mobile robot. The cameras provide a color or grayscale image of the world within a limited or full field of view, whereas the LiDAR sensor creates a point cloud representing the ranges to objects all around the robot. Larger, more sophisticated robots (e.g., an on-road autonomous vehicle) would most likely have a range of other sensing modalities, including radar, sonar, wheel encoders, GPS, and IMUs.

and acceleration, analogous to some extent to the vestibular system in animals. High-end, expensive IMUs, such as those used on submarines, can provide accurate tracking information over long periods of time spanning many days, without the need for any positional correction from other sources. This level of accuracy is beyond any natural system currently known.

In general, the sensors used in most robot navigation systems have significant differences in capability (both superior and inferior) compared to most sensing modalities in nature. Broad similarities can be drawn, such as whether a sensor is an active one that emits energy into the environment or a passive sensor, like a camera. Sensors can explicitly measure the range to objects in the environment around the robot. They do this implicitly, as modern robot vision systems can estimate complete "depth maps" that show the range to every observable part of the environment around them (Figure 3.3).

Sensors enable a variety of navigationally relevant tasks: mapping the environment and estimating the sensor's position relative to parts of the environment, such as in the case of cameras or LiDAR sensors, or estimating motion over time, as in the case of IMUs. Some sensors can be used for multiple tasks. LiDAR sensors, for example, can be used to estimate motion directly by analyzing how scans of the environment change over time, such as when a robot moves closer to a wall. Multiple sensing modalities can be combined in a variety of ways, including *sensor fusion*, to improve robustness and reliability.

(a)

(b)

Figure 3.3 Modern computer vision and AI can estimate depth in an image with purely passive sensing; that is, without need for an active sonar, radar, or LiDAR sensor: (a) daytime image and (b) associated depth mask created using AI techniques. Figure adapted with permission from Milford et al. (2015); copyright 2015 IEEE.

Despite advancements in sensor technology, several challenges remain in ensuring accurate perception for robotic navigation. Noise in sensor data can lead to inaccuracies in mapping and localization, particularly in environments with reflective surfaces or interference. Occlusion presents another significant challenge, as important environmental features may be hidden behind obstacles, limiting the robot's ability to detect and avoid them. For example, an autonomous vehicle stuck in heavy traffic may need to use other vehicles as a source of information to estimate location (Hausler et al. 2023). In addition, environmental variability (e.g., changes in lighting, weather conditions, or terrain) can affect the performance of sensors, particularly cameras (Figure 3.4). Overcoming these challenges requires sophisticated algorithms capable of filtering noise, predicting hidden objects, and adapting to changing environmental conditions to ensure that robots can navigate safely and effectively in a wide range of scenarios.

One of the primary limitations of sensors used to track robot motion over time is *drift*: imperfections in the process of estimating the velocity or acceleration of the robot, and hence the gradual accumulation of errors in the overall position estimate for the robot. This challenge is central to the robot navigation research field and has

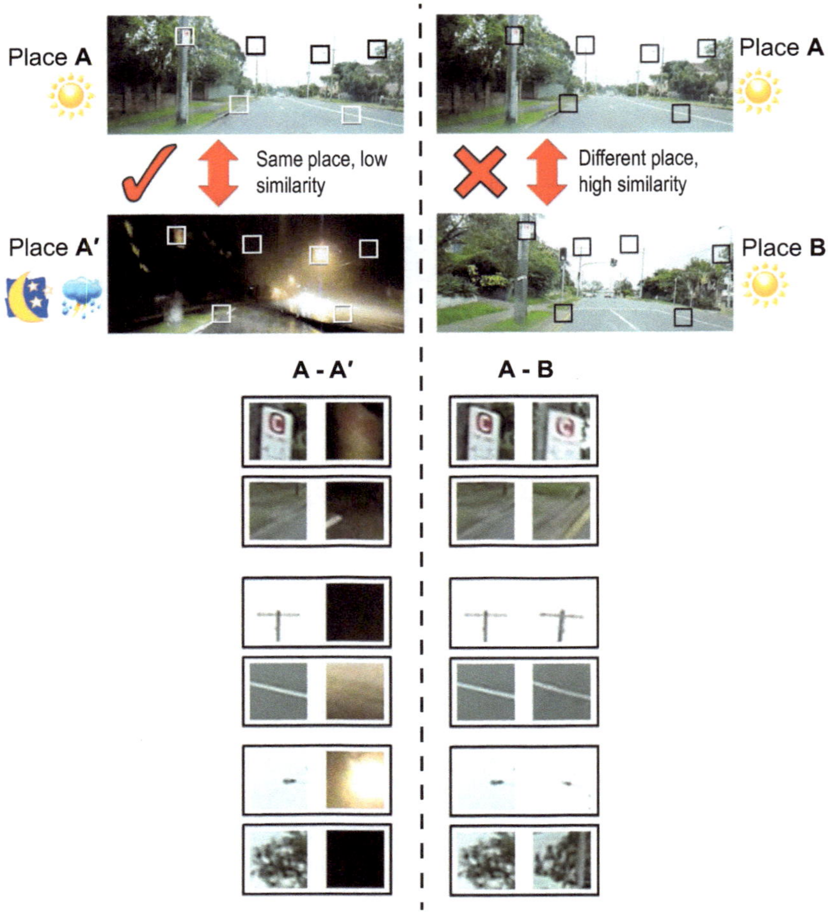

Figure 3.4 Perceptual change is a key challenge for navigating robots. Here, the two images on the left are from the same location, but they appear very different perceptually. In contrast, the two images on the right are of geographically different locations but appear very similar perceptually. Figure adapted with permission from Milford and Wyeth (2012); copyright 2012 IEEE.

driven a range of solutions: from pragmatic work-arounds (e.g., installing artificial beacons of known location in the environment) to complex solutions, such as in SLAM, one of the largest research fields in robotics (Cadena et al. 2016).

3.2 Representations

In robotic navigation, representations are essential for understanding and modeling the environment in which a robot operates. These representations function as internal maps that enable robots to localize themselves, plan movements, and avoid obstacles. Representations can take various forms, from metric maps that accurately describe the environment's geometry to topological maps that represent connections between different places. The choice of representation depends on the specific task, the environment's complexity, and the available computational resources.

Mapping and *localization* are fundamental processes in autonomous robotic navigation. Mapping refers to the construction of an internal model of the environment based on sensor data, whereas localization is the robot's ability to determine its position within this map. Together, these processes enable the robot to navigate through its surroundings. SLAM is widely used to enable robots to build maps of unknown environments while simultaneously tracking their location. This capability is essential for robots that must operate in dynamic or unstructured environments, where prebuilt maps may not be available or may rapidly become outdated due to environmental changes (Cadena et al. 2016). A key component of SLAM has revolved around the notion that such a solution should be *probabilistic* in nature. That is, it should be able to represent uncertainty about the estimate of the robot's position in the world, including representing and independently propagating multiple hypotheses of its position in the world over time (Thrun et al. 2005).

Spatial representations are also critical for robotic navigation, as they provide the foundation for path planning and decision making. Grid maps are one of the most common types of spatial representation used in robotics (Figure 3.5b). They divide the environment into a grid, where each cell represents the presence or absence of obstacles, allowing robots to plan movements on a fine-grained level. Grids may be two or three dimensional; their formal structure can be relaxed somewhat by having grids of variable densities utilizing methods like octrees (Vespa et al. 2018). Other representations of space include (a) point clouds, where surfaces of objects are represented by points in 3D space, forming a cloud that can be dense or sparse; (b) neural radiance fields (Rosinol et al. 2023); and (c) Gaussian splatting techniques (Matsuki et al. 2024), which encode the physical layout of space in the weights of a neural network and through Gaussian "blobs," respectively.

Topological maps, in contrast, represent the environment as a network of interconnected nodes connected by edges or links, each node corresponding to a key location or decision point (Figure 3.5a). Edges connecting nodes can convey information such as traversability, distance, or relative orientation. Topological maps, which are derived from the physical layout and dimensions of the environment, are often used in applications where high-level planning is required, such as navigating large-scale environments, of which the London tube map is a famous example. Hybrid approaches combine (at least theoretically) the strengths of both grid and topological maps, enabling robots to operate efficiently at multiple levels of abstraction (Kuipers 2000).

Roboticists encounter several challenges, as they seek to create accurate and reliable representations of the environment. One of the primary difficulties involves uncertainty in sensor data, which can result from noise, occlusions, or dynamic changes in the environment. In addition, environments can vary dramatically in complexity, ranging from structured indoor settings to highly dynamic outdoor scenes. As the complexity increases, so does the challenge of maintaining an accurate map while ensuring that the robot remains localized within it. Another key challenge involves scalability: representations that work well in small environments may become computationally expensive or impractical in larger or more complex areas. Developing algorithms that can efficiently handle large-scale environments without sacrificing accuracy is a continuing area of research (Cadena et al. 2016).

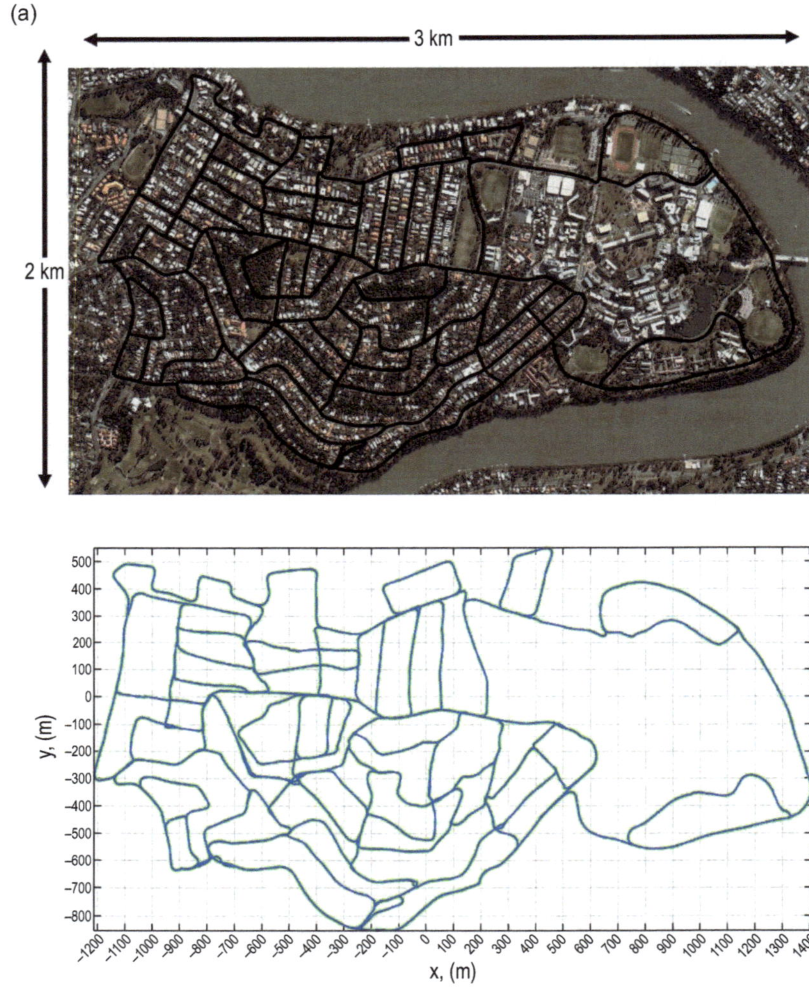

Figure 3.5a Example of a topological map for a robot mapping system (adapted with permission from Milford and Wyeth 2008; copyright 2008 IEEE).

3.3 Path Planning and Execution

As critical components of autonomous robotic navigation, *path planning* refers to the process of determining a route for the robot to follow from its current position to a designated goal, whereas *execution* involves carrying out that plan in real time. Effective path planning must account for static obstacles (e.g., walls or furniture) as well as dynamic elements (e.g., people or moving vehicles). In addition, the robot must continuously monitor its progress and adapt its path, as needed, to respond to unforeseen changes in the environment. The integration of path planning and execution is a core aspect of autonomous systems, ensuring that robots not only generate an optimal route but also follow it accurately and safely in dynamic, real-world conditions.

(b)

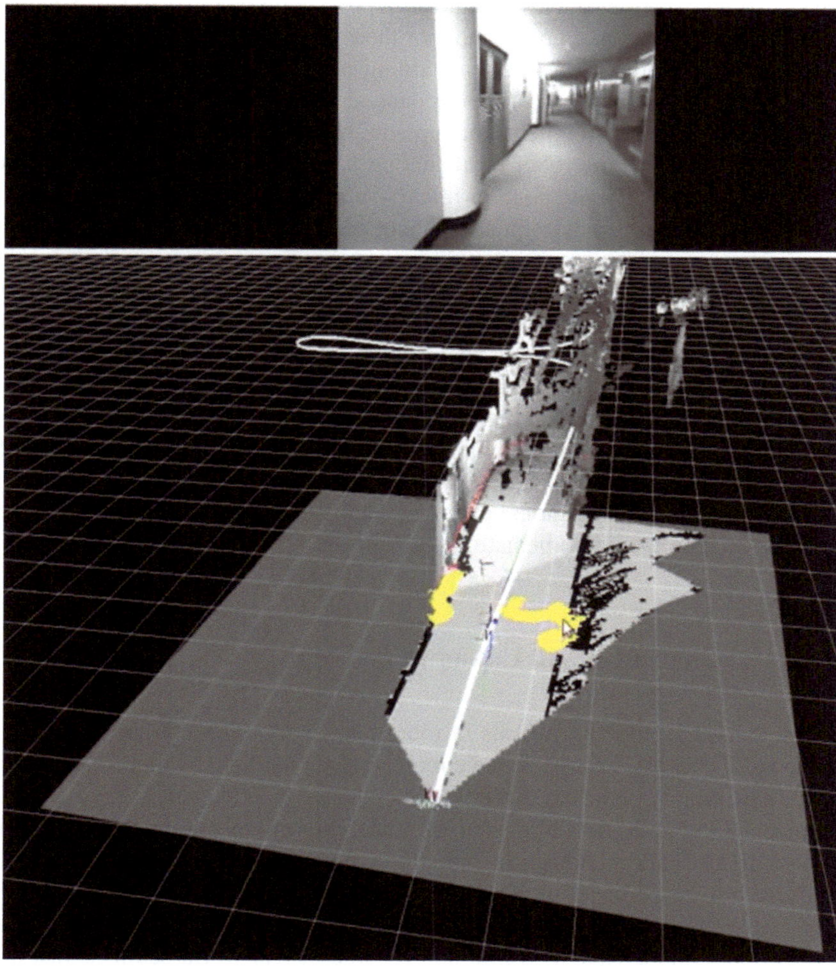

Figure 3.5b Example of an occupancy grid map representation for a robot mapping system.

Several well-established *algorithms* for path planning are used in robotic systems, each with strengths suited to different types of environments and tasks (Zhang et al. 2018). A* (A-star) is a widely used search algorithm that finds the shortest path between two points by evaluating neighboring nodes and estimating the cost to reach the goal. It is known for its efficiency in relatively simple, structured environments. Dijkstra's algorithm, which is similar to A*, evaluates all possible paths to determine the shortest route, making it more comprehensive but also computationally expensive. The rapidly exploring random tree (RRT) algorithm is particularly well-suited for high-dimensional spaces and complex, unstructured environments (Karaman et al. 2011). RRT and its variants work by incrementally building a tree of possible paths, quickly exploring large areas of the environment. This makes it a good choice for dynamic or uncertain environments, with the trade-off being that paths may be longer than is theoretically achievable.

Real-time path execution presents numerous challenges, especially in *dynamic* environments where obstacles can move unpredictably (Mohanan and Salgoankar 2018). Robots must continuously monitor their surroundings and adjust their movements to avoid collisions while maintaining progress toward their goal. This requires the processing of robust sensor data and fast decision-making capabilities. In highly dynamic environments, path planning and execution often need to occur simultaneously, with the robot recalculating its path as it navigates. This process is known as *dynamic replanning* or *reactive planning*; it ensures that the robot can adapt to changes in its environment without coming to a stop or requiring external intervention.

The integration of planning and perception is essential for successful robotic navigation, as the robot must continuously perceive its environment and update its plans accordingly. Integration ensures that the robot can respond to changes in its surroundings, such as new obstacles or altered terrain. A common integration approach involves combining high-level path planning with low-level reactive behaviors, enabling the robot to switch between long-term planning and immediate responses to environmental changes. For example, a robot navigating through a cluttered space may use a global planner to define its overall route while relying on a local planner to avoid immediate obstacles.

4 Grand Challenges and Unknown Questions

Despite significant advancements in robotic navigation, several major challenges related to the issues discussed above remain, spanning both technical and nontechnical (e.g., ethical) domains. Robust navigation in dynamic and unstructured environments, long-term autonomy, and the scalability of navigation systems continue to pose significant difficulties for roboticists. Furthermore, although great progress has been made, roboticists remain uncertain about which types of map representations, as well as which path planning and execution approaches, are truly necessary to make robot systems useful and reliable in real-world settings. Just as the debate over whether insects form cognitive maps has raged for decades (Dhein 2023), we do not know exactly what a robot requires (and by extension does not require) to execute useful, navigation-reliant tasks. Since the release of ChatGPT in November 2022, the popularization and spread of large language models and related approaches has added to this uncertainty, as the exact extent to which these systems store or learn spatial information is somewhat unclear, even though they appear capable of performing navigationally relevant tasks (Rana et al. 2023).

Furthermore, deployment and ethical considerations, such as ensuring safe interactions between robots and humans, are becoming increasingly important as autonomous systems are deployed in more complex and diverse environments. Substantial new research will be needed to progress past the status quo, combining refinement of existing technologies with the exploration of new approaches, such as bio-inspired solutions and advanced learning techniques.

4.1 Fragility, Lack of Generalization, and Unintuitive Behavior

The largest, overarching challenge in robot navigation today is that its systems are relatively specific; they work well (theoretically) in an operational design domain that is narrow and constrained. Take a robot that is engineered to suit one environment (and set of tasks reliant on navigation) to a different setting and it will likely fail catastrophically. A logistics robot that shifts product shelving around in a highly controlled logistics warehouse will not function in an autonomous off-road vehicle, just like a drone navigation system will likely not function well on a submersible robot.

As a partial explanation for this phenomenon, it should be noted that to meet commercial interests, these systems were designed for a very specific purpose. Thus, the lack of generality is not surprising; commercially, generality is often not needed. This design intent, however, obscures a more general issue: it is incredibly difficult to build general-purpose, highly adaptable robot navigation systems. This characteristic is in vast contrast to many natural systems such as the brown rat, which for centuries has demonstrated its ability to be highly effective and resilient in an incredibly diverse range of habitats around the world (Buckle 2015). Such versatility is still a long way from being achieved in robotic navigation systems.

4.2 Robustness to Environmental Variability

A key component of the challenge posed by generality is robustness to environmental variability. Environments constantly change at different timescales, from long-term climate shifts to a wave crashing on the shore. Changes in lighting, weather, or the physical layout of an environment can significantly affect a robot's ability to navigate. For instance, a robot navigating through a city may encounter diverse lighting conditions (e.g., from bright sunlight to deep shadows) as well as changes in weather (e.g., rain or fog), which can obscure sensors like cameras and LiDAR. To address this challenge, researchers have developed algorithms that enable robots to adapt to these variations by improving sensor fusion techniques, which combine data from multiple types of sensors and incorporate environmental context into their models. For example, deep-learning methods have been used to help robots recognize and adapt to different environmental conditions by learning patterns in data collected from diverse environments (Bresson et al. 2017). Modern visual approaches to navigation learn visual features that are incredibly resilient to appearance change (Hausler et al. 2021); some are even capable of predicting how an environment will appear under different conditions (Kurzman et al. 2019). Even with these advancements, however, environmental variability remains a major source of failure in robotic systems, particularly in outdoor settings where conditions can change unpredictably.

In addition, sensor-specific challenges arise under different environmental conditions. Cameras, for example, often struggle with glare or poor lighting, while LiDAR sensors may experience reduced performance in adverse weather conditions, such as rain or snow, which scatter laser beams and degrade range measurements.

Techniques such as sensor fusion and the related concept of sensor redundancy (i.e., multiple sensors are employed to measure the same feature) have been developed to mitigate these issues. However, ensuring that robots can reliably navigate across a wide range of environments, from structured indoor settings to unpredictable outdoor terrains, remains a critical goal in the field of robotics.

4.3 Viewpoint Sensitivity

Navigating robots that use map-based approaches reconcile their current sensory snapshot of the world against a stored map to calculate their likely position. However, robots that navigate through a previously visited (and mapped) environment do not always do so along the exact same path. Consequently, their sensory snapshots may include locations that they have not specifically visited but which lie near those they have. For example, an autonomous vehicle that has only mapped a roadway traveling in one direction may revisit the same roadway from the opposite direction (Figure 3.6). The robot revisits the place along a trajectory that is offset laterally by five or ten meters from its last visit and rotated 180 degrees relative to its previous pass. Much research in robot localization systems has gone into making them as viewpoint-invariant as possible; that is, capable of estimating their location even when they are not revisiting a mapped environment along the same path or orientation (Lowry et al. 2016). Approaches to this problem generally involve designing or learning “viewpoint invariant” visual features (for vision-based mapping systems) that enable recognition of a familiar place from varying viewpoints, as well as geometric approaches that can reason about how the environment would appear from an arbitrary pose. A more straightforward approach is to use sensors that have panoramic fields of view, which radically reduces the difficulty related specifically to viewing orientation changes (but not changes in location like lateral shift). A larger field of view can also be achieved, even with a limited field of view sensor, by employing a turn-back-and-look behavior. When the challenge of viewpoint variation is combined with the challenge of appearance variation, the robot localization problem becomes very difficult (Garg et al. 2022).

4.4 Handling Unstructured Environments

Unlike structured environments (e.g., warehouses or factories), where the layout is predictable and known, unstructured environments (e.g., forests, deserts, or marine areas) are dynamic and lack predefined paths. To navigate in these environments, robots must be capable of detecting and interpreting a wide range of obstacles (e.g., irregular terrain, vegetation, or debris) and must often make decisions in real time based on incomplete or noisy sensor data. This requires advanced perception algorithms capable of detecting obstacles as well as classifying and responding to them appropriately (Cadena et al. 2016). Extreme cases of unstructured environments include navigation inside the body in surgical or medical applications as well as navigation on or under the ocean.

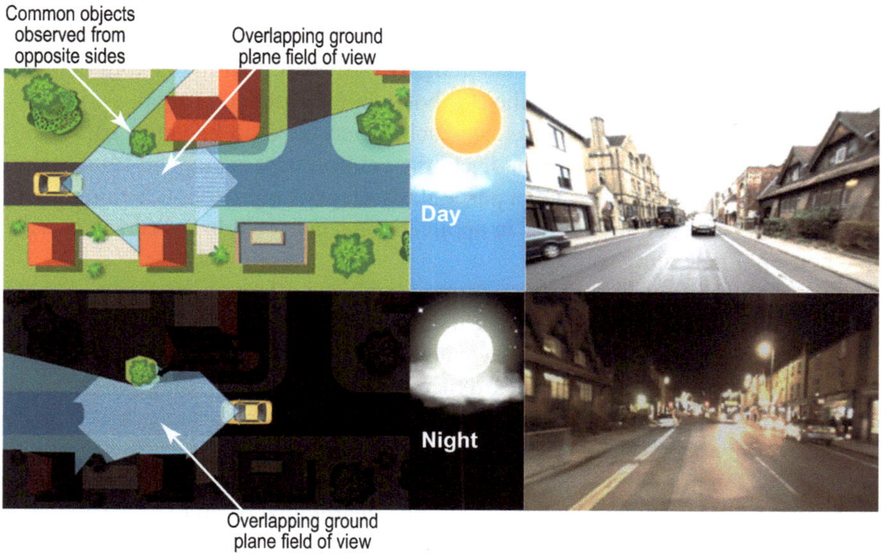

Figure 3.6 Viewpoint variation combined with appearance variation poses a major challenge for robot localization systems. Here, an autonomous vehicle revisits a previously mapped environment from the opposite direction, with a lateral shift from one road lane to the next under different light conditions (day vs. night). Left panel from Garg et al. (2022), reprinted by permission of Sage Publications.

Ground-based robots operating in unstructured environments must contend with uneven terrain, which can impact their mobility and stability. Algorithms for real-time terrain analysis, combined with robust locomotion systems, are critical for successful navigation under such conditions (Borges et al. 2022). For example, robots like Boston Dynamics' Atlas utilize sophisticated balancing systems and perception algorithms to traverse complex terrains. Researchers are also exploring bio-inspired approaches, such as modeling the navigation strategies of animals like ants, which combine dead reckoning with visual landmarks to navigate effectively even in unstructured environments (Dupeyroux et al. 2019).

4.5 Long-Term Autonomy and Adapting to Change

Long-term autonomy is one of the most significant challenges in robotic navigation. Like animals, autonomous robots must be able to operate over extended periods without human intervention, continuously perceiving their environment, making decisions, and executing tasks. This requires not only robust navigation systems but also the ability to adapt to changing environments over time. Unlike short-term operations where conditions may be relatively static, long-term autonomy often involves coping with a range of changes in the environment, such as shifting obstacles, fluctuating weather patterns, or the wear and tear of physical infrastructure. These changes can degrade the performance of traditional mapping and navigation systems if they are not designed to learn and adapt over time.

Techniques like continuous SLAM, where maps are updated dynamically as the environment evolves, offer the potential for maintaining autonomy in the long term. The problem is sufficiently challenging, however, that many commercially deployed approaches (e.g., household robot vacuum cleaners) rely to varying extents on a "reset and remap" type approach: when the map becomes sufficiently out of date or corrupted, it is deleted, and the environment is mapped from scratch (Figure 3.7). Although this is not a particularly elegant approach, it is functionally sufficient and acceptable to users in applications where robots are currently used, like household cleaning robots.

One approach to achieving long-term autonomy is through life-long learning, where robots continuously refine their models of the environment and update their internal representations based on new data. This can involve improving the robot's understanding of the environment's dynamics, such as anticipating when certain obstacles might appear or disappear, or adjusting navigation strategies based on experience. Many ways have been proposed to do this, from pruning maps over time (Figure 3.8) (Milford and Wyeth 2010) to the explicit tracking of dynamic objects in the environment (Yu et al. 2018). Continual learning adds an extra challenge of map certification. How can the map's correctness be assured when the map is being modified autonomously without any human supervision? This is an especially relevant challenge in safety-critical operations, such as autonomous vehicles in mines (Jacobson et al. 2021).

4.6 Maintaining Navigation Accuracy over Extended Periods

In autonomous systems, maintaining navigation accuracy over long periods poses a fundamental challenge, especially in dynamic or large-scale environments. Over time, sensor drift, environmental changes, and map inaccuracies can lead to significant errors in localization. For example, in outdoor environments where GPS signals may be unreliable or obstructed, robots must rely on alternative navigation methods such as visual place recognition or inertial navigation systems. These systems, however, can also degrade over time due to changes in appearance or sensor noise. To counteract this, advanced techniques, such as loop closure detection, are used to correct accumulated errors by recognizing previously visited places and resetting the robot's position within the map. This process is central to SLAM.

In addition to correcting errors, maintaining navigation accuracy requires constant monitoring of sensor reliability and recalibration. This is particularly important in long-term autonomous systems that may experience sensor degradation or encounter new environmental conditions. The integration of multiple sensors (e.g., combining visual, inertial, and LiDAR data) can improve the robustness of navigation by providing redundancy and allowing the system to cross-validate sensor readings. This multisensor fusion approach is vital for ensuring that navigation remains accurate over extended periods and across a variety of environments (Alatise and Hancke 2020).

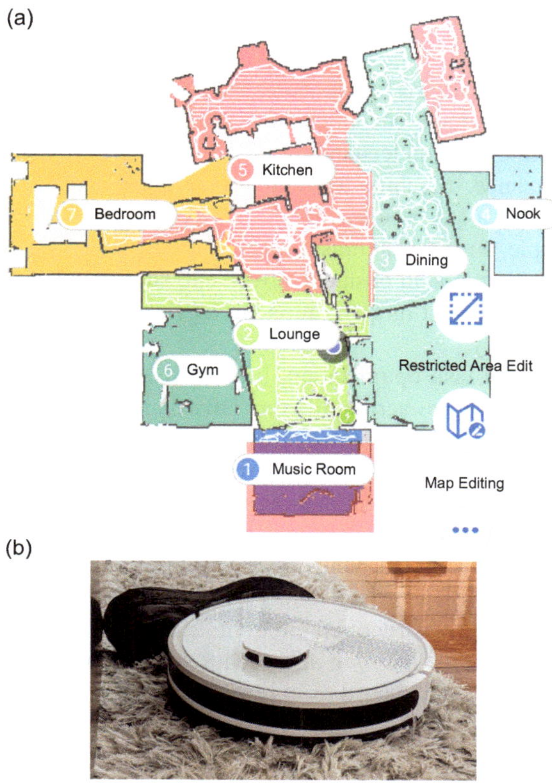

Figure 3.7 To ensure long-term navigational reliability, details of the robot's function and environment must be available. For many robot vacuum cleaners, an occasional remap from scratch of the area to be serviced has been determined to be the best compromise of user tolerance and technical difficulty. Here, the robot (a) has learned two separate but overlapping representations of the environment (b), rotated by about twenty degrees from each other. Without a reset and remapping procedure, the robot's performance will likely be impaired.

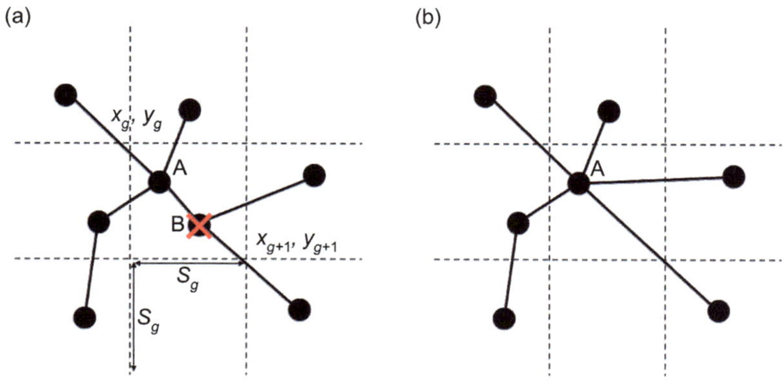

Figure 3.8 A simple example of pruning a topological map: the nodes and edges of the graph that make up the map are pruned so as to not exceed a certain density per grid square, and new graph edges are formed to maintain the overall connectivity of the graph. From Milford and Wyeth (2010), reprinted with permission of Sage Publications.

4.7 Learning and Adapting from Experience

To achieve long-term success, autonomous robots must be able to learn and adapt from experience. As robots encounter new environments or repeated tasks, they would ideally refine their understanding of the world and adjust their behavior to optimize performance. Machine-learning techniques, particularly reinforcement learning, could play a key role in enabling robots to adapt based on past experiences. For example, a robot that must navigate around a warehouse may be able to learn to recognize patterns in obstacle movement, such as the paths of forklifts or human workers, and adjust its navigation strategy accordingly to avoid collisions and improve efficiency.

Beyond individual task optimization, lifelong learning would enable robots to generalize their experiences across different environments and tasks. This capacity for adaptation is especially important in dynamic environments where fixed rules and preprogrammed behaviors may be insufficient. For instance, in search-and-rescue missions, robots need to adapt their strategies based on the specific conditions of the disaster site, such as blocked paths or unstable terrain. By learning from their experiences and using that knowledge to inform future decisions, robots can improve their resilience and effectiveness over time, enabling them to tackle increasingly complex and unstructured environments. Learning does not need to be a singular affair: *federated learning* approaches share information across multiple robots (or agents) and can be used to improve robot mapping and navigation systems (Li et al. 2019).

4.8 Scalability of Navigation Systems

As robotic applications expand to larger and more complex environments, the scalability of navigation systems becomes a critical challenge. Traditional navigation systems, while effective in small, structured spaces, often struggle to maintain performance when applied to vast or highly dynamic environments. Scalability involves not only the ability to map and localize in larger areas but also the capacity to handle increasing amounts of sensor data and more complex obstacle configurations. Without scalability, the operational effectiveness of autonomous systems in industries such as logistics, agriculture, and urban transportation will remain limited. Although computational hardware (e.g., graphics processing units) continues to improve rapidly, large-scale mapping problems (e.g., mapping an entire city or country at a fine resolution) are still computationally daunting.

In *complex and varied environments*, robots face the challenge of managing extensive and diverse maps. The robot's internal maps must cover a vast area, often with intricate layouts and numerous dynamic obstacles. For example, an autonomous vehicle operating in a city must contend not only with streets and intersections but also pedestrians, other vehicles, and varying traffic patterns. Some of these challenges are only relevant at the local level, whereas others (e.g., traffic) must be accounted for at larger spatial scales to enable effective navigation decisions.

Similarly, drones or autonomous ground robots in agricultural fields must process large-scale maps while accounting for complex terrain and environmental changes. To tackle these challenges, hierarchical mapping techniques, such as dividing environments into smaller sub-maps or using topological maps alongside metric maps, are often employed to reduce the computational burden.

Cloud-based navigation systems have also emerged as a potential solution for scaling robotic operations. By offloading intensive mapping and processing tasks to remote servers, robots can access updated maps and environmental models in real time without being constrained by onboard computational limitations. This approach, however, introduces additional challenges, such as ensuring reliable communication links and managing latency in time-critical operations, where outages can ground fleets (Templeton 2023). As robots increasingly operate in larger, more complex environments, the scalability of both hardware and software solutions remains a central focus of ongoing research. "Offloading to the cloud" is another example of an engineering approach to navigation in robots that bears little direct resemblance to any natural system.

Resource-constrained robots (e.g., small drones, mobile service robots, or robots operating in extreme environments) face the dual challenges of maintaining high levels of autonomy while operating under strict limits on processing power, memory, and energy. Efficient computation is essential for these robots to successfully navigate and perform tasks. To address this challenge, researchers have developed lightweight algorithms that reduce the computational load without sacrificing accuracy. For example, algorithms that approximate solutions to computationally expensive path-planning problems, such as RRTs, have proven effective in resource-limited systems (Karaman et al. 2011).

Specialized hardware accelerators, (e.g., graphics processing units, field-programmable gate arrays, or neuromorphic hardware) provide another approach to improve efficiency in resource-constrained robots (Sandamirskaya et al. 2022). These accelerators can handle computationally demanding tasks, like image processing or neural network inference, allowing the robot to achieve faster and more efficient performance (Hussaini et al. 2024). In addition, sensor fusion techniques can optimize the use of multiple low-cost sensors to reduce reliance on expensive or high-power sensors like LiDAR.

5 Multi-Robot Navigation

Multi-robot navigation involves coordinating the movements of multiple robots in shared environments and is a growing area of research due to its potential for enhancing efficiency and productivity across various fields, from logistics to search-and-rescue operations. In such systems, each robot must not only navigate individually but also work collaboratively to achieve a common goal. This requires both sophisticated communication and coordination mechanisms. Multi-robot systems provide significant advantages, such as covering larger areas in less time, distributing tasks, and sharing sensory data to improve the collective perception of the environment. These systems, however, also introduce challenges, particularly in terms of ensuring that robots can navigate safely without interfering with each other or causing collisions. This problem has been mostly solved in highly constrained environments, such as warehouse logistics (Jain and Sharma 2017), but is more challenging in less constrained environments, like underground settings (Ebadi et al. 2023).

The complexity of multi-robot navigation increases with the number of robots involved (sometimes exponentially) and the dynamic nature of the environment. Algorithms for multi-robot coordination, such as decentralized control strategies, enable each robot to make independent decisions while still adhering to overall mission objectives (Bayer et al. 2023). These decentralized approaches are particularly valuable in environments where communication between robots may be limited or unreliable, such as in underwater locations, disaster zones, or thick-walled urban environments. To assign tasks and monitor the positions of all robots, centralized control strategies can still be used when communication bandwidth allows. Both distributed and centralized approaches must be fault tolerant, so that the failure of one robot does not compromise the success of the entire team.

Coordination among multiple robots optimizes task allocation, reduces redundancy, and ensures smooth operation in shared environments. Task partitioning divides the mission into subtasks based on each robot's capabilities and positions, requiring real-time communication for adaptability to changes, such as obstacles or new priorities (Parker 2007). Shared localization and mapping enhance navigation by improving environmental understanding and reducing redundant efforts. Dynamic task reassignment also enables robots to take over tasks when others encounter failures or obstacles, maintaining continuous operation.

6 Introspection in Robot Navigation

In the context of robot navigation, introspection refers to the ability of a system to assess its own state, recognize its limitations, and determine when it lacks sufficient information to navigate reliably. This capability is crucial in any operationally or safety critical scenario (e.g., when an autonomous vehicle must navigate around pedestrians) as well as in dynamic environments where navigation decisions must adapt to real-time changes. A robot equipped with introspection can detect when its sensors or mapping are unreliable, such as in low-visibility conditions or when GPS has dropped out, and take corrective action, such as slowing down, reevaluating its path, or handing control over to a human operator. By integrating introspection, navigation systems can become more robust, avoiding errors caused by faulty data or uncertain decisions and enabling adherence to established ethical norms (Claxton et al. 2024). Introspection enhances communication and collaboration with other navigating agents, including robots that support each other in navigating difficult environments. Effective communication with human operators is facilitated, enabling robots to alert humans when they need help to ensure safe navigation in human-centric spaces.

Introspection can help with other, sometimes more nefarious challenges. Navigating robots do not always operate in friendly, passive environments: other agents in the environment may accidentally or intentionally interfere with their navigational processes. Two common examples are jamming by global navigation satellite systems (i.e., removing access to a satellite-based positioning signal) and spoofing, which does not remove the signal but instead intercepts and changes it in a way designed to be undetectable by the robot or robot operators. Currently, ongoing work is addressing this challenge, and the ability to introspect is likely to be key (Claxton et al. 2024).

7 Financial and Energy Cost Considerations

The cost of autonomous robotic systems is a major factor affecting their adoption, and a substantial aspect of this cost is related to navigation. Expenses arise from sensors, computational hardware, and materials. Thus, balancing performance with affordability is essential for industries like logistics, agriculture, and health care. Lower costs help democratize the technology, making it accessible to smaller businesses and more widely available on a global scale. Cost reduction often involves using mass-produced components and minimizing expensive sensors, but this must be done without compromising functionality or safety. As always, this is a challenging trade-off.

One major contributor to the cost of autonomous robots is their bulk and size, which affects practicality, mobility, and energy consumption. Larger robots are more capable but they are generally more expensive to manufacture and operate; smaller robots are cheaper but may lack the necessary sensor fidelity and computing capability to address complex or large-scale navigation tasks (Nagy et al. 2023).

Sensor choice represents another significant cost factor, as high-precision technologies like LiDAR incur a substantial expense. Reducing reliance on expensive sensors through sensor fusion offers a promising solution. Advances in cloud and edge computing also help offload tasks, reducing the need for costly onboard processing. Improving computational efficiency, such as using lightweight neural networks, helps balance performance and cost (Khaliq et al. 2019).

8 Ethical and Safety Considerations

As robots become more integrated into human environments, ethical and safety considerations emerge and must be addressed. Autonomous navigation raises different ethical issues, such as how decision making is reached under dangerous situations as well as privacy concerns. Robots must balance competing objectives, such as speed and safety, while adhering to ethical standards set to prioritize safe, ethical interactions with humans, including mechanisms to prevent accidents and minimize harm. Embedding ethical decision making into algorithms is essential, particularly in morally ambiguous situations, such as self-driving cars which must make split-second navigational choices (Arkin 2008; Goodall 2014).

Ensuring safe interaction with humans requires navigating robots to recognize and predict human behavior and act accordingly. Human-aware navigation algorithms combine sensor data with models of human movement to avoid collisions (Kruse et al. 2013). Safety also includes the psychological comfort of humans: robots should be designed to behave predictably and communicate their intentions clearly. This is particularly important in mixed robot-human environments, such as roads, where nonverbal communication between vehicles is prevalent.

Ethical and privacy considerations may be explicit, such as in military contexts, or more subtle, as with the privacy implications of visual mapping systems used to map cities (Burdon et al. 2024). Data collected by robots need to comply with privacy laws, a growing challenge as robots (e.g., autonomous vehicles and drones) become more widespread. Transparency is crucial for accountability and public trust (Arkin 2008). Addressing these concerns requires collaboration between engineers, ethicists, and policymakers to develop responsible frameworks.

9 Interdisciplinary Opportunities for Future Research

Robotic navigation continues to face major challenges, including environmental variability, scalability, multi-robot coordination, and the integration of perception, mapping, localization, and path planning. Although technological advancements have improved system performance, significant hurdles remain in enhancing generality, adaptability, and efficiency across diverse, real-world scenarios. A comprehensive and detailed treatment of all these challenges and associated research opportunities is beyond the scope of this chapter. Here, I wish to highlight key opportunities that are particularly relevant to human, animal, or organism navigation researchers, drawing attention to significant shortcomings in current robotic approaches and suggesting opportunities to close the loop back to nature.

9.1 Adaptability and Generality

A major focus for future robot navigation systems will be to achieve greater generality and adaptability, inspired by the resilience observed in natural systems. Unlike current robotic systems, which often falter outside their narrowly defined environments, biological organisms, such as animals, exhibit remarkable adaptability as they seamlessly navigate across a wide range of unpredictable conditions. To mirror this, robotic systems need long-term resilience through continuous learning and dynamic adaptation, similar to how animals adjust their behavior based on experience. Techniques such as life-long learning and continuous SLAM will be essential for enabling robots to adapt their perception and navigation strategies as their environments evolve. This generality and adaptability will be key to the development of more general-purpose robots capable of performing a broad range of tasks and operating across diverse environments.

How exactly do animals achieve this remarkable adaptability and generality? We have some, but by no means all, of the clues. For example, recordings taken over minutes or hours have provided valuable insight into the neural underpinnings of spatial representations in the mammalian brain. To make further progress and potentially provide nature-inspired breakthroughs in robot navigation, much more will be needed. Of particular interest to roboticists will be combining behavioral and neural analysis with more sophisticated, long-term and larger-scale experimental paradigms, particularly when organisms navigate over extended periods of time and across diverse environments.

9.2 Introspection

Introspection in robotic navigation, akin to the self-assessment capabilities seen in neural and natural systems, is another critical development. Just as animals constantly assess their surroundings and adjust their actions based on confidence in their perceptions, robots must learn to recognize when they lack sufficient information to proceed safely. By integrating introspective mechanisms, robots can detect unreliable sensor inputs or decision-making processes, enabling them to mitigate risks and prevent failures, akin to how biological systems err on the side of caution in uncertain situations.

The concept of uncertainty in natural systems has been studied both neurally and behaviorally, but much remains unknown. The probabilistic representations—methods underpinned by Kalman and particle filters, and the application of the Bayes' theorem—have been central to robot mapping and navigation systems for the past 25 years but are unlikely to be exactly replicated in the brain. Bio-inspired mapping systems, such as RatSLAM and NeuroSLAM, have demonstrated that capable robot mapping and navigation performance is achievable without these exact "optimal" probabilistic representations; they have, however, been driven heavily by engineering practicalities rather than detailed neural considerations. In a navigational context, the extent to which uncertainty encoding and introspective capabilities are achieved in the brain, and how, remain of great interest to roboticists.

9.3 Efficiency

In both energy and computational terms, efficiency is important to the future of robotic navigation technologies. Natural systems, particularly neural processes in animals, demonstrate a high degree of energy efficiency in complex tasks such as navigation. The human brain notably uses only around 20 watts (Madhavan 2023), compared to the kilowatt-range consumption of onboard computing systems in autonomous vehicles, and the far larger consumption involved in training these systems. Developing lightweight algorithms, much like nature's energy-efficient neural systems, will enable robots to achieve scalable, robust navigation in resource-constrained environments.

Optimization in artificial systems is a very different process to that which occurs in natural systems. Rather than evolving gradually over time, current artificial navigation systems are typically developed with performance first in mind (e.g., navigation accuracy); thereafter, computing efficiency is investigated. The efficiency of learning in natural systems is potentially of even greater interest. Many current state-of-the-art navigation techniques require exhaustive training on huge datasets, which stands in stark contrast to many animals and insects that begin to navigate competently only hours or days after being born.

9.4 Closing the Loop

The opportunities discussed above have involved advances in our understanding of natural organisms informing robotics development, but there are opportunities for benefits in the reverse direction (Webb 2000). Perhaps most significantly, researchers have near-complete visibility of the mechanisms behind robot navigation systems—in large part because they have created and overseen their training. Thus, they can explicitly set a robot's intent, something that can only be inferred when studying natural organisms. Causality, in general, is also much easier to determine. This means that navigating robots provide a potent mechanism for studying theories of navigation in the natural world, even though exact, faithful replication of natural navigation systems in computational models is difficult and necessarily involves abstraction.

10 Conclusion

As robotic systems evolve, the ability to emulate adaptability, introspective awareness, and efficiency of biological systems will be key to overcoming current challenges. Achieving these outcomes will enable robots to navigate with greater autonomy, robustness, and resilience, driving their successful integration into a much wider range of more complex and dynamic environments. In solving these challenges in robots, we may also generate new insights into the biological systems from which we draw inspiration, advancing our knowledge of the incredible navigation systems of the natural world.

Acknowledgments This research was partially supported by funding from an ARC Laureate Fellowship FL210100156 to MM, and we acknowledge the support of the Ernst Strüngmann Forum.

References

Ackerman E (2024) Boston Dynamics and Toyota Research Team up on Robots. https://spectrum.ieee.org/boston-dynamics-toyota-research

Alatise MB, and Hancke GP (2020) A Review on Challenges of Autonomous Mobile Robot and Sensor Fusion Methods. IEEE Access 8:39830–39846. https://doi.org/10.1109/ACCESS.2020.2975643

Arkin RC (2008) Governing Lethal Behavior: Embedding Ethics in a Hybrid Deliberative/Reactive Robot Architecture. In: 3rd ACM/IEEE International Conference on Human Robot Interaction. ACM, New York, pp 121–128

Banino A, Barry C, Uria B, et al. (2018) Vector-Based Navigation Using Grid-Like Representations in Artificial Agents. Nature 557:429–433. https://doi.org/10.1038/s41586-018-0102-6

Bayer J, Čížek P, and Faigl J (2023) Autonomous Multi-Robot Exploration with Ground Vehicles in DARPA Subterranean Challenge Finals. Field Robot 3:266–300. https://doi.org/10.55417/fr.2023008

Borges PV, Peynot T, Liang S, et al. (2022) A Survey on Terrain Traversability Analysis for Autonomous Ground Vehicles: Methods, Sensors, and Challenges. Field Robot 2:1567–1627. https://doi.org/10.55417/fr.2022049

Bresson G, Alsayed Z, Yu L, and Glaser S (2017) Simultaneous Localization and Mapping: A Survey of Current Trends in Autonomous Driving. IEEE Trans Intell Vehicl 2 (3):194–220. https://doi.org/10.1109/TIV.2017.2749181

Brooks R (1986) A Robust Layered Control System for a Mobile Robot. IEEE J Robot Autom 2:14–23. https://doi.org/10.1109/JRA.1986.1087032

Buckle A, and Smith R (2015) Rodent Pests and Their Control. CABI Publ, Wallingford, UK

Burdon M, Cohen T, Buckley J, and Milford M (2024) From Object Obfuscation to Contextually-Dependent Identification: Enhancing Automated Privacy Protection in Street-Level Image Platforms (Slips). Inform Commun Technol Law 33 (2):198–221. https://dx.doi.org/10.2139/ssrn.4585387

Cadena C, Carlone L, Carrillo H, et al. (2016) Past, Present, and Future of Simultaneous Localization and Mapping: Toward the Robust-Perception Age. IEEE Trans Robot 32 (6):1309–1332. https://doi.org/10.1109/TRO.2016.2624754

Claxton O, Carson H, Ford JJ, et al. (2024) Improving Visual Place Recognition Based Robot Navigation by Verifying Localization Estimates. IEEE Robot Autom Lett 9 (12):11098–11105. https://doi.org/10.1109/LRA.2024.3483045

Dhein K (2023) The Cognitive Map Debate in Insects: A Historical Perspective on What Is at Stake. Stud Hist Philos Sci 98:62–79. https://doi.org/10.1016/j.shpsa.2022.12.008

Dijk Tv, De Wagter C, and de Croon G (2024) Visual Route Following for Tiny Autonomous Robots. Sci Robot 9 (92):eadk0310. https://doi.org/10.1126/scirobotics.adk0310

Dupeyroux J, Serres JR, and Viollet S (2019) AntBot: A Six-Legged Walking Robot Able to Home Like Desert Ants in Outdoor Environments. Sci Robot 4 (27):eaau0307. https://doi.org/10.1126/scirobotics.aau0307

Ebadi K, Bernreiter L, Biggie H, et al. (2023) Present and Future of SLAM in Extreme Environments: The DARPA SubT Challenge. IEEE Trans Robot 39 (1):936–959. https://doi.org/10.1109/TRO.2023.3323938

Fiete I, Burak Y, and Brookings T (2008) What Grid Cells Convey about Rat Location. J Neurosci 28 (27):6858–6871. https://doi.org/10.1523/jneurosci.5684-07.2008

Floreano D, and Wood RJ (2015) Science, Technology and the Future of Small Autonomous Drones. Nature 521 (7553):460–466. https://doi.org/10.1038/nature14542

Garg S, Suenderhauf N, and Milford M (2022) Semantic-Geometric Visual Place Recognition: A New Perspective for Reconciling Opposing Views. Int J Robot Res 41 (6):573–598. https://doi.org/10.1177/0278364919839761

Gasparetto A, and Scalera L (2018) From the Unimate to the Delta Robot: The Early Decades of Industrial Robotics. In: 2018 HMM IFToMM Symposium on History of Machines and Mechanisms. Springer, Cham, pp 284–295

Gates B (2008) A Robot in Every Home. Sci Am 298 (1):58–65. https://www.scientificamerican.com/article/a-robot-in-every-home-2008-02/

Goodall NJ (2014) Machine Ethics and Automated Vehicles. In: Meyer G, Beiker S (eds) Road Vehicle Automation. Lecture Notes in Mobility. Springer, Cham, pp 93–102

Hausler S, Garg S, Chakravarty P, et al. (2023) Locking On: Leveraging Dynamic Vehicle-Imposed Motion Constraints to Improve Visual Localization. In: IEEE/RSJ International Conference on Intelligent Robots and Systems (IROS). IEEE, New York, pp 5258–5265

Hausler S, Garg S, Xu M, Milford M, and Fischer T (2021) Patch-NetVLAD: Multi-Scale Fusion of Locally-Global Descriptors for Place Recognition. In: IEEE Conference on Computer Vision and Pattern Recognition. IEEE, New York, pp 14136–14147

Hussaini S, Milford M, and Fischer T (2024) Applications of Spiking Neural Networks in Visual Place Recognition. IEEE Trans Robot 40 (2):345–359. https://doi.org/10.1109/TRO.2024.3508053

Jacobson A, Zeng F, Smith D, et al. (2021) What Localizes Beneath: A Metric Multisensor Localization and Mapping System for Autonomous Underground Mining Vehicles. J Field Robot 38 (1):5–27. https://doi.org/10.1002/rob.21978

Jain D, and Sharma Y (2017) Adoption of Next Generation Robotics: A Case Study on Amazon. Perspect Case Res 3:9–23. https://www.sibmbengaluru.edu.in/wp-content/uploads/2023/08/2_Adoption-of-next-generation-robotics.pdf

Karaman S, Walter M, Perez A, Frazzoli E, and Teller S (2011) Anytime Motion Planning Using the RRT*. In: IEEE International Conference on Robotics and Automation. IEEE, New York, pp 1478–1483

Khaliq A, Ehsan S, Chen Z, Milford M, and McDonald-Maier K (2019) A Holistic Visual Place Recognition Approach Using Lightweight CNNs for Significant Viewpoint and Appearance Changes. IEEE Trans Robot 36 (2):561–569. https://doi.org/10.1109/TRO.2019.2956352

Kruse T, Pandey A, Alami R, and Kirsch A (2013) Human-Aware Robot Navigation: A Survey. Rob Auton Syst 61 (12):1726–1743. https://doi.org/10.1016/j.robot.2013.05.007

Kuipers B (2000) The Spatial Semantic Hierarchy. Artif Intell 119 (1-2):191–233. https://doi.org/10.1016/S0004-3702(00)00017-5

Kurzman L, Vazquez D, and Laradji I (2019) Class-Based Styling: Real-Time Localized Style Transfer with Semantic Segmentation. In: IEEE/CVF International Conference on Computer Vision Workshops. IEEE, New York, pp 3189–3192

Li Z, Wang L, Jiang L, and Xu C-Z (2019) FC-SLAM: Federated Learning Enhanced Distributed Visual-LiDAR SLAM in Cloud Robotic System. In: IEEE International Conference on Robotics and Biomimetics (ROBIO). IEEE, New York, pp 1995–2000

Lowry S, Sünderhauf N, Newman P, et al. (2016) Visual Place Recognition: A Survey. IEEE Trans Robot 32 (1):1–19. https://doi.org/10.1109/TRO.2015.2496823

Madhavan A (2023) Brain-Inspired Computing Can Help Us Create Faster, More Energy-Efficient Devices—If We Win the Race. https://www.nist.gov/blogs/taking-measure/brain-inspired-computing-can-help-us-create-faster-more-energy-efficient

Mangan M, Floreano D, Yasui K, et al. (2023) A Virtuous Cycle between Invertebrate and Robotics Research: Perspective on a Decade of Living Machines Research. Bioinspir Biomim 18 (3):035005. https://doi.org/10.1088/1748-3190/acc223

Matsuki H, Murai R, Kelly PH, and Davison AJ (2024) Gaussian Splatting SLAM. In: IEEE/CVF Conference on Computer Vision and Pattern Recognition. IEEE, New York, pp 18039–18048

Milford M, Shen C, Lowry S, et al. (2015) Sequence Searching with Deep-Learnt Depth for Condition- and Viewpoint-Invariant Route-Based Place Recognition. In: IEEE Conference on Computer Vision and Pattern Recognition Workshops. IEEE, New York, pp 18–25

Milford M, and Wyeth G (2008) Mapping a Suburb with a Single Camera Using a Biologically Inspired SLAM System. IEEE Trans Robot 24 (5):1038–1053. https://doi.org/10.1109/TRO.2008.2004520

——— (2010) Persistent Navigation and Mapping Using a Biologically Inspired SLAM System. Int J Robot Res 29 (9):1131–1153. https://doi.org/10.1177/0278364909340592

——— (2012) SeqSLAM: Visual Route-Based Navigation for Sunny Summer Days and Stormy Winter Nights. In: IEEE International Conference on Robotics and Automation. IEEE, New York, pp 1643–1649

Milford MJ, Wyeth GF, and Prasser D (2004) RatSLAM: A Hippocampal Model for Simultaneous Localization and Mapping. In: IEEE International Conference on Robotics and Automation. IEEE, New York, pp 403–408

Mohanan M, and Salgoankar A (2018) A Survey of Robotic Motion Planning in Dynamic Environments. Rob Auton Syst 100:171–185. https://doi.org/10.1016/j.robot.2017.10.011

Nagy M, Lăzăroiu G, and Valaskova K (2023) Machine Intelligence and Autonomous Robotic Technologies in the Corporate Context of SMEs: Deep Learning and Virtual Simulation Algorithms, Cyber-Physical Production Networks, and Industry 4.0-Based Manufacturing Systems. Appl Sci 13 (3):1681. https://doi.org/10.3390/app13031681

Ohnsman A (2024) Waymo Logging 150,000 Robotaxi Rides and 1 Million Miles a Week. https://www.forbes.com.au/news/innovation/waymo-logging-150000-robotaxi-rides-and-1-million-miles-a-week/

Parker L (2007) Distributed Intelligence: Overview of the Field and Its Application in Multi-Robot Systems. In: AAAI Fall Symposium: Regarding the Intelligence in Distributed Intelligent Systems. AAAI, Washington, DC, pp 1–6

Pomerleau DA (1988) ALVINN: An Autonomous Land Vehicle in a Neural Network. In: NIPS 1988: Neural Information Processing Systems. Morgan Kaufmann, San Francisco, pp 305–313

Rana K, Haviland J, Garg S, et al. (2023) SayPlan: Grounding Large Language Models Using 3D Scene Graphs for Scalable Robot Task Planning. In: Conference on Robot Learning (CoRL). PMLR, Cambridge, MA, pp 23–72

Rosinol A, Leonard JJ, and Carlone L (2023) NeRF-SLAM: Real-Time Dense Monocular Slam with Neural Radiance Fields. In: IEEE/RSJ International Conference on Intelligent Robots and Systems (IROS) IEEE, New York, pp 3437–3444

Sandamirskaya Y, Kaboli M, Conradt J, and Celikel T (2022) Neuromorphic Computing Hardware and Neural Architectures for Robotics. Sci Robot 7 (67):eabl8419. https://doi.org/10.1126/scirobotics.abl8419

Templeton B (2023) GM's Cruise Falls Down During Concert Cell Overload; Here's How to Fix It. https://www.forbes.com/sites/bradtempleton/2023/08/14/gms-cruise-falls-down-during-concert-cell-overload-heres-how-to-fix-it/

Thrun S, Burgard W, and Fox D (2005) Probabilistic Robotics. MIT Press, Cambridge, MA

Vespa E, Nikolov N, Grimm M, et al. (2018) Efficient Octree-Based Volumetric SLAM Supporting Signed-Distance and Occupancy Mapping. IEEE Robot Autom Lett 3 (2):1144–1151. https://doi.org/10.1109/LRA.2018.2792537

Webb B (2000) What Does Robotics Offer Animal Behaviour? Anim Behav 60 (5):545–558. https://doi.org/10.1006/anbe.2000.1514

Wu F, Shang J, Yu Y, and Milford M (2019) NeuroSLAM: A Brain-Inspired SLAM System for 3D Environments. Biol Cybern 113 (5):515–545. https://doi.org/10.1007/s00422-019-00806-9

Yu C, Liu Z, Liu X, et al. (2018) DS-SLAM: A Semantic Visual SLAM Towards Dynamic Environments. In: IEEE/RSJ International Conference on Intelligent Robots and Systems. IEEE, New York, pp 1168–1174

Zhang H-Y, Lin W-M, and Chen A-X (2018) Path Planning for the Mobile Robot: A Review. Symmetry 10 (10):450. https://doi.org/10.3390/sym10100450

4

Universal Principles of Navigation across Species and Spatial Scales

Sylvia Wirth, Ken Cheng, Pauline Fleischman, Kenneth J. Lohmann, Michael Milford, Henrik Mouritsen, David J. Pritchard, Benjamin Robira, Nachum Ulanovsky, and Kirsty Wan

Abstract Many organisms share the essential need to move through their environment to find resources, shelters, and mates as well as to escape predation. These movements are highly diverse across species, in terms of patterns and mechanisms, and are shaped by various ecological pressures. Despite this diversity, there are universal principles of navigation that apply across species and spatial scales. Integrating knowledge across organisms, this chapter provides an overarching framework underpinning the many ways species implement navigation across scales and goals. What common mechanism(s) could possibly unify the seemingly stochastic wandering of a unicellular organism, the foraging trip of a desert ant, and the more deterministic long-distance migration of vertebrates across land or ocean? Here, common principles are put forward that organize navigation across scales through different phases, transitioning from large-scale to fine-grained spatial resolution. Furthermore, a universal functional model of navigational systems is presented that operates at each of these scales. This model describes a set of modules functioning within a servomechanistic loop, whose completion underpins navigation at each phase. Enabled by species-specific sensorimotor systems and shaped by the navigational context, the model relies on a central integrator module. This integrator facilitates control of the motor system, which implements navigation behavior through decision making, planning, and continuous error correction. This model encourages a neuroethological approach to navigation, offering precise hypotheses to test at both behavioral and neuroscience levels.

Keywords Comparative model, behavioral strategy, navigation, neuroethology

Sylvia Wirth (✉)
Institut des Sciences Cognitives Marc Jeannerod, CNRS 5229, Bron, France
Email: sylvia.wirth@isc.cnrs.fr

Coauthor affiliations are available in the List of Contributors.

N. S. Newcombe and K. Cheng (eds.), *Challenges in Navigation Research*,
Strüngmann Forum Reports, https://doi.org/10.1007/978-3-032-20563-6_4

Group photos (top left to bottom right) Sylvia Wirth, Ken Cheng, Pauline Fleischman, Nachum Ulanovsky, Kenneth Lohmann, Benjamin Robira, Michael Milford, David Pritchard, Kenneth Lohmann, Nachum Ulanovsky, Henrik Mouritsen, Kirsty Wan, Benjamin Robira, Michael Milford, Ken Cheng, Sylvia Wirth, Pauline Fleischman, Kirsty Wan

1 Introduction: Navigational Diversity across Species

Many organisms across the entire range of life (from bacteria to archaea or single-celled eukaryotes, to animals with and without brains) travel to specific locations in space across many orders of magnitude in spatial scale relative to body size (Buehlmann et al. 2020; Freas and Cheng 2022; Grob et al. 2021; Lohmann et al. 2007; Mouritsen 2018, 2022; Putman 2018). Bacteria (Berg and Brown 1972), archaea (Quax et al. 2018), and single-celled eukaryotes, such as *Paramecium caudatum* (Van Houten 1978) and *Euplotes vannus* (Laeverenz-Schlogelhofer and Wan 2024), climb up favorable chemical gradients and down unfavorable ones, over spatial scales of centimeters. Insects with nests, such as ants and honeybees, routinely forage and then navigate back to their home over scales of meters or kilometers (Fleischmann et al. 2016; Freas and Cheng 2022; Knaden and Graham 2016; Srinivasan 2011). Rodents such as wild rats roam and find places within an outdoor enclosure of ~10 m^2 (Calhoun 1963) as well as over areas of several kilometers, such as on an island or in agricultural areas (Russell et al. 2010; Taylor 1978). Primates roam large areas to meet their daily energetic requirement, scaling with their size: whereas the tiny Microcebus covers an area no larger than a couple of hectares (i.e., ca. 150 m^2), the home range of large western gorillas spans ca. 10 km^2 (Judson et al. 2024; Robira et al. 2023b; Scarry et al. 2023). Fruit bats find their way around a home range of tens of kilometers and can home readily when displaced within or shortly outside their home range (Tsoar et al. 2011). On a longer, global scale, sea turtles navigate across entire ocean basins (i.e., thousands of kilometers) using magnetic cues for guidance (Lohmann and Lohmann 2019; Lohmann et al. 2001, 2012), and migratory birds traverse the globe using, for example, celestial and magnetic cues (Emlen 1975; Mouritsen 2018; Wynn et al. 2020, 2022). Bar-tailed godwits can find their way from Alaska to the shores of New Zealand or Australia in a single nonstop flight that is over more than 10,000 km (Gill et al. 2009). These examples reflect just some of the taxa with which we, as coauthors, are familiar, spanning all domains of life and seven orders of magnitude in terms of the size of organisms (Figure 4.1).

Embedded within such diversity is a multitude of strategies and mechanisms for reaching navigational goals. Our task at the Forum was to go beyond this potpourri of diversity and search for common characteristics that hold across all navigating organisms. Reviews of organismal navigation are plentiful (Freas and Cheng 2022; Grob et al. 2021; Lohmann et al. 2008c, 2022; Mouritsen 2018, 2022; Wadhwa and Berg 2022; Wiltschko and Wiltschko 2023), but the drawing of common themes has rarely been attempted (Menti et al. 2023; Parra-Barrero et al. 2023). Thus, we set out to identify commonalities and distill universal themes, drawing on existing knowledge from a wide array of organisms. Purposely, we focus on the behavioral aspects of navigation, as we hope this will be beneficial to the wide array of disciplines that study this topic (e.g., neuroscience, behavioral ecology, comparative psychology, microbiology, artificial systems, and robotics; see Milford, this volume).

We begin by sketching out some common phases of navigational journeys across a range of spatial scales, phases that typically rely on different sensory cues and mechanisms and involve phase transitions. Thereafter, we highlight how a general scheme of navigation can describe the inner working process of these navigational

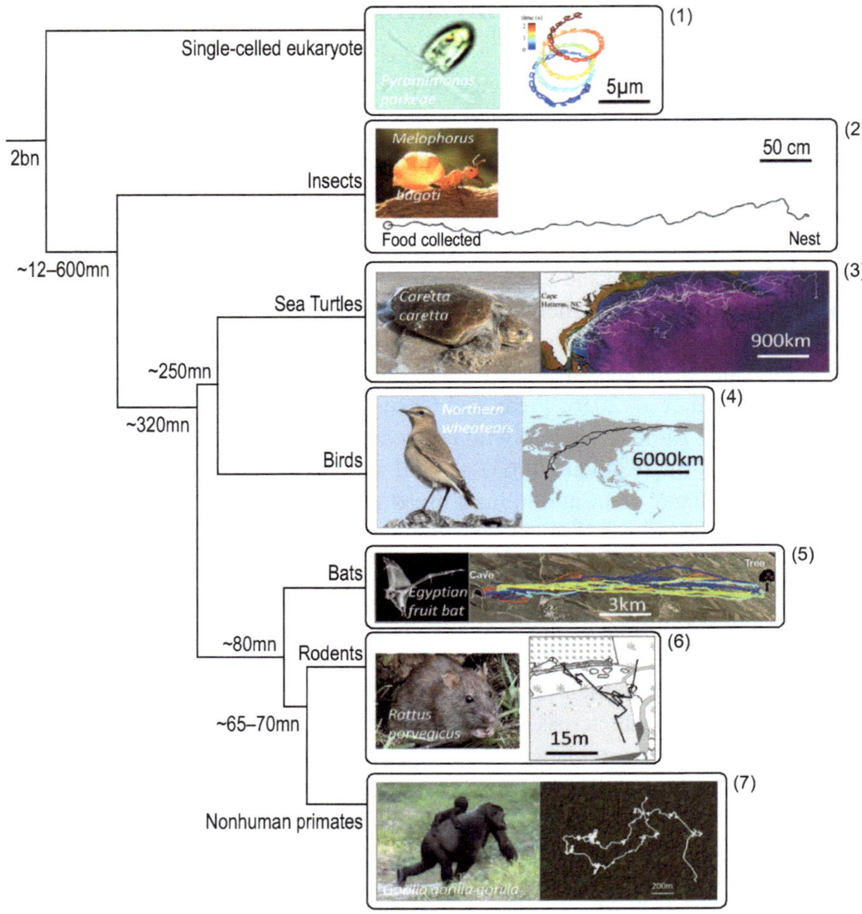

Figure 4.1 Phylogenetic tree of species represented in this chapter (Chiari et al. 2012; Erwin and Davidson 2002; Foley et al. 2023), with examples of navigational trajectories, top to bottom.

(1) Trajectory of a single-celled flagellate green alga *Pyramimonas parkeae* (adapted from Wan 2020; http://creativecommons.org/licenses/by/4.0/).
(2) Homing trajectory of a *Melophorus bagoti* ant (data from Freas et al. 2024). Photo credit: Patrick Schultheiss.
(3) Satellite telemetry tracks of young loggerhead turtles from Florida, USA, during the first few months of their multiyear, transoceanic migration in which they gradually circle the Sargasso Sea before returning to North America (from Mansfield et al. 2014; https://creativecommons.org/licenses/by/3.0/). Photo credit: Kenneth J. Lohman.
(4) Cross-hemisphere trajectory of a northern wheatear (Bairlein et al. 2012; used with permission from the Royal Society Great Britain). Photo credit: Heiko Schmaljohann.
(5) Flight trajectory of an Egyptian fruit bat foraging from tree to cave (adapted from Geva-Sagiv et al. 2015; used with permission from Springer Nature). Photo credit: Nachum Ulanovsky.
(6) Foraging trajectory of a wild rat from burrow to feed sites on a farm. Redrawn from Montes De Oca et al. (2017); used with the permission of Commonwealth Scientific and Industrial Research, or Australian Academy of Science. Photo credit: Reg McKenna, https:/commons.wikimedia.org/wiki/File:WildRat.jpg
(7) A daily foraging path of a western lowland gorilla group in the Dzanga-Sangha Forest Special Reserve (Lara Nellissen, Shelly Masi, and Klaus Zuberbühler, pers. comm.) Photo credit: Lara Nellissen.

phases through integrated modules, independent of scale. We illustrate how the modules operate in different species and robotic systems, demonstrating both the commonalities and the diversity of solutions to similar challenges. As with any descriptive framework, there are exceptions. Rather than forcing every exemplar to fit, our aim was to model what is generally applicable with meaningful detail; that is, without resorting to excessive abstraction or simplification for the sake of universality. In conclusion, we offer perspectives on future research that could build upon the framework presented here.

2 Multiple Navigational Phases during One Navigation Journey

For many organisms, short-distance movement through space follows external gradients, particularly chemical gradients in the case of chemotaxis, a seemingly stochastic random process whose state transitions can nevertheless be characterized as servomechanisms (Cheng 2022, 2023). Depending on the comparison between a current state and a target state, an error correction mechanism may precisely adjust the course of travel until the desired state is reached. This simple principle may suffice to support many orienting mechanisms, including alternation between exploration and exploitation states, driving movement, and transitions, depending on internal and external cues. We propose that across species, navigation behavior universally follows a principle of state transitions, involving multiple state or behavioral transitions, whether moving or orienting at a short or longer distance. As illustrated in Figure 4.2a, transition switches are triggered by different sets of cues that recruit different mechanisms underpinning these behaviors.

For example, even at the microscopic scale, motile organisms routinely choose between random exploration and goal-directed behaviors. Under nutrient-rich conditions, the amoeba *Dictyostelium* preferentially remains in the unicellular form as cells randomly crawl on the substrate. However, upon sensing of cyclic adenosine monophosphate, cells rapidly transition to directed navigation and aggregate to form a multicellular slug. The slug then moves in a directed manner toward a food source (Marée and Hogeweg 2001).

When organisms travel over long distances relative to their body size, navigational behavior generally involves multiple phases, especially, though not exclusively, in long-distance migrating animals (Lohmann et al. 1999, 2008a; Mouritsen 2018, 2022). Three phases were defined by Mouritsen (2018): In a *long-distance phase*, an animal typically relies on global cues such as Earth's magnetic field (Wiltschko and Wiltschko 1995; Zapka et al. 2009), the Sun (Schmidt-Koenig et al. 1991; Zapka et al. 2009), and/or celestial UV polarization patterns (Wehner 2020), and the stars (Emlen 1975; Zapka et al. 2009). During *a narrowing-in* or *homing phase*, a variety of learned or innate local maps, routes, or cues are used (Brothers and Lohmann 2015; Mouritsen 2018; Putman et al. 2013; Wynn et al. 2020, 2022). In a *pinpointing-the-goal phase*, the animal usually bases navigation on localized familiar sensory cues that emanate from the vicinity of the goal (Bonadonna et al. 2003; Gémard et al. 2019; Zeil 2025) (Figure 4.2b).

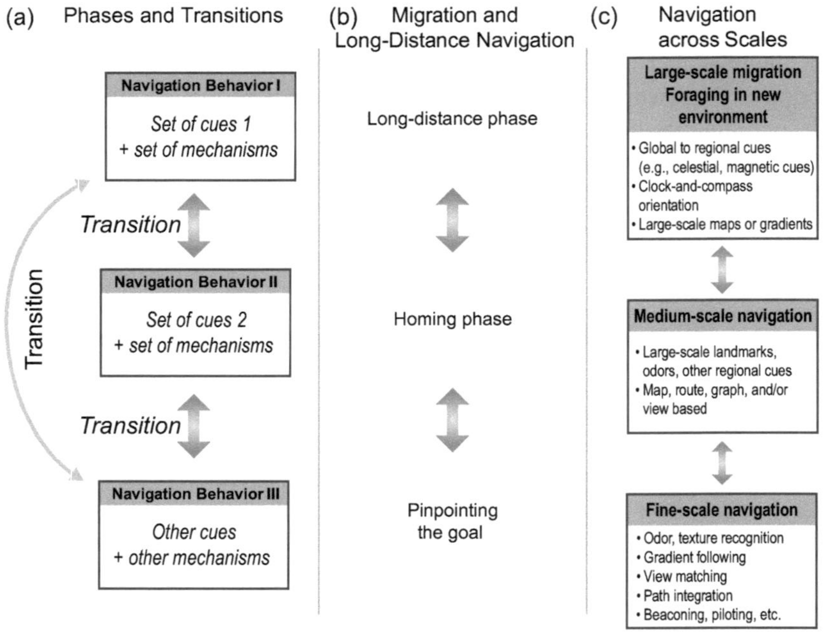

Figure 4.2 Navigational phases: (a) Transition between navigation behaviors depends on current sensory cues and the organism's adapted mechanisms. (b) Typical navigational phases in migrating animals and other animals navigating back to a known goal over long distances relative to their body size. (c) Common framework of navigational phases for all animals. The model developed by Mouritsen (2018) to account for bird migration was adapted here to emphasize the different navigational behaviors across scales in other species.

The phases defined in animals navigating long distances compared to their body size (Figure 4.2b) can be generalized to many organisms (Figure 4.2c), depending on the resolution from large to fine scale:

- Long-distance phases in *unfamiliar* areas (e.g., migration or foraging in unknown territories); when there are repeated migrations, this phase can also take place in areas where the animals have previously been.
- Medium-scale phases in *familiar* areas (e.g., map-based or route-based navigation in a familiar area).
- Fine-scale navigation phase locating the goal in a *very familiar* and very local area around the goal (e.g., finding a hole in the ground, a specific sleeping perch, or a specific feeding site).

During a given journey, organisms might switch back and forth between these phases several times, such as when specific intermediate, familiar locations are targeted in sequence or when changes in internal motivation motivate a local foraging trip (Figure 4.2a). Phase transitions may not be sharply defined, and multiple phases may be active in parallel, either through some form of weighted combination or rapidly alternating mechanism. Thus, in multigoal trips, these phases can be recycled several times.

Each phase is underpinned by different cues, strategies, and mechanisms (Figure 4.2a), and this has several important consequences (Dittman and Quinn

1996; Lohmann et al. 2008a; Mouritsen 2018, 2022). First, no single cue, sense, or mechanism can completely account for how organisms navigate. Even during a single phase, an organism must weigh and integrate multiple sensory cues to obtain the most reliable and robust guidance when deciding in which direction to move. Second, knowing which phase an organism is in is crucial when investigating underlying mechanisms, as apparent contradictions may arise if the same organism is inadvertently tested in different navigational phases while assuming that it remained in the same phase. Third, transitions between the navigational phases must occur behaviorally and/or neuronally.

To illustrate that the occurrence of several distinct phases of navigational journeys is quite universal across taxa, we provide examples that demonstrate the recurrent principle of phase transitioning for the classes of organism represented in Figure 4.1. In addition, from a robotic perspective, we illustrate that navigation is often implemented through subgoals, making the conditional robotic "behavior" more effective.

2.1 Microorganisms

In microorganisms, motile cells employ multiple modes of movement that are often enacted by the same motor apparatus, including (but not limited to) cilia and flagella which generate motility. In the "run and tumble" strategy, coordinated activity of appendages leads to steady swimming according to a specific gait, whereas transients cause unbundling or asynchrony of flagella and reorientations, sometimes following a sharp distribution in turning angles (Berg and Brown 1972; Polin et al. 2009; Wan and Goldstein 2018). Environmental cues (e.g., chemical gradients) can alter the relative frequencies of these distinct states. Small organisms can chain these different behaviors together to achieve some overall goal, such as chemotaxis or chemokinesis, typically associated with changes in the speed of the run phase or the frequency of turns (Berg and Brown 1972; Bray 2002; Van Houten 1978).

2.2 Insects

Insects use a combination of strategies when navigating that depend on spatial scales. Many more insects than previously thought perform directed, long-distance migrations (Chapman et al. 2015). In the most famous long-distance migratory insect, the Monarch butterfly, the long-distance phase in autumn is dominated by sun-compass orientation in an inherited mean migratory direction (Mouritsen and Frost 2002), which is aided by landscape features (not flying over water or mountains) to help them narrow in on their final goal (Freas and Cheng 2022; Mouritsen et al. 2013b; Reppert et al. 2010). The narrowing-in phase is less well understood but could involve searching for the right altitude and/or microclimate, whereas the pinpointing-the-goal phase most likely involves olfactory cues and detecting the presence of a large number of conspecifics (Mouritsen et al. 2013b). Desert ants use a celestial compass to get to the vicinity of the nest when they have traveled far away from it (Wehner 2020). They then utilize the memory of the panoramic view around their nest to home (Zeil 2025), followed by a visual and/or odor-based mechanism (Steck et al. 2009; Zeil 2025), sometimes combined with a systematic search pattern, to pinpoint the nest entrance (Wehner 2020). Hence, these examples illustrate insects switching reliance from broad to narrow cues to reach their final destinations with precision.

2.3 Marine Animals

In the marine realm, animals such as sea turtles and salmon engage in natal homing, a behavioral pattern in which animals leave the area where they began life, migrate long distances, and then return to reproduce in their area of origin (Lohmann and Lohmann 2019; Lohmann et al. 2008c). In the open ocean, animals moving toward a distant goal are likely to navigate using positional information in the Earth's magnetic field combined with magnetic and celestial compasses (Avens and Lohmann 2003; Lohmann and Lohmann 2019; Lohmann et al. 2004; Luschi et al. 2007). As they near the target area of coastline, salmon detect the distinctive chemical cues associated with their home river and use this information to move upstream toward the spawning area (Dittman and Quinn 1996). Similarly, turtles nearing islands or continental coastlines, where they nest, are likely to exploit numerous short-range cues, including waterborne and airborne chemical cues, changes in wave patterns near land, the sounds of waves breaking, and visual landmarks (Endres and Lohmann 2013; Endres et al. 2016; Lohmann et al. 1999, 2008b). How turtles select nesting sites is not known, but in at least some populations, female turtles appear to imprint on the magnetic signatures of their home beach and return to locations near the learned magnetic fields (Brothers and Lohmann 2015, 2018). Similarly, turtles can learn the magnetic signatures that correspond to feeding areas (Goforth et al. 2025), providing them with a means to reach the vicinity of these targets, close enough for local cues to guide them to the goal.

2.4 Birds

Young night-migratory songbirds are born with the information to look for rotating light dots in the sky for several weeks and to interpret the center of celestial rotation as north (Emlen 1975; Michalik et al. 2014). They also imprint on the magnetic inclination angle of where they were born (Wynn et al. 2020, 2022). To reach their wintering range for the first time, they use a simple clock-and-compass strategy (i.e., a time and direction program offering no or limited feedback on where they actually are) (Berthold 1991; Mouritsen and Mouritsen 2000; Mouritsen and Næsbye Larsen 1998). During the return migratory journey of experienced, night-migratory songbirds, magnetic gradient maps and compasses combined with celestial cues are primarily employed during the long-distance phase (Kishkinev et al. 2021; Mouritsen 2018; Packmor et al. 2024; Pakhomov et al. 2018; Wynn et al. 2022). During the homing phase, medium-scale landmarks, such as rivers, mountains, landscape structures, and medium-scale gradient maps (e.g., odor-scapes; Gagliardo 2013) in several sensory dimensions, are used to narrow in on the final goal. The magnetic inclination angle they imprinted on the year prior is also used as an approximate stop signal (Wynn et al. 2020, 2022). During the pinpointing-the-goal phase, highly localized cues in several sensory dimensions are used to locate the nesting site or sleeping perch (Mouritsen 2018, 2022).

2.5 Bats

In nonmigratory fruit bats, which dwell in caves, a nightly foraging journey consists of multiple phases. First, the bat uses route following inside the cave to reach the cave entrance (Geva-Sagiv et al. 2015). Second, it uses visual information to navigate to its preferred tree over several tens of kilometers (Tsoar et al. 2011). This is an example of medium-scale navigation. Third, it narrows in on a fruit tree by relying, most likely, on a combination of olfactory, visual, and echolocation cues (Geva-Sagiv et al. 2015). Finally, it repeats these same three phases in reverse order to navigate back to the cave at the end of the night.

2.6 Rodents

Much less is known about how rodents navigate naturally outdoors. It has been suggested that rats employ map-based navigation over medium scales of up to one kilometer in agricultural habitats and then pinpoint their burrow using direct sensory cues (e.g., olfaction, touch, and vision) that emanate from the burrow (Calhoun 1963; Geva-Sagiv et al. 2015; Taylor 1978).

2.7 Primates

Across a wide range of habitats, depending on species, primates often establish a home range that suits their metabolic or reproductive needs. These ranges are bordered by natural barriers (e.g., woodland edges, cliffs, shores) or by the presence of conspecifics, both of which may serve as navigational cues for medium- to local-scale navigation (Edwards et al. 2023; Hansen et al. 2020; Pearce et al. 2013). Within their home range, navigation can vary from local- to larger-terrain exploitation, depending on resource scarcity (Hansen et al. 2020; Robira et al. 2023a) or variation in resource types (Janmaat et al. 2006, 2016, 2021; Robira et al. 2024). Since most are visual foragers, landmark perception is crucial for reaching out-of-sight targets (Trapanese et al. 2019) and often involves following specific paths or habitual routes (Asensio et al. 2011; De Guinea et al. 2021). Closer to the goal, the transition to local guidance may be triggered by specific cues, such as recognizable fruit patches (Janmaat 2019). Thus, reaching the current goal may rely on particular visual, acoustic, or olfactory guidance (Kaas et al. 2022).

2.8 Robots

Robot navigation systems fall along a spectrum of navigational phases. Monolithic, end-to-end learning-type navigation systems have no separate phases. Instead, all navigational functions are combined into a single system module. At the other extreme, explicit navigation hierarchies have separate phases. For example, a global navigation phase based on getting to the approximate goal location may be followed by a local homing or visual guiding stage that achieves the exact goal location. Phase transitions can be discrete, but multiple navigation phases can also run in parallel. Autonomous aerial vehicles typically switch from a global navigation strategy to a local homing strategy when landing in a constricted space such as a helipad on a rocking ship.

2.9 Summary

The behavioral perspective described here illustrates how most navigational journeys consist of several distinct phases across species and even in artificial systems. Programs sensitive to different cues appear to enable each phase, and transitions between phases signify a universal principle of navigation. Showcasing the relevance of multiple phases in applied engineering, subprograms prompted by specific variable switches are usefully implemented in robots, but in the living organisms, how these transitions occur has hardly been studied. This inevitably leads to open questions, the answers of which would considerably enhance our understanding of the cognitive machinery at play in the navigation of organisms:

- Are the transitions between phases abrupt or gradual?
- What triggers the transition from one phase to the next?
- How could these transitions be reflected in the nervous system (where applicable)?

We hypothesize that in most animals, the switching between navigational phases or strategies is performed by the central executive brain area of each species: the prefrontal cortex of mammals (Friedman and Robbins 2022; Rushworth et al. 2004; Smith and Jonides 1999), the nidopallium caudolaterale (NCL) in birds (Mouritsen et al. 2016; Stacho et al. 2020), the central complex in insects (El Jundi et al. 2014; Heinze et al. 2018; Honkanen et al. 2019), or the homologous or analogous brain areas in other species. Many aspects of these claims, however, still lack fundamental empirical demonstration, thus opening the door for future neuroethological studies (see also Sections 3.4 and 5). An intriguing, open issue is how cognitive behavioral switching in the context of navigation can occur in "simple" organisms without nervous systems.

Further questions arise regarding the differential processing of navigational cues during different navigational phases. Are the cues needed for each phase continuously sensed and processed, or sensed only depending on the current state? The *relative importance* of directional cues and *the means by which* particular cues are used during different navigation phases can shift with environmental, motivational, and contextual changes, and may vary dramatically across and between phases

(Lohmann et al. 2008c; Mouritsen 2018; Wehner 2020; see also Section 4). Despite the diversity of phase and navigational behaviors, how cues are processed to enable navigation can be universally described along a modular system.

3 Modules of Navigation

Whether an organism navigates an environment to forage from food patch to food patch, to migrate to a distant location, or to pursue a moving prey, we propose that the same machinery can operate on the spatial information to support navigation (Figure 4.3); that is, the overall process of *where* to go and *how* to get there based on *what* information and *why* (Nathan et al. 2008). As illustrated in Figure 4.3, this machinery is rooted in largely universal modules (i.e., processes) that operate sequentially, often iteratively, along a directed movement path:

- A *sensory* module perceives spatial information, addressing "which cues" are used for navigation.
- A *direction* module addresses the "where-to" or heading component of navigation. (Although this direction module refers to a heading, it may rely on absolute or relative compasses.)
- A *position* module estimates the organism's position referring to the "where-am-I" component.
- A *central* module integrates information to address the "why and how" component of navigation, encompassing its objectives and means, applying higher cognitive processes to make navigational decisions depending on context (Section 4), and continuously performing error correction.
- A *locomotion* module leads ultimately to movement, which generates new sensory input, completing a servomechanistic loop (Cheng 2022, 2023).

Conceptually, we consider that these modules constitute the first functional (computational) level of the so-called Marr's three *levels of computation* (Marr 1982), and thus define the different types of problem needed to be solved to achieve navigation behavior. The modules' algorithms and implementation for accomplishing the task (second and third levels, i.e., the transformational operations performed and their biological realization) depend on the nature of sensory information available and the organism's abilities to process it (e.g., the brain hardware) and is thus likely to vary across species. Yet, despite the diversity of navigational behavior across species, we propose that these components are invariant across species and have a mechanistic parallel in robots. Not all modules, however, are necessarily used during navigation, and the modularity of the navigational machinery gives rise to four emerging properties that are probably invariant across species.

First, this modularity directly induces robustness of the navigational system to movement scales, provided that all modules are coherent with one another and that the resolution of the information at which the module operates is suitable for the intended task (Benhamou 2014). For example, long-distance navigation may operate on cues at coarse resolution, whereas locally constrained navigation may imply the use of finer-grained cues.

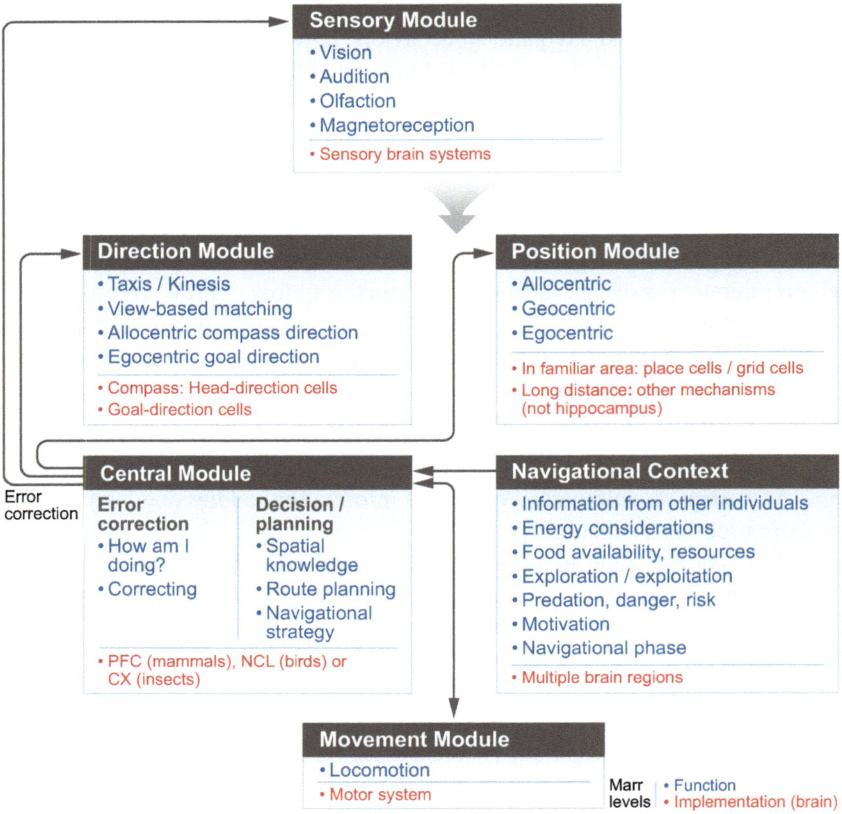

Figure 4.3 A set of universal modules underpin navigation across species. Each module is defined according to its function (in blue) and its neural implementation (in red), constituting the first and third levels of the three computational levels of Marr (1982). CX: central complex; NCL: nidopallium caudolaterale; PFC: prefrontal cortex.

Second, as the modules may possibly operate independently of each other (e.g., an organism in a particular phase may only rely on direction, and not position, and thus only use the direction module and not the position module), the whole navigational process becomes coherently (functionally) scalable. This is important as not all organisms may possess all modules and not all navigational movements may involve all modules. As such, this system can be plastic (i.e., robust) to changes in the navigational context (see Section 4) by modulating *independently* only the modules at play or the mechanism(s) that underpin these modules.

Third, while the implementation of these modules (i.e., the mechanisms by which they operate) may be diverse across species, leading to the emergence of the myriad patterns of movement observed in nature (Mouritsen 2018; Riotte-Lambert and Matthiopoulos 2020; Shaw 2020; Teitelbaum and Mueller 2019; Wisnoski and Lennon 2023), their function may be essentially invariant across the tree of life (Figure 4.1).

Fourth, as a consequence, this system is easily evolvable, as each of these modules can have a genetic origin, such as the inherited migratory abilities in the Monarch butterfly (Zhan et al. 2014) or night-migratory songbirds (Helbig 1991).

As such, new modules emerging over evolutionary timescales can be built onto those that already exist (e.g., in many cases the position module relies on the direction module, and thus likely emerged after the latter in evolutionary history), or old modules can be adapted to rely on new mechanisms as the environment changes or the organisms' abilities evolve.

As a next step, we will characterize the functions of modules underpinning navigation by addressing questions such as "What are they used for?" and "What are the organisms' abilities supporting the function?" while showcasing the species considered in this chapter. We will end by introducing possible implementations at the neuronal level, highlighting the functional properties of the neuronal populations that could underpin these functions. Yet, in many cases, we need more *causal* evidence linking specific cell types to navigational performance. Providing such causal links remains an important challenge for future work. Thus, the neural perspectives ending each section provide grounds for future research to test a potential causal link between the presence of specific cell types and a specific function.

3.1 Sensory Module

Depending on their ecological niches, all organisms have evolved species-specific sensory systems and biological mechanisms to process relevant cues based on their state, context, and navigational goal. Animals might be guided by visual cues, such as celestial cues (e.g., Sun, stars, moon, and polarization patterns) (Dacke et al. 2003, 2013, 2021; Emlen 1975; Mouritsen and Frost 2002; Mouritsen and Larsen 2001; Schmidt-Koenig et al. 1991; Warrant et al. 2024; Wehner 2020), or visual landmarks, such as features in the environment (e.g., mountain chain, river, coastline, the panorama, or more discrete objects like a tree or a rock) (Dreyer et al. 2018; Fleischmann et al. 2016, 2018b; Grob et al. 2021; Kürten et al. 2022). Acoustic cues for navigation may also be extracted from the environment, such as the sound of the sea against the coastline or a frog chorus at a pond, or even from their own sounds like echolocation in bats and dolphins (Goldshtein et al. 2024; Thomas and Moss 2004). Olfaction is also an important modality for navigational tasks, such as scent following (e.g., in ants, Steck 2012), gradient-based navigation (pigeons, Gagliardo 2013), or pinpointing a fruit tree (bats, Gonzalez-Terrazas et al. 2016; mice, Liu et al. 2020; rats, Wallace et al. 2002).

More surprising and yet shared across many organisms is the reliance on Earth's magnetic field. The simplest organisms that exploit magnetic cues are magnetotactic bacteria, which contain crystals of magnetite (Barr et al. 2022; Uebe and Schüler 2016). It is possible, though not certain, that magnetite also provides the basis for a magnetic compass in several animals, including ants (Fleischmann et al. 2018a, 2020; Grob et al. 2024b) and lobsters (Ernst and Lohmann 2016; Lohmann et al. 1995a). In birds, a quantum mechanical magnetic compass sense is strongly supported by recent data (Engels et al. 2014; Hore and Mouritsen 2016; Leberecht et al. 2023; Schulten et al. 1978; Timmer et al. 2023; Xu et al. 2021). Whatever the underlying mechanism, magnetic sensitivity is common among long-distance migrants such as birds (Mouritsen 2018; Wiltschko and Wiltschko 1995, 2023; Zapka et al.

2009), sea turtles (Goforth et al. 2025; Lohmann et al. 2001, 2004), fish (Bottesch et al. 2016; Putman et al. 2013, 2014), and migratory insects (Bogong moths, Dreyer et al. 2018; monarchs, Guerra et al. 2014; cf. Stalleicken et al. 2005). The prevalence of the magnetic sense among many animals may relate to its pervasiveness in the absence of other cues (visual or olfactory) that may not be consistent across environments (air, ocean, desert). Some animals also use mechanosenses to detect cues provided by ocean waves (Lohmann and Fittinghoff Lohmann 1992; Lohmann et al. 1995b), wind (Dacke et al. 2021; Müller and Wehner 2007), gravity (Angelaki et al. 2020; Laurens and Angelaki 2018; Takahashi et al. 2021; Vowles 1954), turbulence (Reynolds et al. 2010, 2016), and temperature (Reynolds 1977) for orientation in space. Below, we account for the incredible diversity of sensory processing developed by living organisms, emphasizing the relationship between sensing and moving across different environments, and combining multiple sensory cues.

3.1.1 Microorganisms

Due to their small sizes, sensory modules are typically embodied in many unicellular organisms. For example, the entire cell membrane of many protists acts as a sensor, containing a myriad of ion channels that enable excitable responses, chemosensitivity, and mechanosensitivity (Wan and Jékely 2021). The membranes are involved in prey detection and its subsequent phagocytosis. Additional specialized sensory modules can be located at specific positions around the organism's body, such as algal "eyespots," which in the biflagellate cell *Chlamydomonas* is located 45° in advance from the plane of the flagella (Witman 1993). These primitive photosensors have a visual sensitivity sufficient to detect single photons of light from the environment (Hegemann and Marwan 1988; Witman 1993). Many protists also use appendages like cilia for locomotion and flow sensing. For example, transient receptor potential channels embedded in a cilium at proximal location, activated during bending, trigger behavioral modification and backward swimming (Arias-Darraz et al. 2015). In some protist species, specialized sensory cilia (not involved in motility) protrude into the surrounding fluid to detect weak water currents and changes in the hydrodynamic landscape signaling potential predators (Marshall and Nonaka 2006; Oshima et al. 2023; Wan and Goldstein 2018).

3.1.2 Insects

The sensory systems of insects vary greatly across species' ecologies. Focusing on ants only, cue reliance depends on local availability and reliability. In temperate climates, many species rely on olfactory cues for navigation and lay pheromone trails to recruit and guide their nest mates toward a food source or back to the nest (Hölldobler and Wilson 1990; p. 270 ff). In contrast, desert ants combine multiple nonolfactory cues, most likely because chemical trails may be too volatile in a desert environment. These ants rely on magnetic cues during initial learning walks and acquire the visual panorama around their nest progressively in time (Fleischmann et al. 2016) and space (Fleischmann et al. 2018b) before going off to forage (Fleischmann et al. 2018b; Grob et al. 2024a, b). In contrast, when confronted with a new panoramic view, experienced foragers perform relearning walks without relying on the geomagnetic field as a reference system; instead, they rely on other, most likely visual, cues (Fleischmann et al. 2022). Further, desert ants evolved a specialized area, the "dorsal rim area," which detects the polarization pattern in the sky's ultraviolet (UV) spectrum (Wehner 2020; p. 132 ff) and provides directional information. Long-distance migratory insects (e.g., Monarchs and Bogong moths) use global cues such as the Sun (Mouritsen and Frost 2002; Stalleicken et al. 2005), the Milky Way (Warrant et al. 2024), and/or the magnetic field (Dreyer et al. 2018) during the long-distance navigation phase. Some insects are masters of olfactory cue detection and are able to detect single pheromone molecules released by a potential mate (Kaissling and Priesner 1970). Thus, insects' sensory systems are incredibly diverse and adapted to detect both local and global cues.

3.1.3 Marine Animals

Given the relative scarcity of cues for navigation in the open sea, it is perhaps not surprising that sensitivity to Earth's magnetic field appears to be widespread among migratory marine animals (Lohmann et al. 2008a). Relative to other sensory cues, the geomagnetic field is a pervasive environmental feature, present day and night, largely unaffected by weather and season, and present at all depths of the ocean. Numerous marine animals use Earth's magnetic field in orientation and navigation, including invertebrates (Lohmann and Ernst 2014; Lohmann and Willows 1987; Wyeth et al. 2021), fishes (Bottesch et al. 2016; Naisbett-Jones and Lohmann 2022; Putman et al. 2020), sea turtles (Lohmann and Lohmann 2019), and possibly marine mammals (Granger et al. 2020). At different scales and distances during the completion of long migrations, magnetic cues are supplemented by celestial, chemical, auditory, and hydrodynamic cues (Avens and Lohmann 2003; Endres and Lohmann 2012, 2013; Lohmann and Fittinghoff Lohmann 1992; Lohmann et al. 1995b).

3.1.4 Birds

Many birds possess remarkable visual abilities based on high-resolution tetrachromatic color vision, species-specific adapted eye placement (forward or lateral facing), and retinal topography (e.g., the number and location of fovea, and the presence of other sensitive areas such as visual streaks) (Martin 2017). Their sense of vision also underlies their time-compensated sun compass (Schmidt-Koenig et al. 1991; Zapka et al. 2009) and is used to detect the center of rotation of the starry sky for their star compass (Emlen 1975; Michalik et al. 2014; Zapka et al. 2009). Surprisingly, the visual system of at least some birds also hosts their magnetic compass (Zapka et al. 2009), which seems to be based on a radical-pair mechanism (Hore and Mouritsen 2016) in cryptochrome 4 proteins (Xu et al. 2021) located in their double-cone photoreceptor cells (Chetverikova et al. 2022; Günther et al. 2018, 2024). Due to the electron-spin-based quantum mechanism, night-migratory songbirds are highly sensitive to radiofrequency electromagnetic noise between ca. 0.4–116 kHz (Leberecht et al. 2023). In addition to vision, olfaction also plays a key role in navigation, in particular in oceanic seabirds and homing pigeons (Bonadonna et al. 2003; Gagliardo 2013; Gémard et al. 2019), allowing them to form a representation of an olfactory landscape.

3.1.5 Bats

Four sensory systems are thought to contribute primarily to bat navigation: vision, echolocation, olfaction, and magnetosensation (reviewed in Geva-Sagiv et al. 2015). Visual acuity varies across bat species; some bat species, especially old-world fruit bats, possess excellent vision that has much better acuity than laboratory rodents (Geva-Sagiv et al. 2015). Several bat species have been shown experimentally to rely heavily on vision for navigating over spatial scales of tens of kilometers (Tsoar et al. 2011; Ulanovsky and Moss 2008; Yovel and Ulanovsky 2017). Echolocation is a sense that is rather unique to bats (among land mammals), whereby the animal emits a brief ultrasonic pulse and uses the returning echoes to estimate the distance, direction, and physical properties of objects in the environment (Ulanovsky and Moss 2008). Echolocation was recently shown to be crucial for navigation in an insectivorous bat, Kuhl's pipistrelle (Goldshtein et al. 2024). Olfaction is important for pinpointing-the-goal phase in fruit bats, which can detect fruit-associated odors at very low concentrations (Brokaw et al. 2021; Laska 1990). Finally, magnetoreception was shown to exist in several bat species, in particular in long-distance migrants, and to contribute to their navigation (Holland et al. 2006; Schneider et al. 2023).

3.1.6 Rodents

Rodents' sensory perception involves vision, audition, olfaction, and somatosensation (Burn 2008). As primarily nocturnal animals, their vision is adapted to dim light with a retina lacking a fovea and consisting almost exclusively of rods (Kaas et al. 2022). This results in poor visual acuity compared to primates (Prusky et al. 2002), especially when exposed to bright light (Koskela et al. 2020). Yet, rodent vision can still locate visual cues at a distance, provided these bear a salient contrast (Hoy et al. 2016; Leinonen and Tanila 2018). To support their poor visual acuity, rodents have highly developed olfaction (Doty 1986; Shepherd 2006), which allows them to locate odorants without moving their heads, by comparing internostril odor intensity and timing differences (Burn 2008; Rajan et al. 2006). Further, rodents detect changes in odor intensity with a sensitivity approximately ten times greater than that of humans (Slotnick and Ptak 1977). They also possess more than twice the number of genes coding for chemoreceptors, likely accounting for higher olfactory discrimination abilities (Niimura and Nei 2006). Indeed, olfactory cues can support locating food sources or conspecifics when navigating, sometimes providing a chemical trail that animals can follow when pinpointing the goal (Wallace et al. 2002). Rats also detect a wide range of sounds, including ultrasound frequencies, which they use for social communication and can thereby also provide local acoustic cues when homing back to the nest (Brudzynski 2021). Somatosensation is supported by rodents' extremely sensitive vibrissae (Carvell and Simons 1990; Davis et al. 1989), allowing fine-grained discriminations, and permitting whisking-based navigation in their complex underground burrow systems (Calhoun 1963). Finally, vestibular and proprioceptive systems provide information relative to one's self-motion, which can be particularly useful during path integration and the retracing of steps in nocturnal environments (Cullen 2012).

3.1.7 Primates

The sensory system of primates prioritizes vision over other senses, such as olfaction (Kaas et al. 2022), diverging from some but not all other animals, and likely resulting from their mostly diurnal activities (cf. Tagg et al. 2018). Compared to many other mammals (e.g., rodents, bovines, equines), a reorientation of the eye position in the skull from lateral to front facing enabled depth perception (Land 2015). Further, primates evolved a central region of the retina, the fovea, that is densely equipped with cones and provides colorful, dense visual resolution (Bringmann et al. 2018). This system is particularly efficient for navigating space at a distance as it enables elements of interest to be identified (e.g., landmarks, food, conspecifics), even in a dense environment, before movement is made toward them. Supporting vision, auditory cues are processed continuously as primates are particularly sensitive to sudden noises indicative of danger. Further, primates also developed long-range acoustic communication to inform others of food locations (Fischer 2021) and maintain contact in dense environments, therefore also serving as guiding signals during navigation (Morrill et al. 2013; Waser and Brown 1986). Finally, primates share with other mammals a vestibular system that supports tracking one's own body and head movement in space using gravity-sensitive receptors (Laurens and Angelaki 2018).

3.1.8 Robots

A key point of differentiation for robotic sensors such as light detection and ranging (LiDAR) (Li et al. 2022) is that they were deliberately created for both robotic navigation and related task purposes, rather than going through a long process of natural evolution (although their artificial creation could be framed as a type of evolution). Robots combine external sensors that sample the environment around the robot using cameras, LiDAR, sonar, and radar sensors. Cameras are passive sensors in that they do not typically emit energy into the environment (versus the echolocation system of a bat) and do not explicitly infer geometry and layout of the scene. Conversely, LiDAR, sonar, and radar are active energy emitting sensors and directly sample range to objects and parts of the physical environment. In contrast, internal sensors (e.g., an accelerometer, gyroscope, or magnetometer) measure state and changes in state. Finally, sensor readings are often fused together for navigational purposes, in particular because different sensing modalities have different utilities for different components of navigation (Alatise and Hancke 2020). An autonomous vehicle typically combines cameras, LiDAR, radar, wheel encoders, inertial sensors, and GPS to safely achieve its navigational goals on roads (Bresson et al. 2017). It is important to note that the origin of these various sensors varies: some were designed and optimized for non-navigational tasks. For example, LiDAR has its heritage in lasers used to ensure safety in a manufacturing environment by detecting when objects or people entered dangerous "no-go" zones. GPS was always designed to provide positional information but is a "platform" capability in that it is not particularly specialized to each individual robotic platform.

3.1.9 Neural Perspective

Each sensory system in the brain is processed through a series of specialized brain areas, all of which have been studied in detail in sensory neuroscience, across many species, and summarized in many reviews (Baars and Gage 2018; Boyd et al. 2009; Cullen 2012; Fulton et al. 2024; Kaas 2013; O'Connor et al. 2021; Rauschecker 2015; Sinakevitch et al. 2018; Sumner-Rooney 2023). The only exception is the magnetic sense, which is the only major sense for which the most important neural pathways remain to be discovered. Only in night-migratory songbirds are we beginning to get an understanding of the pathways involved (Mouritsen et al. 2016). Light-dependent magnetic compass information is most likely detected in double-cone photoreceptor cells of the retina (Günther et al. 2018, 2024). From the eye, magnetic information is transmitted via the thalamofugal pathway (Heyers et al. 2007) to Cluster N, a forebrain area that is part of the visual wulst (Heyers et al. 2022; Zapka et al. 2009), which is further connected to the hippocampus (Heyers et al. 2022). Magnetic information essential for positional information requires input from the ophthalmic branch of the trigeminal nerve, which is processed in the two hindbrain regions (Heyers et al. 2010). From there the information is sent to the telencephalic frontal nidopallium (Kobylkov et al. 2020). In almost any other animal, nothing is known about the magnetosensory pathways in the brain. Thus, this is a major open question for further research.

In mammals, most sensory information streams are integrated into higher-order *multisensory* brain areas, often called association areas (Eichenbaum 2000; Felleman and Van Essen 1991; Olcese et al. 2013), through a high degree of anatomical convergence. These areas then transmit the multisensory information to the brain regions responsible for computing the animal's direction and position, specifically the direction and position modules.

In sum, while there is an incredible diversity of sensory systems evolved by animals to perceive their environment, a common principle shared by both living organisms and artificial robotic systems is that real-life navigation cannot rely on a single set of sensory cues. Instead, multiple sources of information are combined, which together enable the determination of position and direction at the behavioral and neural levels, providing crucial data at both large and fine scales.

3.2 Direction Module

A key requirement for successful navigation is simply traveling in an appropriate direction, whether heading toward a goal, moving away from danger, or just maintaining a consistent trajectory (Grob et al. 2021; Mouritsen and Mouritsen 2000). Depending on the available cues and the scale at which navigation takes place, accurately ascertaining direction can therefore be a challenge for many different navigators. Global direction cues can reliably provide directional information across very long distances, even when a navigator is traveling through unknown territory (Mouritsen and Mouritsen 2000). For example, the east-to-west trajectory of the Sun is accompanied by patterns of polarized light that fill the sky (Guilford and Taylor 2014; Wehner 2020) even at night via the moonlight's reflection (Dacke et al. 2013;

Freas et al. 2024). Providing a more stable signal, Earth's magnetic field continuously provides orientation information to those capable of sensing it even in conditions of low visibility (Bottesch et al. 2016; Dreyer et al. 2018; Fleischmann et al. 2020; Lohmann et al. 2022, 2023; Wiltschko and Wiltschko 1995, 2023; Zapka et al. 2009).

Local directional cues can come from familiar features in the world, such as a mountain range, seashore, or river, or by directly sensing the goal. A navigator traveling through the world could, for example, orient themselves by remembering how the world looks when heading along a route and maintain a bearing relative to a salient landmark, by keeping a noisy factory or a swamp's frog chorus to the left, or by heading toward an enticing smell.

Finally, cues generated by a navigator's own movement, such as optic flow (Srinivasan et al. 2000) or vestibular information (Cullen 2012), can also provide directional feedback as an organism moves and turns in order to maintain a consistent sense of direction. This section illustrates how direction behaviorally operates across species. We note that here we do not distinguish between the sense of absolute allocentric direction (a "compass") and egocentric direction relative to a goal (goal direction; see Figure 4.3).

3.2.1 Unicellular Organisms

Unicellular organisms have an acute directional sense. They can extract directional information from the chemical, mechanical, and light-based cues from their specialized sensory mechanisms. For example, ion channels in their membrane can detect chemical gradients (And and Hellung-Larsen 1992; Wolfe 2000) and sensory cilia can detect the direction of fluid currents (Bezares-Calderón et al. 2020), whereas algal eyespots are sensitive to light direction (Foster and Smyth 1980). These directional responses often trigger adjustments in movement, such as flagellar rotation or ciliary bending, allowing the organism to adapt its directed trajectory (Cortese and Wan 2021; Tsang et al. 2018). This integrated sensory system enables precise and adaptive taxis within their immediate surroundings, differentiating between stochastic and deterministic forms of directed migration depending on the nature of the cue.

3.2.2 Insects

Insects and other arthropods have developed many different strategies to determine direction (Chapman et al. 2015; Grob et al. 2021; Reynolds et al. 2016). Desert ants, for example, use a celestial compass system (Wehner 2020), the distant panorama (Fleischmann et al. 2018b; Wystrach et al. 2011, 2012), a magnetic compass (Fleischmann et al. 2016; Grob et al. 2024b), and a wind compass (Müller and Wehner 2007) to determine directions (Wehner 2020). Importantly, the celestial compass must be calibrated at the beginning of the ants' outside activities (Grob et al. 2019; Rössler et al. 2023) to be time-compensated and thus useful. Recent studies reveal that ants must experience a rotating celestial UV polarization pattern to undergo synaptic changes in mushroom bodies (the higher-order integration center for learning and memory) and central complex (the navigational center in the insect brain) (Fleischmann et al. 2022; Grob et al. 2017).

3.2.3 Marine Animals

As described previously, marine animals can maintain consistent headings using Earth's magnetic field, celestial cues, and ocean waves (Avens and Lohmann 2003; Bottesch et al. 2016; Lohmann et al. 1990, 2008a; Mott and Salmon 2011; Mouritsen et al. 2013a; Spiecker et al. 2022). A few animals that inhabit shallow, clear water are able to exploit sky polarization patterns when orienting (Patel and Cronin 2020).

3.2.4 Birds

Birds traveling over short distances can rely on many of the same cues used over much longer journeys. Homing pigeons, for example, respond to shifts in the sun compass even within sight of their loft (Armstrong et al. 2013). Traveling in a familiar area also enables birds to learn and use familiar landmarks. These landmarks can provide directional information by themselves, for example, as a remembered panoramic view (Schwarz et al. 2024), or can be combined with a sun compass (Guilford and Biro 2014). Clark's nutcrackers can adaptively switch between strategies (Kelly et al. 2019).

Across long distances, migrating birds may use at least three different compasses—sun, star, and magnetic (Mouritsen 2018; Wiltschko and Wiltschko 1995, 2023)—to determine and keep flying in the right direction after they have determined their position using their map (the classical map and compass concept of Kramer 1953). When juveniles migrate for the first time, an inherited time and direction program (clock-and-compass or vector navigation mechanism) supports navigation to the destination (Berthold 1991; Holland et al. 2006; Mouritsen 1998a; Mouritsen and Mouritsen 2000).

3.2.5 Bats

Bats were shown to extract directional information from visual inputs, such as the sight of a distal mountain range in Trinidad (Williams et al. 1966), as well as from celestial cues, such as the post-sunset glow and the stars (Buchler and Childs 1982; Childs and Buchler 1981). Some bat species were proposed to use distant sound sources (e.g., frog choruses) for directional orientation (Buchler and Childs 1981). Finally, long-distance migratory bats were proposed to use Earth's magnetic field as a magnetic compass for directional orientation (Lindecke et al. 2021; Schneider et al. 2023).

3.2.6 Rodents

Most evidence of a rodent's ability to orient in particular directions comes from laboratory experiments performed in mice and rats. A detailed behavioral analysis during orientation suggests the presence of sequential mechanisms when learning a new environment, with a progressively expanding exploration (Gire et al. 2016; Maaswinkel and Whishaw 1999). Initial behavior is directed to local cues (Alyan and Jander 1994), which could be an olfactory gradient leading to food source (Howard et al. 1968) or back to their nest (Gire et al. 2016; Khan et al. 2012). Next, an examination of their travel path suggests that the homing mouse adds path integration derived from locomotion when it runs (Alyan and Jander 1994). This homing vector through path integration can be formed without olfactory cues, providing directional cues in case the olfactory trail vanishes (Whishaw and Gorny 1999). Finally, manipulations of distal visual landmarks' positions or visibility suggest they provide a spatial directional reference for navigation. For example, blindfolded rats were incapable of finding an escape platform in the water, proving that animals relied on visual cues in the absence of reliable odors (Maaswinkel and Whishaw 1999; Whishaw and Maaswinkel 1998). Neural support for directional signals has been well studied in the rat, as described further in Section 3.2.9.

3.2.7 Primates

In primates, immediate directional cues are provided by visual optic flow (expanding, contracting, or moving laterally) and induce a perception of self-motion (Laurens and Angelaki 2018; Laurens et al. 2017; Longuet-Higgins and Prazdny 1980; Peltier et al. 2020). In addition, internal sensors in the vestibular system respond to linear and angular acceleration, allowing the primate to sense the direction of its movement (Angelaki and Dickman 2000; Laurens et al. 2017). The presence of neural systems associated with the processing of visual or vestibular response evoked by self-motion has been widely demonstrated in humans and other primates (Cullen and Taube 2017; Laurens and Angelaki 2018). Yet as shown in humans, these signals alone do not suffice to maintain direction over long distances (Souman et al. 2009). Rather, proprioceptive and vestibular information are likely combined with view-based information to support maintenance of travel direction (Yoder et al. 2011). While there is no systematic analysis of the visual cues used by primates in the wild, semi-fieldwork (Rosati 2017) and laboratory work show that primates orient themselves using local and distant salient visual cues, appropriately termed landmarks (Trapanese et al. 2019). When such landmarks are rare, or unknown, as in unfamiliar environments, primates may rely on natural compasses, such as the sun compass. This has been demonstrated by BaYaka people, who inhabit the tropical forest of the Congo Basin (Jang et al. 2019). For western gorillas, the sun compass may not be the primary navigational tool due to their high site fidelity and hence familiarity with the environment (Robira et al. 2024).

3.2.8 Robots

For many robot navigational tasks, directional accuracy is more important than exact locational accuracy, because errors in direction typically translate immediately into errors in location. By contrast, errors in location are not always catastrophic, at least for periods of time, such as when large open spaces are traversed. Robots employ external directional cues like a compass and global navigation satellite systems, as well as tracking orientation as part of their dead reckoning/path-integration process. Orientation can also be determined by localizing within a prior map of the environment. Robot gliders are a type of submersible robot that travel thousands of kilometers deep underwater, where orientation accuracy is key to navigation. All errors are not equally significant when considering orientation: random zero-mean noise can be compensated for well in many navigational situations, whereas consistent bias is often more problematic.

3.2.9 Neural Perspective

Perhaps the most well-documented evidence for the presence of a neural compass in the brain is found in *head-direction cells*, located in several brain areas (Ranck 1985; Taube 2007; Taube et al. 1990). This neuronal redundancy (e.g., thalamus, subiculum, retrosplenial cortex, entorhinal cortex) suggests that the sense of direction is a fundamental building block of navigation, as it is also in robotics.

Head-direction neurons are tuned to a particular allocentric direction of the head (i.e., where the head is pointing toward), thus encoding the animal's direction. This signal is independent of the animal's position, such that a cell which responds when the animal is pointing to the north, does so for many locations, as long as its head points toward the same direction. Thus, for the majority of these cells, direction is independent from positional signals. In rodents and primates in laboratory settings, it was shown that head-direction cells depend on both self-motion (idiothetic) cues and external visual landmarks. The cell's tuning is present and stable in blind rats, showing that olfactory or self-motion cues suffice to drive them (Asumbisa et al. 2022). Yet, in sighted rats, their direction appears to be primarily set by visual cues (Asumbisa et al. 2022; Goodridge et al. 1998; Goodridge and Taube 1995). In the absence of visual and olfactory cues, the cells respond to self-motion but drift over time (Asumbisa et al. 2022), showing that the cells need constant calibration by external cues (Cullen and Taube 2017).

Interestingly, the sensitivity of head-direction cells to self-motion and visual cues parallels the way direction is computed in robots, in which path integration is combined with the use of external cues, highlighting again common engineering solutions.

Head-direction cells are found across a large range of taxa: rodents, bats (Finkelstein et al. 2015), primates (Laurens et al. 2016), birds (Ben-Yishay et al. 2021), and, surprisingly, fish (Petrucco et al. 2023) and insects (*Drosophila*, Hulse and Jayaraman 2020). Remarkably, the neural mechanisms in insects, fish, and mammals seem to be highly similar (Petrucco et al. 2023; Seelig and Jayaraman 2015) and instantiates the theoretical ring attractor that was originally proposed to explain rodents' head-direction cells (Hulse and Jayaraman 2020; Skaggs et al. 1994). This similarity might arise through conserved evolutionary origins or through common learning processes across species (i.e., cases of convergent evolution). Importantly, head-direction cells were recently found in bats flying outdoors on an oceanic island; these neurons maintained their directionality over the entire geographical extent of the island (Palgi et al. 2024). Thus, head-direction cells might be true neural compasses for outdoor navigation, serving as a neural mechanism for the sense of direction, including in the wild.

Goal-direction cells are another type of directional neuron that may be important for goal-directed navigation. These neurons were found in bats and rodents (Campagner et al. 2023; Sarel et al. 2017), and they encode the animal's egocentric direction relative to the goal, the "goal-direction angle" (goal-centric direction). By minimizing the goal-direction angle to zero, an animal can potentially keep a direct path toward the navigational goal and thus reach the goal. Notably, however, goal-direction cells have not yet been studied in the wild, so it remains to be seen if they might serve as a navigational mechanism during real-life navigation outdoors.

An interesting question is how the allocentric directional information encoded by head-direction cells is transformed from an allocentric directional reference frame to an egocentric reference frame, relative to the goal, to produce goal-direction cells and vice versa. In robotics, directional systems often rely on external reference frames provided, for example, by GPS, and thus directional coordinates are absolute. In the parietal cortex of primates, cells encode the presence of stimuli in a self-referenced (egocentric) reference frame relative to the retina, head, body, or hand, thus providing information necessary for the control of actions in space such as grasping objects (Andersen et al. 1993; Ben Hamed et al. 2001; Duhamel et al. 1997). Similar egocentric signals are found in the retrosplenial cortex of the rat (Alexander et al. 2020) and in the parietal cortex (Wilber et al. 2014). The transformation of cues' position or direction into multiple reference frames is thought to take place in the parietal cortex or retrosplenial cortex, brain areas in mammals that have been suggested to be important for egocentric-to-allocentric transformations (Alexander et al. 2023b; Clark et al. 2018; Wilber et al. 2014). The exact process underlying these transformations remains, however, to be understood and may involve gain field modulation ultimately combining multiple cue sources (Salinas and Sejnowski 2001).

3.3 Position Module

The position module informs the organism about locations, including both self and objects' positions, and it may be used to compute the distance between locations and select appropriate trajectories. In a broad sense, position knowledge can be referenced from various origins (Grob et al. 2021; Holland 2014). For example, the current position could be relative to distal landmarks, a starting position during self-motion, a "home" such as a cave, a foraging goal, or a distant migratory destination. In that sense, the position module provides a map that can inform an animal about a location along a familiar or unfamiliar route. Although many animals mostly navigate flat terrain, positional information can also include a vertical dimension for animals that fly, swim, or climb.

The position module provides information that is mostly tied to distances (relative to a start or relative to a goal), which can be used to compute a directional vector toward the goal, including both direction and distance. Thus, in many animals, the position module interacts closely with the direction module, with the animal periodically assessing its location relative to the goal and then using the direction module to maintain an appropriate heading.

If an animal can determine its position and correct for its displacement even when moved far away from its route to unfamiliar locations, it is said to have performed "true navigation." True navigation has been demonstrated in experienced migratory birds that can reach a final goal with a precision of centimeters after traveling thousands of kilometers (Holland 2014; Mouritsen 2018, 2022). How could this be achieved?

In many migratory animals, including lobsters, sea turtles, salmon, and birds, positional information is underpinned by a sensitivity to Earth's magnetic field, providing a way to assess approximate instantaneous geographic location relative to a goal (Boles and Lohmann 2003; Brothers and Lohmann 2015, 2018; Kishkinev et al. 2021; Lohmann et al. 2004, 2008b, 2008c, 2022; Pakhomov et al. 2018; Putman et al. 2013; Wynn et al. 2020, 2022). Further, several magnetic parameters have been shown to be useful to some animals: magnetic inclination, the angle between the magnetic field lines and the Earth's surface (Lohmann and Lohmann 1994; Putman and Lohmann 2008; Putman et al. 2013; Wynn et al. 2020, 2022), magnetic intensity (Lohmann et al. 2022; Lohmann and Lohmann 1996; Lohmann et al. 2007, 2008a; Putman et al. 2014), and magnetic declination, the difference in directions between geographic and magnetic north (Chernetsov et al. 2017).

Signposts, such as a specific magnetic inclination, can function as a simple stop signal (Wynn et al. 2020, 2022) or to help marine animals recognize locations along a coastline (Brothers and Lohmann 2018; Lohmann and Lohmann 2019; Lohmann et al. 2008c, 2023). Magnetic cues can also be part of a gradient map, allowing for true navigation even after displacements far from the familiar area (Kishkinev et al. 2021; Lohmann et al. 2022; Packmor et al. 2024). Alternately, distance tracking allows an agent to update its position relative to the starting point and compare it to a target distance of travel (ants, Voegeli et al. 2024).

Magnetic information is used by some animals as a positional cue during long-distance movements, but such cues are not used universally. In particular, most mammalian species do not exhibit magnetosensation. On a smaller scale, many animals, including most mammals and many birds, use sets of distal landmarks as positional map cues (Geva-Sagiv et al. 2015; Lipp et al. 2004).

Below, we provide examples of the ways many animals (except the migratory ones described above) assess position. These implementational mechanisms for the position module vary greatly among species and with the spatial scale of movement.

3.3.1 Unicellular Organisms

Unicellular organisms are unlikely to possess any positional sense in a map-like fashion. Nonetheless, many single-celled organisms can detect gravity, which is perhaps the axis of most relevance (Roberts 2006). The exact mechanism is disputed, as is whether it is active or passive. A popular theory suggests that motile flagellates are bottom heavy, with the chloroplast located in the bottom half of the cell (Yoshimura et al. 2003). Due to the offset between the cell's geometric and gravitational centers, a gravitational torque acts to align the cell to promote upward swimming, ensuring that the cells migrate upward toward the surface of the water column where they can gain access to light. Intrinsic morphological asymmetries in the shape of the organism may also play a key role in mediating the gravitactic response of unicellular organisms (Kage et al. 2020).

3.3.2 Insects

Like bats and other mammals, insects are known to use path integration (Müller and Wehner 1988; Srinivasan et al. 2000; Voegeli et al. 2024; Wehner 2020; Wehner and Srinivasan 2003). Recent modeling suggests that they compute a vector to their starting point as they travel, comprising the distance and direction to the starting point (Heinze et al. 2018; Lyu et al. 2022; Stone et al. 2017).

Hymenopterans living in habitats with surrounding views also use views as a position sense. For ants, it is not thought that views give a map sense, but rather that views provide instructions for what to do to reach various locations (Hoinville and Wehner 2018; Wehner 2020). A map-like representation has been suggested for honeybees (Cheeseman et al. 2014; Menzel et al. 2005), but this claim is disputed (Cheung et al. 2014; Hoinville and Wehner 2018; Wehner 2020). Hoinville and Wehner (2018) could account for Cheeseman et al.'s (2014) data based on a model without map coordinates.

3.3.3 Birds

When determining their position in a local area, nonmigratory birds may need to be very accurate if they are to find their nest hole or recover a buried cache. To achieve this, many birds learn locations with reference to visual landmarks, either extracting spatial properties, such as distances or bearings (Kelly et al. 2008), or matching what they currently perceive with a remembered view (Pritchard et al. 2018). Some birds (e.g., Clark's nutcrackers) can also encode relative position, such as halfway between two landmarks, and generalize this to novel distances (Kamil and Jones 1997).

3.3.4 Bats

The position module of bats includes several mechanisms. Bats exhibit map-like navigational abilities, including showing novel shortcuts, which is considered the hallmark of a *cognitive map* (Harten et al. 2020; Toledo et al. 2020; Tsoar et al. 2011). Bats often fly along linear elements in the terrain, including along underground cave passages, vegetation hedges, and forest roads, thus exhibiting *route following* (reviewed in Schnitzler et al. 2003; Ulanovsky and Moss 2008; Yovel and Ulanovsky 2017). Finally, bats have also been shown to exhibit path integration, estimating the distance flown based on self-movement (Aharon et al. 2017). Together, the bat's position module exhibits a diversity of locational mechanisms.

3.3.5 Rodents

Numerous experiments show rodents' robust ability to navigate in space from one place to another (Hok et al. 2016; Paul et al. 2009; Vorhees and Williams 2014). Under debate, however, is whether these navigational abilities rely on a cognitive map, based on Euclidian metrics (Benhamou 1996; Wallace et al. 2002). Such a map could, in principle, allow animals to take shortcuts from locations they have never visited before, without relying on other directional cues such as salient visual landmarks, which provide beacon guidance, or cues like scent trails. Regardless, experimental evidence shows that rodents have a representation of the location of food sources, burrows, or nests (Alyan and Jander 1994; Schaeffer et al. 2023) and often follow the most efficient trajectories goal locations (Blaser and Ginchansky 2012; Jackson et al. 2020). In Section 3.3.8, we describe the idea that a representation of self-position must exist in the rodent brain in the form of hippocampal place cells (O'Keefe and Nadel 1978) and entorhinal grid cells (Hafting et al. 2005).

3.3.6 Primates

In primates, the position module is not necessarily represented in a precise cartographic map (Warren 2019). Instead, primates may rely on knowledge of "routes sequences" tied to specific sites, such as landmarks or odors (Abreu et al. 2021), or to route intersections (Presotto et al. 2018). This is akin to a labeled topological graph which does not contain exact Euclidean positions (De Guinea et al. 2021; De Raad and Hill 2019; Presotto et al. 2018), but rather an incomplete set of distances, directions, and connections linking memorized locations (Warren 2019). As a result, tracking position and distance from a start or goal is likely to be carried out by anchoring routes with visual cues in a familiar environment. Further, "positions" could be defined using specific vantage points (a landmark or tree seen from a particular location) combined with stereovision, motion parallax information, and perspective. These visual cues along with local odors may correctly guide oriented trajectories, or route selection, from one position to another, incorporating distance to goal (Xavier et al. 2024).

3.3.7 Robots

Robot positioning is often referred to as localization in the robotics field. Simultaneous localization and mapping (SLAM) is a type of mapping and positioning system that builds a map of a newly explored environment while simultaneously keeping track of the robot's position within that environment (Cadena et al. 2016). Robot positioning performance is typically described by parameters like localization error, worse-case localization performance, how bad can that error be, and availability: over what fraction of the robot's journey can it be expected to have a "good" estimate of its position (Hausler et al. 2022). A surgical robot needs highly precise positioning, often to within a millimeter, to be able to perform very fine-grained navigation, like drilling a hole in a bone. It can achieve this using techniques like stereo-camera calibration of tools (achieved through multiple cameras or by moving a single camera) with respect to references "landmarks" on the bone (Windsor et al. 2024).

3.3.8 Neural Perspective

In the rodent brain, there are many neurons that encode one's self-position or the position of landmarks in space. These include place and grid cells (Hafting et al. 2005; O'Keefe and Dostrovsky 1971). Place cells are active when an animal is in a specific location in the environment, the "place field." Each cell has a different place field; thus, many place cells tile the entire environment. Next, the grid cells have location-specific activity as well, but their activity has a spatial periodicity, such that the multiple place fields of one cell tile the environment like a hexagonal grid. First identified in rodents, place cells were later found in birds (Agarwal et al. 2023; Payne et al. 2021) and have been reported in fish (Yang et al. 2024). In primates, the "format" of position selectivity seems to be referenced not to the animal's current location, as in rodents, but rather to the direction the animal is looking (Georges-Francois 1999). Thus, the hippocampal cells in primates provide a map of the visual space for all the objects that can be seen from a place (Mao et al. 2021; Piza et al. 2024). Place cells are, broadly speaking, ubiquitous across species and may thus form a mechanism for allocentric self-position across species. Importantly, place cells are thought to represent the animal's position in a *familiar* environment, as they are found in the hippocampus—a brain area well known to be important for learning and memory. They also show "remapping" when the animal is placed into a new environment (Leutgeb et al. 2005; Muller and Kubie 1987). Thus, akin to SLAM in robotics, place cells permit tracking one's own position in an environment and building a map suitable for one environment at a time. Place cells also accumulate around reward locations (Jarzebowski et al. 2022) or goal locations (Hollup et al. 2001), suggesting that they signal salient navigational cues. Thus, place cells can be relevant for the medium-scale navigational phase in familiar environments and for pinpointing a goal, but *not* for long-distance navigation outside of the familiar area—that is, locations the animal has never been before, such as during first-year migration (see Section 2). Finally, we note that place cells were recently found in the hippocampus of bats navigating outdoors on an island (Ray et al. 2025), strengthening the possibility that place cells may indeed serve as a neural mechanism for positional representation during real-life navigation in the wild.

3.4 Central Module: Information Integration for Decision and Error Correction

Knowing where one is (position module) and where one is heading in space (direction module) is necessary but not sufficient for navigation. Successful navigation also requires additional processes that are crucial for reaching the goal:

- Planning the route before starting
- Choosing the navigational strategy to use or switching to a new strategy (e.g., map-based or route-based navigation, self-centered or goal-centered)
- Making decisions during navigation (e.g., Should I now veer left or right? Which path should I take at an intersection?)
- Updating the plan along the way and correcting errors

We propose that a central module integrates information and implements complex navigation behavior depending on scales, goals, and memory. This module may underlie "cognitive" functions, qualified broadly as a set of operations involving integrating information, deciding, planning, and error correction. The module integrates information from other modules, and depending on the current evidence, selects a navigational plan. Further, if the information is degraded from one module, the integration of multiple sources may allow switching cue-dependent strategies. To this end, we propose that the central module contains a continuous error-monitoring process which enables corrections along the way. Indeed, assessing how well one is performing through error monitoring allows for maintaining appropriate trajectory to the goal through error correction (Figure 4.3). This error-monitoring stage applies to all previous modules and permits the organism to adjust, based on the relative importance that the organism gives to one module or the other depending on current evidence:

- Sensory module: Assess whether multisensory integration of one's own sensory information (via Bayesian weighing or other weighing) reflects the true reliability of sensory inputs and correct accordingly (e.g., if fog sets in, less weight should be placed on visual information and more on other senses).
- Direction module: Assess whether one's own direction estimate reflects true direction and correct accordingly. This may apply to the organism's estimate of absolute allocentric direction (heading north) as well as to the egocentric direction relative to a goal (goal-centric direction).
- Position module: Assess whether one's own position estimate reflects true position and correct accordingly. This applies both to estimates of absolute allocentric position and egocentric distance from the start or goal.

Notably, all these processes (planning, decision making, error correction) need to take into account the *navigational context* (see Section 4), including reward valuation, energy expenditure, and predation risk (e.g., navigating in a non-straight line to avoid climbing uphill, to access food resources along the way, or to avoid danger zones).

Neuronally, we hypothesize that all these steps (planning/switching between navigational strategies/decision making/error correction) are performed by the central executive brain area of each species; that is, the prefrontal cortex in mammals, the NCL in birds, and the central complex in insects. Below, we illustrate navigational behaviors across species in which information would be integrated to support decisions and error corrections.

3.4.1 Unicellular Organisms

At the scale of an individual cell, the world is very noisy and can be influenced by thermal fluctuations. While bacteria can have little control over their orientation (heading) due to rotational diffusion, algae (or larger protists) and eukaryotic cells (e.g., immune cells) have finer control over direction and trajectory (Wan and Jékely 2021). Membrane potential or other signaling molecules (e.g., calcium or cAMP) trigger voltage-gated ion channels, which can change the dynamics of the motor effector (e.g., flagella, cilia). Indeed, coupling motion with sensors saves time and avoids reliance on centralized processes. Cells are also able to optimize their location depending on circumstances by moving toward low or moderate light intensity (positive phototaxis) or away from intense light (negative phototaxis), thus evidencing the ability to make choices. These choices also depend on a sense of "internal state," such as previous exposure to light or circadian rhythm (perhaps analogous to the notion of motivational state in animals explored above).

Correcting errors in travel direction is a huge step in the evolution of navigation in small organisms, marking the difference between *kinesis* and *taxis* proper. In bacterial chemokinesis, only the rate of turning, in an uncontrolled random direction, is adjusted (Cheng 2022, 2023). The rate of turning is reduced when the gradient is more favorable than it was a short time earlier. In contrast, vectorial cues such as light allow even single-celled organisms to adjust travel direction so that a better, rather than random, direction is selected (Ros-Rocher and Brunet 2023). In flagellated algae, light is detected by directional photosensors on the cell (embodied again) and transduced to the motor apparatus. The cell swims in helices as a form of trial and error; the response magnitude is a function of the difference in orientation between stimulus and body. Further, the response time for the response of the cilia is optimized by evolution to correspond to the rate of rotation of the cell (Drescher et al. 2010; Yoshimura and Kamiya 2001). Once the cell successfully reorients toward the light, it will maintain its heading, thus demonstrating successful error correction. As another example, the network of syncytial slime mold *Physarum* will slowly adapt and prune itself over time to strengthen certain connections to optimize flow transport, a variant form of error correction (Alim 2018; Reid 2023).

3.4.2 Insects

Ants can rely on multiple cues to orient. When these cues are in conflict, the selection of navigation cues taken by *Cataglyphis fortis* depends on their experience history (Wehner 2020). Young ants that did not yet acquire the landmark panorama tend to rely on a home vector to return to the nest. In contrast, experienced ants, which had performed learning walks on a large spatial scale prior to testing (i.e., to learn the surrounding landmark panorama), relied on panorama information to find their way back to the nest (Fleischmann et al. 2016, 2018b). Many other cue-conflict experiments showed that ants often use both cues in conflict, striking an intermediate path between the commands of each cue set (Wehner et al. 2016). Some ants can use mechanosensors in their joints to sense the direction of the grip while clutching the ground to resist the wind; then, when displaced by the wind, they correct toward the right orientation by accounting for wind displacement (Wystrach and Schwarz 2013). Thus, ants possess multiple mechanisms allowing them to correct their heading.

3.4.3 Marine Animals

Research on information integration and decision making in marine animals remains limited, primarily because tracking wide-ranging individuals in marine environments is challenging. Nevertheless, existing evidence indicates "decisions" about which goals to pursue are often made based on some combination of environmental information and internal state. In a temperate-zone population of green turtles and loggerhead turtles, for example, turtles displaced from a capture site attempted to home back to the area during summer and early fall; by contrast, when turtles were displaced during the autumn migratory season, they oriented southward, consistent with their migratory direction (Avens and Lohmann 2004). In leatherback turtles that inhabit Canadian waters, a mix of environmental factors appear to affect when turtles decide to initiate a southward migration, including water temperature and the chlorophyll concentration of the water (Sherrill-Mix et al. 2008). Finally, in mass-nesting turtles, such as the olive ridley, turtles first gather offshore of the nesting area by the thousands, then emerge synchronously on the beach to nest together in enormous numbers—a process that requires numerous turtles to move toward shore at the same time, typically at night. How the collective decision to move landward is made, and how this synchronized directional movement is achieved, remains enigmatic (Bézy et al. 2020).

3.4.4 Birds

Birds often need to choose between different destinations, or when foraging, they need to choose the order in which to visit different goals. Hungry pigeons released in unfamiliar territory, for example, charted a course toward the location of a known feeder, whereas sated birds flew home (Blaser et al. 2013). Pigeons traveling short distances between different feeders rapidly learn the most efficient route, apparently planning multiple steps ahead (Gibson et al. 2012). Similar behavior is seen in rufous hummingbirds, which not only rapidly learn the most efficient route between flowers (Tello-Ramos et al. 2015), but also adapt this route as they acquire new information (Tello-Ramos et al. 2022) and learn when flowers refill, so they can prioritize recently refilled flowers (Henderson et al. 2006).

3.4.5 Bats

Bats exhibit highly sophisticated decision making during foraging and navigation in the wild, incorporating into their decision process a multitude of ecological factors as well as the behavior of conspecifics (reviewed in Prat and Yovel 2020). For example, insectivorous bats eavesdrop on the echolocation "feeding buzzes" of other bats and change their own behavior accordingly (Cvikel et al. 2015). Foraging fruit bats choose which fruit tree to navigate to according to the season of the year and the tree's location, thus demonstrating a well-developed *planning* capability (Harten et al. 2024).

3.4.6 Rodents

Evidence that rodents evaluate information before making decisions comes from an extensive corpus of behavioral research that builds upon Tolman's work and the development of the concept of cognitive map (Tolman 1948). Selecting from a large literature of rats in mazes reviewed by Wijnen et al. (2024), when tested on an equivalent of the traveling salesman challenge adapted to rats, animals converge on a short route between an array of rewards, minimizing distance, thus showing the ability to plan ahead and minimize efforts. At each location, the animal's next chosen location is biased by the current orientation of the animal, suggesting that animals integrate current position to plan ahead. Further, reward history on previous explorations also modulates the decision-making process and goal selection, with rats selecting the most rewarded sites (De Jong et al. 2011).

Another example is the observation that rats often stop at intersections, a behavior termed "vicarious trial and error" (Tolman 1948), as if animals evaluated possible outcomes of trajectories. This behavior is observed with higher frequency during early exposure to novel environments and increases with choice difficulty (Redish 2016), suggesting a role in outcome evaluation. Further, in the case of maze configuration that can be solved via two strategies, there were more stop episodes, as if the animal was weighing which strategy to rely upon (Schmidt et al. 2013).

When animals are closer to the goal, error correction may rely on place recognition, with animals orienting until finding the right sets of cues (visual, olfactory, tactile) matching with the memory of the goal.

3.4.7 Primates

There are abundant examples of selection of navigational strategies depending on current information in primates. Wild chimpanzees appear to anticipate the distance they have to travel for the next day (Janmaat et al. 2014) based on the current evaluation of available food sources, using cues indicative of the qualitative (sweetness of the fruit) and quantitative (trunk size) nature of the tree (Ban et al. 2014, 2016). Evidence suggests that after evaluating their current position, chimpanzees move to a night location in the vicinity of trees providing large food quantities the next morning (Janmaat et al. 2014). Primates may also be sensitive to abiotic cues. Wild mangabeys (an old world monkey found in Africa) are influenced by past weather conditions (solar radiation) when deciding to revisit a tree known to contain fruits on previous visits (Janmaat et al. 2006). This suggests that these animals form a knowledge of the location and ripeness of fruits depending on past history and use this to predict food availability and state as well as to form their next movement decisions. Gorillas' foraging decisions also appear to depend on their current position and goal. Travel paths to long-distance targets (e.g., a swamp of mineral-rich water plants) were more direct than daily foraging searches, likely minimizing efforts for long journeys compared to routine foraging for fruits and leaves (Robira et al. 2023a). For such long journeys, which likely traverse over unfamiliar terrain, nonhuman primates tend to use cues that are more effective across large scales (e.g., a sun compass), while they depend on local cues for navigating daily in their familiar environment. A similar strategy has been documented in the BaYaka people (Jang et al. 2019). Thus, navigating in familiar or unfamiliar places may affect cues and the overall navigation strategy and travel patterns (Jang et al. 2019). Finally, for smaller primates, much like any prey species, their overall space use is shaped by predation risk (Willems and Hill 2009). Thus their movement decisions, in particular the chosen travel path, are a function of both immediate energetic and predation pressures (Gee et al. 2024).

3.4.8 Robots

Robot navigational decisions are linked both to the core navigational task and the higher-level goal being pursued, such as finding a charging station. Robots employ many decision-making approaches, from the subsumption architecture of Brooks (1986), which prioritizes certain tasks over others, to behavior trees (Colledanchise and Ögren 2017). In most architectures, decisions are made based on imprecise information: probabilistic approaches help manage this process, including quantification of the level of uncertainty in the data being used to make the decision, such as sensor data (Thrun et al. 2005). A consumer robot vacuum cleaner has a simple but effective subsumption architecture: catastrophic problems (e.g., jams) are prioritized over returning to the charging base when low on charge, which is prioritized over finishing the vacuuming run of the floor being cleaned.

Many navigating robots employ discrete or continuous error correction processes. A key component of error correction and compensation in robotics is the ability to autonomously recognize that there is an error, a process known as *introspection* (Claxton et al. 2024). Once an error, or likely error, is identified, robots can then employ a number of responses including a fail-safe behavior, switching to an alternative navigational modality, or recruiting the help of another robot or human collaborator. Introspection capabilities can compensate for the relatively poor performance of a navigational system. For example, an autonomous vehicle with a positioning system that is 99% accurate but lacks introspection will unknowingly navigate based on incorrect data 1% of the time. By contrast, a system that is 90% accurate but correctly introspects its uncertainty 99% of the time will do so only 0.1% of the time—a tenfold improvement (Carson et al. 2022).

3.4.9 Neural Perspective

Due to the complexity of navigation processes, several brain regions are likely involved, integrating inputs from multiple other brain regions, such as information on reward, sensory cues, and state-dependent goals (e.g., hunger, thirst). Because of their high interconnections with other brain areas, the prefrontal and retrosplenial cortices in mammals are well poised to exhibit such high-level control functions. The prefrontal cortex is considered the central executive brain area in mammals (Friedman and Robbins 2022; Rushworth et al. 2004; Smith and Jonides 1999). In birds, this would most likely be the brain area NCL, the avian homologue of prefrontal cortex (Kobylkov et al. 2022; Stacho et al. 2020). In mammals, the prefrontal cortex has been shown to have a strong role in learning rules and actions associated with specific items, depending on context, by integrating reward (or nonreward) outcomes (Levy 2024). A role in rule learning has also been described for the NCL in birds (Veit and Nieder 2013). Thus, the prefrontal cortex and NCL could provide a map of actions and rewards that can guide behavior to maximize reward and avoid nonreward. This is reminiscent of the decisional tree that can be built in robotics, which guides actions prompted by contextual cues. In insects, the "executive" work could be performed by the central complex region of the insects' brain (El Jundi et al. 2014; Heinze et al. 2018; Honkanen et al. 2019).

A series of interesting questions arise: How are navigational plans or trajectories represented? How is the strategy choice and the switching between navigational strategies being performed at the circuit level? Is this an abrupt switch, or is it gradual? Further, how is error correction implemented mechanistically?

We note that the processes identified as introspection in the robot may be paralleled at the neural level in brain areas subserving the comparison between a current input and a learned expectation. Such representation of the violation of an expectation (mismatch) exists in multiple brain areas and is underpinned by a neural architecture supporting comparisons across cortical layers or across different cortical inputs. Thus, uncertainty and error signals have been found in many brain areas, such as the prefrontal cortex (Alexander and Brown 2018), cingulate (Carter et al. 1998), and retrosplenial cortices (Alexander et al. 2023a). In fact, error signals can also integrate consequences of actions (absence of reward, Genovesio et al. 2014) and are followed by several mechanisms aiming to reduce error or uncertainty, and hence participating in learning. Thus, error monitoring and error correction are likely implemented at all stages of navigation, from the sensorimotor level (which may involve the cerebellum) to higher-level map representations (hippocampus, entorhinal cortex, prefrontal cortex). Both are essential for memory-based map acquisition and play a pivotal role in navigation learning. Unlike in robotics (where error monitoring enables switching between learned programs) but like in artificial intelligence (where error back-propagation supports learning), error monitoring and error correction in living organisms are central to behavioral expression and memory formation.

3.5 Movement Module

Movement is the ultimate expression of navigation enabling behaviors such as foraging, mating, and escape from predators. The mechanisms of locomotion are inherently tied to the anatomy of the organism, with appendages adapted for motion ranging from microscopic cilia and flagella to macroscopic structures such as legs, wings, fins, and tentacles. These anatomical adaptations are further shaped by the physics in which the organism resides, whether terrestrial, avian, or aquatic.

Movement operates across different scales and is subject to distinct physical regimens, depending on the environment, whether swimming through fluid, flying through air, or locomoting on a dry surface. At smaller scales, motion is dominated by viscosity, as seen in microorganisms that navigate through viscous fluids. At larger scales, inertia becomes a primary consideration, influencing the dynamics of flight, swimming, or terrestrial locomotion in larger organisms. These differences are encapsulated in the concept of the Reynolds number (Re), a dimensionless quantity that characterizes the relative influence of inertial to viscous forces (Purcell 1977). The Re spans several orders of magnitude across the biological world, from microscopic swimmers operating in low Re environments to large animals in high Re regimes (Klotsa 2019), highlighting the role of the physical environment in dictating movement strategies. Understanding these dynamics requires integrating anatomical adaptations with the physics of fluid mechanics (e.g., viscosity, drag, lift) to appreciate how organisms navigate their environments at vastly different scales.

Animals often move using multiple appendages (legs, wings, fins) which need to work together in a coordinated way. This coordination might involve a nervous system, but not necessarily. Simple animals like *Trichoplax* can move without a brain or nervous system, deciding their overall direction spontaneously. However, some cells exhibit more directed movement and use front-back actin polarization to move toward a chemical signal (eukaryotic chemotaxis). Algae and ciliates coordinate multiple flagella to propel themselves through fluids. Coordination of appendages is a prerequisite, such as wing beats for flights and leg moves for walks. Insects even need to coordinate six, eight, or more legs. Movement can also happen in groups, like slime molds spreading out or birds flying in flocks to save energy by reducing air resistance.

Movement type and corresponding effectors can be adjusted depending on the navigation phase (Section 2) and require central control. For example, there can be drastic changes in behavior even when the same appendage is used. When a bird takes off or lands, the movement of its wing is drastically different, implying a sensorimotor loop (Figure 4.3) to adjust movements depending on sensory input. The link between movement effectors and the biological environment in which an organism moves is discussed further in Section 4, where solutions to overcome or even exploit constraints in different organisms will be highlighted.

3.5.1 Unicellular Organisms

Due to the physical constraints of needing to break time-reversal symmetry at low Re, unicellular organisms need a specific aspect ratio if designing an appendage. At these scales, movement relies on oscillatory dynamics of such appendages (beating, twirling, or rotating), for which oscillations are necessary (Wan 2023). However, convergent evolution of different appendage-based movements in bacteria, archaea, and ciliates can be classified as either boundary actuated, with a single degree of freedom/rotation from the base, or distributively actuated, with many degrees of freedom actuation along the length of the filament. Basal coupling for control of multiple cilia within the same organism takes place. Then membrane potentiation and other signaling pathways can affect the overall movement mode. With multiple cilia, coordination among cilia is needed, much like the limbs of animals. It is thought that both external factors of hydrodynamics and internal, physiological factors contribute to coordination (Hamilton et al. 2020; Wan and Goldstein 2016).

3.5.2 Insects

Insect gaits exemplify the working of oscillators and oscillations at multiple levels. Each of the six legs must lift, swing, and plant in a cycle, requiring six oscillators to drive them (Gallistel 1980; Wilson 1966a, b). These six oscillators must be coordinated or, technically speaking, coupled to produce sensible locomotion. In the most common gait, the front and back legs of one side lift and plant together (approximately) with the middle leg on the opposite side in the tripod gait (Pfeffer et al. 2019; Tross et al. 2021; Wahl et al. 2015; Wilson 1966a, b). Recent studies suggest that in navigation, the tripod gait does not drive straight movement in the chosen direction; rather, the ant "meanders" or oscillates from left to right in its path (meat ants and day-active bull ants, Clement et al. 2023; Murray et al. 2019). These continuous oscillatory adjustments serve to keep ants traveling in an optimum direction.

3.5.3 Marine Animals

Locomotion of aquatic animals is shaped by a need to minimize friction during displacements and to harness underwater currents effectively. Despite a significant evolutionary gap, fishes and cetaceans evolved similar adaptations, with streamlined bodies and fins, to optimize swimming efficiency. Fins serve diverse functions such as propulsion, steering, and stability. In some marine animals (sea turtles), legs have evolved into flippers to enable both terrestrial and aquatic locomotion. Additionally, overall body hydrodynamical shape reduces drag over long distances or during rapid movements.

3.5.4 Birds

The motor output in navigating birds can involve walking, flying, swimming, and paddling. Within flight, birds might use forward flapping flight, soar on thermals, hover in position, or dive toward the ground. Each of these sets of movements can involve different power requirements (Tobalske 2007), made more complicated by variation in, for example, the turbulence of the environment through which a bird is flying (Chapman et al. 2011). To fit their movements to the environment, many flying birds use visual motion cues to control their speed, avoid obstacles, and maneuver through gaps (Altshuler and Srinivasan 2018).

3.5.5 Bats

Bats are the only mammals that exhibit powered flight, the dominant locomotion mode of all bat species (Las and Ulanovsky 2024). Their flapping flight is energetically efficient (Hedenström and Johansson 2015), allowing bats to fly nonstop for over 100 km without stopping to rest. This movement capacity enables foraging flights of >100 km in some species and multi-night migration flights of >2,500 km (Egert-Berg et al. 2018; Ulanovsky and Moss 2008).

3.5.6 Rodents

Rodent species exhibit a wide range of locomotion modes adapted to diverse ecological niches, including arboreal (squirrels), semiaquatic, fully subterranean (moles), or mixed. Many rodent species exhibit high habitat diversity, which is thought to stem from a high behavioral flexibility rather than morphological specializations (Hedrick et al. 2020), with the exception of underground and desert species. Many of their locomoting behavior stems from strategies used to escape predators. While most rodents are quadrupedal, desert gerbils evolved a bipedal locomotion through facilitated gait transitions, which enhances their evasion ability (Moore et al. 2017). Locomotion in rats and mice is expressed in mixed terrestrial and underground habitats that fit the predator-escaping strategy, with a morphology (small limbs and flexible small body) that also matches this need (Hedrick et al. 2020).

3.5.7 Nonhuman Primates

Some primates are almost exclusively arboreal, with specialized morphological adaptations like clasping feet and hands (e.g., marmosets and colobus monkeys) or even prehensile tails (e.g., howler and spider monkeys) that have evolved to maintain a strong grip on branches. While predominantly arboreal, few species are strictly confined to trees, as they may occasionally descend for water, food, or social interaction, exhibiting a mixed terrestrial-arboreal environment exploitation (e.g., chimpanzees, gorillas, macaques). The locomotion modes associated with this mixed habitat include climbing, walking (arboreal, terrestrial, knuckle walking), galloping, and jumping between branches (Schmidt 2011). These activities require precise grasping and landing adjustments, along with effective gait and postural adjustments. To achieve this, primates have evolved clasping hands and feet, sometimes further equipped with claws for enhanced grip. Additionally, a wide range of eye movements supports and compensates for body motion. The vestibulo-occulomotor reflex counteracts distortions caused by frequent head movements, helping to maintain a stable visual field during locomotion (Cullen 2012; Fetter 2007). Saccades and fixations allow primates to explore their surroundings at a distance and to focus on landing targets during jumps or objects they intend to grasp (Kaas et al. 2022). This coordination requires eye movements to always accompany movements of the rest of the body, facilitating seamless and efficient navigation.

3.5.8 Robots

Robot movement modalities vary widely, depending on the type of navigational systems suitable for a particular platform. Broadly speaking, robotics platforms and movement modalities break down into air, land, and sea domains. Aerial robots move using rotors, gliding, jets, rockets, and thermals; land robots move using wheels, legs (often biped, quadruped, and more), and tracks; sea surface and submersible robots use propellers, jets, and ocean currents (Jahn et al. 2020). Key aspects of how movement modalities affect navigation systems are the speed and stability of the movement and the observability of the environment caused by that movement: these affect optimal sensing choice, as well as navigational techniques and phases used. A humanoid robot walks approximately like a human (Tong et al. 2024). Navigational sensors are typically chosen to maximize the ability to measure the robot's gait, and hence movement, and are positioned near the head to maximize visibility of the environment.

3.5.9 Neural Perspective

In mammals, movement is controlled and coordinated by a number of brain areas, including the basal ganglia (Graybiel et al. 1994), motor cortex (Kaas 2004), and cerebellum, which is particularly important for online correction of gait and gaze control (Sokolov et al. 2017). These motor areas receive feedback from sensory areas of the brain, forming a sensorimotor loop (Frigon et al. 2021). Importantly, no effective locomotion can be performed without integrating external sensory input as well as self-pose information through proprioception and vestibular system. Hence, several mechanisms implement postural control through vestibulo-motor loops (e.g., for gaze or gait stabilization) in regions of the brain including motor cortex, basal ganglia, and cerebellum (Takakusaki 2017). Functionally equivalent regions for motor control in birds also include cerebellum (Gutiérrez-Ibáñez et al. 2023), avian striatal complex, and pallium (Kumar and Iyengar 2018; Reiner 2002). In insects, motor control is performed in local premotor networks within the ventral nerve cord on which multimodal information converges (Gebehart and Büschges 2024).

4 The Ecological Context in which Predation, Mating, and Resource Decisions Are Made

Navigating implies processing myriad information continuously because nothing is ever fixed: the environment changes as the organism moves, and so does the organism itself. The interaction between environmental conditions and the internal state of the organism creates a dynamic "context" that may shape the incentive to move and redefine the current goal as well as the underlying inner working processes. Further, an environment is multifaceted and reflects several intertwined landscapes (Kashetsky et al. 2021). It comprises not only the abiotic (e.g., topography) and biotic (e.g., the resource) environmental conditions (Abrahms et al. 2021) but also the organism's perception and knowledge of the conditions, such as familiarity with some areas (Piper 2011). The abiotic properties of landscapes may present some spatial variability (Kolasa and Pickett 1991) and thus are likely to impact immediate travel plans depending on various locations in space. Moreover, while most landscapes are generally stable during the movement itself, acute deviations from long-term trends can occur over time (Palmer et al. 2022). For example, for a prey species, predation may be represented in a "landscape of fear" built through a history of long-term trends (Laundre et al. 2010). However, chase-and-escape behaviors of predators and preys will inevitably impact their next movement (Gee et al. 2024; Vanderlocht et al. 2025). Further, depending on their current goal, animals need to make a cost-benefit assessment between predation avoidance and resource acquisition. Their decisions need to take into account trade-offs based on a spatial "landscape of resource" and the dynamic "landscape of fear" (Gallagher et al. 2017). Moreover, the organism may even anticipate changes in the resource landscape in the near future, which may affect navigational decisions from possible anticipated payoffs (Ban et al. 2014). Thus, while the navigational context changes, so may the cues and clues that are likely to be integrated by the different navigation modules. As such, the overall navigational strategy, and the machinery that underpins it, is driven by the degree of heterogeneity and predictability of the cues within an environment, (Figure 4.1a) (Riotte-Lambert and Matthiopoulos 2020).

Changes in an organism's external and internal states may alter navigation processes in four aspects. First, these changes can impact which modules are used (i.e., the overall architecture of the modules). For example, depending on available cues, ants may prioritize path integration (and associated celestial/magneto-compass use in the orientation module) or landmark-based views (in the position module) (Hoinville and Wehner 2018).

Second, the changes may shape how they are used (i.e., the algorithm that underlies each module). For example, the positional system that underpins human navigation gradually matures during childhood, switching from a prevalent self-referencing at a young age to a stronger referencing to external objects at more mature ages (Newcombe et al. 1998). Thus, the nature of spatial processing in children's early years is radically different from adults (Pullano and Foti 2022).

Third, these changes can affect how module output are integrated to make a decision. An instructive example comes from collective navigation. For large soaring birds, finding thermal lift is a prerequisite for energy-efficient flights (Duriez et al. 2014). In migratory wild storks, experienced older individuals tend to move more independently of conspecifics, whereas younger, less experienced individuals tend to follow other individuals (Brønnvik et al. 2024). In other words, the "weaker" individuals rely more on social rather than individual knowledge. In contrast, in griffon vultures, nondominant (but fully experienced) vultures are more likely to explore the environment alone, whereas dominant individuals are more likely to follow them (Sassi et al. 2024), certainly to minimize possible subsequent competition for food at the found carcasses. As such, the overall navigational strategy and associated toolbox not only emerges from the species constraints, but also on the current condition an organism faces at a given time.

Fourth, the variations in context perceived by the organism can also stimulate the use of an error-correction submodule (Figure 4.3), leading potentially to the recalibration of the orientation and navigation modules. This, in turn, provides the information-integrating central module with new information. Such an error-correction mechanism is, for example, a prerequisite for robots navigating in the real world, as their navigational systems are optimized for cost and energy consumption at the cost of performance like positional accuracy (Ghosh et al. 2021). Most likely, these robotic rules find resonance in natural systems for which the cognitive machinery is usually bent to a speed-accuracy trade-off (Chittka et al. 2009).

4.1 Unicellular Organisms

The critical importance of navigational context extends to microscopic organisms. Gliding diatoms are known to choose between when to search for food and when to seek out a mating partner depending on the generational age of the culture (Bondoc et al. 2019), whereas some microorganisms such as ciliates and flagellates constantly choose between feeding and swimming (Esteban and Fenchel 2020) because the same apparatus is often used for both functions therefore preventing them from swimming and feeding at the same time (Fung et al. 2023). There are also fundamental trade-offs between feeding, swimming, and predation risk associated with the arrangement of cilia and flagella (Nielsen and Kiørboe 2021). Rosette-forming choanoflagellates become colonial when certain bacteria are added to the medium, but remain as solitary swimmers when resources are lacking (Woznica et al. 2016).

4.2 Insects

Navigating hymenopteran insects often face the choice of going home or going to a foraging location. Both navigational context and individual differences drive this decision. Navigational context supports the retrieval of correct memories to meet the goal (for honeybees, Cheng 2005; Collett et al. 2003), and the individual circumstances (e.g., going home or foraging) help the retrieval of correct memories for the job. In the choice of going to a foraging spot versus going home, one key driver used in numerous studies is food: give a foraging ant a piece of food to carry and it becomes a homing "machine" driven by the sole goal of getting home. The same can be said of a foraging honeybee that has filled her crop. Adverse or unexpected circumstances on a foraging trip, often perpetrated by experimenters, can make an ant or bee go home. After displacement, some honeybees eventually head to a known feeder (after flying off a vector) whereas others head home (Menzel et al. 2005). Long-lived bull ants are "sensitive" to experimental manipulations. After encountering a barrier on their foraging tree, some ants turned back and went home, whereas others struggled to find a way around the barrier (Islam et al. 2023).

4.3 Marine Animals

Navigational context affects the movement of numerous marine animals. Many fish alter their behavior to diminish risks of predation, for example, by avoiding surface waters during daylight hours when they may be readily visible from below (Helfman 1986). Indeed, the widespread vertical migration of plankton, in which plankton ascend to the food-rich surface at night only to descend to safer areas during the day (Brierley 2014), can be viewed as movement driven by a fear landscape and navigational context.

Another interesting example is the northern elephant seal, which undertakes a seven-month foraging migration during which it must balance the need to feed with the need to avoid predators. Recent findings indicate that the seals change strategies as their body condition improves, increasingly prioritizing safety over foraging (Beltran et al. 2021). Elephant seals also transition from foraging behavior, where animals frequently change direction while searching for prey in the open ocean, to directed migratory behavior, where the animals swim directly toward distant beaches to reproduce (Beltran et al. 2022).

4.4 Birds

Birds traveling in their local area navigate for many reasons. Blue tits will head off on more than twenty foraging trips per day to find and bring back caterpillars for their chicks (Iserbyt et al. 2019). Outside of the breeding season, many birds find their way to preferred roosts as night falls (Laughlin et al. 2014). Even in socially monogamous species, one partner might visit other territories to find extra-pair mates (Griffith et al. 2002). These diverse motivations are accompanied by constraints that determine how birds move in their environment. The risk of predation, for example, is associated with multiple behavior and even morphological changes in prey species, such as small songbirds, influencing where they travel (Abbey-Lee et al. 2016; Suhonen 1993). Small migratory birds also strongly increase their flight altitude during the day, possibly to avoid predation (Sjöberg et al. 2021).

4.5 Bats

Bats incorporate multiple types of navigational context into their navigational decision making. One example is the onset of daylight—a dangerous time for bats as they can be hunted by hawks. Homing experiments in GPS-tracked Egyptian fruit bats showed that when translocated ~80 km away from their cave, bats homed straight back to their cave (Tsoar et al. 2011). However, if the bats were released after translocation at a late hour of the night, just before sunrise, they initially flew straight in the direction of their cave but then they abruptly changed direction at the precise moment of sunrise, and flew to a closer cave, where another colony of fruit bats resided, thus avoiding the danger of daytime predation (Tsoar et al. 2011). This demonstrates a time-of-day-dependent predation awareness by bats, the *landscape of fear* concept. Another example of navigational context in fruit bats comes from a study which showed that these bats switch their navigation goal—the fruit tree to which they fly to forage—according to the time of year (fruit season) and location of the seasonal fruit trees (Harten et al. 2024). Through careful manipulation experiments, this study demonstrated that fruit bats map the spatiotemporal distribution of food resources in their environment, the *resource landscape*.

4.6 Rodents

In rodents, decisions need to combine the need for a fast escape route to shelter in the face of predation with the need to explore beyond the home to acquire resources. In experimental setups, mice quickly learn to choose shortest trajectories to shelter to minimize effort and facilitate quick escape (Claudi et al. 2022). Rats consider time costs when deciding to keep collecting from one food source before switching to a different food source (Neville et al. 2024). When enabled to explore outside their home cage (with their home cage being brought to a new place), rats make only small explorative trips until the environment becomes familiar (Eilam and Golani 1989; Whishaw et al. 2006), likely increasing security by securing knowledge of escape routes. Further, increasing familiarity with the environment not only modifies exploration but also benefits resource exploitation (Fagan et al. 2013) for these central-place foragers. Finally, rodents are generally social animals that live in groups. When they become adults, they face the "decision" of being philopatric (remaining where they are) or dispersing to a new area or joining a new social group or population (Nunes 2008).

Dispersal may be a response to several factors (e.g., competition for food, nest sites, mates) and has the adaptive value of reducing inbreeding; there are, however, some gender-based differences, with females being more philopatric (Nunes 2008). Finally, when an adult rat remains in a group, social bonds influence spatial behavior, even in males: active exploratory patterns are inhibited by the establishment of social relationships among adult rats fostering some kind of philopatry (Arakawa 2006). Dynamics of social interaction modulate the group's explorative pattern, with social-companionship seeking a variable acting on spatial cognition (Dorfman et al. 2021; Weiss et al. 2017). In sum, spatial and "migratory" navigation in rodents is strongly influenced by navigational context, and different navigational behaviors may be triggered by specific socioenvironmental cues unraveling different consequences depending on current sex-based developmental state.

4.7 Primates

In many primates, navigation involves confronting daily challenges that arise from the interplay between internal states, social group dynamics, and the difficulties of the terrain (Garcia et al. 2023). For example, in some species, individual and group foraging strategies must adapt to reduced food availability, creating a trade-off between the protection provided by the group and access to resources (Grove 2012; Reyna-Hurtado et al. 2018). Within-group competition requires an individual to choose a foraging site that offers food with little predation risk, while minimizing the risk of displacement by a more dominant animal (Arseneau-Robar et al. 2022). Animals may also limit their explorative foraging trips to maintain some distance with neighbors and avoid territorial conflicts with other troops (Ceccarelli et al. 2019; Seiler et al. 2017). Difficulties inherent in the terrain may factor into a cost-benefit evaluation that considers both the energetic costs associated with distance and challenge of reaching the goal (Green et al. 2020; Villioth et al. 2023). Further, seasonal variability impacts the current diet upon which animals may rely and adapt their foraging strategies accordingly (Janmaat et al. 2006; Matsuda et al. 2022). In sum, primate navigation strategies must adapt to a dynamic context that requires balancing challenges of shared social space with fluctuating resource availability.

4.8 Robots

Robot navigational choice is heavily influenced by the ecological context in which an agent operates. Navigation systems for robots in safety-critical environments (e.g., autonomous vehicles or surgical robots) must combine high performance, reliability, and introspection to minimize catastrophic outcomes. Robots that operate in noncritical or highly redundant environments (e.g., low-cost aerial drones or marine robots) often use navigational systems optimized for cost and energy consumption at the cost of performance indices such as positional accuracy. In adversarial contexts, a robot may prioritize navigational aspects around stealth and survivability, mirroring the context for many navigating prey animals. Robot teams, such as the DARPA Subterranean Challenge robot teams, may coordinate to maximize coverage and hence spatial mapping of a previously unexplored environment while maintaining sufficient contact for inter-robot communication channels (Hudson et al. 2022).

4.9 Neural Perspective

How is navigational context represented in the brain? How and where does the brain represent the landscape for food resources or predation risk? Ecologically relevant variables such as the nature and probability of food resources are thought to be represented in the prefrontal cortex, in areas such as the orbitofrontal cortex in rodents (Basu et al. 2021) and primates (Ballesta et al. 2020; Setogawa et al. 2019). Reciprocal connections between the orbitofrontal cortex and the hippocampus likely integrate the goals in a joint representation of space (Wikenheiser and Schoenbaum 2016) or contribute to the representation of goals and rewards within the hippocampus and entorhinal cortex (Gauthier and Tank 2018; Sosa and Giocomo 2021). Another candidate likely involved in the representation of the landscape of fear is the amygdala (Choi and Kim 2010). Its neurons represent predation risk in a semi-naturalistic foraging task (Amir et al. 2015) and are recruited during the classic fear-conditioned response task (Phillips and LeDoux 1992). The tight neuroanatomical relationship between the amygdala and the hippocampus supports a joint representation of the landscape of fear across the amygdala and the hippocampus (Kong et al. 2021; Moita et al. 2003; Phillips and LeDoux 1992). In sum, memories of the physical properties of an environment, likely represented by the hippocampus, are integrated with representations of goals and threats, generating a representation of space adapted to behavioral needs.

Finally, while the brain may represent social stimuli in different formats (auditory, visual, olfactory) in different regions, the hippocampus has been found to contain information representing other individuals in primates, rodents, and bats based on multiple cues (Kong et al. 2023; Omer et al. 2018; Tyree et al. 2023). In bats, the hippocampus represents both the allocentric position and the egocentric direction and distance to other individuals, including their identity (Omer et al. 2018; Ray et al. 2025). Encoding of the position of other individuals was also shown in the prefrontal cortex (Murugan et al. 2017). These neuronal representations could potentially form the mechanistic basis for behaviors such as eavesdropping or following a group leader during navigation. In sum, despite long-standing interest in studying contextual variables for navigation, the perspective that connects neuroscience with ecology has only recently emerged to propose that spatial variables dynamically integrate nonspatial variables to allow real-world navigation.

5 Conclusion and Future Outlook

By focusing on the diverse behavioral expressions of navigation through a unique comparative approach, we have identified common principles of navigation across vast differences in organismal size and complexity. A central element of our proposal is that navigation is underpinned by a series of sub-behaviors, which we refer to as "phases," that can be implemented at different spatial and temporal scales. These phases are driven by context-specific cues and mechanisms and are often completed via feedback loops within and across phases. Within each phase, a set of universal modules supports the completion of a navigational behavior, from sensory uptake to motor execution, forming a mechanistic loop, with the final serial steps of execution depending on context. From our proposals, the following points arise.

5.1 Mechanisms Supporting Phase Completion and Phase Transition

We propose that navigation consists of multiple phases and transitions between them (Mouritsen 1998a). Expanding on the ideas of Cheng (2005, 2022, 2023), navigational phases fit into a servomechanistic loop consisting of sensory evaluation, the determination of a direction of travel, a module determining the traveler's place in the world or its distance to the goal, and an information integrator ("cognitive" central module) that enables a decision and generates the action plan (Section 3, Figure 4.3). Once an action plan is selected, action must be taken in the form of coordinated movements of effectors, often requiring oscillators driving locomotion. The action taken provides new sensory feedback to be evaluated, completing a loop. Variations on this theme are plentiful, showcasing diversity in a unifying scheme. Both the multiphase navigation scheme and the expanded servomechanistic scheme are highly sensitive to what we have termed *navigational context*, a broad concept encompassing at least the motivation of the organism, the physical context of travel (e.g., weather, time of day, time of year, and the Re for tiny creatures), available food resources, predation risk, and more (Section 4). The processing of contextual information can lead to self-organized behavior, where simple mechanisms switch actions based on cues, such as approaching food when hungry and stopping once satiated, until the need arises again.

In this scheme, one key open question for all navigating organisms is how transitions from one phase to another take place. Many of these transitions depend on, for example, scale and/or time of year. Might these transitions in some organisms happen randomly over time? Are various options evaluated continuously, with the most valued option driving the navigation at any moment, once it outweighs other options? Are transitions sharp or gradual? Do they depend on the nature of the threshold implemented (accumulation or winner takes all) and the amount of conflict between cues? These types of questions can be fruitfully studied both at a fundamental behavioral level and at the neuronal level. In seeking generalizations and commonalities across natural forms of navigation, it will be instructive to explore strategies employed by particularly "simple" organisms as they make such choices, in some cases in the absence of any nervous systems control. For example, during ontogeny, ants transition from relying on a magnetic compass to using a celestial compass, yet it is unknown how they transition from one compass (using one set of cues) to the other (using completely different cues). Further, how ants transition from a "home" learning phase to an exploitative foraging phase may depend on "self-evaluation" of one's own ability to return home (Fleischmann et al. 2017), paralleling a process that has been described in rats (Redish 2016). This apparent confidence-based evaluation requires an organism to access a distribution of its performances and recognize a mismatch between an expected input and the actual input, resulting from a self-generated error. Therefore, error monitoring and error correction are pervasive mechanisms in navigation. This can be highly consequential when it comes to data analysis of animal movements. To analyze spatial behaviors of wild animals, many ecologists consider their path as a time series of locations (measured, e.g., using GPS devices). Because these locations are generally logged with low spatial resolution (cf. Munden et al. 2021), inaccurate conclusions may result (Avgar et al. 2016). Thus, a better understanding of the servomechanistic loop could considerably enhance the identification of decision points, the inferences of the movement models, and the conclusions of the animal's cognition and information-processing abilities, which underpin movement choices.

5.2 Introspection, Error Monitoring, and Correction as a Prerequisite to Any Navigational Behavior

According to Milford and colleagues (Carson et al. 2022), a key characteristic for navigating robots is *introspection*—an estimate of how well the performance of the navigational task is going. We propose that living organisms have a similar self-evaluation scheme (Figure 4.3) and suggest that it will be important to elucidate how this self-evaluation process affects navigational behavior and performance, and how it is implemented neuronally.

For example, organisms often "scan" the environment to determine the best direction to head in and adjust their behavior accordingly. Most microswimmers generate intrinsically helical trajectories as they swim to enable various embodied sensors to scan their environment periodically so they can perform dynamic course-correction by integrating the error between the swimming and stimulus direction (Cortese and Wan 2021; Drescher et al. 2010; Jékely 2009). Ants often stop and look around the environment at the start of a trip, stopping and turning in various directions with pauses (Deeti et al. 2023; Fleischmann et al. 2017; Wystrach et al. 2014). Similar scanning behavior has been found in rodents (Redish 2016) and birds (Mouritsen et al. 2004). Homing pigeons (Schiffner and Wiltschko 2009) and night-migratory songbirds (Mouritsen 1998b), when released at an unfamiliar location, often circle around in the air before heading straight in some direction. The nematode *Caenorhabditis elegans* behaves as if it has sampled different directions until it locates a direction that climbs a favorable gradient (Tanimoto and Kimura 2019). Such exploration can arise and persist from a behavior focused on seeking a match, which continues as long as mismatches are encountered. How error monitoring and error correction are implemented at the biological or neuronal levels remains an important challenge. This is a nontrivial problem as both may depend on the current behavior expressed by the animal and hence on the fine-grained anatomy of each system.

In our proposal, we implemented direct error correction within the central module (Figure 4.3) because decisions and adjustments of behavior can be a direct consequence of errors, whether these arise from graded mismatches or from a binary error. However, it remains an open question whether error correction is truly part of the planning/decision-making module, or whether it forms a separate module involving separate neural circuits.

5.3 Modularity of Direction, Position, and Memory

Our proposal of module composition stems from functional levels that define navigation as purposefully aiming for goals or resources. Within this framework, the modules address key navigational questions: Where is it? Where am I? How do I get from here to there? The navigational context provides the motivation. While sensorimotor loops, represented through sensory and movement modules, are central to interacting with the world, the composition of other modules may be more flexible. Yet, the proposed direction module is strongly justified by the reliance on directional signals in both biological and artificial systems. Migrating and nonmigrating species alike demonstrate the importance of compass-like directional information. Such information also plays a crucial role in robotics, where precise directional signals are essential for maintaining course. At the neuronal level, directional signals are prominently represented by head- and goal-direction cells, further supporting the existence of a separate direction module.

In contrast, the nature and pervasiveness of the position module are less clear. While there is ample evidence for place cells coding self-position in the brain, behavioral data often reflect positions as knowledge about locations in space rather than direct representations of self-position. Some organisms may navigate effectively without a position module. For example, a movement strategy called guidance, where salient objects or "beacons" toward which the animal navigates are targeted sequentially, may not necessarily involve an explicit positioning level; namely, the animal does not need to know its absolute location in space. Instead, the sequence of beacons is memorized without any spatial embedding (Waller and Lippa 2007), a form of what has been called the taxon system (O'Keefe and Nadel 1978). Moreover, some modeling and robotic studies show that navigation can be implemented by relying on directional bearing, without using position (Gaussier et al. 1997; Krajník et al. 2010; LaChance and Taube 2023).

The position module also encodes information about angles and distances between locations, implying the need for a memory-based mechanism to support learning of locations in space, integrating their nature, relative distance, and so on. The hippocampus likely plays a key role here, integrating spatial cues to form an experience-dependent map. Memory, a pervasive ability, supports the operation of all the modules in many organisms. Simple signpost values (e.g., stop migrating when magnetic inclination angle x° has been reached) and relative local readings of such values relative to that of a goal could be the basis of a simple form of relative position without specific distances being needed (Mouritsen 2003; Wynn et al. 2020, 2022).

Further, stemming from the error-correction mechanism described above, memory may support error monitoring. This could occur in two ways. First, memory may serve as a reference against which incoming information is compared. This would allow the organism to detect matches or mismatches with memory-based predictions. Second, memory may shape behavior in response to feedback, helping to refine actions in new contexts and consolidate new navigation strategies. We suggest that memory can pervasively influence all modules by biasing cue selection, sharpening tuning of position or direction modules, forming the basis of context, and more. Future research is needed to understand how exactly memory alters the functioning of the various modules.

5.4 The Mystery of Magnetic Sense and the Compass

As seen in this review, many animals possess a magnetic sense, which may provide the most continuous source of directional and positional information. Yet further knowledge is needed on how the magnetic sense works on the biophysical, biochemical, neural, and behavioral levels. This investigation is nontrivial as magnetosensation mechanisms appear to vary across species (Lohmann et al. 2022; Mouritsen 2018, 2022), and at least three different types of basic sensors for magnetoreception have been supported by experimental evidence. Some examples include magnetite chains, as used in magnetotactic bacteria (Barr et al. 2022; Uebe and Schüler 2016); a quantum mechanical, electron-spin-based mechanism based on radical pairs in cryptochromes, which seems to underlie the light-dependent magnetic compass in night-migratory songbirds (Hore and Mouritsen 2016; Leberecht et al. 2023; Xu et al. 2021); and electromagnetic induction in saltwater fish (Paulin 1995), recently suggested for birds (Nimpf et al. 2019). Recent studies have also revealed that desert ants possess a polarity-sensitive magnetic compass, likely based on magnetite particles (Grob et al. 2024b). While altered magnetic fields were found to affect synaptic plasticity in the ant brain (Grob et al. 2024a), the location of the magnetite sensors and the integration of magnetic information for navigational decisions remain unresolved. One hypothesis suggests that the sensors are in the ants' antennae and that magnetoreception involves active sensing that enables differences across time to be integrated (Fleischmann et al. 2020; Grob et al. 2021). In all of these sensory systems across species, there are many exciting open questions, compounded further by the fact that several animals seem to possess at least two of these magnetoreception mechanisms side-by-side (Mouritsen 2018). Finding the underlying mechanism of magnetosensation and establishing its similarities and differences across species is an interesting future research direction.

5.5 Toward a Neuroethology of Spatial Navigation

How the brain subserves adaptive navigation during naturalistic behavior remains a challenge. Most experiments on the neural basis of navigation are performed in laboratory settings and focus on a single region. Single-cell neural recording techniques have now advanced to permit recordings to be made from many individual neurons in animals performing navigation tasks over relatively long distances, including outdoors (Eliav et al. 2021; Palgi et al. 2024; Ray et al. 2025; Sarel et al. 2022). A key approach that is needed to understand the underlying neural mechanisms is to simultaneously record from multiple brain areas while animals perform complex navigational tasks in the wild. In the coming decade, we believe that it will be possible to record from the brains of multiple animal species during natural navigation outdoors, including in long-distance migrating species. Based on the many navigation behaviors described herein, such neural recording methods will open the door for a host of fascinating future questions.

Regarding the neural implementation of direction: Do head-direction cells form a genuine neural compass during outdoor navigation? Preliminary experiments suggest that they do, based on recordings of head-direction cells in bats navigating outdoors (Palgi et al. 2024). Do head-direction cells form a global compass across the entire navigational range of the animal, or do they form different compasses in different local regions, reflecting local frames of reference? In animals sensitive to Earth's magnetic field, what are the neural circuits supporting the magnetic compass?

Questions pertaining to the representation of position arise from two challenges that most navigating animals face: the processing of large scales and multiple environments. This leads to a host of mechanistic questions: In mammals and birds, do place cells form a cognitive map of the animal's familiar area during outdoor navigation, similar to place cells recorded indoors? On which scale do place cells map? Are place cells specific to the pinpointing-the-goal phase? Or can place cells also code space on a much larger scale during the long-distance and/or homing phases? Do place cells form a global unified map outdoors or many local maps? If so, how are these local maps "stitched" together?

How is navigational context represented in the brain? For example, how and where does the brain represent the landscape for food resource or predation risk? How are navigational strategies (or phases) chosen by neural circuits; that is, how are the switches between strategies implemented? How do various motivational factors come into play? How does the brain's reward system interact with motivational factors, food-resource maps, and fear maps? Where and how are navigational goals represented in the brain?

Further questions arise from translating engineering solutions for robotic navigation to real navigation in animals. Robotic navigation systems can range from very lightweight, "simple" systems to extremely complex. Lightweight systems demonstrate that effective navigation can emerge from basic principles such as error correction, efficient cue integration, and modular processing. This prompted us to put forward the need for self-monitoring (introspection). Yet how this is neuronally implemented in all modules remains unknown and is an important question, for both natural and artificial navigation systems. Further, while robotic systems can inform about fundamental mechanisms, their navigational generality is still currently limited compared to many animal and insect species: shifting a system to a totally different environment type typically requires the development or deployment of a new approach to address new types of challenges. Hence, discovering the implementation of natural flexibility in animal navigation remains a goal for anyone interested in both artificial and natural navigation. With the rise of the use of modern artificial intelligence (AI) approaches and the search for solutions that can generalize across problems, this is a key area for future research. One interesting perspective to take into account when mitigating between AI- and robotics-based solutions lies in the fact that unlike AI, the field of robotics (at least classical approaches) is underpinned by operant solutions in which computational and power efficiency and real-time capability are prioritized. This results in solutions that are less costly to compute and which use less energy than many brute-force AI methods. An interesting way forward may be to develop hybrid systems, where a compositional approach is adopted: an overarching AI that organizes bio-inspired robotic modules. This may enhance adaptability and efficiency leveraging on the modular architecture we proposed to ground biological navigation (Figure 4.3).

These and other fascinating questions stand at the interface between the neuroscience of navigation, the ethology and ecology of navigation, and robotics. In sum, in this chapter, we have presented a unified scheme for researchers studying navigation, complementing yet going beyond previously proposed schemes (Cheng 2022, 2023; Mouritsen 2018). We have proposed ways to conceptualize how navigation works, outlined its central functional modules, and examined key mechanisms across species. We hope these insights will foster future research on the algorithms and mechanisms of navigation.

References

Abbey-Lee RN, Mathot KJ, and Dingemanse NJ (2016) Behavioral and Morphological Responses to Perceived Predation Risk: A Field Experiment in Passerines. Behav Ecol 27 (3):857–864. https://doi.org/10.1093/beheco/arv228

Abrahms B, Aikens EO, Armstrong JB, et al. (2021) Emerging Perspectives on Resource Tracking and Animal Movement Ecology. Trends Ecol Evol 36 (4):308–320. https://doi.org/10.1016/j.tree.2020.10.018

Abreu F, Garber PA, Souto A, Presotto A, and Schiel N (2021) Navigating in a Challenging Semiarid Environment: The Use of a Route-Based Mental Map by a Small-Bodied Neotropical Primate. Anim Cogn 24 (3):629–643. https://doi.org/10.1007/s10071-020-01465-1

Agarwal A, Sarel A, Derdikman D, Ulanovsky N, and Gutfreund Y (2023) Spatial Coding in the Hippocampus and Hyperpallium of Flying Owls. PNAS 120 (5):e2212418120. https://doi.org/10.1073/pnas.2212418120

Aharon G, Sadot M, and Yovel Y (2017) Bats Use Path Integration Rather Than Acoustic Flow to Assess Flight Distance Along Flyways. Curr Biol 27 (23):3650–3657.e3653. https://doi.org/10.1016/j.cub.2017.10.012

Alatise MB, and Hancke GP (2020) A Review on Challenges of Autonomous Mobile Robot and Sensor Fusion Methods. IEEE Access 8:39830–39846. https://doi.org/10.1109/ACCESS.2020.2975643

Alexander AS, Carstensen LC, Hinman JR, et al. (2020) Egocentric Boundary Vector Tuning of the Retrosplenial Cortex. Sci Adv 6 (8):eaaz2322. https://doi.org/10.1126/sciadv.aaz2322

Alexander AS, Place R, Starrett MJ, Chrastil ER, and Nitz DA (2023a) Rethinking Retrosplenial Cortex: Perspectives and Predictions. Neuron 111 (2):150–175. https://doi.org/10.1016/j.neuron.2022.11.006

Alexander AS, Robinson JC, Stern CE, and Hasselmo ME (2023b) Gated Transformations from Egocentric to Allocentric Reference Frames Involving Retrosplenial Cortex, Entorhinal Cortex, and Hippocampus. Hippocampus 33 (5):465–487. https://doi.org/10.1002/hipo.23513

Alexander WH, and Brown JW (2018) Frontal Cortex Function as Derived from Hierarchical Predictive Coding. Sci Rep 8 (1):3843. https://doi.org/10.1038/s41598-018-21407-9

Alim K (2018) Fluid Flows Shaping Organism Morphology. Philos Trans R Soc Lond B Biol Sci 373 (1747):20170112. https://doi.org/10.1098/rstb.2017.0112

Altshuler DL, and Srinivasan MV (2018) Comparison of Visually Guided Flight in Insects and Birds. Front Neurosci 12:157. https://doi.org/10.3389/fnins.2018.00157

Alyan S, and Jander R (1994) Short-Range Homing in the House Mouse, *Mus musculus*: Stages in the Learning of Directions. Anim Behav 48 (2):285–298. https://doi.org/10.1006/anbe.1994.1242

Amir A, Lee S-C, Headley DB, Herzallah MM, and Pare D (2015) Amygdala Signaling during Foraging in a Hazardous Environment. J Neurosci 35 (38):12994–13005. https://doi.org/10.1523/JNEUROSCI.0407-15.2015
And VL, and Hellung-Larsen P (1992) Chemosensory Behaviour of Tetrahymena. BioEssays 14 (1):61–66. https://doi.org/10.1002/bies.950140113
Andersen RA, Snyder LH, Li C-S, and Stricanne B (1993) Coordinate Transformations in the Representation of Spatial Information. Curr Opin Neurobiol 3 (2):171–176. https://doi.org/10.1016/0959-4388(93)90206-E
Angelaki DE, and Dickman JD (2000) Spatiotemporal Processing of Linear Acceleration: Primary Afferent and Central Vestibular Neuron Responses. J Neurophysiol 84 (4):2113–2132. https://doi.org/10.1152/jn.2000.84.4.2113
Angelaki DE, Ng J, Abrego AM, et al. (2020) A Gravity-Based Three-Dimensional Compass in the Mouse Brain. Nat Commun 11 (1):1855. https://doi.org/10.1038/s41467-020-15566-5
Arakawa H (2006) Changes in the Pattern of Exploratory Behavior Are Associated with the Emergence of Social Dominance Relationships in Male Rats. Dev Psychobiol 48 (1):39–47. https://doi.org/10.1002/dev.20114
Arias-Darraz L, Cabezas D, Colenso CK, et al. (2015) A Transient Receptor Potential Ion Channel in *Chlamydomonas* Shares Key Features with Sensory Transduction-Associated TRP Channels in Mammals. Plant Cell 27 (1):177–188. https://doi.org/10.1105/tpc.114.131862
Armstrong C, Wilkinson H, Meade J, et al. (2013) Homing Pigeons Respond to Time-Compensated Solar Cues Even in Sight of the Loft. PLoS ONE 8 (5):e63130. https://doi.org/10.1371/journal.pone.0063130
Arseneau-Robar TJM, Anderson KA, Vasey EN, Sicotte P, and Teichroeb JA (2022) Think Fast!: Vervet Monkeys Assess the Risk of Being Displaced by a Dominant Competitor When Making Foraging Decisions. Front Ecol Evol 10:775288. https://doi.org/10.3389/fevo.2022.775288
Asensio N, Brockelman WY, Malaivijitnond S, and Reichard UH (2011) Gibbon Travel Paths Are Goal Oriented. Anim Cogn 14 (3):395–405. https://doi.org/10.1007/s10071-010-0374-1
Asumbisa K, Peyrache A, and Trenholm S (2022) Flexible Cue Anchoring Strategies Enable Stable Head Direction Coding in Both Sighted and Blind Animals. Nat Commun 13 (1):5483. https://doi.org/10.1038/s41467-022-33204-0
Avens L, and Lohmann KJ (2003) Use of Multiple Orientation Cues by Juvenile Loggerhead Sea Turtles *Caretta caretta*. J Exp Biol 206 (23):4317–4325. https://doi.org/10.1242/jeb.00657
——— (2004) Navigation and Seasonal Migratory Orientation in Juvenile Sea Turtles. J Exp Biol 207 (11):1771–1778. https://doi.org/10.1242/jeb.00946
Avgar T, Potts JR, Lewis MA, and Boyce MS (2016) Integrated Step Selection Analysis: Bridging the Gap between Resource Selection and Animal Movement. Methods Ecol Evol 7 (5):619–630. https://doi.org/10.1111/2041-210X.12528
Baars BJ, and Gage NM (2018) Fundamentals of Cognitive Neuroscience: A Beginner's Guide. Second edition edn. Academic Press, San Diego
Bairlein F, Norris DR, Nagel R, et al. (2012) Cross-Hemisphere Migration of a 25 G Songbird. Biol Lett 8 (4):505–507. https://doi.org/10.1098/rsbl.2011.1223
Ballesta S, Shi W, Conen KE, and Padoa-Schioppa C (2020) Values Encoded in Orbitofrontal Cortex Are Causally Related to Economic Choices. Nature 588 (7838):450–453. https://doi.org/10.1038/s41586-020-2880-x
Ban SD, Boesch C, and Janmaat KRL (2014) Taï Chimpanzees Anticipate Revisiting High-Valued Fruit Trees from Further Distances. Anim Cogn 17 (6):1353–1364. https://doi.org/10.1007/s10071-014-0771-y

Ban SD, Boesch C, N'Guessan A, et al. (2016) Taï Chimpanzees Change Their Travel Direction for Rare Feeding Trees Providing Fatty Fruits. Anim Behav 118:135–147. https://doi.org/10.1016/j.anbehav.2016.05.014

Barr CR, Bedrossian M, Lohmann KJ, and Nealson KH (2022) Magnetotactic Bacteria: Concepts, Conundrums, and Insights from a Novel *in Situ* Approach Using Digital Holographic Microscopy (DHM). J Comp Physiol A Neuroethol Sens Neural Behav Physiol 208 (1):107–124. https://doi.org/10.1007/s00359-022-01543-4

Basu R, Gebauer R, Herfurth T, et al. (2021) The Orbitofrontal Cortex Maps Future Navigational Goals. Nature 599 (7885):449–452. https://doi.org/10.1038/s41586-021-04042-9

Beltran RS, Kendall-Bar JM, Pirotta E, et al. (2021) Lightscapes of Fear: How Mesopredators Balance Starvation and Predation in the Open Ocean. Sci Adv 7 (12):eabd9818. https://doi.org/10.1126/sciadv.abd9818

Beltran RS, Yuen AL, Condit R, et al. (2022) Elephant Seals Time Their Long-Distance Migrations Using a Map Sense. Curr Biol 32 (4):R156–R157. https://doi.org/10.1016/j.cub.2022.01.031

Ben-Yishay E, Krivoruchko K, Ron S, et al. (2021) Directional Tuning in the Hippocampal Formation of Birds. Curr Biol 31 (12):2592–2602.e2594. https://doi.org/10.1016/j.cub.2021.04.029

Ben Hamed S, Duhamel JR, Bremmer F, and Graf W (2001) Representation of the Visual Field in the Lateral Intraparietal Area of Macaque Monkeys: A Quantitative Receptive Field Analysis. Exp Brain Res 140 (2):127–144. 10.1007/s002210100785

Benhamou S (1996) No Evidence for Cognitive Mapping in Rats. Anim Behav 52 (1):201–212. https://doi.org/10.1006/anbe.1996.0165

——— (2014) Of Scales and Stationarity in Animal Movements. Ecol Lett 17 (3):261–272. https://doi.org/10.1111/ele.12225

Berg HC, and Brown DA (1972) Chemotaxis in *Escherichia coli* Analysed by Three-Dimensional Tracking. Nature 239 (5374):500–504. https://doi.org/10.1038/239500a0

Berthold P (1991) Spatiotemporal Programmes and Genetics of Orientation. In: Berthold P (ed) Orientation in Birds, vol 60. Birkhäuser, Basel, pp 86–105

Bezares-Calderón LA, Berger J, and Jékely G (2020) Diversity of Cilia-Based Mechanosensory Systems and Their Functions in Marine Animal Behaviour. Philos Trans R Soc Lond B Biol Sci 375 (1792):20190376. https://doi.org/10.1098/rstb.2019.0376

Bézy VS, Putman NF, Umbanhowar JA, et al. (2020) Mass-Nesting Events in Olive Ridley Sea Turtles: Environmental Predictors of Timing and Size. Anim Behav 163:85–94. https://doi.org/10.1016/j.anbehav.2020.03.002

Blaser N, Dell'Omo G, Dell'Ariccia G, Wolfer DP, and Lipp HP (2013) Testing Cognitive Navigation in Unknown Territories: Homing Pigeons Choose Different Targets. J Exp Biol 216 (16):3123–3131. https://doi.org/10.1242/jeb.083246

Blaser RE, and Ginchansky RR (2012) Route Selection by Rats and Humans in a Navigational Traveling Salesman Problem. Anim Cogn 15 (2):239–250. https://doi.org/10.1007/s10071-011-0449-7

Boles LC, and Lohmann KJ (2003) True Navigation and Magnetic Maps in Spiny Lobsters. Nature 421 (6918):60–63. https://doi.org/10.1038/nature01226

Bonadonna F, Cunningham GB, Jouventin P, Hesters F, and Nevitt GA (2003) Evidence for Nest-Odour Recognition in Two Species of Diving Petrel. J Exp Biol 206 (20):3719–3722. https://doi.org/10.1242/jeb.00610

Bondoc KGV, Lembke C, Lang SN, et al. (2019) Decision-Making of the Benthic Diatom *Seminavis robusta* Searching for Inorganic Nutrients and Pheromones. ISME J 13 (2):537–546. https://doi.org/10.1038/s41396-018-0299-2

Bottesch M, Gerlach G, Halbach M, et al. (2016) A Magnetic Compass That Might Help Coral Reef Fish Larvae Return to Their Natal Reef. Curr Biol 26 (24):R1266–R1267. https://doi.org/10.1016/j.cub.2016.10.051

Boyd JD, Khaytin I, and Casagrande VA (2009) Evolution of the Visual System in Mammals – Comparative Evolutionary Aspects across Orders. In: Binder MD, Hirokawa N, Windhorst U (eds) Encyclopedia of Neuroscience. Springer, Berlin, pp 1448–1455

Bray D (2002) Bacterial Chemotaxis and the Question of Gain. PNAS 99 (1):7–9. https://doi.org/10.1073/pnas.022641699

Bresson G, Alsayed Z, Yu L, and Glaser S (2017) Simultaneous Localization and Mapping: A Survey of Current Trends in Autonomous Driving. IEEE Trans Intell Vehicl 2 (3):194–220. https://doi.org/10.1109/TIV.2017.2749181

Brierley AS (2014) Diel Vertical Migration. Curr Biol 24 (22):R1074–R1076. https://doi.org/10.1016/j.cub.2014.08.054

Bringmann A, Syrbe S, Görner K, et al. (2018) The Primate Fovea: Structure, Function and Development. Prog Retin Eye Res 66:49–84. https://doi.org/10.1016/j.preteyeres.2018.03.006

Brokaw AF, Davis E, Page RA, and Smotherman M (2021) Flying Bats Use Serial Sampling to Locate Odour Sources. Biol Lett 17 (10):20210430. https://doi.org/10.1098/rsbl.2021.0430

Brønnvik H, Nourani E, Fiedler W, and Flack A (2024) Experience Reduces Route Selection for Conspecifics by the Collectively Migrating White Stork. Curr Biol 34 (9):2030–2037.e2033. https://doi.org/10.1016/j.cub.2024.03.052

Brooks R (1986) A Robust Layered Control System for a Mobile Robot. IEEE J Robot Autom 2:14–23. https://doi.org/10.1109/JRA.1986.1087032

Brothers JR, and Lohmann KJ (2015) Evidence for Geomagnetic Imprinting and Magnetic Navigation in the Natal Homing of Sea Turtles. Curr Biol 25 (3):392–396. https://doi.org/10.1016/j.cub.2014.12.035

——— (2018) Evidence That Magnetic Navigation and Geomagnetic Imprinting Shape Spatial Genetic Variation in Sea Turtles. Curr Biol 28 (8):1325–1329.e1322. https://doi.org/10.1016/j.cub.2018.03.022

Brudzynski SM (2021) Biological Functions of Rat Ultrasonic Vocalizations, Arousal Mechanisms, and Call Initiation. Brain Sci 11 (5):605. https://doi.org/10.3390/brainsci11050605

Buchler ER, and Childs SB (1981) Orientation to Distant Sounds by Foraging Big Brown Bats (*Eptesicus fuscus*). Anim Behav 29 (2):428–432. https://doi.org/10.1016/S0003-3472(81)80102-9

——— (1982) Use of the Post-Sunset Glow as an Orientation Cue by Big Brown Bats (*Eptesicus fuscus*). J Mammal 63 (2):243–247. https://doi.org/10.2307/1380633

Buehlmann C, Mangan M, and Graham P (2020) Multimodal Interactions in Insect Navigation. Anim Cogn 23 (6):1129–1141. https://doi.org/10.1007/s10071-020-01383-2

Burn CC (2008) What Is It Like to Be a Rat? Rat Sensory Perception and Its Implications for Experimental Design and Rat Welfare. Appl Anim Behav Sci 112 (1-2):1–32. https://doi.org/10.1016/j.applanim.2008.02.007

Cadena C, Carlone L, Carrillo H, et al. (2016) Past, Present, and Future of Simultaneous Localization and Mapping: Toward the Robust-Perception Age. IEEE Trans Robot 32 (6):1309–1332. https://doi.org/10.1109/TRO.2016.2624754

Calhoun JB (1963) The Ecology and Sociology of the Norway Rat. US Dept. of Health, Education, and Welfare, Public Health Service, Bethesda

Campagner D, Vale R, Tan YL, et al. (2023) A Cortico-Collicular Circuit for Orienting to Shelter during Escape. Nature 613 (7942):111–119. https://doi.org/10.1038/s41586-022-05553-9

Carson H, Ford JJ, and Milford M (2022) Predicting to Improve: Integrity Measures for Assessing Visual Localization Performance. IEEE Robot Autom Lett 7 (4):9627–9634. https://doi.org/10.1109/LRA.2022.3191205

Carter CS, Braver TS, Barch DM, et al. (1998) Anterior Cingulate Cortex, Error Detection, and the Online Monitoring of Performance. Science 280 (5364):747–749. https://doi.org/10.1126/science.280.5364.747

Carvell GE, and Simons DJ (1990) Biometric Analyses of Vibrissal Tactile Discrimination in the Rat. J Neurosci 10 (8):2638–2648. https://doi.org/10.1523/JNEUROSCI.10-08-02638.1990

Ceccarelli E, Negrín AR, Coyohua-Fuentes A, Canales-Espinosa D, and Dias PAD (2019) An Exploration of the Factors Influencing the Spatial Behavior of Mantled Howler Monkeys (*Alouatta palliata*). Int J Primatol 40 (2):197–213. https://doi.org/10.1007/s10764-018-0075-1

Chapman J, Klaassen R, Drake VA, et al. (2011) Animal Orientation Strategies for Movement in Flows. Curr Biol 21 (20):R861–R870. https://doi.org/10.1016/j.cub.2011.08.014

Chapman JW, Reynolds DR, and Wilson K (2015) Long-Range Seasonal Migration in Insects: Mechanisms, Evolutionary Drivers and Ecological Consequences. Ecol Lett 18 (3):287–302. https://doi.org/10.1111/ele.12407

Cheeseman JF, Millar CD, Greggers U, et al. (2014) Way-Finding in Displaced Clock-Shifted Bees Proves Bees Use a Cognitive Map. PNAS 111 (24):8949–8954. https://doi.org/10.1073/pnas.1408039111

Cheng K (2005) Context Cues Eliminate Retroactive Interference Effects in Honeybees *Apis mellifera*. J Exp Biol 208 (6):1019–1024. https://doi.org/10.1242/jeb.01499

——— (2022) Oscillators and Servomechanisms in Orientation and Navigation, and Sometimes in Cognition. Proc R Soc Lond B Biol Sci 289 (1974):20220237. https://doi.org/10.1098/rspb.2022.0237

——— (2023) From Representations to Servomechanisms to Oscillators: My Journey in the Study of Cognition. Anim Cogn 26 (1):73–85. https://doi.org/10.1007/s10071-022-01677-7

Chernetsov N, Pakhomov A, Kobylkov D, et al. (2017) Migratory Eurasian Reed Warblers Can Use Magnetic Declination to Solve the Longitude Problem. Curr Biol 27 (17):2647–2651.e2642. https://doi.org/10.1016/j.cub.2017.07.024

Chetverikova R, Dautaj G, Schwigon L, Dedek K, and Mouritsen H (2022) Double Cones in the Avian Retina Form an Oriented Mosaic Which Might Facilitate Magnetoreception and/or Polarized Light Sensing. J R Soc Interface 19 (189):20210877. https://doi.org/10.1098/rsif.2021.0877

Cheung A, Collett M, Collett TS, et al. (2014) Still No Convincing Evidence for Cognitive Map Use by Honeybees. PNAS 111 (42):E4396–E4397. https://doi.org/10.1073/pnas.1413581111

Chiari Y, Cahais V, Galtier N, and Delsuc F (2012) Phylogenomic Analyses Support the Position of Turtles as the Sister Group of Birds and Crocodiles (Archosauria). BMC Biol 10 (1):65. https://doi.org/10.1186/1741-7007-10-65

Childs SB, and Buchler ER (1981) Perception of Simulated Stars by *Eptesicus fuscus* (Vespertilionidae): A Potential Navigational Mechanism. Anim Behav 29 (4):1028–1035. https://doi.org/10.1016/S0003-3472(81)80056-5

Chittka L, Skorupski P, and Raine NE (2009) Speed–Accuracy Tradeoffs in Animal Decision Making. Trends Ecol Evol 24 (7):400–407. https://doi.org/10.1016/j.tree.2009.02.010

Choi J-S, and Kim JJ (2010) Amygdala Regulates Risk of Predation in Rats Foraging in a Dynamic Fear Environment. PNAS 107 (50):21773–21777. https://doi.org/10.1073/pnas.1010079108

Clark BJ, Simmons CM, Berkowitz LE, and Wilber AA (2018) The Retrosplenial-Parietal Network and Reference Frame Coordination for Spatial Navigation. Behav Neurosci 132 (5):416–429. https://doi.org/10.1037/bne0000260

Claudi F, Campagner D, and Branco T (2022) Innate Heuristics and Fast Learning Support Escape Route Selection in Mice. Curr Biol 32 (13):2980–2987.e2985. https://doi.org/10.1016/j.cub.2022.05.020

Claxton O, Malone C, Carson H, et al. (2024) Improving Visual Place Recognition Based Robot Navigation by Verifying Localization Estimates. IEEE Robot Autom Lett 9 (12):11098–11105. https://doi.org/10.1109/LRA.2024.3483045

Clement L, Schwarz S, and Wystrach A (2023) An Intrinsic Oscillator Underlies Visual Navigation in Ants. Curr Biol 33 (3):411–422.e415. https://doi.org/10.1016/j.cub.2022.11.059

Colledanchise M, and Ögren P (2017) Behavior Trees in Robotics and AI: An Introduction. CRC Press, Boca Raton. https://doi.org/10.48550/ARXIV.1709.00084

Collett TS, Fauria K, and Dale K (2003) Contextual Cues and Insect Navigation. In: Jeffery KJ (ed) The Neurobiology of Spatial Behaviour. Oxford Univ Press, pp 67–82

Cortese D, and Wan KY (2021) Control of Helical Navigation by Three-Dimensional Flagellar Beating. Phys Rev Lett 126 (8):088003. https://doi.org/10.1103/PhysRevLett.126.088003

Cullen KE (2012) The Vestibular System: Multimodal Integration and Encoding of Self-Motion for Motor Control. Trends Neurosci 35 (3):185–196. https://doi.org/10.1016/j.tins.2011.12.001

Cullen KE, and Taube JS (2017) Our Sense of Direction: Progress, Controversies and Challenges. Nat Neurosci 20 (11):1465–1473. https://doi.org/10.1038/nn.4658

Cvikel N, Egert Berg K, Levin E, et al. (2015) Bats Aggregate to Improve Prey Search but Might Be Impaired When Their Density Becomes Too High. Curr Biol 25 (2):206–211. https://doi.org/10.1016/j.cub.2014.11.010

Dacke M, Baird E, Byrne M, Scholtz Clarke H, and Warrant Eric J (2013) Dung Beetles Use the Milky Way for Orientation. Curr Biol 23 (4):298–300. https://doi.org/10.1016/j.cub.2012.12.034

Dacke M, Baird E, El Jundi B, Warrant EJ, and Byrne M (2021) How Dung Beetles Steer Straight. Annu Rev Entomol 66 (1):243–256. https://doi.org/10.1146/annurev-ento-042020-102149

Dacke M, Nilsson D-E, Scholtz CH, Byrne M, and Warrant EJ (2003) Insect Orientation to Polarized Moonlight. Nature 424 (6944):33–33. https://doi.org/10.1038/424033a

Davis H, MacKenzie KA, and Morrison S (1989) Numerical Discrimination by Rats (*Rattus norvegicus*) Using Body and Vibrissal Touch. J Comp Psychol 103 (1):45–53. https://doi.org/10.1037/0735-7036.103.1.45

Deeti S, Cheng K, Graham P, and Wystrach A (2023) Scanning Behaviour in Ants: Interplay between Random-Rate Processes and Oscillators. J Comp Phys A Neuroethol Sens Neural Behav Physiol 209 (4):625–639. https://doi.org/10.1007/s00359-023-01628-8

De Guinea M, Estrada A, Nekaris KA-I, and Van Belle S (2021) Cognitive Maps in the Wild: Revealing the Use of Metric Information in Black Howler Monkey Route Navigation. J Exp Biol 224 (15):jeb242430. https://doi.org/10.1242/jeb.242430

De Jong LW, Gereke B, Martin GM, and Fellous J-M (2011) The Traveling Salesrat: Insights into the Dynamics of Efficient Spatial Navigation in the Rodent. J Neural Eng 8 (6):065010. https://doi.org/10.1088/1741-2560/8/6/065010

De Raad AL, and Hill RA (2019) Topological Spatial Representation in Wild Chacma Baboons (*Papio ursinus*). Anim Cogn 22 (3):397–412. https://doi.org/10.1007/s10071-019-01253-6

Dittman AH, and Quinn TP (1996) Homing in Pacific Salmon: Mechanisms and Ecological Basis. J Exp Biol 199 (1):83–91. https://doi.org/10.1242/jeb.199.1.83

Dorfman A, Weiss O, Hagbi Z, Levi A, and Eilam D (2021) Social Spatial Cognition. Neurosci Biobehav Rev 121:277–290. https://doi.org/10.1016/j.neubiorev.2020.12.023

Doty RL (1986) Odor-Guided Behavior in Mammals. Experientia 42 (3):257–271. https://doi.org/10.1007/BF01942506

Drescher K, Goldstein RE, and Tuval I (2010) Fidelity of Adaptive Phototaxis. PNAS 107 (25):11171–11176. https://doi.org/10.1073/pnas.1000901107

Dreyer D, Frost B, Mouritsen H, et al. (2018) The Earth's Magnetic Field and Visual Landmarks Steer Migratory Flight Behavior in the Nocturnal Australian Bogong Moth. Curr Biol 28 (13):2160–2166.e2165. https://doi.org/10.1016/j.cub.2018.05.030

Duhamel J-R, Bremmer F, Ben Hamed S, and Graf W (1997) Spatial Invariance of Visual Receptive Fields in Parietal Cortex Neurons. Nature 389 (6653):845–848. https://doi.org/10.1038/39865

Duriez O, Kato A, Tromp C, et al. (2014) How Cheap Is Soaring Flight in Raptors? A Preliminary Investigation in Freely-Flying Vultures. PLoS ONE 9 (1):e84887. https://doi.org/10.1371/journal.pone.0084887

Edwards WI, Popescu VD, Guma N, et al. (2023) Living at the Edge: Home Range Patterns of the Buraiga Chimpanzee Community, Kibale National Park, Uganda. Front Ecol Evol 11:1267688. https://doi.org/10.3389/fevo.2023.1267688

Egert-Berg K, Hurme ER, Greif S, et al. (2018) Resource Ephemerality Drives Social Foraging in Bats. Curr Biol 28 (22):3667–3673.e3665. https://doi.org/10.1016/j.cub.2018.09.064

Eichenbaum H (2000) A Cortical-Hippocampal System for Declarative Memory. Nat Rev Neurosci 1:41–50. https://doi.org/10.1038/35036213

Eilam D, and Golani I (1989) Home Base Behavior of Rats (*Rattus norvegicus*) Exploring a Novel Environment. Behav Brain Res 34 (3):199–211. https://doi.org/10.1016/S0166-4328(89)80102-0

Eliav T, Maimon SR, Aljadeff J, et al. (2021) Multiscale Representation of Very Large Environments in the Hippocampus of Flying Bats. Science 372 (6545):eabg4020. https://doi.org/10.1126/science.abg4020

El Jundi B, Pfeiffer K, Heinze S, and Homberg U (2014) Integration of Polarization and Chromatic Cues in the Insect Sky Compass. J Comp Physiol A Neuroethol Sens Neural Behav Physiol 200:575–589. https://doi.org/10.1007/s00359-014-0890-6

Emlen ST (1975) The Stellar-Orientation System of a Migratory Bird. Sci Am 233 (2):102–111. https://doi.org/10.1038/scientificamerican0875-102

Endres CS, and Lohmann KJ (2012) Perception of Dimethyl Sulfide (DMS) by Loggerhead Sea Turtles: A Possible Mechanism for Locating High-Productivity Oceanic Regions for Foraging. J Exp Biol 215 (20):3535–3538. https://doi.org/10.1242/jeb.073221

——— (2013) Detection of Coastal Mud Odors by Loggerhead Sea Turtles: A Possible Mechanism for Sensing Nearby Land. Mar Biol 160 (11):2951–2956. https://doi.org/10.1007/s00227-013-2285-6

Endres CS, Putman NF, Ernst DA, et al. (2016) Multi-Modal Homing in Sea Turtles: Modeling Dual Use of Geomagnetic and Chemical Cues in Island-Finding. Front Behav Neurosci 10:00019. https://doi.org/10.3389/fnbeh.2016.00019

Engels S, Schneider N-L, Lefeldt N, et al. (2014) Anthropogenic Electromagnetic Noise Disrupts Magnetic Compass Orientation in a Migratory Bird. Nature 509 (7500):353–356. https://doi.org/10.1038/nature13290

Ernst DA, and Lohmann KJ (2016) Effect of Magnetic Pulses on Caribbean Spiny Lobsters: Implications for Magnetoreception. J Exp Biol 219 (12):1827–1832. https://doi.org/10.1242/jeb.136036

Erwin DH, and Davidson EH (2002) The Last Common Bilaterian Ancestor. Development 129 (13):3021–3032. https://doi.org/10.1242/dev.129.13.3021

Esteban GF, and Fenchel TM (2020) Ecology of Protozoa: The Biology of Free-Living Phagotrophic Protists. Springer, Cham

Fagan WF, Lewis MA, Auger-Méthé M, et al. (2013) Spatial Memory and Animal Movement. Ecol Lett 16 (10):1316–1329. https://doi.org/10.1111/ele.12165

Felleman PS, and Van Essen P (1991) Distributed Hierarchical Processing in the Primate Cerebral Cortex. Cereb Cortex 1 (1):1–1. https://doi.org/10.1093/cercor/1.1.1-a

Fetter M (2007) Vestibulo-Ocular Reflex. In: Straube A, Büttner U (eds) Developments in Ophthalmology, vol 40. Karger, Basel, pp 35–51

Finkelstein A, Derdikman D, Rubin A, et al. (2015) Three-Dimensional Head-Direction Coding in the Bat Brain. Nature 517 (7533):159–164. https://doi.org/10.1038/nature14031

Fischer J (2021) Primate Vocal Communication and the Evolution of Speech. Curr Dir Psychol Sci 30 (1):55–60. https://doi.org/10.1177/0963721420979580

Fleischmann PN, Christian M, Müller VL, Rössler W, and Wehner R (2016) Ontogeny of Learning Walks and the Acquisition of Landmark Information in Desert Ants, *Cataglyphis fortis*. J Exp Biol 219:3137–3145. https://doi.org/10.1242/jeb.140459

Fleischmann PN, Grob R, Müller VL, Wehner R, and Rössler W (2018a) The Geomagnetic Field Is a Compass Cue in *Cataglyphis* Ant Navigation. Curr Biol 28 (9):1440–1444. e1442. https://doi.org/10.1016/j.cub.2018.03.043

Fleischmann PN, Grob R, and Rössler W (2020) Magnetoreception in Hymenoptera: Importance for Navigation. Anim Cogn 23 (6):1051–1061. https://doi.org/10.1007/s10071-020-01431-x

——— (2022) Magnetosensation during Re-Learning Walks in Desert Ants (*Cataglyphis nodus*). J Comp Physiol A Neuroethol Sens Neural Behav Physiol 208 (1):125–133. https://doi.org/10.1007/s00359-021-01511-4

Fleischmann PN, Grob R, Wehner R, and Rössler W (2017) Species-Specific Differences in the Fine Structure of Learning Walk Elements in *Cataglyphis* Ants. J Exp Biol 220 (13):2426–2435. https://doi.org/10.1242/jeb.158147

Fleischmann PN, Rössler W, and Wehner R (2018b) Early Foraging Life: Spatial and Temporal Aspects of Landmark Learning in the Ant *Cataglyphis noda*. J Comp Physiol A Neuroethol Sens Neural Behav Physiol 204 (6):579–592. https://doi.org/10.1007/s00359-018-1260-6

Foley NM, Mason VC, Harris AJ, et al. (2023) A Genomic Timescale for Placental Mammal Evolution. Science 380 (6643):eabl8189. https://doi.org/10.1126/science.abl8189

Foster KW, and Smyth RD (1980) Light Antennas in Phototactic Algae. Microbiol Rev 44 (4):572–630. https://doi.org/10.1128/mr.44.4.572-630.1980

Freas CA, and Cheng K (2022) The Basis of Navigation across Species. Annu Rev Psychol 73 (1):217–241. https://doi.org/10.1146/annurev-psych-020821-111311

Freas CA, Narendra A, Murray T, and Cheng K (2024) Polarised Moonlight Guides Nocturnal Bull Ants Home. eLife 13:RP97615. https://doi.org/10.7554/eLife.97615.4

Friedman NP, and Robbins TW (2022) The Role of Prefrontal Cortex in Cognitive Control and Executive Function. Neuropsychopharmacol 47 (1):72–89. https://doi.org/10.1038/s41386-021-01132-0

Frigon A, Akay T, and Prilutsky BI (2021) Control of Mammalian Locomotion by Somatosensory Feedback. In: Prakash YS (ed) Comprehensive Physiology. Wiley, Hoboken, pp 2877–2947

Fulton KA, Zimmerman D, Samuel A, Vogt K, and Datta SR (2024) Common Principles for Odour Coding across Vertebrates and Invertebrates. Nat Rev Neurosci 25 (7):453–472. https://doi.org/10.1038/s41583-024-00822-0

Fung L, Konkol A, Ishikawa T, et al. (2023) Swimming, Feeding, and Inversion of Multicellular Choanoflagellate Sheets. Phys Rev Lett 131 (16):168401. https://doi.org/10.1103/PhysRevLett.131.168401

Gagliardo A (2013) Forty Years of Olfactory Navigation in Birds. J Exp Biol 216 (12):2165–2171. https://doi.org/10.1242/jeb.070250

Gallagher AJ, Creel S, Wilson RP, and Cooke SJ (2017) Energy Landscapes and the Landscape of Fear. Trends Ecol Evol 32 (2):88–96. https://doi.org/10.1016/j.tree.2016.10.010

Gallistel CR (1980) The Organization of Action: A New Synthesis. Lawrence Erlbaum, Hillsdale

Garcia C, Biro D, Janmaat K, and Prat S (2023) Editorial: Cognition, Foraging, and Energetics in Extant and Extinct Primates. Front Ecol Evol 11:1192061. https://doi.org/10.3389/fevo.2023.1192061

Gaussier P, Joulain C, Zrehen S, Banquet JP, and Revel A (1997) Visual Navigation in an Open Environment without Map. In: 1997 IEEE/RSJ International Conference on Intelligent Robot and Systems. Innovative Robotics for Real-World Applications. IROS '97. IEEE, New York, pp 545–550

Gauthier JL, and Tank DW (2018) A Dedicated Population for Reward Coding in the Hippocampus. Neuron 99 (1):179–193 e177. https://doi.org/10.1016/j.neuron.2018.06.008

Gebehart C, and Büschges A (2024) The Processing of Proprioceptive Signals in Distributed Networks: Insights from Insect Motor Control. J Exp Biol 227 (1):jeb246182. https://doi.org/10.1242/jeb.246182

Gee M, Gonzalez-Granda N, Joshi S, et al. (2024) Navigating the Landscape of Fear. bioRxiv Preprint:608007. https://doi.org/10.1101/2024.08.15.608007
Gémard C, Aubin T, and Bonadonna F (2019) Males' Calls Carry Information about Individual Identity and Morphological Characteristics of the Caller in Burrowing Petrels. J Avian Biol 50 (12):jav.02270. https://doi.org/10.1111/jav.02270
Genovesio A, Tsujimoto S, Navarra G, Falcone R, and Wise SP (2014) Autonomous Encoding of Irrelevant Goals and Outcomes by Prefrontal Cortex Neurons. J Neurosci 34 (5):1970–1978. https://doi.org/10.1523/JNEUROSCI.3228-13.2014
Georges-Francois P (1999) Spatial View Cells in the Primate Hippocampus: Allocentric View Not Head Direction or Eye Position or Place. Cereb Cortex 9 (3):197–212. https://doi.org/10.1093/cercor/9.3.197
Geva-Sagiv M, Las L, Yovel Y, and Ulanovsky N (2015) Spatial Cognition in Bats and Rats: From Sensory Acquisition to Multiscale Maps and Navigation. Nat Rev Neurosci 16 (2):94–108. https://doi.org/10.1038/nrn3888
Ghosh B, Chinchali S, and Duggirala PS (2021) Interpretable Trade-Offs between Robot Task Accuracy and Compute Efficiency. In: 2021 IEEE/RSJ International Conference on Intelligent Robots and Systems (IROS). IEEE, New York, pp 5364–5371
Gibson B, Wilkinson M, and Kelly D (2012) Let the Pigeon Drive the Bus: Pigeons Can Plan Future Routes in a Room. Anim Cogn 15 (3):379–391. https://doi.org/10.1007/s10071-011-0463-9
Gill RE, Tibbitts TL, Douglas DC, et al. (2009) Extreme Endurance Flights by Landbirds Crossing the Pacific Ocean: Ecological Corridor Rather Than Barrier? Proc R Soc Lond B Biol Sci 276 (1656):447–457. https://doi.org/10.1098/rspb.2008.1142
Gire DH, Kapoor V, Arrighi-Allisan A, Seminara A, and Murthy VN (2016) Mice Develop Efficient Strategies for Foraging and Navigation Using Complex Natural Stimuli. Curr Biol 26 (10):1261–1273. https://doi.org/10.1016/j.cub.2016.03.040
Goforth KM, Lohmann CMF, Gavin A, et al. (2025) Learned Magnetic Map Cues and Two Mechanisms of Magnetoreception in Turtles. Nature 638 (8052):1015–1022. https://doi.org/10.1038/s41586-024-08554-y
Goldshtein A, Chen X, Amichai E, et al. (2024) Acoustic Cognitive Map–Based Navigation in Echolocating Bats. Science 386 (6721):561–567. https://doi.org/10.1126/science.adn6269
Gonzalez-Terrazas TP, Koblitz JC, Fleming TH, et al. (2016) How Nectar-Feeding Bats Localize Their Food: Echolocation Behavior of *Leptonycteris yerbabuenae* Approaching Cactus Flowers. PLoS ONE 11 (9):e0163492. https://doi.org/10.1371/journal.pone.0163492
Goodridge JP, Dudchenko PA, Worboys KA, Golob EJ, and Taube JS (1998) Cue Control and Head Direction Cells. Behav Neurosci 112 (4):749–761. https://doi.org/10.1037/0735-7044.112.4.749
Goodridge JP, and Taube JS (1995) Preferential Use of the Landmark Navigational System by Head Direction Cells in Rats. Behav Neurosci 109 (1):49–61. https://doi.org/10.1037/0735-7044.109.1.49
Granger J, Walkowicz L, Fitak R, and Johnsen S (2020) Gray Whales Strand More Often on Days with Increased Levels of Atmospheric Radio-Frequency Noise. Curr Biol 30 (4):R155–R156. https://doi.org/10.1016/j.cub.2020.01.028
Graybiel AM, Aosaki T, Flaherty AW, and Kimura M (1994) The Basal Ganglia and Adaptive Motor Control. Science 265 (5180):1826–1831. https://doi.org/10.1126/science.8091209
Green SJ, Boruff BJ, Bonnell TR, and Grueter CC (2020) Chimpanzees Use Least-Cost Routes to Out-of-Sight Goals. Curr Biol 30 (22):4528–4533.e4525. https://doi.org/10.1016/j.cub.2020.08.076
Griffith SC, Owens IPF, and Thuman KA (2002) Extra Pair Paternity in Birds: A Review of Interspecific Variation and Adaptive Function. Mol Ecol 11 (11):2195–2212. https://doi.org/10.1046/j.1365-294X.2002.01613.x

Grob R, El Jundi B, and Fleischmann PN (2021) Towards a Common Terminology for Arthropod Spatial Orientation. Ethol Ecol Evol 33 (3):338–358. https://doi.org/10.1080/03949370.2021.1905075

Grob R, Fleischmann PN, Grübel K, Wehner R, and Rössler W (2017) The Role of Celestial Compass Information in *Cataglyphis* Ants during Learning Walks and for Neuroplasticity in the Central Complex and Mushroom Bodies. Front Behav Neurosci 11:226. https://doi.org/10.3389/fnbeh.2017.00226

Grob R, Fleischmann PN, and Rössler W (2019) Learning to Navigate – How Desert Ants Calibrate Their Compass Systems. Neuroforum 25 (2):109–120. https://doi.org/10.1515/nf-2018-0011

Grob R, Müller VL, Grübel K, Rössler W, and Fleischmann PN (2024a) Importance of Magnetic Information for Neuronal Plasticity in Desert Ants. PNAS 121 (8):e2320764121. https://doi.org/10.1073/pnas.2320764121

Grob R, Wegmann JW, Rössler W, and Fleischmann PN (2024b) *Cataglyphis* Ants Have a Polarity-Sensitive Magnetic Compass. Curr Biol 34 (24):5833–5838.e5832. https://doi.org/10.1016/j.cub.2024.11.012

Grove M (2012) Space, Time, and Group Size: A Model of Constraints on Primate Social Foraging. Anim Behav 83 (2):411–419. https://doi.org/10.1016/j.anbehav.2011.11.011

Guerra PA, Gegear RJ, and Reppert SM (2014) A Magnetic Compass Aids Monarch Butterfly Migration. Nat Commun 5:4164. https://doi.org/10.1038/ncomms5164

Guilford T, and Biro D (2014) Route Following and the Pigeon's Familiar Area Map. J Exp Biol 217 (2):169–179. https://doi.org/10.1242/jeb.092908

Guilford T, and Taylor GK (2014) The Sun Compass Revisited. Anim Behav 97:135–143. https://doi.org/10.1016/j.anbehav.2014.09.005

Günther A, Einwich A, Sjulstok E, et al. (2018) Double-Cone Localization and Seasonal Expression Pattern Suggest a Role in Magnetoreception for European Robin Cryptochrome 4. Curr Biol 28 (2):211–223.e214. https://doi.org/10.1016/j.cub.2017.12.003

Günther A, Haverkamp S, Irsen S, et al. (2024) Species–Specific Circuitry of Double Cone Photoreceptors in Two Avian Retinas. Commun Biol 7 (1):992. https://doi.org/10.1038/s42003-024-06697-2

Gutiérrez-Ibáñez C, Wylie DR, and Altshuler DL (2023) From the Eye to the Wing: Neural Circuits for Transforming Optic Flow into Motor Output in Avian Flight. J Comp Physiol A Neuroethol Sens Neural Behav Physiol 209 (5):839–854. https://doi.org/10.1007/s00359-023-01663-5

Hafting T, Fyhn M, Molden S, Moser M-B, and Moser EI (2005) Microstructure of a Spatial Map in the Entorhinal Cortex. Nature 436 (7052):801–806. https://doi.org/10.1038/nature03721

Hamilton E, Pellicciotta N, Feriani L, and Cicuta P (2020) Motile Cilia Hydrodynamics: Entrainment versus Synchronization When Coupling through Flow. Philos Trans R Soc Lond B Biol Sci 375 (1792):20190152. https://doi.org/10.1098/rstb.2019.0152

Hansen MF, Ellegaard S, Moeller MM, et al. (2020) Comparative Home Range Size and Habitat Selection in Provisioned and Non-Provisioned Long-Tailed Macaques (*Macaca fascicularis*) in Baluran National Park, East Java, Indonesia. Contrib Zool 89 (4):393–411. https://doi.org/10.1163/18759866-bja10006

Harten L, Chen X, De Marcas L, et al. (2024) Time-Mapping and Future-Oriented Behavior in Free-Ranging Wild Fruit Bats. Curr Biol 34 (13):3005–3010.e3004. https://doi.org/10.1016/j.cub.2024.05.046

Harten L, Katz A, Goldshtein A, Handel M, and Yovel Y (2020) The Ontogeny of a Mammalian Cognitive Map in the Real World. Science 369 (6500):194–197. https://doi.org/10.1126/science.aay3354

Hausler S, Garg S, Chakravarty P, et al. (2022) Improving Worst Case Visual Localization Coverage via Place-Specific Sub-Selection in Multi-Camera Systems. In: IEEE Robotics and Automation Letters. IEEE, New York, pp 10112–10119.

Hedenström A, and Johansson LC (2015) Bat Flight: Aerodynamics, Kinematics and Flight Morphology. J Exp Biol 218 (5):653–663. https://doi.org/10.1242/jeb.031203
Hedrick BP, Dickson BV, Dumont ER, and Pierce SE (2020) The Evolutionary Diversity of Locomotor Innovation in Rodents Is Not Linked to Proximal Limb Morphology. Sci Rep 10 (1):717. https://doi.org/10.1038/s41598-019-57144-w
Hegemann P, and Marwan W (1988) Single Photons Are Sufficient to Trigger Movement Responses in *Chlamydomonas reinhardtii*. Photochem Photobiol 48 (1):99–106. https://doi.org/10.1111/j.1751-1097.1988.tb02793.x
Heinze S, Narendra A, and Cheung A (2018) Principles of Insect Path Integration. Curr Biol 28 (17):R1043–R1058. https://doi.org/10.1016/j.cub.2018.04.058
Helbig AJ (1991) Inheritance of Migratory Direction in a Bird Species: A Cross-Breeding Experiment with SE- and SW-Migrating Blackcaps (*Sylvia atricapilla*). Behav Ecol Sociobiol 28 (1):9–12. https://doi.org/10.1007/BF00172133
Helfman GS (1986) Fish Behaviour by Day, Night and Twilight. In: Pitcher TJ (ed) The Behaviour of Teleost Fishes. Springer, Boston, pp 366–387
Henderson J, Hurly TA, Bateson M, and Healy SD (2006) Timing in Free-Living Rufous Hummingbirds, *Selasphorus rufus*. Curr Biol 16 (5):512–515. https://doi.org/10.1016/j.cub.2006.01.054
Heyers D, Manns M, Luksch H, Güntürkün O, and Mouritsen H (2007) A Visual Pathway Links Brain Structures Active during Magnetic Compass Orientation in Migratory Birds. PLoS ONE 2 (9):e937. https://doi.org/10.1371/journal.pone.0000937
Heyers D, Musielak I, Haase K, et al. (2022) Morphology, Biochemistry and Connectivity of Cluster N and the Hippocampal Formation in a Migratory Bird. Brain Struct Funct 227 (8):2731–2749. https://doi.org/10.1007/s00429-022-02566-y
Heyers D, Zapka M, Hoffmeister M, Wild JM, and Mouritsen H (2010) Magnetic Field Changes Activate the Trigeminal Brainstem Complex in a Migratory Bird. PNAS 107 (20):9394–9399. https://doi.org/10.1073/pnas.0907068107
Hoinville T, and Wehner R (2018) Optimal Multiguidance Integration in Insect Navigation. PNAS 115 (11):2824–2829. https://doi.org/10.1073/pnas.1721668115
Hok V, Poucet B, Duvelle É, Save É, and Sargolini F (2016) Spatial Cognition in Mice and Rats: Similarities and Differences in Brain and Behavior. WIRES Cogn Sci 7 (6):406–421. https://doi.org/10.1002/wcs.1411
Holland RA (2014) True Navigation in Birds: From Quantum Physics to Global Migration. J Zool 293 (1):1–15. https://doi.org/10.1111/jzo.12107
Holland RA, Thorup K, Vonhof MJ, Cochran WW, and Wikelski M (2006) Bat Orientation Using Earth's Magnetic Field. Nature 444 (7120):702–702. https://doi.org/10.1038/444702a
Hölldobler B, and Wilson EO (1990) The Ants. Belknap Press, Cambridge, MA
Hollup SA, Molden S, Donnett JG, Moser M-B, and Moser EI (2001) Accumulation of Hippocampal Place Fields at the Goal Location in an Annular Watermaze Task. J Neurosci 21 (5):1635–1644. https://doi.org/10.1523/jneurosci.21-05-01635.2001
Honkanen A, Adden A, Da Silva Freitas J, and Heinze S (2019) The Insect Central Complex and the Neural Basis of Navigational Strategies. J Exp Biol 222 (Suppl_1):jeb188854. https://doi.org/10.1242/jeb.188854
Hore PJ, and Mouritsen H (2016) The Radical-Pair Mechanism of Magnetoreception. Annu Rev Biophys 45:299–344. https://doi.org/10.1146/annurev-biophys-032116-094545
Howard WE, Marsh RE, and Cole RE (1968) Food Detection by Deer Mice Using Olfactory Rather Than Visual Cues. Anim Behav 16 (1):13–17. https://doi.org/10.1016/0003-3472(68)90100-0
Hoy JL, Yavorska I, Wehr M, and Niell CM (2016) Vision Drives Accurate Approach Behavior during Prey Capture in Laboratory Mice. Curr Biol 26 (22):3046–3052. https://doi.org/10.1016/j.cub.2016.09.009

Hudson N, Talbot F, Cox M, et al. (2022) Heterogeneous Ground and Air Platforms, Homogeneous Sensing: Team CSIRO Data61's Approach to the DARPA Subterranean Challenge. Field Robot 2 (1):595–636. https://doi.org/10.55417/fr.2022021

Hulse BK, and Jayaraman V (2020) Mechanisms Underlying the Neural Computation of Head Direction. Annu Rev Neurosci 43 (1):31–54. https://doi.org/10.1146/annurev-neuro-072116-031516

Iserbyt A, Griffioen M, Eens M, and Müller W (2019) Enduring Rules of Care within Pairs: How Blue Tit Parents Resume Provisioning Behaviour after Experimental Disturbance. Sci Rep 9 (1):2776. https://doi.org/10.1038/s41598-019-39139-9

Islam M, Deeti S, Mahmudah Z, Kamhi JF, and Cheng K (2023) Detouring While Foraging up a Tree: What Bull Ants (*Myrmecia midas*) Learn and Their Reactions to Novel Sensory Cues. J Comp Psychol 137 (1):4–15. https://doi.org/10.1037/com0000333

Jackson BJ, Fatima GL, Oh S, and Gire DH (2020) Many Paths to the Same Goal: Balancing Exploration and Exploitation during Probabilistic Route Planning. eNeuro 7 (3):ENEURO.0536–0519.2020. https://doi.org/10.1523/ENEURO.0536-19.2020

Jahn U, Heß D, Stampa M, et al. (2020) A Taxonomy for Mobile Robots: Types, Applications, Capabilities, Implementations, Requirements, and Challenges. Robotics 9 (4):109. https://doi.org/10.3390/robotics9040109

Jang H, Boesch C, Mundry R, Kandza V, and Janmaat KRL (2019) Sun, Age and Test Location Affect Spatial Orientation in Human Foragers in Rainforests. Proc R Soc Lond B Biol Sci 286 (1907):20190934. https://doi.org/10.1098/rspb.2019.0934

Janmaat KRL (2019) What Animals Do Not Do or Fail to Find: A Novel Observational Approach for Studying Cognition in the Wild. Evol Anthropol 28 (6):303–320. https://doi.org/10.1002/evan.21794

Janmaat KRL, Boesch C, Byrne R, et al. (2016) Spatio-Temporal Complexity of Chimpanzee Food: How Cognitive Adaptations Can Counteract the Ephemeral Nature of Ripe Fruit. American J Primatol 78 (6):626–645. https://doi.org/10.1002/ajp.22527

Janmaat KRL, Byrne RW, and Zuberbühler K (2006) Primates Take Weather into Account When Searching for Fruits. Curr Biol 16 (12):1232–1237. https://doi.org/10.1016/j.cub.2006.04.031

Janmaat KRL, De Guinea M, Collet J, et al. (2021) Using Natural Travel Paths to Infer and Compare Primate Cognition in the Wild. iScience 24 (4):102343. https://doi.org/10.1016/j.isci.2021.102343

Janmaat KRL, Polansky L, Ban SD, and Boesch C (2014) Wild Chimpanzees Plan Their Breakfast Time, Type, and Location. PNAS 111 (46):16343–16348. https://doi.org/10.1073/pnas.1407524111

Jarzebowski P, Hay YA, Grewe BF, and Paulsen O (2022) Different Encoding of Reward Location in Dorsal and Intermediate Hippocampus. Curr Biol 32 (4):834–841.e835. https://doi.org/10.1016/j.cub.2021.12.024

Jékely G (2009) Evolution of Phototaxis. Philos Trans R Soc Lond B Biol Sci 364 (1531):2795–2808. https://doi.org/10.1098/rstb.2009.0072

Judson K, Sanz C, Ebombi TF, et al. (2024) Socioecological Factors Influencing Intraspecific Variation in Ranging Dynamics of Western Lowland Gorillas (*Gorilla gorilla gorilla*) in Ndoki Forest. American J Primatol 86 (2):e23586. https://doi.org/10.1002/ajp.23586

Kaas JH (2004) Evolution of Somatosensory and Motor Cortex in Primates. Anat Rec A Discov Mol Cell Evol Biol 281A (1):1148–1156. https://doi.org/10.1002/ar.a.20120

——— (2013) The Evolution of Brains from Early Mammals to Humans. Wiley Interdiscip Rev Cogn Sci 4 (1):33–45. https://doi.org/10.1002/wcs.1206

Kaas JH, Qi H-X, and Stepniewska I (2022) Escaping the Nocturnal Bottleneck, and the Evolution of the Dorsal and Ventral Streams of Visual Processing in Primates. Philos Trans R Soc Lond B Biol Sci 377 (1844):20210293. https://doi.org/10.1098/rstb.2021.0293

Kage A, Omori T, Kikuchi K, and Ishikawa T (2020) The Shape-Effect of Flagella Is More Important Than Bottom-Heaviness on Passive Gravitactic Orientation in *Chlamydomonas reinhardtii*. J Exp Biol 223 (5):jeb.205989. https://doi.org/10.1242/jeb.205989

Kaissling K-E, and Priesner E (1970) Die Riechschwelle des Seidenspinners. Naturwiss 57 (1):23–28. https://doi.org/10.1007/BF00593550
Kamil AC, and Jones JE (1997) The Seed-Storing Corvid Clark's Nutcracker Learns Geometric Relationships among Landmarks. Nature 390 (6657):276–279. https://doi.org/10.1038/36840
Kashetsky T, Avgar T, and Dukas R (2021) The Cognitive Ecology of Animal Movement: Evidence from Birds and Mammals. Front Ecol Evol 9:724887. https://doi.org/10.3389/fevo.2021.724887
Kelly DM, Cheng K, Balda R, and Kamil AC (2019) Effects of Sun Compass Error on Spatial Search by Clark's Nutcrackers. Integr Zool 14 (2):172–181. https://doi.org/10.1111/1749-4877.12302
Kelly DM, Kippenbrock S, Templeton J, and Kamil AC (2008) Use of a Geometric Rule or Absolute Vectors: Landmark Use by Clark's Nutcrackers (*Nucifraga columbiana*). Brain Res Bull 76 (3):293–299. https://doi.org/10.1016/j.brainresbull.2008.02.008
Khan AG, Sarangi M, and Bhalla US (2012) Rats Track Odour Trails Accurately Using a Multi-Layered Strategy with near-Optimal Sampling. Nat Commun 3 (1):703. https://doi.org/10.1038/ncomms1712
Kishkinev D, Packmor F, Zechmeister T, et al. (2021) Navigation by Extrapolation of Geomagnetic Cues in a Migratory Songbird. Curr Biol 31 (7):1563–1569.e1564. https://doi.org/10.1016/j.cub.2021.01.051
Klotsa D (2019) As above, So Below, and Also in Between: Mesoscale Active Matter in Fluids. Soft Matter 15 (44):8946–8950. https://doi.org/10.1039/C9SM01019J
Knaden M, and Graham P (2016) The Sensory Ecology of Ant Navigation: From Natural Environments to Neural Mechanisms. Annu Rev Entomol 61 (1):63–76. https://doi.org/10.1146/annurev-ento-010715-023703
Kobylkov D, Musielak I, Haase K, et al. (2022) Morphology of the "Prefrontal" Nidopallium Caudolaterale in the Long-Distance Night-Migratory Eurasian Blackcap (*Sylvia atricapilla*). Neurosci Lett 789:136869. https://doi.org/10.1016/j.neulet.2022.136869
Kobylkov D, Schwarze S, Michalik B, et al. (2020) A Newly Identified Trigeminal Brain Pathway in a Night-Migratory Bird Could Be Dedicated to Transmitting Magnetic Map Information. Proc R Soc Lond B Biol Sci 287 (1919):20192788. https://doi.org/10.1098/rspb.2019.2788
Kolasa J, and Pickett STA (eds) (1991) Ecological Heterogeneity, vol 86. Ecological Studies. Springer, New York
Kong E, Lee K-H, Do J, Kim P, and Lee D (2023) Dynamic and Stable Hippocampal Representations of Social Identity and Reward Expectation Support Associative Social Memory in Male Mice. Nat Commun 14 (1):2597. https://doi.org/10.1038/s41467-023-38338-3
Kong MS, Kim EJ, Park S, et al. (2021) "Fearful-Place" Coding in the Amygdala-Hippocampal Network. eLife 10:e72040. https://doi.org/10.7554/elife.72040
Koskela S, Turunen T, and Ala-Laurila P (2020) Mice Reach Higher Visual Sensitivity at Night by Using a More Efficient Behavioral Strategy. Curr Biol 30 (1):42–53.e44. https://doi.org/10.1016/j.cub.2019.11.021
Krajník T, Faigl J, Vonásek V, et al. (2010) Simple yet Stable Bearing-Only Navigation. J Field Robot 27 (5):511–533. https://doi.org/10.1002/rob.20354
Kramer G (1953) Wird die Sonnehöhe bei der Heimfindeorientierung verwertet? J Ornithol 94:201–219. https://doi.org/10.1007/BF01922508
Kumar S, and Iyengar S (2018) Avian Striatal Complex. In: Vonk J, Shackelford T (eds) Encyclopedia of Animal Cognition and Behavior. Springer, Cham, pp 1–6
Kürten N, Schmaljohann H, Bichet C, et al. (2022) High Individual Repeatability of the Migratory Behaviour of a Long-Distance Migratory Seabird. Mov Ecol 10 (1):5. https://doi.org/10.1186/s40462-022-00303-y
LaChance PA, and Taube JS (2023) A Model for Transforming Egocentric Views into Goal-Directed Behavior. Hippocampus 33 (5):488–504. https://doi.org/10.1002/hipo.23510

Laeverenz-Schlogelhofer H, and Wan KY (2024) Bioelectric Control of Locomotor Gaits in the Walking Ciliate *Euplotes*. Curr Biol 34 (4):697–709.e696. https://doi.org/10.1016/j.cub.2023.12.051

Land MF (2015) Eye Movements of Vertebrates and Their Relation to Eye Form and Function. J Comp Physiol A Neuroethol Sens Neural Behav Physiol 201 (2):195–214. https://doi.org/10.1007/s00359-014-0964-5

Las L, and Ulanovsky N (2024) Bats. Nat Methods 21 (7):1135–1137. https://doi.org/10.1038/s41592-024-02330-6

Laska M (1990) Olfactory Discrimination Ability in Short-Tailed Fruit Bat, *Carollia perspicillata* (Chiroptera: Phyllostomatidae). J Chem Ecol 16 (12):3291–3299. https://doi.org/10.1007/BF00982099

Laughlin AJ, Sheldon DR, Winkler DW, and Taylor CM (2014) Behavioral Drivers of Communal Roosting in a Songbird: A Combined Theoretical and Empirical Approach. Behav Ecol 25 (4):734–743. https://doi.org/10.1093/beheco/aru044

Laundre JW, Hernandez L, and Ripple WJ (2010) The Landscape of Fear: Ecological Implications of Being Afraid. Open Ecol J 3:1–7. https://www.predatordefense.org/docs/research_LandscapeOfFear_Laundre_2010.pdf

Laurens J, and Angelaki DE (2018) The Brain Compass: A Perspective on How Self-Motion Updates the Head Direction Cell Attractor. Neuron 97:275–289. https://doi.org/10.1016/j.neuron.2017.12.020

Laurens J, Kim B, Dickman JD, and Angelaki DE (2016) Gravity Orientation Tuning in Macaque Anterior Thalamus. Nat Neurosci 19 (12):1566–1568. https://doi.org/10.1038/nn.4423

Laurens J, Liu S, Yu X-J, et al. (2017) Transformation of Spatiotemporal Dynamics in the Macaque Vestibular System from Otolith Afferents to Cortex. eLife 6:e20787. https://doi.org/10.7554/eLife.20787

Leberecht B, Wong SY, Satish B, et al. (2023) Upper Bound for Broadband Radiofrequency Field Disruption of Magnetic Compass Orientation in Night-Migratory Songbirds. PNAS 120 (28):e2301153120. https://doi.org/10.1073/pnas.2301153120

Leinonen H, and Tanila H (2018) Vision in Laboratory Rodents—Tools to Measure It and Implications for Behavioral Research. Behav Brain Res 352:172–182. https://doi.org/10.1016/j.bbr.2017.07.040

Leutgeb S, Leutgeb JK, Barnes CA, et al. (2005) Independent Codes for Spatial and Episodic Memory in Hippocampal Neuronal Ensembles. Science 309 (5734):619–623. https://doi.org/10.1126/science.1114037

Levy R (2024) The Prefrontal Cortex: From Monkey to Man. Brain 147 (3):794–815. https://doi.org/10.1093/brain/awad389

Li N, Ho CP, Xue J, et al. (2022) A Progress Review on Solid-State LiDAR and Nanophotonics-Based LiDAR Sensors. Laser Photon Rev 16 (11):2100511. https://doi.org/10.1002/lpor.202100511

Lindecke O, Holland RA, Pētersons G, and Voigt CC (2021) Corneal Sensitivity Is Required for Orientation in Free-Flying Migratory Bats. Commun Biol 4 (1):522. https://doi.org/10.1038/s42003-021-02053-w

Lipp H-P, Vyssotski AL, Wolfer DP, et al. (2004) Pigeon Homing Along Highways and Exits. Curr Biol 14 (14):1239–1249. https://doi.org/10.1016/j.cub.2004.07.024

Liu A, Papale AE, Hengenius J, et al. (2020) Mouse Navigation Strategies for Odor Source Localization. Front Neurosci 14:218. https://doi.org/10.3389/fnins.2020.00218

Lohmann K, Putman N, Johnsen S, and Lohmann C (2023) Misconceptions about Magnetic Maps. zenodo Preprint. https://doi.org/10.5281/ZENODO.10052196

Lohmann KJ, Cain SD, Dodge SA, and Lohmann CMF (2001) Regional Magnetic Fields as Navigational Markers for Sea Turtles. Science 294 (5541):364–366. https://doi.org/10.1126/science.1064557

Lohmann KJ, and Ernst DA (2014) The Geomagnetic Sense of Crustaceans and Its Use in Orientation and Navigation. In: Derby CD, Thiel M (eds) Crustacean Nervous Systems and Control of Behavior. Oxford Univ Press, New York, pp 321–336

Lohmann KJ, and Fittinghoff Lohmann CM (1992) Orientation to Oceanic Waves by Green Turtle Hatchlings. J Exp Biol 171 (1):1–13. https://doi.org/10.1242/jeb.171.1.1

Lohmann KJ, Goforth KM, Mackiewicz AG, Lim DS, and Lohmann CMF (2022) Magnetic Maps in Animal Navigation. J Comp Physiol A Neuroethol Sens Neural Behav Physiol 208 (1):41–67. https://doi.org/10.1007/s00359-021-01529-8

Lohmann KJ, Hester JT, and Lohmann CMF (1999) Long-Distance Navigation in Sea Turtles. Ethol Ecol Evol 11 (1):1–23. https://doi.org/10.1080/08927014.1999.9522838

Lohmann KJ, and Lohmann CMF (1994) Detection of Magnetic Inclination Angle by Sea Turtles: A Possible Mechanism for Determining Latitude. J Exp Biol 194 (1):23–32. https://doi.org/10.1242/jeb.194.1.23

——— (1996) Detection of Magnetic Field Intensity by Sea Turtles. Nature 380 (6569):59–61. https://doi.org/10.1038/380059a0

——— (2019) There and Back Again: Natal Homing by Magnetic Navigation in Sea Turtles and Salmon. J Exp Biol 222 (Suppl_1):jeb184077. https://doi.org/10.1242/jeb.184077

Lohmann KJ, Lohmann CMF, Ehrhart LM, Bagley DA, and Swing T (2004) Geomagnetic Map Used in Sea-Turtle Navigation. Nature 428 (6986):909–910. https://doi.org/10.1038/428909a

Lohmann KJ, Lohmann CMF, and Endres CS (2008a) The Sensory Ecology of Ocean Navigation. J Exp Biol 211 (11):1719–1728. https://doi.org/10.1242/jeb.015792

Lohmann KJ, Lohmann CMF, and Putman NF (2007) Magnetic Maps in Animals: Nature's GPS. J Exp Biol 210 (21):3697–3705. https://doi.org/10.1242/jeb.001313

Lohmann KJ, Luschi P, and Hays GC (2008b) Goal Navigation and Island-Finding in Sea Turtles. J Exp Mar Biol Ecol 356 (1-2):83–95. https://doi.org/10.1016/j.jembe.2007.12.017

Lohmann KJ, Pentcheff ND, Nevitt GA, et al. (1995a) Magnetic Orientation of Spiny Lobsters in the Ocean: Experiments with Undersea Coil Systems. J Exp Biol 198 (10):2041–2048. https://doi.org/10.1242/jeb.198.10.2041

Lohmann KJ, Putman NF, and Lohmann CMF (2012) The Magnetic Map of Hatchling Loggerhead Sea Turtles. Curr Opin Neurobiol 22 (2):336–342. https://doi.org/10.1016/j.conb.2011.11.005

——— (2008c) Geomagnetic Imprinting: A Unifying Hypothesis of Long-Distance Natal Homing in Salmon and Sea Turtles. PNAS 105 (49):19096–19101. https://doi.org/10.1073/pnas.0801859105

Lohmann KJ, Salmon M, and Wyneken J (1990) Functional Autonomy of Land and Sea Orientation Systems in Sea Turtle Hatchlings. Biol Bull 179 (2):214–218. https://doi.org/10.2307/1541772

Lohmann KJ, Swartz AW, and Lohmann CMF (1995b) Perception of Ocean Wave Direction by Sea Turtles. J Exp Biol 198 (5):1079–1085. https://doi.org/10.1242/jeb.198.5.1079

Lohmann KJ, and Willows AOD (1987) Lunar-Modulated Geomagnetic Orientation by a Marine Mollusk. Science 235 (4786):331–334. https://doi.org/10.1126/science.3798115

Longuet-Higgins HC, and Prazdny K (1980) The Interpretation of a Moving Retinal Image. Proc R Soc Lond B Biol Sci 208 (1173):385–397. https://doi.org/10.1098/rspb.1980.0057

Luschi P, Benhamou S, Girard C, et al. (2007) Marine Turtles Use Geomagnetic Cues during Open-Sea Homing. Curr Biol 17 (2):126–133. https://doi.org/10.1016/j.cub.2006.11.062

Lyu C, Abbott LF, and Maimon G (2022) Building an Allocentric Travelling Direction Signal via Vector Computation. Nature 601 (7891):92–97. https://doi.org/10.1038/s41586-021-04067-0

Maaswinkel H, and Whishaw IQ (1999) Homing with Locale, Taxon, and Dead Reckoning Strategies by Foraging Rats: Sensory Hierarchy in Spatial Navigation. Behav Brain Res 99 (2):143–152. https://doi.org/10.1016/S0166-4328(98)00100-4

Mansfield KL, Wyneken J, Porter WP, and Luo J (2014) First Satellite Tracks of Neonate Sea Turtles Redefine the "Lost Years" Oceanic Niche. Proc R Soc Lond B Biol Sci 281 (1781):20133039. https://doi.org/10.1098/rspb.2013.3039
Mao D, Avila E, Caziot B, et al. (2021) Spatial Modulation of Hippocampal Activity in Freely Moving Macaques. Neuron 109 (21):3521–3534.e3526. https://doi.org/10.1016/j.neuron.2021.09.032
Marée AFM, and Hogeweg P (2001) How Amoeboids Self-Organize into a Fruiting Body: Multicellular Coordination in *Dictyostelium discoideum*. PNAS 98 (7):3879–3883. https://doi.org/10.1073/pnas.061535198
Marr D (1982) Vision: A Computational Investigation into the Human Representation and Processing of Visual Information. MIT Press, Cambridge, MA
Marshall WF, and Nonaka S (2006) Cilia: Tuning in to the Cell's Antenna. Curr Biol 16 (15):R604–R614. https://doi.org/10.1016/j.cub.2006.07.012
Martin GR (2017) The Sensory Ecology of Birds. Oxford Univ Press, Oxford
Matsuda I, Hashimoto C, Ihobe H, et al. (2022) Dietary Choices of a Foregut-Fermenting Primate, Colobus guereza: A Comprehensive Approach Including Leaf Chemical and Mechanical Properties, Digestibility and Abundance. Front Ecol Evol 10:795015. https://doi.org/10.3389/fevo.2022.795015
Menti GM, Meda N, Zordan MA, and Megighian A (2023) Towards a Unified Vision on Animal Navigation. Eur J Neurosci 57 (12):1980–1997. https://doi.org/10.1111/ejn.15881
Menzel R, Greggers U, Smith A, et al. (2005) Honey Bees Navigate According to a Map-Like Spatial Memory. PNAS 102 (8):3040–3045. https://doi.org/10.1073/pnas.0408550102
Michalik A, Alert B, Engels S, Lefeldt N, and Mouritsen H (2014) Star Compass Learning: How Long Does It Take? J Ornithol 155 (1):225–234. https://doi.org/10.1007/s10336-013-1004-x
Moita MAP, Rosis S, Zhou Y, LeDoux JE, and Blair HT (2003) Hippocampal Place Cells Acquire Location-Specific Responses to the Conditioned Stimulus during Auditory Fear Conditioning. Neuron 37 (3):485–497. https://doi.org/10.1016/S0896-6273(03)00033-3
Montes De Oca DP, Lovera R, and Cavia R (2017) Where Do Norway Rats Live? Movement Patterns and Habitat Selection in Livestock Farms in Argentina. Wildl Res 44 (4):324. https://doi.org/10.1071/WR16219
Moore TY, Rivera AM, and Biewener AA (2017) Vertical Leaping Mechanics of the Lesser Egyptian Jerboa Reveal Specialization for Maneuverability Rather Than Elastic Energy Storage. Front Zool 14 (1):32. https://doi.org/10.1186/s12983-017-0215-z
Morrill RJ, Thomas AW, Schiel N, Souto A, and Miller CT (2013) The Effect of Habitat Acoustics on Common Marmoset Vocal Signal Transmission. American J Primatol 75 (9):904–916. https://doi.org/10.1002/ajp.22152
Mott CR, and Salmon M (2011) Sun Compass Orientation by Juvenile Green Sea Turtles (*Chelonia mydas*). Chelonian Conserv Biol 10 (1):73–81. https://doi.org/10.2744/CCB-0888.1
Mouritsen H (1998a) Modelling Migration: The Clock-and-Compass Model Can Explain the Distribution of Ringing Recoveries. Anim Behav 56 (4):899–907. https://doi.org/10.1006/anbe.1998.0826
——— (1998b) Redstarts, *Phoenicurus phoenicurus*, Can Orient in a True-Zero Magnetic Field. Anim Behav 55 (5):1311–1324. https://doi.org/10.1006/anbe.1997.0696
——— (2003) Spatiotemporal Orientation Strategies of Long-Distance Migrants. In: Berthold P, Gwinner E, Sonnenschein E (eds) Avian Migration. Springer, Berlin, Heidelberg, pp 493–513
——— (2018) Long-Distance Navigation and Magnetoreception in Migratory Animals. Nature 558 (7708):50–59. https://doi.org/10.1038/s41586-018-0176-1
——— (2022) Sturkie's Avian Physiology. Academic Press, London
Mouritsen H, Atema J, Kingsford MJ, and Gerlach G (2013a) Sun Compass Orientation Helps Coral Reef Fish Larvae Return to Their Natal Reef. PLoS ONE 8 (6):e66039. https://doi.org/10.1371/journal.pone.0066039

Mouritsen H, Derbyshire R, Stalleicken J, et al. (2013b) An Experimental Displacement and over 50 Years of Tag-Recoveries Show That Monarch Butterflies Are Not True Navigators. PNAS 110 (18):7348–7353. https://doi.org/10.1073/pnas.1221701110

Mouritsen H, Feenders G, Liedvogel M, and Kropp W (2004) Migratory Birds Use Head Scans to Detect the Direction of the Earth's Magnetic Field. Curr Biol 14 (21):1946–1949. https://doi.org/10.1016/j.cub.2004.10.025

Mouritsen H, and Frost BJ (2002) Virtual Migration in Tethered Flying Monarch Butterflies Reveals Their Orientation Mechanisms. PNAS 99 (15):10162–10166. https://doi.org/10.1073/pnas.152137299

Mouritsen H, Heyers D, and Güntürkün O (2016) The Neural Basis of Long-Distance Navigation in Birds. Annu Rev Physiol 78 (1):133–154. https://doi.org/10.1146/annurev-physiol-021115-105054

Mouritsen H, and Larsen ON (2001) Migrating Songbirds Tested in Computer-Controlled Emlen Funnels Use Stellar Cues for a Time-Independent Compass. J Exp Biol 204 (22):3855–3865. https://doi.org/10.1242/jeb.204.22.3855

Mouritsen H, and Mouritsen O (2000) A Mathematical Expectation Model for Bird Navigation Based on the Clock-and-Compass Strategy. J Theor Biol 207 (2):283–291. https://doi.org/10.1006/jtbi.2000.2171

Mouritsen H, and Næsbye Larsen O (1998) Migrating Young Pied Flycatchers *Ficedula hypoleuca* Do Not Compensate for Geographical Displacements. J Exp Biol 201 (21):2927–2934. https://doi.org/10.1242/jeb.201.21.2927

Müller M, and Wehner R (1988) Path Integration in Desert Ants, *Cataglyphis fortis*. PNAS 85 (14):5287–5290. https://doi.org/10.1073/pnas.85.14.5287

——— (2007) Wind and Sky as Compass Cues in Desert Ant Navigation. Naturwiss 94 (7):589–594. https://doi.org/10.1007/s00114-007-0232-4

Muller R, and Kubie J (1987) The Effects of Changes in the Environment on the Spatial Firing of Hippocampal Complex-Spike Cells. J Neurosci 7 (7):1951–1968. https://doi.org/10.1523/JNEUROSCI.07-07-01951.1987

Munden R, Börger L, Wilson RP, et al. (2021) Why Did the Animal Turn? Time-Varying Step Selection Analysis for Inference between Observed Turning-Points in High Frequency Data. Methods Ecol Evol 12 (5):921–932. https://doi.org/10.1111/2041-210X.13574

Murray T, Kocsi Z, Dahmen H, et al. (2019) The Role of Attractive and Repellent Scene Memories in Ant Homing (*Myrmecia croslandi*). J Exp Biol 223 (3):jeb.210021. https://doi.org/10.1242/jeb.210021

Murugan M, Jang HJ, Park M, et al. (2017) Combined Social and Spatial Coding in a Descending Projection from the Prefrontal Cortex. Cell 171 (7):1663–1677.e1616. https://doi.org/10.1016/j.cell.2017.11.002

Naisbett-Jones LC, and Lohmann KJ (2022) Magnetoreception and Magnetic Navigation in Fishes: A Half Century of Discovery. J Comp Physiol A Neuroethol Sens Neural Behav Physiol 208 (1):19–40. https://doi.org/10.1007/s00359-021-01527-w

Nathan R, Getz WM, Revilla E, et al. (2008) A Movement Ecology Paradigm for Unifying Organismal Movement Research. PNAS 105 (49):19052–19059. https://doi.org/10.1073/pnas.0800375105

Neville V, Finnegan E, Paul ES, et al. (2024) You Are How You Eat: Foraging Behavior as a Potential Novel Marker of Rat Affective State. Affec Sci 5 (3):232–245. https://doi.org/10.1007/s42761-024-00242-4

Newcombe N, Huttenlocher J, Drummey AB, and Wiley JG (1998) The Development of Spatial Location Coding: Place Learning and Dead Reckoning in the Second and Third Years. Cogn Dev 13 (2):185–200. https://doi.org/10.1016/S0885-2014(98)90038-7

Nielsen LT, and Kiørboe T (2021) Foraging Trade-Offs, Flagellar Arrangements, and Flow Architecture of Planktonic Protists. PNAS 118 (3):e2009930118. https://doi.org/10.1073/pnas.2009930118

Niimura Y, and Nei M (2006) Evolutionary Dynamics of Olfactory and Other Chemosensory Receptor Genes in Vertebrates. J Hum Genet 51 (6):505–517. https://doi.org/10.1007/s10038-006-0391-8

Nimpf S, Nordmann GC, Kagerbauer D, et al. (2019) A Putative Mechanism for Magnetoreception by Electromagnetic Induction in the Pigeon Inner Ear. Curr Biol 29 (23):4052–4059.e4054. https://doi.org/10.1016/j.cub.2019.09.048

Nunes S (2008) Dispersal and Philopatry. In: Rodent Societies: An Ecological and Evolutionary Perspective. Univ Chicago Press, Chicago

O'Connor DH, Krubitzer L, and Bensmaia S (2021) Of Mice and Monkeys: Somatosensory Processing in Two Prominent Animal Models. Prog Neurobiol 201:102008. https://doi.org/10.1016/j.pneurobio.2021.102008

O'Keefe J, and Dostrovsky J (1971) The Hippocampus as a Spatial Map: Preliminary Evidence from Unit Activity in the Freely Moving Rat. Brain Res 34 (1):171–175. https://doi.org/10.1016/0006-8993(71)90358-1

O'Keefe J, and Nadel L (1978) The Hippocampus as a Cognitive Map. Oxford Univ Press, London

Olcese U, Iurilli G, and Medini P (2013) Cellular and Synaptic Architecture of Multisensory Integration in the Mouse Neocortex. Neuron 79 (3):579–593. https://doi.org/10.1016/j.neuron.2013.06.010

Omer DB, Maimon SR, Las L, and Ulanovsky N (2018) Social Place-Cells in the Bat Hippocampus. Science 359 (6372):218–224. https://doi.org/10.1126/science.aao3474

Oshima D, Yoshida M, Saga K, et al. (2023) Mechanoresponses Mediated by the TRP11 Channel in Cilia of *Chlamydomonas reinhardtii*. iScience 26 (10):107926. https://doi.org/10.1016/j.isci.2023.107926

Packmor F, Kishkinev D, Zechmeister T, Mouritsen H, and Holland RA (2024) Migratory Birds Can Extract Positional Information from Magnetic Inclination and Magnetic Declination Alone. Proc R Soc Lond B Biol Sci 291 (2034):rspb20241363. https://doi.org/10.1098/rspb.2024.1363

Pakhomov A, Anashina A, Heyers D, et al. (2018) Magnetic Map Navigation in a Migratory Songbird Requires Trigeminal Input. Sci Rep 8 (1):11975. https://doi.org/10.1038/s41598-018-30477-8

Palgi S, Ray S, Maimon S, et al. (2025) Head-direction cells as a neural compass in bats navigating outdoors on a remote oceanic island. Science 390:eadw6202 https://doi.org/10.1126/science.adw6202

Palmer MS, Gaynor KM, Becker JA, et al. (2022) Dynamic Landscapes of Fear: Understanding Spatiotemporal Risk. Trends Ecol Evol 37 (10):911–925. https://doi.org/10.1016/j.tree.2022.06.007

Parra-Barrero E, Vijayabaskaran S, Seabrook E, Wiskott L, and Cheng S (2023) A Map of Spatial Navigation for Neuroscience. Neurosci Biobehav Rev 152:105200. https://doi.org/10.1016/j.neubiorev.2023.105200

Patel RN, and Cronin TW (2020) Mantis Shrimp Navigate Home Using Celestial and Idiothetic Path Integration. Curr Biol 30 (11):1981–1987.e1983. https://doi.org/10.1016/j.cub.2020.03.023

Paul C-M, Magda G, and Abel S (2009) Spatial Memory: Theoretical Basis and Comparative Review on Experimental Methods in Rodents. Behav Brain Res 203 (2):151–164. https://doi.org/10.1016/j.bbr.2009.05.022

Paulin MG (1995) Electroreception and the Compass Sense of Sharks. J Theor Biol 174 (3):325–339. https://doi.org/10.1006/jtbi.1995.0102

Payne HL, Lynch GF, and Aronov D (2021) Neural Representations of Space in the Hippocampus of a Food-Caching Bird. Science 373 (6552):343–348. https://doi.org/10.1126/science.abg2009

Pearce F, Carbone C, Cowlishaw G, and Isaac NJB (2013) Space-Use Scaling and Home Range Overlap in Primates. Proc Biol Sci 280 (1751):20122122. https://doi.org/10.1098/rspb.2012.2122

Peltier NE, Angelaki DE, and DeAngelis GC (2020) Optic Flow Parsing in the Macaque Monkey. J Vis 20 (10):8. https://doi.org/10.1167/jov.20.10.8

Petrucco L, Lavian H, Wu YK, et al. (2023) Neural Dynamics and Architecture of the Heading Direction Circuit in Zebrafish. Nat Neurosci 26 (5):765–773. https://doi.org/10.1038/s41593-023-01308-5

Pfeffer SE, Wahl VL, Wittlinger M, and Wolf H (2019) High-Speed Locomotion in the Saharan Silver Ant, *Cataglyphis bombycina*. J Exp Biol 222 (20):jeb198705. https://doi.org/10.1242/jeb.198705

Phillips RG, and LeDoux JE (1992) Differential Contribution of Amygdala and Hippocampus to Cued and Contextual Fear Conditioning. Behav Neurosci 106 (2):274–285. https://doi.org/10.1037/0735-7044.106.2.274

Piper WH (2011) Making Habitat Selection More "Familiar": A Review. Behav Ecol Sociobiol 65 (7):1329–1351. https://doi.org/10.1007/s00265-011-1195-1

Piza DB, Corrigan BW, Gulli RA, et al. (2024) Primacy of Vision Shapes Behavioral Strategies and Neural Substrates of Spatial Navigation in Marmoset Hippocampus. Nat Commun 15 (1):4053. https://doi.org/10.1038/s41467-024-48374-2

Polin M, Tuval I, Drescher K, Gollub JP, and Goldstein RE (2009) *Chlamydomonas* Swims with Two "Gears" in a Eukaryotic Version of Run-and-Tumble Locomotion. Science 325 (5939):487–490. https://doi.org/10.1126/science.1172667

Prat Y, and Yovel Y (2020) Decision Making in Foraging Bats. Curr Opin Neurobiol 60:169–175. https://doi.org/10.1016/j.conb.2019.12.006

Presotto A, Verderane MP, Biondi L, et al. (2018) Intersection as Key Locations for Bearded Capuchin Monkeys (*Sapajus libidinosus*) Traveling within a Route Network. Anim Cogn 21 (3):393–405. https://doi.org/10.1007/s10071-018-1176-0

Pritchard DJ, Hurly TA, and Healy SD (2018) Wild Hummingbirds Require a Consistent View of Landmarks to Pinpoint a Goal Location. Anim Behav 137:83–94. https://doi.org/10.1016/j.anbehav.2018.01.014

Prusky GT, Harker KT, Douglas RM, and Whishaw IQ (2002) Variation in Visual Acuity within Pigmented, and between Pigmented and Albino Rat Strains. Behav Brain Res 136 (2):339–348. https://doi.org/10.1016/S0166-4328(02)00126-2

Pullano L, and Foti F (2022) The Development of Human Navigation in Middle Childhood: A Narrative Review through Methods, Terminology, and Fundamental Stages. Brain Sci 12 (8):1097. https://doi.org/10.3390/brainsci12081097

Purcell EM (1977) Life at Low Reynolds Number. Am J Phys 45 (1):3–11. https://doi.org/10.1119/1.10903

Putman N (2018) Marine Migrations. Curr Biol 28 (17):R972–R976. https://doi.org/10.1016/j.cub.2018.07.036

Putman NF, and Lohmann KJ (2008) Compatibility of Magnetic Imprinting and Secular Variation. Curr Biol 18 (14):R596–R597. https://doi.org/10.1016/j.cub.2008.05.008

Putman NF, Lohmann KJ, Putman EM, et al. (2013) Evidence for Geomagnetic Imprinting as a Homing Mechanism in Pacific Salmon. Curr Biol 23 (4):312–316. https://doi.org/10.1016/j.cub.2012.12.041

Putman NF, Scanlan MM, Billman EJ, et al. (2014) An Inherited Magnetic Map Guides Ocean Navigation in Juvenile Pacific Salmon. Curr Biol 24 (4):446–450. https://doi.org/10.1016/j.cub.2014.01.017

Putman NF, Williams CR, Gallagher EP, and Dittman AH (2020) A Sense of Place: Pink Salmon Use a Magnetic Map for Orientation. J Exp Biol:jeb.218735. https://doi.org/10.1242/jeb.218735

Quax TEF, Albers S-V, and Pfeiffer F (2018) Taxis in Archaea. Emerg Top Life Sci 2 (4):535–546. https://doi.org/10.1042/ETLS20180089

Rajan R, Clement JP, and Bhalla US (2006) Rats Smell in Stereo. Science 311 (5761):666–670. https://doi.org/10.1126/science.1122096

Ranck JB (1985) Head Direction Cells in the Deep Cell Layer of Dorsal Presubiculum in Freely Moving Rats. Akademiai Kiado, Budapest

Rauschecker JP (2015) Auditory and Visual Cortex of Primates: A Comparison of Two Sensory Systems. Eur J Neurosci 41 (5):579–585. https://doi.org/10.1111/ejn.12844

Ray S, Yona I, Elami N, et al. (2025) Hippocampal Coding of Identity, Sex, Hierarchy, and Affiliation in a Social Group of Wild Fruit Bats. Science 387 (6733):eadk9385. https://doi.org/10.1126/science.adk9385

Redish AD (2016) Vicarious Trial and Error. Nat Rev Neurosci 17 (3):147–159. https://doi.org/10.1038/nrn.2015.30

Reid CR (2023) Thoughts from the Forest Floor: A Review of Cognition in the Slime Mould *Physarum polycephalum*. Anim Cogn 26 (6):1783–1797. https://doi.org/10.1007/s10071-023-01782-1

Reiner A (2002) Functional Circuitry of the Avian Basal Ganglia: Implications for Basal Ganglia Organization in Stem Amniotes. Brain Res Bull 57 (3-4):513–528. https://doi.org/10.1016/S0361-9230(01)00667-0

Reppert SM, Gegear RJ, and Merlin C (2010) Navigational Mechanisms of Migrating Monarch Butterflies. Trends Neurosci 33 (9):399–406. https://doi.org/10.1016/j.tins.2010.04.004

Reyna-Hurtado R, Teichroeb JA, Bonnell TR, et al. (2018) Primates Adjust Movement Strategies Due to Changing Food Availability. Behav Ecol 29 (2):368–376. https://doi.org/10.1093/beheco/arx176

Reynolds AM, Reynolds DR, Sane SP, Hu G, and Chapman JW (2016) Orientation in High-Flying Migrant Insects in Relation to Flows: Mechanisms and Strategies. Philos Trans R Soc Lond B Biol Sci 371 (1704):20150392. https://doi.org/10.1098/rstb.2015.0392

Reynolds AM, Reynolds DR, Smith AD, and Chapman JW (2010) A Single Wind-Mediated Mechanism Explains High-Altitude "Non-Goal Oriented" Headings and Layering of Nocturnally Migrating Insects. Proc R Soc Lond B Biol Sci 277 (1682):765–772. https://doi.org/10.1098/rspb.2009.1221

Reynolds WW (1977) Temperature as a Proximate Factor in Orientation Behavior. J Fish Res Bd Can 34 (5):734–739. https://doi.org/10.1139/f77-114

Riotte-Lambert L, and Matthiopoulos J (2020) Environmental Predictability as a Cause and Consequence of Animal Movement. Trends Ecol Evol 35 (2):163–174. https://doi.org/10.1016/j.tree.2019.09.009

Roberts AM (2006) Mechanisms of Gravitaxis in *Chlamydomonas*. Biol Bull 210 (2):78–80. https://doi.org/10.2307/4134597

Robira B, Benhamou S, Bayanga EO, Breuer T, and Masi S (2023a) How Do Primates Decide Where to Feed? Insights from Wild Western Gorillas. Anim Behav 204:25–36. https://doi.org/10.1016/j.anbehav.2023.07.012

Robira B, Benhamou S, Fuh TN, and Masi S (2023b) Do Seasonal Frugivory and Cognition Shape Foraging Movements in Wild Western Gorillas? In: Reyna-Hurtado R, Chapman CA, Melletti M (eds) Movement Ecology of Afrotropical Forest Mammals. Springer, Cham, pp 151–170

Robira B, Benhamou S, Obeki Bayanga E, Breuer T, and Masi S (2024) Changes in Movement Patterns in Relation to Sun Conditions and Spatial Scales in Wild Western Gorillas. Anim Cogn 27 (1):37. https://doi.org/10.1007/s10071-024-01871-9

Ros-Rocher N, and Brunet T (2023) What Is It Like to Be a Choanoflagellate? Sensation, Processing and Behavior in the Closest Unicellular Relatives of Animals. Anim Cogn 26 (6):1767–1782. https://doi.org/10.1007/s10071-023-01776-z

Rosati AG (2017) Foraging Cognition: Reviving the Ecological Intelligence Hypothesis. Trends Cogn Sci 21 (9):691–702. https://doi.org/10.1016/j.tics.2017.05.011

Rössler W, Grob R, and Fleischmann PN (2023) The Role of Learning-Walk Related Multisensory Experience in Rewiring Visual Circuits in the Desert Ant Brain. J Comp Physiol A Neuroethol Sens Neural Behav Physiol 209 (4):605–623. https://doi.org/10.1007/s00359-022-01600-y

Rushworth M, Walton M, Kennerley S, and Bannerman D (2004) Action Sets and Decisions in the Medial Frontal Cortex. Trends Cogn Sci 8 (9):410–417. https://doi.org/10.1016/j.tics.2004.07.009
Russell JC, McMorland AJC, and MacKay JWB (2010) Exploratory Behaviour of Colonizing Rats in Novel Environments. Anim Behav 79 (1):159–164. https://doi.org/10.1016/j.anbehav.2009.10.020
Salinas E, and Sejnowski TJ (2001) Gain Modulation in the Central Nervous System: Where Behavior, Neurophysiology, and Computation Meet. Neuroscientist 7 (5):430–440. https://doi.org/10.1177/107385840100700512
Sarel A, Finkelstein A, Las L, and Ulanovsky N (2017) Vectorial Representation of Spatial Goals in the Hippocampus of Bats. Science 355 (6321):176–180. https://doi.org/10.1126/science.aak9589
Sarel A, Palgi S, Blum D, et al. (2022) Natural Switches in Behaviour Rapidly Modulate Hippocampal Coding. Nature 609 (7925):119–127. https://doi.org/10.1038/s41586-022-05112-2
Sassi Y, Nouzières B, Scacco M, et al. (2024) The Use of Social Information in Vulture Flight Decisions. Proc R Soc Lond B Biol Sci 291 (2018):20231729. https://doi.org/10.1098/rspb.2023.1729
Scarry CJ, Salmi R, Lodwick J, and Doran-Sheehy DM (2023) Long-Term Home Range Stability Provides Foraging Benefits in Western Gorillas. Am J Biol Anthropol 181 (2):296–311. https://doi.org/10.1002/ajpa.24743
Schaeffer EA, Campbell N, Sampson H, et al. (2023) Cue Polarization and Representation in Mouse Home Base Behaviors. Anim Cogn 26 (3):861–883. https://doi.org/10.1007/s10071-022-01729-y
Schiffner I, and Wiltschko R (2009) Point of Decision: When Do Pigeons Decide to Head Home? Naturwiss 96 (2):251–258. https://doi.org/10.1007/s00114-008-0476-7
Schmidt-Koenig K, Ganzhorn JU, and Ranvaud R (1991) Orientation in Birds: The Sun Compass. EXS 60:1–15. https://pubmed.ncbi.nlm.nih.gov/1838509/
Schmidt B, Papale A, Redish AD, and Markus EJ (2013) Conflict between Place and Response Navigation Strategies: Effects on Vicarious Trial and Error (VTE) Behaviors. Learn Mem 20 (3):130–138. https://doi.org/10.1101/lm.028753.112
Schmidt M (2011) Locomotion and Postural Behaviour. Adv Sci Res 5 (1):23–39. https://doi.org/10.5194/asr-5-23-2010
Schneider WT, Holland RA, Keišs O, and Lindecke O (2023) Migratory Bats Are Sensitive to Magnetic Inclination Changes during the Compass Calibration Period. Biol Lett 19 (11):20230181. https://doi.org/10.1098/rsbl.2023.0181
Schnitzler H-U, Moss CF, and Denzinger A (2003) From Spatial Orientation to Food Acquisition in Echolocating Bats. Trends Ecol Evol 18 (8):386–394. https://doi.org/10.1016/S0169-5347(03)00185-X
Schulten K, Swenberg CE, and Weller A (1978) A Biomagnetic Sensory Mechanism Based on Magnetic Field Modulated Coherent Electron Spin Motion. Z Physikal Chem 111 (1):1–5. https://doi.org/10.1524/zpch.1978.111.1.001
Schwarz S, Wystrach A, Cheng K, and Kelly DM (2024) Landmarks, Beacons, or Panoramic Views: What Do Pigeons Attend to for Guidance in Familiar Environments? Learn Behav 52 (1):69–84. https://doi.org/10.3758/s13420-023-00610-3
Seelig JD, and Jayaraman V (2015) Neural Dynamics for Landmark Orientation and Angular Path Integration. Nature 521 (7551):186–191. https://doi.org/10.1038/nature14446
Seiler N, Boesch C, Mundry R, Stephens C, and Robbins MM (2017) Space Partitioning in Wild, Non-Territorial Mountain Gorillas: The Impact of Food and Neighbours. R Soc Open Sci 4 (11):170720. https://doi.org/10.1098/rsos.170720
Setogawa T, Mizuhiki T, Matsumoto N, et al. (2019) Neurons in the Monkey Orbitofrontal Cortex Mediate Reward Value Computation and Decision-Making. Commun Biol 2 (1):126. https://doi.org/10.1038/s42003-019-0363-0

Shaw AK (2020) Causes and Consequences of Individual Variation in Animal Movement. Mov Ecol 8 (1):12. https://doi.org/10.1186/s40462-020-0197-x
Shepherd GM (2006) Smells, Brains and Hormones. Nature 439 (7073):149–151. https://doi.org/10.1038/439149a
Sherrill-Mix SA, James MC, and Myers RA (2008) Migration Cues and Timing in Leatherback Sea Turtles. Behav Ecol 19 (2):231–236. https://doi.org/10.1093/beheco/arm104
Sinakevitch I, Bjorklund GR, Newbern JM, Gerkin RC, and Smith BH (2018) Comparative Study of Chemical Neuroanatomy of the Olfactory Neuropil in Mouse, Honey Bee, and Human. Biol Cybern 112 (1-2):127–140. https://doi.org/10.1007/s00422-017-0728-8
Sjöberg S, Malmiga G, Nord A, et al. (2021) Extreme Altitudes during Diurnal Flights in a Nocturnal Songbird Migrant. Science 372 (6542):646–648. https://doi.org/10.1126/science.abe7291
Skaggs W, Knierim J, Kudrimoti H, and McNaughton B (1994) A Model of the Neural Basis of the Rat's Sense of Direction. Adv Neural Inf Process Syst 7:173–180
Slotnick BM, and Ptak JE (1977) Olfactory Intensity-Difference Thresholds in Rats and Humans. Physiol Behav 19 (6):795–802. https://doi.org/10.1016/0031-9384(77)90317-1
Smith EE, and Jonides J (1999) Storage and Executive Processes in the Frontal Lobes. Science 283 (5408):1657–1661. https://doi.org/10.1126/science.283.5408.1657
Sokolov AA, Miall RC, and Ivry RB (2017) The Cerebellum: Adaptive Prediction for Movement and Cognition. Trends Cogn Sci 21 (5):313–332. https://doi.org/10.1016/j.tics.2017.02.005
Sosa M, and Giocomo LM (2021) Navigating for Reward. Nat Rev Neurosci 22 (8):472–487. https://doi.org/10.1038/s41583-021-00479-z
Souman JL, Frissen I, Sreenivasa MN, and Ernst MO (2009) Walking Straight into Circles. Curr Biol 19 (18):1538–1542. https://doi.org/10.1016/j.cub.2009.07.053
Spiecker L, Laurien M, Dammann W, et al. (2022) Juvenile Atlantic Herring (*Clupea harengus*) Use a Time-Compensated Sun Compass for Orientation. J Exp Biol 225 (18):jeb244607. https://doi.org/10.1242/jeb.244607
Srinivasan MV (2011) Honeybees as a Model for the Study of Visually Guided Flight, Navigation, and Biologically Inspired Robotics. Physiol Rev 91 (2):413–460. https://doi.org/10.1152/physrev.00005.2010
Srinivasan MV, Zhang S, Altwein M, and Tautz J (2000) Honeybee Navigation: Nature and Calibration of the "Odometer." Science 287 (5454):851–853. https://doi.org/10.1126/science.287.5454.851
Stacho M, Herold C, Rook N, et al. (2020) A Cortex-Like Canonical Circuit in the Avian Forebrain. Science 369 (6511):eabc5534. https://doi.org/10.1126/science.abc5534
Stalleicken J, Mukhida M, Labhart T, et al. (2005) Do Monarch Butterflies Use Polarized Skylight for Migratory Orientation? J Exp Biol 208 (12):2399–2408. https://doi.org/10.1242/jeb.01613
Steck K (2012) Just Follow Your Nose: Homing by Olfactory Cues in Ants. Curr Opin Neurobiol 22 (2):231–235. https://doi.org/10.1016/j.conb.2011.10.011
Steck K, Hansson BS, and Knaden M (2009) Smells Like Home: Desert Ants, *Cataglyphis fortis*, Use Olfactory Landmarks to Pinpoint the Nest. Front Zool 6 (1):5. https://doi.org/10.1186/1742-9994-6-5
Stone T, Webb B, Adden A, et al. (2017) An Anatomically Constrained Model for Path Integration in the Bee Brain. Curr Biol 27 (20):3069–3085.e3011. https://doi.org/10.1016/j.cub.2017.08.052
Suhonen J (1993) Predation Risk Influences the Use of Foraging Sites by Tits. Ecology 74 (4):1197–1203. https://doi.org/10.2307/1940490
Sumner-Rooney L (2023) "Distributed" Vision and the Architecture of Animal Visual Systems. J Exp Biol 226 (23):jeb245392. https://doi.org/10.1242/jeb.245392
Tagg N, McCarthy M, Dieguez P, et al. (2018) Nocturnal Activity in Wild Chimpanzees (*Pan troglodytes*): Evidence for Flexible Sleeping Patterns and Insights into Human Evolution. American J Phys Anthropol 166 (3):510–529. https://doi.org/10.1002/ajpa.23478

Takahashi K, Takahashi H, Furuichi T, et al. (2021) Gravity Sensing in Plant and Animal Cells. NPJ Microgravity 7 (1):2. https://doi.org/10.1038/s41526-020-00130-8
Takakusaki K (2017) Functional Neuroanatomy for Posture and Gait Control. J Mov Disord 10 (1):1–17. https://doi.org/10.14802/jmd.16062
Tanimoto Y, and Kimura KD (2019) Neuronal, Mathematical, and Molecular Bases of Perceptual Decision-Making in *C. elegans*. Neurosci Res 140:3–13. https://doi.org/10.1016/j.neures.2018.10.010
Taube JS (2007) The Head Direction Signal: Origins and Sensory-Motor Integration. Annu Rev Neurosci 30:181–207. https://doi.org/10.1146/annurev.neuro.29.051605.112854
Taube JS, Muller R, and Ranck J (1990) Head-Direction Cells Recorded from the Postsubiculum in Freely Moving Rats. I. Description and Quantitative Analysis. J Neurosci 10 (2):420–435. https://doi.org/10.1523/jneurosci.10-02-00420.1990
Taylor KD (1978) Range of Movement and Activity of Common Rats (*Rattus norvegicus*) on Agricultural Land. J Appl Ecol 15 (3):663. https://www.jstor.org/stable/2402767?origin=crossref
Teitelbaum CS, and Mueller T (2019) Beyond Migration: Causes and Consequences of Nomadic Animal Movements. Trends Ecol Evol 34 (6):569–581. https://doi.org/10.1016/j.tree.2019.02.005
Tello-Ramos MC, Hurly TA, Barclay M, and Healy SD (2022) Hummingbirds Modify Their Routes to Avoid a Poor Location. Learn Behav 50 (1):89–98. https://doi.org/10.3758/s13420-021-00476-3
Tello-Ramos MC, Hurly TA, and Healy SD (2015) Traplining in Hummingbirds: Flying Short-Distance Sequences among Several Locations. Behav Ecol 26 (3):812–819. https://doi.org/10.1093/beheco/arv014
Thomas JA, and Moss C (2004) Echolocation in Bats and Dolphins. Univ Chicago Press, Chicago
Thrun S, Burgard W, and Fox D (2005) Probabilistic Robotics. MIT Press, Cambridge, MA
Timmer D, Frederiksen A, Lünemann DC, et al. (2023) Tracking the Electron Transfer Cascade in European Robin Cryptochrome 4 Mutants. J Am Chem Soc 145 (21):11566–11578. https://doi.org/10.1021/jacs.3c00442
Tobalske BW (2007) Biomechanics of Bird Flight. J Exp Biol 210 (18):3135–3146. https://doi.org/10.1242/jeb.000273
Toledo S, Shohami D, Schiffner I, et al. (2020) Cognitive Map-Based Navigation in Wild Bats Revealed by a New High-Throughput Tracking System. Science 369 (6500):188–193. https://doi.org/10.1126/science.aax6904
Tolman EC (1948) Cognitive Maps in Rats and Men. Psychol Rev 55 (4):189–208. https://doi.org/10.1037/h0061626
Tong Y, Liu H, and Zhang Z (2024) Advancements in Humanoid Robots: A Comprehensive Review and Future Prospects. IEEE/CAA J Autom Sinica 11 (2):301–328. https://doi.org/10.1109/JAS.2023.124140
Trapanese C, Meunier H, and Masi S (2019) What, Where and When: Spatial Foraging Decisions in Primates. Biological Reviews 94 (2):483–502. https://doi.org/10.1111/brv.12462
Tross J, Wolf H, and Pfeffer SE (2021) Allometry in Desert Ant Locomotion (*Cataglyphis albicans* and *Cataglyphis bicolor*): Does Body Size Matter? J Exp Biol 224 (18):jeb242842. https://doi.org/10.1242/jeb.242842
Tsang ACH, Lam AT, and Riedel-Kruse IH (2018) Polygonal Motion and Adaptable Phototaxis via Flagellar Beat Switching in the Microswimmer *Euglena gracilis*. Nature Phys 14 (12):1216–1222. https://doi.org/10.1038/s41567-018-0277-7
Tsoar A, Nathan R, Bartan Y, et al. (2011) Large-Scale Navigational Map in a Mammal. PNAS 108 (37):E718–E724. https://doi.org/10.1073/pnas.1107365108
Tyree TJ, Metke M, and Miller CT (2023) Cross-Modal Representation of Identity in the Primate Hippocampus. Science 382 (6669):417–423. https://doi.org/10.1126/science.adf0460

Uebe R, and Schüler D (2016) Magnetosome Biogenesis in Magnetotactic Bacteria. Nat Rev Microbiol 14 (10):621–637. https://doi.org/10.1038/nrmicro.2016.99
Ulanovsky N, and Moss CF (2008) What the Bat's Voice Tells the Bat's Brain. PNAS 105 (25):8491–8498. https://doi.org/10.1073/pnas.0703550105
Vanderlocht C, Robira B, Corradini A, et al. (2025) Proactive and Reactive Movement Behaviours Shape the Antipredator Sequence in a Large Herbivore. Mov Ecol 13 (1):57. https://doi.org/10.1186/s40462-025-00584-z
Van Houten J (1978) Two Mechanisms of Chemotaxis in *Paramecium*. J Comp Physiol 127 (2):167–174. https://doi.org/10.1007/BF01352301
Veit L, and Nieder A (2013) Abstract Rule Neurons in the Endbrain Support Intelligent Behaviour in Corvid Songbirds. Nat Commun 4 (1):2878. https://doi.org/10.1038/ncomms3878
Villioth J, Zuberbühler K, and Newton-Fisher NE (2023) Discrete Choices: Understanding the Foraging Strategies of Wild Chimpanzees. Anim Behav 200:209–219. https://doi.org/10.1016/j.anbehav.2023.04.003
Voegeli B, Sommer S, Knaden M, and Wehner R (2024) Vector-Based Navigation in Desert Ants: The Significance of Path-Integration Vectors. J Comp Physiol A Neuroethol Sens Neural Behav Physiol 211:209–220. https://doi.org/10.1007/s00359-024-01725-2
Vorhees CV, and Williams MT (2014) Assessing Spatial Learning and Memory in Rodents. ILAR J 55 (2):310–332. https://doi.org/10.1093/ilar/ilu013
Vowles DM (1954) The Orientation of Ants. J Exp Biol 31 (3):356–375. https://doi.org/10.1242/jeb.31.3.356
Wadhwa N, and Berg HC (2022) Bacterial Motility: Machinery and Mechanisms. Nat Rev Microbiol 20 (3):161–173. https://doi.org/10.1038/s41579-021-00626-4
Wahl V, Pfeffer SE, and Wittlinger M (2015) Walking and Running in the Desert Ant *Cataglyphis fortis*. Journal of Comparative Physiology A, Neuroethology, Sensory, Neural, and Behavioral Physiology 201 (6):645–656. https://doi.org/10.1007/s00359-015-0999-2
Wallace DG, Gorny B, and Whishaw IQ (2002) Rats Can Track Odors, Other Rats, and Themselves: Implications for the Study of Spatial Behavior. Behav Brain Res 131 (1-2):185–192. https://doi.org/10.1016/S0166-4328(01)00384-9
Waller D, and Lippa Y (2007) Landmarks as Beacons and Associative Cues: Their Role in Route Learning. Mem Cogn 35 (5):910–924. https://doi.org/10.3758/BF03193465
Wan KY (2020) Synchrony and Symmetry-Breaking in Active Flagellar Coordination. Philos Trans R Soc Lond B Biol Sci 375 (1792):20190393. https://doi.org/10.1098/rstb.2019.0393
——— (2023) Active Oscillations in Microscale Navigation. Anim Cogn 26 (6):1837–1850. https://doi.org/10.1007/s10071-023-01819-5
Wan KY, and Goldstein RE (2016) Coordinated Beating of Algal Flagella Is Mediated by Basal Coupling. PNAS 113 (20):E2784–E2793. https://doi.org/10.1073/pnas.1518527113
——— (2018) Time Irreversibility and Criticality in the Motility of a Flagellate Microorganism. Phys Rev Lett 121 (5):058103. https://doi.org/10.1103/PhysRevLett.121.058103
Wan KY, and Jékely G (2021) Origins of Eukaryotic Excitability. Philos Trans R Soc Lond B Biol Sci 376 (1820):20190758. https://doi.org/10.1098/rstb.2019.0758
Warrant E, Dreyer D, Adden A, et al. (2024) Bogong Moths Use a Stellar Compass for Long-Distance Navigation at Night. Res Sq Preprint. https://doi.org/10.21203/rs.3.rs-4394686/v1
Warren WH (2019) Non-Euclidean Navigation. J Exp Biol 222:jeb187971. https://doi.org/10.1242/jeb.187971
Waser PM, and Brown CH (1986) Habitat Acoustics and Primate Communication. American J Primatol 10 (2):135–154. https://doi.org/10.1002/ajp.1350100205
Wehner R (2020) Desert Navigator: The Journey of an Ant. Belknap Press, Cambridge, MA

Wehner R, Hoinville T, Cruse H, and Cheng K (2016) Steering Intermediate Courses: Desert Ants Combine Information from Various Navigational Routines. J Comp Physiol A Neuroethol Sens Neural Behav Physiol 202 (7):459–472. https://doi.org/10.1007/s00359-016-1094-z

Wehner R, and Srinivasan MV (2003) Path Integration in Insects. In: Jeffery KJ (ed) The Neurobiology of Spatial Behaviour. Oxford Univ Press, Oxford, pp 9–30

Weiss O, Segev E, and Eilam D (2017) Social Spatial Cognition in Rat Tetrads: How They Select Their Partners and Their Gathering Places. Anim Cogn 20 (3):409–418. https://doi.org/10.1007/s10071-016-1063-5

Whishaw I, Gharbawie O, Clark B, and Lehmann H (2006) The Exploratory Behavior of Rats in an Open Environment Optimizes Security. Behav Brain Res 171 (2):230–239. https://doi.org/10.1016/j.bbr.2006.03.037

Whishaw IQ, and Gorny B (1999) Path Integration Absent in Scent-Tracking Fimbria-Fornix Rats: Evidence for Hippocampal Involvement in "Sense of Direction" and "Sense of Distance" Using Self-Movement Cues. J Neurosci 19 (11):4662–4673. https://doi.org/10.1523/JNEUROSCI.19-11-04662.1999

Whishaw IQ, and Maaswinkel H (1998) Rats with Fimbria-Fornix Lesions Are Impaired in Path Integration: A Role for the Hippocampus in "Sense of Direction." J Neurosci 18 (8):3050–3058. https://doi.org/10.1523/JNEUROSCI.18-08-03050.1998

Wijnen K, Genzel L, and van der Meij J (2024) Rodent Maze Studies: From Following Simple Rules to Complex Map Learning. Brain Struct Funct 229 (4):823–841. https://doi.org/10.1007/s00429-024-02771-x

Wikenheiser AM, and Schoenbaum G (2016) Over the River, through the Woods: Cognitive Maps in the Hippocampus and Orbitofrontal Cortex. Nat Rev Neurosci 17 (8):513–523. https://doi.org/10.1038/nrn.2016.56

Wilber AA, Clark BJ, Forster TC, Tatsuno M, and McNaughton BL (2014) Interaction of Egocentric and World-Centered Reference Frames in the Rat Posterior Parietal Cortex. J Neurosci 34 (16):5431–5446. https://doi.org/10.1523/JNEUROSCI.0511-14.2014

Willems EP, and Hill RA (2009) Predator-Specific Landscapes of Fear and Resource Distribution: Effects on Spatial Range Use. Ecology 90 (2):546–555. https://doi.org/10.1890/08-0765.1

Williams TC, Williams JM, and Griffin DR (1966) The Homing Ability of the Neotropical Bat *Phyllostomus hastatus*, with Evidence for Visual Orientation. Anim Behav 14 (4):468–473. https://doi.org/10.1016/S0003-3472(66)80047-7

Wilson DM (1966a) Central Nervous Mechanisms for the Generation of Rhythmic Behaviour in Arthropods. Symp Soc Exp Biol 20:199–228. https://pubmed.ncbi.nlm.nih.gov/5958362/

——— (1966b) Insect Walking. Annu Rev Entomol 11 (1):103–122. https://doi.org/10.1146/annurev.en.11.010166.000535

Wiltschko R, and Wiltschko W (1995) Magnetic Orientation in Animals. Zoophysiology. Springer, Berlin

——— (2023) Animal Navigation: How Animals Use Environmental Factors to Find Their Way. Eur Phys J Spec Top 232 (2):237–252. https://doi.org/10.1140/epjs/s11734-022-00610-w

Windsor M, Fontan A, Pivonka P, and Milford MJ (2024) Forward Prediction of Target Localization Failure through Pose Estimation Artifact Modelling. IEEE Robot Autom Lett 9 (5):4591–4598. https://doi.org/10.1109/LRA.2024.3382528

Wisnoski NI, and Lennon JT (2023) Scaling up and Down: Movement Ecology for Microorganisms. Trends Microbiol 31 (3):242–253. https://doi.org/10.1016/j.tim.2022.09.016

Witman GB (1993) *Chlamydomonas* Phototaxis. Trends Cell Biol 3 (11):403–408. https://doi.org/10.1016/0962-8924(93)90091-E

Wolfe G (2000) The Chemical Defense Ecology of Marine Unicellular Plankton: Constraints, Mechanisms, and Impacts. Biol Bull 198 (2):225–244. https://doi.org/10.2307/1542526

Woznica A, Cantley AM, Beemelmanns C, et al. (2016) Bacterial Lipids Activate, Synergize, and Inhibit a Developmental Switch in Choanoflagellates. PNAS 113 (28):7894–7899. https://doi.org/10.1073/pnas.1605015113

Wyeth RC, Holden T, Jalala H, and Murray JA (2021) Rare-Earth Magnets Influence Movement Patterns of the Magnetically Sensitive Nudibranch *Tritonia exsulans* in Its Natural Habitat. Biol Bull 240 (2):105–117. https://doi.org/10.1086/713663

Wynn J, Padget O, Mouritsen H, et al. (2022) Magnetic Stop Signs Signal a European Songbird's Arrival at the Breeding Site after Migration. Science 375 (6579):446–449. https://doi.org/10.1126/science.abj4210

Wynn J, Padget O, Mouritsen H, Perrins C, and Guilford T (2020) Natal Imprinting to the Earth's Magnetic Field in a Pelagic Seabird. Curr Biol 30 (14):2869–2873.e2862. https://doi.org/10.1016/j.cub.2020.05.039

Wystrach A, Beugnon G, and Cheng K (2011) Landmarks or Panoramas: What Do Navigating Ants Attend to for Guidance? Front Zool 8 (1):21. https://doi.org/10.1186/1742-9994-8-21

——— (2012) Ants Might Use Different View-Matching Strategies on and Off the Route. J Exp Biol 215 (1):44–55. https://doi.org/10.1242/jeb.059584

Wystrach A, Philippides A, Aurejac A, Cheng K, and Graham P (2014) Visual Scanning Behaviours and Their Role in the Navigation of the Australian Desert Ant *Melophorus bagoti*. J Comp Physiol A Neuroethol Sens Neural Behav Physiol 200 (7):615–626. https://doi.org/10.1007/s00359-014-0900-8

Wystrach A, and Schwarz S (2013) Ants Use a Predictive Mechanism to Compensate for Passive Displacements by Wind. Curr Biol 23 (24):R1083–R1085. https://doi.org/10.1016/j.cub.2013.10.072

Xavier DP, Abreu F, Souto A, and Schiel N (2024) Choosing the Best Way: How Wild Common Marmosets Travel to Efficiently Exploit Resources. Anim Cogn 27 (1):20. https://doi.org/10.1007/s10071-024-01864-8

Xu J, Jarocha LE, Zollitsch T, et al. (2021) Magnetic Sensitivity of Cryptochrome 4 from a Migratory Songbird. Nature 594 (7864):535–540. https://doi.org/10.1038/s41586-021-03618-9

Yang C, Mammen L, Kim B, et al. (2024) A Population Code for Spatial Representation in the Zebrafish Telencephalon. Nature 634 (8033):397–406. https://doi.org/10.1038/s41586-024-07867-2

Yoder RM, Clark BJ, and Taube JS (2011) Origins of Landmark Encoding in the Brain. Trends Neurosci 34 (11):561–571. https://doi.org/10.1016/j.tins.2011.08.004

Yoshimura K, and Kamiya R (2001) The Sensitivity of *Chlamydomonas* Photoreceptor Is Optimized for the Frequency of Cell Body Rotation. Plant Cell Physiol 42 (6):665–672. https://doi.org/10.1093/pcp/pce084

Yoshimura K, Matsuo Y, and Kamiya R (2003) Gravitaxis in *Chlamydomonas reinhardtii* Studied with Novel Mutants. Plant Cell Physiol 44 (10):1112–1118. https://doi.org/10.1093/pcp/pcg134

Yovel Y, and Ulanovsky N (2017) Bat Navigation. In: Learning and Memory: A Comprehensive Reference. Elsevier, Amsterdam, pp 333–345

Zapka M, Heyers D, Hein CM, et al. (2009) Visual but Not Trigeminal Mediation of Magnetic Compass Information in a Migratory Bird. Nature 461 (7268):1274–1277. https://doi.org/10.1038/nature08528

Zeil J (2025) Coming Home: How Visually Navigating Ants (*Myrmecia* spp.) Pinpoint Their Nest. J Exp Biol 228 (2):JEB249499. https://doi.org/10.1242/jeb.249499

Zhan S, Zhang W, Niitepõld K, et al. (2014) The Genetics of Monarch Butterfly Migration and Warning Colouration. Nature 514 (7522):317–321. https://doi.org/10.1038/nature13812

Linking Cells and Circuits to Brains and Behavior

5

What Are the Contributions of Modeling to Bridging the Gaps between Methods and across Species?

Andrej Bicanski

Abstract Modeling has strongly influenced our understanding of biological navigation, spanning diverse model organisms, neural subsystems, and varying levels of abstraction. Yet to compare models and methods across species requires nontrivial assumptions about how nervous systems of different species are related to each other. Aided by evolutionary considerations and a framing of neural systems according to Marr's three levels of analysis, the relations between different organisms can be rendered explicit and will be explored here through two case studies. First, head-direction coding in flies, rodents, and potentially humans highlights likely commonalities between phylogenetically distant organisms. Second, different neural implementations of vector navigation in flies and rodents highlight likely differences. Marr's three levels can also be leveraged to cope with the increasing complexity and fewer constraining data for neural-level models of human navigation. A way forward could lie in the use of large-scale systems-level models, which allow us to explore interactions among multiple neural subsystems during successful navigation and, crucially, idiosyncratic failures. These models may also lead to a mechanistic understanding of how individual differences and the effects of aging across subsystems might manifest in behavior, bringing to the forefront the need for robust, biologically plausible, and gracefully degrading neural models.

Keywords Animal navigation, cross-species comparison, heading representation, vector navigation

Andrej Bicanski (✉)
Neural Computation Group, Max Planck Institute for Human Cognitive and Brain Sciences, Dept. of Psychology, 04103 Leipzig, Germany
Email: bicanski@cbs.mpg.de

N. S. Newcombe and K. Cheng (eds.), *Challenges in Navigation Research*, Strüngmann Forum Reports, https://doi.org/10.1007/978-3-032-20563-6_5

1 Introduction

In 1973, three years before the first report of place cells (O'Keefe 1976), the influential geneticist, Theodosius Dobzhansky, famously stated that "nothing in biology makes sense except in the light of evolution" (Dobzhansky 1973). Without the concept of evolution, the observations that make up the "pile of sundry facts" about biology cannot be related to each other in a meaningful picture of the whole. The same holds for the subset of biology we call brains, the navigational abilities supported by these brains, and models thereof.

Here, we explore how our understanding of animal navigation (human and non-human) might benefit from collaborative work across species, with a specific emphasis on how the modeling of neural systems can contribute. Dobzhansky's words are particularly helpful in this endeavor: evolutionary considerations, together with an appropriate framing of the modeling of neural systems, may help us assess the potential for productive exchange (or "cross talk") across methods and species and explain why it is possible in the first place.

Before diving into this issue, let us consider an obvious example in the realm of navigation, around which I will construct some of my arguments. Recent work (Giraldo et al. 2018; Green et al. 2017; Kim et al. 2017) has beautifully shown how theoretical models of heading representation via head-direction (HD) cells (Taube 2007) transfer from rodents to insects, and likely all arthropods, which share the brain area referred to as the central complex (the locus of HD cells in arthropods). Moreover, the comparatively smaller and simpler brains of insects allow for direct visualization of the attractor ring. This has led to a comprehensive characterization of the neurons involved and their species-specific connectivity (Green et al. 2017, 2019), arguably surpassing knowledge about the equivalent rodent system with regard to the details of the neural implementation. Thus, although we have a wealth of connectivity data in rodents, the precise functional connectivity that allows for the emergence of a specific firing pattern given some input is not known.

This provides a clear example of productive exchange across species and methods, and specifically modeling, which I will discuss in further detail below. However, before doing so, let us recall the words of Dobzhansky and consider what changes in the neural underpinnings of behavior could have accompanied the gradual transformation of phenotypes by asking: How can cross talk between researchers who study different species (with partially incongruent methods) be productive if the evolutionary lines of animals being compared have diverged from common ancestors in the distant past? This is the case for insects and rodents. In addition, how can we productively compare animals that are more closely related but occupy radically different environmental niches? Say, rats, bats, and monkeys. How do we need to formulate our ideas about brains in general, and how should we frame models of the neural circuits underlying navigation to accommodate these cases? Answers to these questions should help us to gauge where we can reasonably expect fruitful collaboration across species and methods as well as a transfer of models, as we can observe for rodents and insects.

Below I will articulate these problems in more detail, making recourse to Marr's three levels of analysis (Marr and Poggio 1976) and later to Braitenberg's law of uphill analysis and downhill invention (Braitenberg 1986). Aided by the latter, I will comment specifically on how we might bridge the gap between invasive animal research (with access to single cells and connectivity) and work with humans, which yields predominantly (though not exclusively) behavioral and meso-/macroscale data based on noninvasive imaging methods.

To clarify terminology, we need to analyze what actually constitutes productive cross talk or exchanges. In the realm of theory and modeling, productive cross talk is given when the models and theories from one model organism can be adapted and used to study another. In other words, they enable us to explain some aspect of behavior or neural data and/or generate predictions for a species other than the one used in the original model. In the realm of experimental neuroscience, productive cross talk is given when one experimental technique, behavioral task, or analysis method can be transferred between model organisms; for instance, spatially selective cells (e.g., grid cells) recorded directly in epilepsy patients, analogous classical rodent experiments (Jacobs et al. 2013), or the use of virtual reality environments with human participants akin to a Morris water maze from classical rodent studies (Astur et al. 2002; Morris 1984). Similarly, to mimic the comparison between healthy animals and animal models of Alzheimer disease (Ying et al. 2022) or animals with hippocampal lesions, aging or mildly cognitively impaired individuals with (early onset) Alzheimer disease could be compared to healthy human participants (Howett et al. 2019). I consider such cases to be successful transfers between model organisms and methods. With these loose definitions of theoretical and experimental cross talk in mind, we must understand why we are justified in assuming that existing models and experimental designs developed in one domain can transfer to another. Using the ideas of Marr, let us lay out the necessary conceptual groundwork.

1.1 Marr

Marr's three levels of analysis distinguish the computational, algorithmic, and implementation levels of a neural system (Marr and Poggio 1976) and are best illustrated with a concrete example (Figure 5.1a). In the case of neural oscillations in the spinal cord, which generate muscle activation patterns (Grillner and Wallen 1985), we can formulate the computational-level problem as generating (or "computing" from more primitive, tonic brain stem inputs) the oscillatory muscle activation patterns for locomotion (e.g., a traveling wave along the spinal cord for anguilliform swimming). The algorithmic level is concerned with the general algorithm that should address the problem. For instance, we could chain oscillators composed of pacemakers operating via release from inhibition (an oscillator terminates on its own) or population-based oscillators that allow for escape from inhibition (the silent oscillator can activate and overpower an active one due to some fatigue mechanism) (cf. Figure 5.1a). That is, more than one algorithm can solve the same computational problem. In this case, the implementation level concerns the exact composition of ionic channels (or lack thereof, depending on the neuron model) of the neurons in the network, their neuronal parameters (e.g., time constants and input resistances), the types of synapses used (e.g., AMPA vs. NMDA, GABA vs.

glycine), the learning rule if learning is involved, the specific form of any external inputs for the network that is to be simulated, and the details of static connectivity.

A more navigation-focused example would be that of heading direction, where the computational problem involves representing heading, updating, and keeping it in register with the external world via sensory feedback. The algorithmic solution in multiple species—from insects to rodents, likely in humans and possibly fish—is the use of some form of one-dimensional attractor network/ring attractor (Green et al. 2017; Kim et al. 2017; Shine et al. 2016; Taube 2007; Vinepinsky et al. 2020). As with the locomotor circuits described above, the implementation level is usually understood as being concerned with the concrete neural implementation of the chosen algorithm. For instance, an attractor could be implemented as one ring (Zhang 1996), two rings (Bicanski and Burgess 2016; Xie et al. 2002), or three rings (Song and Wang 2005; Bernard et al. 2025), and each of these can be built with different neuronal properties at the implementation level (Figure 5.1b, c). Several variations of HD models have been explored over the past three decades (e.g., Redish et al. 1996; Stringer et al. 2002). Marr's framework can exhibit some ambiguity as to what constitutes an algorithm and what constitutes the implementation level. For instance, one could view the general task to represent heading with one-dimensional continuous ring attractor as the algorithm, and the specific choice of one versus two (or three) rings as the implementation differences. Alternatively, one could view one- and two-ring attractors as two different algorithms (akin to the distinction of escape and release from inhibition for oscillators) and focus on the neuronal and synaptic properties as the implementation level. To the best of my knowledge, there are no hard and precise criteria for what should be considered the algorithm in the description of a neural system, though we may often have strong intuitions.

This fuzziness, however, does not detract from the utility of Marr's framework and could be construed as an advantage, as it allows researchers to define algorithm and implementation levels according to the problem under investigation. Perhaps the most important feature of framing models in terms of Marr's levels of analysis is that it renders explicit how a model can be correct (and thus useful) on one level while wrong on another. Conversely, this also means that a mere implementation difference between model and experiment does not falsify a model. That is, an experimenter cannot simply point to a difference, such as in the sodium conductance (an implementation level difference) or the label of the neural population in the model (area A suggested but area B shown to hold the representation in question), and claim that the model is false if the overall combination of computation, algorithm, and implementation is still a good explanation for the neural processes in question and the associated behavior. It goes without saying that all models are trivially wrong on the ground truth implementation level, as no model is a perfect representation of the actual biological system (Salis 2021), not even networks made of complex multicompartment Hodgkin-Huxley neurons (Bjelkmar et al. 2009).

Marr's perspective allows us to contextualize research from a broad range of species and allows us to establish well-defined commonalities and differences across levels of analysis—commonalities in terms of computation, algorithm, and implementation that extend beyond mere observation of external similarities in (navigation) behavior. This integration across a broad range of species can be made very explicit when Marr's approach is brought into contact with Dobzhansky.

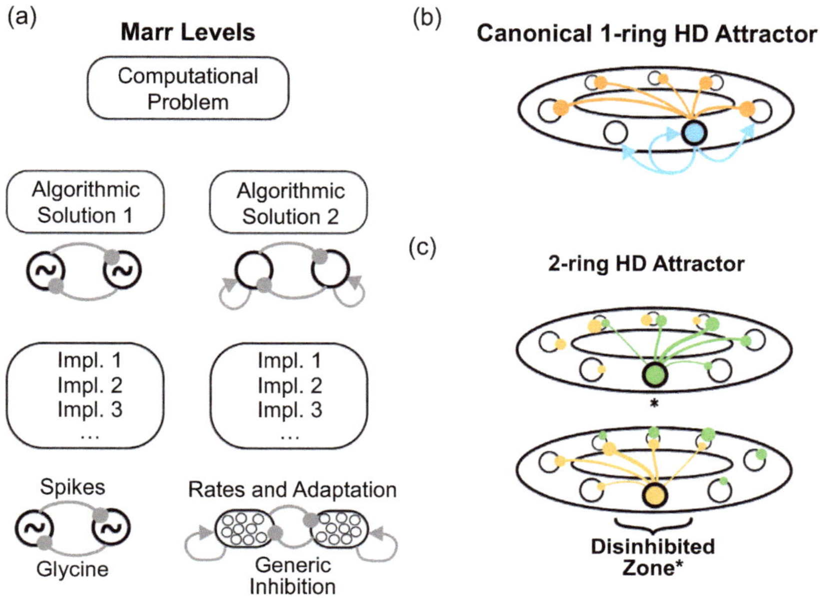

Figure 5.1 Framing computational models according to Marr. (a) In the three levels of analysis (top to bottom), the computational problem constitutes the high-level description of a neural circuit's purpose (i.e., what function/computation is implemented). This problem can be solved using different algorithms, illustrated here with neural oscillations generated by mutually inhibiting pacemakers (intrinsically bursting neurons, solution 1; note the oscillator symbol in the nodes) and an alternative (solution 2) employing neurons incapable of bursting on their own. Both circuits can generate antiphase oscillations and could be implemented in a variety of ways (Impl. 1, 2, 3,...). Bottom: sample implementations are shown for two populations of intrinsically bursting neurons (modeled as spiking neurons) that employ glycinergic inhibition (with corresponding time constant, reversal potential, etc.) and constitute an implementation of algorithm 1. Bottom right: populations of rate-coded neurons with some recurrence, slowly accumulating fatigue (adaptation) and coupled by generic inhibition constitute another implementation of algorithm 2. (b) In the canonical HD circuit developed by Zhang (1996) on theoretical grounds, recurrent excitation (blue) and long-range inhibition (orange) ensure a stable activity packet centered on the blue neuron. The connectivity pattern repeats for all neurons (not shown). An asymmetric connectivity component (not shown) activated by head turns can translate the activity packet around the ring. (c) An alternative implementation/algorithm for HD coding made of two coupled rings with offset inhibitory connections. Connection patterns are shown only for the green and yellow neurons. Thickness of the lines indicates the strength of the inhibition. The connection patterns are offset in different directions and leave a zone of relative disinhibition at which the HD activity packet can form. Both rings receive tonic background drive and the down-regulation of one set of connections (e.g., yellow) leads to a shift of the activity pattern; thus, there is no need for dedicated translation weights (Bicanski and Burgess 2016). This architecture does not require self-excitation (similar to Figure 5.2c), as all neurons receive indiscriminate background drive, presumably originating in the vestibular nuclei (not shown).

1.2 Marr and Evolution

Armed with Marr's terminology, we can assess different models of the navigation system and types of experiments to evaluate the potential for constructive cross talk. To a very first approximation, we can assume that evolutionarily conserved anatomical structures implement similar algorithms (Reznik et al. 2024). Nevertheless, when the evolutionary distance between two species is great, the algorithm may require greater modification—particularly if the species occupy different ecological niches (e.g., terrestrial vs. aquatic). This would likely correspond to weaker and less precise anatomical homology. Alternatively, we might posit convergent evolution (Jarvis et al. 2005), a discussion of which is beyond the scope of this chapter. Conversely, if two animals are closely related in evolutionary terms, similar or identical Marr algorithms might be sufficient and yield stronger homology. However, we need to be prepared to encounter implementation differences, though possibly to a lower degree. Thus, presupposing that two species are faced with the same computational problem, and using the terminology of Marr, we can assume the following:

1. A distant common ancestor increases the probability of differences in algorithm, and common algorithms are likely to differ strongly at the level of implementation (weak homology).
2. More recent common ancestors increase the probability of a conserved algorithm, and implementations are more likely to be similar (strong homology).
3. Organisms with a distant common ancestor and with similar algorithms for a given problem suggest a conserved neural mechanism or convergent evolution.

An additional, often implicitly assumed working hypothesis is that a conserved neural mechanism does not automatically imply similar cognitive capabilities, as is obvious from rodents and insects. Organisms can share some neural subsystems, but such a conserved neural system will be augmented by, and interact with, additional neural machinery in a more complex brain (e.g., expanded cortex in mammals relative to reptiles or between different mammalian species), which is absent from the simpler brains. For a specific navigation-related example, consider that in rodents, the HD system expands (in terms of neuron number) along Papez's Circuit (Taube 2007). In primates, similar spatial cells are found as in rodents (e.g., grid cells, HD cells) (Jacobs et al. 2013; Mao et al. 2021), but many cell types are likely absent in large number from the rodent, expressing view-based coding (Mao et al. 2021; Rolls et al. 1997).

2 Rodents versus Insects versus Primates

As a specific navigation-focused case study, let us compare rodent and insect navigation systems. Here we can find examples of common and divergent algorithms. The same can be said for the implementation level, depending on the definition of the algorithm. The detailed neuronal properties of the cells in the HD system are bound to differ trivially because they share a distant common ancestor and rodents have much larger brains. Hence, it is more useful to define the algorithm as the ability to maintain a stable packet of activity to represent head direction, the ability to update it based on self-motion, and to correct for errors via sensory feedback. How this is implemented (e.g., as one or two rings of neurons) will be treated at the implementation level.

In *Drosophila*, several groundbreaking studies have been able to directly image the activity of the so-called ellipsoid body in the behaving animal, thus providing direct visual evidence of the packet of neural activity that represents heading direction in the animal (Giraldo et al. 2018; Green et al. 2017; Kim et al. 2017). Copies of the activity bump in the ellipsoid body are located in the anatomically adjacent protocerebral bridge (Figure 5.2a). This work has led to a precise characterization of the neurons involved and suggests that the neurons in the protocerebral bridge with excitatory connections target neurons offset from the activity bump in the ellipsoid body, ensuring translation of the activity bump. Moreover, a network that can guarantee menotaxis has also been found in the fruit fly's brain, such that head direction (and thus movement direction) is kept in a fixed relation to some directional cue, thus ensuring goal-directed navigation relative to angular cues (Green et al. 2019). Like the HD attractor itself, this system is also characterized at the level of individual neurons and projections. These findings have also been incorporated into recent modeling studies (Pisokas et al. 2020; Vafidis et al. 2022). It is difficult to determine whether the rodent and insect HD systems are algorithmically similar due to convergent evolution (possibly driven by a very narrow range of viable solutions to the HD problem in neural systems) or whether both derive from an ancient conserved system present in a common ancestor.

In contrast to the fly literature, HD cells were first reported in rodents in 1990 (Taube et al. 1990) and are found throughout an extended network of brain areas (Figure 5.2d), including the lateral mammillary nucleus (LMN) (Blair et al. 1998), the dorsal tegmental nucleus (DTN) (Bassett et al. 2007), the anterior dorsal thalamic nucleus (ADN) (Blair et al. 1999; Peyrache et al. 2019), retrosplenial cortex (Cho and Sharp 2001; Jacob et al. 2017), and postsubicular HD cortex (PoS) (Taube 2007). Although the rodent HD literature is very rich in terms of direct recordings, tracing and lesion studies, and behavioral/cognitive aspects of HD coding, in terms of precise connectivity patterns as well as synaptic and neuronal properties, the rodent HD system is arguably less well characterized than the insect system. Furthermore, unlike in flies, it is difficult to image these cells directly in rodents, as they are located in deep brain regions. In the rodent, HD cell numbers also expand from $O(10^3)$ in the LMN, $O(10^4)$ in the ADN, and come close to $O(10^5)$ in the PoS (Taube 2007). The reason for this expansion is not understood. The insect system does not seem to require it. Lesion studies have established that the LMN and the DTN likely comprise the generative circuitry of the attractor in rodents (Bassett et al. 2007; Blair et al. 1998, 1999). Lesions to the LMN abolish HD coding in the ADN but not vice versa (Taube 2007).

Models of head direction in the insect system have clearly benefited from an extensive preexisting modeling literature on the rodent system, where the canonical HD model—the continuous ring attractor (Figure 5.2b)—was developed on theoretical grounds (Zhang 1996). This transfer is a quintessential example of successful cross-species theory in the area of navigation. It is especially noteworthy that in the fly, cells with neighboring directional preference are co-localized in the ellipsoid body. This is a surprising contrast to the commonly anticipated objection in computational work on the rodent system. There, it is often stipulated (to appease critics) that spatial co-localization of cells is not needed, contrary to the modeling convenience of arranging cells explicitly in a ring. All that is needed is the correct

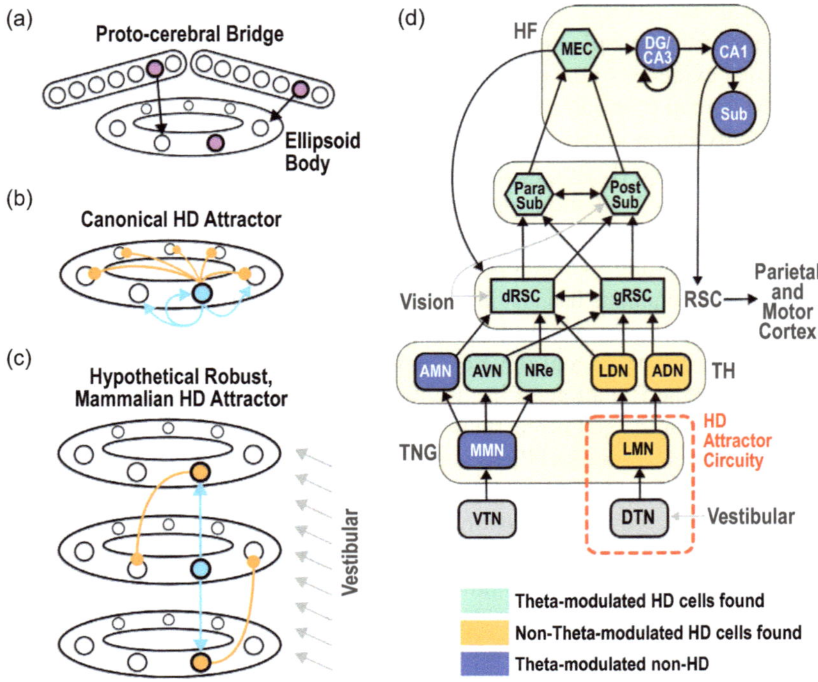

Figure 5.2 Head direction (HD) in insects and rodents. (a) The fly HD system consists of three copies of the current heading estimate (pink cell active) in the ellipsoid body and two protocerebral bridges. The latter targets the ellipsoid body, offset from the active cell to translate the activity packet. The connectivity pattern repeats for all cell triplets (3 pink cells). (b) The canonical HD circuit developed by Zhang (1996) on theoretical grounds. Recurrent excitation (blue) and long-range inhibition (orange) emanate from the same cell. The connectivity pattern repeats for all neurons (not shown). An asymmetric connectivity component (not shown) activated by head turns can translate the activity packet around the ring. (c) A hypothetical mammalian attractor that is robust to neuronal death and noise. Recurrence is removed and the activity level is set by incoming vestibular drive. An excitatory ring activates two inhibitory rings that project back (offset) to the excitatory ring. Modulating the inhibitory connections (or, equivalently, the inhibitory neurons) according to head turns translates the activity packet. The use of three rings parallels some aspects of the fly circuitry, albeit with a different excitation/inhibition mix. (d) The extended HD network in rodents, distinguishing theta-modulated HD cells, non-theta-modulated HD cells, and theta-modulated non-HD cells. It is unknown why HD cell numbers expand from LMN to the thalamus and the postsubiculum. Some connections neglected for clarity (e.g., feedback connections back to the HD attractor, detailed connectivity to parietal and motor areas). Blocks top to bottom: HF (hippocampal formation); RSC (retrosplenial cortex); TH (thalamus); TNG (tegmental nucleus of Gudden). Individual areas: VTN (ventral) and DTN (dorsal) tegmental nuclei; LMN (lateral) and MMN (medial) mammillary nuclei; NRe, nucleus reuniens; AMN (anterior medial), AVN (anterior ventral), LDN (lateral ventral), and ADN (anterior dorsal) thalamic nuclei; dRSC (dysgranular) and gRSC (granular) retrosplenial cortex; ParaSub (parasubiculum) and PostSub (postsubiculum); MEC (medial entorhinal cortex); DG (dentate gyrus) and CA1, 3 (hippocampal subfields); Sub (subiculum).

topology of connections, not the physical co-localization of the neurons. Such pre-emptive answers to anticipated objections betray a lack of knowledge of the precise implementation in the rodent systems. In another example of cross-species interaction, the fly system can perhaps help us refine our models of head directions in mammals.

Flies have a short life span and are unlikely to be concerned by neuronal death and age-related changes to synaptic connectivity (for an extensive review, see Chapter 13). Mammals, on the other hand, can persist long enough to require more robustness from neural circuits. This is in stark contrast to the canonical recurrent continuous ring attractor model of head direction (Zhang 1996), which although theoretically elegant, violates Dale's principle (excitation and inhibition coming from the same neurons) and is very brittle. As little as one dead neuron (or equivalent missing connections) can lead to the collapse of the activity packet. Conversely, a small excess of activity can saturate the network. Both deviations render the canonical model extremely sensitive to parameter variations. As such, it is not a good model of an aging mammalian attractor or a biologically plausible circuit in general. In Marr's terminology, the implementation is unlikely to be a good match to the rodent brain. The detailed connectivity patterns gleaned from the fly brain, however, with its three copies of the activity packet in the protocerebral bridge and the ellipsoid body, is somewhat reminiscent of a preexisting model (Song and Wang 2005) and its architecture was recently adapted to model aging (Bernard et al. 2025). In this biologically plausible model of head direction, direct recurrence is removed and Dale's principle is respected. As a result, this attractor can tolerate substantial amounts of noise and neuronal death compared to the canonical recurrent attractor model. Figure 5.2c illustrates such a ring attractor model, devoid of direct recurrence (cf. Figure 5.1c; see also Bicanski and Burgess 2016). The activity packet is established by combining tonic background drive to the three rings, which are connected similarly to the fly network, albeit with different synaptic valence. Thus, the fly circuit (with its three copies of HD cells) could corroborate the search for a new canonical model of head direction for mammals—one that accounts for the persistence of head direction across the lifespan, is resistant to some neuronal death and synaptic variation, and is thus capable of degrading gracefully in aging to some degree. However, until more detailed data are available, all mammalian HD models must remain provisional with regard to connectivity. As mentioned above, this is not to undercut the vast amount of data that exists for rodents and other mammals. It merely recognizes that detailed connection patterns that determine function/computation remain largely unknown in these species.

To summarize, the HD system provides a beautiful example of successful cross talk across species and across methods. This is possible because, in Marr's terminology, similar algorithms and (to some degree) implementations are at play. In addition, comparing rodents and insects allows us to highlight what is very likely a clear difference between these phylogenetically distant species. While the HD system could be remarkably similar on many levels, the same may not be true for the neural infrastructure that underlies vector navigation. In the fly, the calculation of vectors appears to rely on a phasor code in which neural representations of sine waves with different phases (construed as consecutive readings of heading relative to landmarks), when summed, generate an estimate of the vector to a goal (cf. Figure 5.3a; see also Lyu et al. 2022). For instance, this model is able to update a vector pointing to a goal (e.g., to home) during ongoing navigation. Whether vectors can be computed from memory in the fly brain remains, however, unknown. Conversely in rodents, a promising theory of vector navigation (that allows for interactions with spatial memory) posits that it is supported by grid cells (Hafting et al. 2005; see also

Ginosar et al. 2023). A goal vector can be decoded, given two grid cell population vectors for the current and target position (Bush et al. 2015; Edvardsen 2015, 2017, 2018, 2019; Stemmler et al. 2015). Decoding can be implemented as a nested code (Edvardsen 2015; Stemmler et al. 2015), where smaller spatial scales refine coarse location estimates based on larger grid scales, or as a combinatorial code (Bush et al. 2015; Fiete et al. 2008), where grid cells could implement a residue number system. Goal vectors could be read out by fairly simple network architectures (e.g., see Figure 5.3b and Bush et al. 2015; Edvardsen 2015). Alternative proposals also suggest that candidate trajectories toward a target ("look-ahead probes") could be sampled via grid cells (Erdem and Hasselmo 2012).

This vector navigation system could also potentially be at play in the primate visual system, providing a benefit not only in the spatial domain. Direct (Killian et al. 2012) and indirect evidence (Nau et al. 2018; Staudigl et al. 2018) suggests that grid cells play a role in vision, possibly mapping visual space, and it has been proposed that vector navigation based on grid cells could be employed to calculate eye-movement vectors from memory (see Figure 5.3c and Bicanski and Burgess 2019). This model suggests how complex arrays of visual features (or elements of a scene) can be anchored to a grid map, allowing gaze to be directed to specific locations in the visual field (from memory), including to occluded stimuli/features. If primate visual memory is supported by a grid code in this manner, this would constitute an interesting example of a Marr algorithm (grid code) for the computational problem of vector navigation being adapted to different cognitive domains (from planar environmental navigation to eye-movement navigation from memory) in phylogenetically closer (compared to flies) animals. It would also suggest that an existing (probably) conserved mammalian algorithm has been adapted to another domain of cognition, potentially reflecting the differences in niche between rats and primates. Naturally, this does not exclude the possibility that primates also employ grid cells in the spatial domain (Jacobs et al. 2013), thus raising an intriguing question of how the same grid cells could dynamically be adapted to different domains (e.g., visual vs. spatial). Another example of a rodent algorithm adapted to another mammalian species can be seen in models of head direction for bats (Finkelstein et al. 2015). Here, three-dimensional head direction has been suggested to be coded by a toroidal manifold and not a ring, bridging the literature on HD and grid cells, with the latter supporting the notion that a toroidal neural manifold underlies the observed properties of grid cells (McNaughton et al. 2006).

3 Humans and Large-Scale Models

The above examples explore how models can be applied across a range of species and suggest why we are justified to expect this approach to work based on evolutionary considerations. This notion is rendered more precise by adopting the framing of neural systems according to Marr. However, what these models have in common is that we have access to detailed (albeit incomplete) knowledge about neural representations and tuning curves (e.g., the representational format of head direction), *in conjunction with* more or less detailed connectivity knowledge and

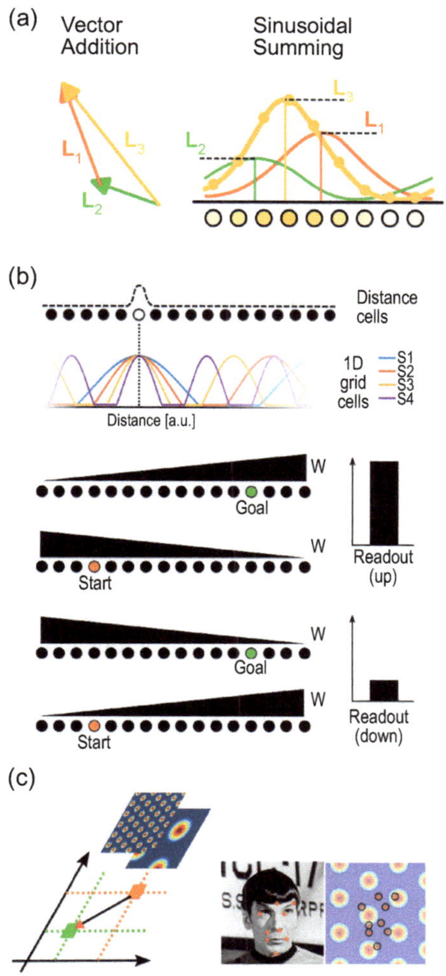

Figure 5.3 Vector calculus for navigation and path integration. (a) In flies (yellow vector), if the length of the heading vectors translates to the amplitude of a sinusoid (L_1, L_2, L_3) and the angle to the phase, adding such sinusoids (red and green) yields another sinusoid that represents the vector to the goal (yellow) (after Lyu et al. 2022). The vector estimate is indicated by the shaded neurons under the sinusoids, firing maximally for the phase of the yellow curve. (b) A simple vector navigation architecture using rodent grid cells (see Figure 5.1d): At any given position, some grid cells have a common phase (presupposing all phases are represented in the grid cell network), shown here for four modules (S1–4). Mapping grid cell population vectors to relative distance cells (black circles) allows for the readout of the relative distance (W, weight profiles to the readout neurons). The difference in firing rate between the readout neurons for "up" and "down" on the grid-sheet (here 1D), indicates the distance and direction to the target location. Additional, alternative implementations can be found in Bush et al. (2015). (c) Left: Repeating the architecture in (b) for two axes yields two-dimensional vector navigation. Such vectors can be computed from memory, given grid cell population vectors for start and goal locations. Thus, straight-line shortcuts can be computed in spatial navigation. Right: Salient features (red crosses) of complex visual stimuli are anchored to a visual grid cell network (at locations indicated on a single representative grid map) and can leverage the same vector calculation architectures to calculate eye-movement vectors from memory (Bicanski and Burgess 2019), potentially suggesting a domain transfer (from spatial navigation to vision) of grid cell-based algorithms in primates. Panels (b) and (c) were modified after Bicanski and Burgess (2019).

sometimes cellular and synaptic properties. Such combined knowledge about structure, function, and dimensionality of the neural representations of interest is often absent or limited in humans, at least at the level of detail needed to constrain and inspire neurocomputational models of biological systems. This poses a major challenge for the development of mechanistic models for humans, including the navigation system.

In most cases, human navigation behavior is studied with noninvasive recordings using tools such as fMRI or MEG, which yield activity distributions smeared out in time and space compared to spiking data from, for example, rodents. Direct single-cell recordings from epilepsy patients, although highly valuable, are rare and limited by the medically determined electrode placement and tasks that human patients can be asked to perform. Nevertheless, such experiments have revealed numerous spatiotemporal correlates, including grid and place codes (Ekstrom et al. 2003; Jacobs et al. 2010, 2013; Kunz et al. 2021; Schonhaut et al. 2023). How can we relate such results to models built for other model organisms, with access to detailed neuronal data? Conversely, can models help us tackle the increased complexity in humans, coupled with the decreased availability of constraining data?

To explore how we might conceptualize and address this increase in complexity, let us take a brief detour and complement Marr's ideas with those of the late neuroscientist Valentino Braitenberg. In his seminal book *Vehicles* (1986), Braitenberg describes thought experiments in which small automatons (the titular vehicles) are built with sensors that drive small motors. A single sensor, for instance, is able to detect light and increase the drive to a single motor. Such a vehicle would approach light sources but could not steer toward or away from them at an angle. Steering toward or away requires two motors, with crossed/parallel connections to two light sensors. More complex vehicles could have sensors in multiple modalities (e.g., heat, nutrients, oxygen, or light), and they monotonically increase or decrease the drive to a motor. When combining two motors and multiple sensors, vehicle behavior becomes much more interesting. A vehicle could, for instance, come to a stop in nutrient-rich parts of an environment (presumably to consume nutrients), move quickly through areas that are too hot, or approach sources of light. Braitenberg also introduced more complex functions for these sensor-motor mappings (akin to tuning curves). The upshot is that with as few as three or four sensors of varying valence and connectivity—some wired in parallel, others crossing when targeting the motors—vehicle behavior becomes highly complex, from an external perspective. If the internal connectivity (structure) is unknown, the behavior (function) may appear unpredictable, even before intermediate processing units (neurons) are included between the sensors and motors, i.e., absent memory, cognitive control, and concepts. Note that here "connectivity" refers to functional connectivity in brains. If at any given time only a subset of all connections is used, then connectomes alone do not suffice to understand the internal structure.

Braitenberg's vehicles provide a scientific fable, a tool for thinking about complexity and the study of brains and behavior. Just by observing the behavior we may overestimate the complexity of the underlying machinery. Yet, knowing how it is constructed (or building the system ourselves) allows us to understand its inner workings and the resultant behavior. Braitenberg called this "the law of uphill analysis and downhill invention," with quite obvious implications for how we can view models and experiments across species. Purely behavioral experimental neuroscience is the uphill analysis, whereas computational models are, quite literally, attempts at "invention," guided by data. Neural data that is noninvasively or directly recorded is necessary to constrain models and generate hypotheses about mechanisms. The results of all approaches—behavior, noninvasive, invasive, models—must somehow converge and align.

Fortunately, invasively acquired primate data (Buzsaki and Moser 2013; Ekstrom et al. 2003; Jacobs et al. 2010, 2013; Kunz et al. 2021; Mao et al. 2021) implies some similarities across mammals (e.g., Ekstrom and Isham 2017). This can be taken as a justification to extrapolate simpler models to humans and engage in constructive attempts at "downhill invention," à la Braitenberg, guided by evolutionary parallels. Importantly, noninvasive experiments corroborate the hypothesis that humans share at least some neural representations relevant to navigation with other mammals (Bellmund et al. 2016; Doeller et al. 2010; Horner et al. 2016; Shine et al. 2016, 2019). This suggests that a worthwhile starting point would be to use known mammalian models and try to adapt them to primates, including humans, necessary modifications notwithstanding. Just as early models of head direction in the rodent system have been developed in the absence of detailed connectivity knowledge in the LMN and DTN, we can suggest models for the human navigation system. Framing such models in Marr's terminology permits us to see how this process can be productive, despite guaranteed errors at the implementation level.

When knowledge about the details of neural properties and local connectivity within microcircuits is limited, what form could "downhill invention" models take? One possibility is to build large-scale models that are constrained by the systems-level interaction of multiple neural populations (neural representations), with individual model components focusing on the algorithmic level according to Marr, rather than the details of the implementation level. This has been attempted through the BB model (Bicanski and Burgess 2018), which relates numerous egocentric and allocentric spatial cells in one coherent framework. Although this model was primarily developed based on rodent data, it straddles the divide between humans and rodents and extrapolates to cognitive processes and pathologies in humans, such as the distinction between (usually co-occurring, though dissociable) representational and perceptual neglect (Becker and Burgess 2001, 2020; Byrne et al. 2007) or mental navigation (the ability to mentally explore a space in visuospatial imagery) (Bellmund et al. 2016; Bicanski and Burgess 2018; Horner et al. 2016). The same model predicted egocentric and allocentric object vector cells as well as HD-modulated versions of these neural responses. A predecessor of the BB model predicted egocentric, allocentric, and HD-modulated boundary vector cells (Becker and Burgess 2001; Byrne et al. 2007). Several of these cell types have subsequently been found (Alexander et al. 2020; Gofman et al. 2019; Hinman et al. 2019; Høydal et al. 2019). More recently, building on this framework, a neural model of episodic memory has been proposed (Bicanski 2025), following the long-standing notion that the neural machinery of spatial memory and navigation can be repurposed to code extended episodes (Buzsaki and Tingley 2018; Eichenbaum and Cohen 2014), with the help of temporal oscillations (Aghajan et al. 2023).

The strength of systems-level accounts lies in the fact that one is forced to specify interactions between multiple cell types, be they predicted or suggested by data.

These interactions constitute hypotheses about neural computation in their own right. For example, the boundary vector cell model of place cell firing (Hartley et al. 2000; O'Keefe and Burgess 1996) suggests some place cells are constructed from allocentric boundary inputs (Lever et al. 2009). In other words, a subset of place cell responses is "computed" from boundary information. Having to specify multiple interactions and computations of this type in one large coherent model increases the potential points of failure when confronted with data, making the model more susceptible to falsification, for example by observing changes due to lesions in the model. Combinatorically there are many more ways to get the interactions wrong than right. For the same reason, if such models are to some degree successful in accounting for experimental data and the generation of predictions, such models can increase our confidence in the hypothesized computations, algorithms, and neural representations, despite the relative lack of detailed knowledge about the implementation level. That is, large-scale models can generate interdependent predictions across brain areas that can be mapped to noninvasively acquired neural data and behavior. From Marr's perspective, this process can be fruitful even in the absence of detailed knowledge about the human brain at the level of precision found in the fly.

It should go without saying that such large-scale systems-level models must be complemented by detailed models to place them on firm footing. Ideally, we can establish a chain of correspondence between systems-level models and more focused models (progressively closer to the ground truth implementation level) that attempt to explain individual subsystems and precise temporal dynamics. The capabilities of large-scale models, and importantly their idiosyncratic errors, should trend in the direction of observed capabilities and errors of humans (or human navigators in the context of the present volume). Specifically, in the case of navigation, we can envision a large-scale model composed of multiple brain areas (Ekstrom et al. 2017; Epstein et al. 2017) to account for the transformation of sensory inputs to allocentric spatial and view-based representations, planning, trajectory coding in, for example parietal areas (Clark et al. 2018; Nitz 2006; Wilber et al. 2014), a cognitive map and vector navigation architecture, a reinforcement system for iterative learning and stimulus response navigation, action selection via the basal ganglia and prefrontal cognitive control (should I take a shortcut or follow landmarks?), and eventual motor output. Across species, individual aspects and some combinations have been modeled repeatedly (e.g., Chersi and Burgess 2015; Chersi and Pezzulo 2012; De Cothi et al. 2022; Dollé et al. 2010; Edvardsen et al. 2020; Erdem and Hasselmo 2012; Koene et al. 2003; Sheynikhovich et al. 2009; Stachenfeld et al. 2017; Stoianov et al. 2018; Yan et al. 2021), including models at the intersection with robotics (Arleo and Gerstner 2000; Burgess et al. 1997; Milford et al. 2004).

A large-scale systems-level model may, for instance, employ a simplified HD circuit (wrong at the implementation level, but correct at Marr's algorithmic level) to keep the large-scale model tractable. The resultant large-scale model would allow for systems-level interactions to be mapped onto human performance. The key conjecture is that the types of behaviors and errors that would be observed in these models will still trend in the same direction as observed in the real organism, despite the need for numerous simplifications. Most importantly, not only should these models allow a simulated agent to navigate successfully, they should also fail in the same manner as the biological system. That is, we can employ large-scale models as tools for research in which the failure modes match biology (Eliasmith 2013). As mentioned above, such large-scale models need to be supplemented by more focused models to explore more details as well as by models that make predictions about how the model activity manifests in noninvasive measures, such as macroscopic grid signals (Doeller et al. 2010; Bin Khalid et al. 2024). In this endeavor, all subdisciplines (e.g., work on flies, rodents, primates including humans, models) interact in a fruitful manner.

3.1 Individual Variation and Environmental Inputs

The systems-level perspective espoused above also lends itself to a meaningful appraisal of individual differences and the effect of environmental inputs. In terms of modeling, individual variation can be assessed much better at the systems level, because the neural processes underlying navigation depend on the interaction of multiple brain areas, each potentially subject to variation. Moreover, different processes (e.g., piloting and vector navigation) engage partially overlapping brain areas but not identical functional networks. Hence, we can envision simulating in the same systems-level model (with some parameter variation) an agent good at piloting and an agent poor at piloting who is a better path integrator (or vector navigator). This also requires models that degrade gracefully, harking back to the biologically plausible attractor for head direction in aging described above (cf. Figure 5.3). This offers an opportunity to study individual or group differences (e.g., young vs. old subjects, Bernard et al. 2025) on a mechanistic level. Such differences can be expressed as variations in the same large-scale architecture.

The same holds for the processing of environmental inputs. A systems-level account may allow us to compare the simulated processing of environmental inputs across multiple processing stages, acting on different navigation subsystems. In this context we must ask whether environmental inputs must be niche specific. More ecologically viable settings may be more conducive to furthering our understanding, especially if we take seriously the notion that evolution has imbued the neural systems of animals with niche-specific inductive biases. Simulations could, for instance, include more realistic environments with variable navigation cues, or environments cluttered with obstacles (Edvardsen et al. 2020; Yan et al. 2021).

3.2 Inductive Biases versus Blank Slates

Is all this talk of evolutionary relations, inherited algorithms, and common ancestors even necessary? Could the brain not just be a blank slate, with insect brains independently learning the same representations as mammalian brains (e.g., head direction) from scratch, because a ring attractor is the optimal solution for neural systems? Could head direction be learned in a blank slate network from experience due to some universal principle such as predictive coding (Huang and Rao 2011)? This would offer an alternative explanation for cross-species commonalities that does not rely on evolution per se, and harkens back to the age-old question of nature and nurture (Locke 1860).

The total absence of any evolutionarily inherited neural structure is, of course, a caricature. Even in learning problems where a match to biological nervous systems is more superficial (and not necessary when solving engineering problems), learning algorithms are still subject to initial assumptions, often referred to as inductive biases (Baxter 2000), which includes architectural constraints (Goyal and Bengio 2022; Zador 2019; Zhang et al. 2024). Still, suppose we let a blank slate architecture that is subject to minimal initial assumptions learn representations that resemble, for instance, HD cells. In terms of *biological* navigation, would such a model reveal something meaningful about brains and how they implement navigational computations? The technical achievement notwithstanding, such approaches face both conceptual and empirical hurdles.

Conceptually, it is not a given that the solution learned in an artificial network—with biologically implausible architectures, units, and learning rules—will produce a network that is informative (to an experimentalist) of the causal, neurobiological processes that underly similar behavior in biological brains, particularly not at the single-cell level. On the other hand (and in keeping with Marr's terminology), one could view such models as offering another algorithmic solution that may share population-level phenomena with brains.

Empirically, insects with short life spans may have little to no time to learn networks (e.g., ring attractors) from experience. This suggests that networks such as ring attractors are likely prewired in development (at least some noisy initial version), and that this developmental setup must be the product of evolution. This notion is strongly supported by the finding that rat pups show prototypical HD coding before eye-opening (Tan et al. 2015), and that the toroidal manifold, which is thought to support grid cell firing, is present in firing rate correlations before grid cells manifest in recordings (Gardner et al. 2022). Thus, across species, we have strong reasons to believe in some pre-wiring: as outlined repeatedly in the literature (e.g., see Zador 2019), animal brains are subject to powerful inductive biases. If useful across ecological niches, these biases should persist throughout evolution and thus explain why we find the same or similar neural algorithms across animals. Despite these empirical findings, we can nevertheless acknowledge that it would be beneficial for brains to be general computational devices, to the greatest extent possible, as this would confer additional adaptability on the neural system.

4 Conclusions

Properly framing models of neural systems can help us identify and understand commonalities in models of navigation systems across species, including humans. Modeling not only accounts for single species data, it can help bridge gaps across species. Marr's three levels of analysis provide a way to frame neural models, evolutionary continuity of neural systems, and complexity (Braitenberg 1986). The examples from rodents, insects, bats, and primates presented in this chapter illustrate the potential for productive dialogue as well as the commonalities and differences between species. I suggest that large-scale models can take us further and help bridge the gap to primate/human research. Such large-scale models have limitations when viewed in isolation, yet when properly construed (e.g., modeling both successful navigation *and* failures) and supported by a broad array of complementary research to link them to more detailed models across species, they can serve as powerful tools to advance mechanistic understanding of the human brain, in general, and navigation, in particular. This approach also provides a meaningful way to study individual differences through modeling. To improve translation across levels and place existing models on firmer ground, new measurement techniques may play a pivotal role: large-scale recordings in behaving monkeys and rodents (e.g., via neuropixel probes) would particularly be useful. Such recordings will undoubtedly support our efforts—and reveal their limits—when extrapolating from other organisms (including other primates) to the human system. However, there are caveats. At present, the generation of data outpaces the development of useful theories and models needed to explain them; "useful" in terms of being truly informative to guide experiments and for understanding brains and brain circuits. Large-scale recordings will intensify this imbalance, but they also provide an opportunity to advance systems-level models of the neural underpinnings of navigation to a new level.

Acknowledgments AB acknowledges funding from the Max Planck Society. AB also wishes to acknowledge all participants at this ES Forum. Numerous discussions helped in the revision of this manuscript.

References

Aghajan ZM, Kreiman G, and Fried I (2023) Minute-Scale Periodicity of Neuronal Firing in the Human Entorhinal Cortex. Cell Rep 42 (11):113271. https://doi.org/10.1016/j.celrep.2023.113271

Alexander AS, Carstensen LC, Hinman JR, et al. (2020) Egocentric Boundary Vector Tuning of the Retrosplenial Cortex. Sci Adv 6 (8):eaaz2322. https://doi.org/10.1126/sciadv.aaz2322

Arleo A, and Gerstner W (2000) Spatial Cognition and Neuro-Mimetic Navigation: A Model of Hippocampal Place Cell Activity. Biol Cybern 83 (3):287–299. https://doi.org/10.1007/s004220000171

Astur RS, Taylor LB, Mamelak AN, Philpott L, and Sutherland RJ (2002) Humans with Hippocampus Damage Display Severe Spatial Memory Impairments in a Virtual Morris Water Task. Behav Brain Res 132 (1):77–84. https://doi.org/10.1016/s0166-4328(01)00399-0

Bassett JP, Tullman ML, and Taube JS (2007) Lesions of the Tegmentomammillary Circuit in the Head Direction System Disrupt the Head Direction Signal in the Anterior Thalamus. J Neurosci 27 (28):7564–7577. https://doi.org/10.1523/jneurosci.0268-07.2007

Baxter J (2000) A Model of Inductive Bias Learning. J Artif Intell Res 12:149–198. https://arxiv.org/pdf/1106.0245

Becker S, and Burgess N (2001) Modelling Spatial Recall, Mental Imagery and Neglect. In: Leen T, Dietterich T, Tresp V (eds) Advances in Neural Information Processing Systems (NIPS 2000). MIT Press, Boston, pp 96–102

Bellmund JL, Deuker L, Navarro Schroder T, and Doeller CF (2016) Grid-Cell Representations in Mental Simulation. eLife 5:17089. https://doi.org/10.7554/eLife.17089

Bernard M, Shine J, Bicanski A et al. (2025) How Human Aging Disrupts the Head Direction Network: Evidence from VR Experiments and Mechanistic Models. bioRxiv https://www.biorxiv.org/content/10.1101/2025.11.24.690256v1

Bicanski A (2025) Breaking the Chains: Toward a Neural-Level Account of Episodic Memory. Psychol Rev Sept. 8. https://doi.org/10.1037/rev0000571

Bicanski A, and Burgess N (2016) Environmental Anchoring of Head Direction in a Computational Model of Retrosplenial Cortex. J Neurosci 36 (46):11601–11618. https://doi.org/10.1523/jneurosci.0516-16.2016

——— (2018) A Neural-Level Model of Spatial Memory and Imagery. eLife 7 (7052):1–3. https://doi.org/10.7554/elife.33752

——— (2019) A Computational Model of Visual Recognition Memory via Grid Cells. Curr Biol 29 (6):979–990. https://doi.org/10.1016/j.cub.2019.01.077

——— (2020) Neuronal Vector Coding in Spatial Cognition. Nat Rev Neurosci 21:453–470. https://doi.org/10.1038/s41583-020-0336-9

Bin Khalid I, Reifenstein ET, Auer N, Kunz L, and Kempter R (2024) Quantitative Modeling of the Emergence of Macroscopic Grid-Like Representations. eLife 13:e85742. https://doi.org/10.7554/eLife.85742

Bjelkmar P, Niemelä PS, Vattulainen I, and Lindahl E (2009) Conformational Changes and Slow Dynamics through Microsecond Polarized Atomistic Molecular Simulation of an Integral Kv1.2 Ion Channel. PLoS Comput Biol 5 (2):e1000289–e1000289. https://doi.org/10.1371/journal.pcbi.1000289

Blair HT, Cho J, and Sharp PE (1998) Role of the Lateral Mammillary Nucleus in the Rat Head Direction Circuit: A Combined Single Unit Recording and Lesion Study. Neuron 21 (6):1387–1397. https://doi.org/10.1016/s0896-6273(00)80657-1

——— (1999) The Anterior Thalamic Head-Direction Signal Is Abolished by Bilateral but Not Unilateral Lesions of the Lateral Mammillary Nucleus. J Neurosci 19 (15):6673–6683. https://doi.org/10.1523/jneurosci.19-15-06673.1999

Braitenberg V (1986) Vehicles: Experiments in Synthetic Psychology. MIT press, Boston

Burgess N, Donnett JG, Jeffery KJ, and O–keefe J (1997) Robotic and Neuronal Simulation of the Hippocampus and Rat Navigation. Philos Trans R Soc Lond B Biol Sci 352 (1360):1535–1543. https://doi.org/10.1098/rstb.1997.0140

Bush D, Barry C, Burgess N, et al. (2015) Using Grid Cells for Navigation. Neuron 87 (3):507–520. https://doi.org/10.1016/j.neuron.2015.07.006

Buzsaki G, and Moser EI (2013) Memory, Navigation and Theta Rhythm in the Hippocampal-Entorhinal System. Nat Neurosci 16 (2):130–138. https://doi.org/10.1038/nn.3304

Buzsaki G, and Tingley D (2018) Space and Time: The Hippocampus as a Sequence Generator. Trends Cogn Sci 22 (10):853–869. https://doi.org/10.1016/j.tics.2018.07.006

Byrne P, Becker S, and Burgess N (2007) Remembering the Past and Imagining the Future: A Neural Model of Spatial Memory and Imagery. Psychol Rev 114 (2):340–375. https://doi.org/10.1037/0033-295X.114.2.340

Chersi F, and Burgess N (2015) The Cognitive Architecture of Spatial Navigation: Hippocampal and Striatal Contributions. Neuron 88 (1):64–77. https://doi.org/10.1016/j.neuron.2015.09.021

Chersi F, and Pezzulo G (2012) Using Hippocampal-Striatal Loops for Spatial Navigation and Goal-Directed Decision-Making. Cogn Process 13:125–129. https://doi.org/10.1007/s10339-012-0475-7

Cho J, and Sharp PE (2001) Head Direction, Place, and Movement Correlates for Cells in the Rat Retrosplenial Cortex. Behav Neurosci 115 (1):3–25. https://doi.org/10.1037/0735-7044.115.1.3

Clark BJ, Simmons CM, Berkowitz LE, and Wilber AA (2018) The Retrosplenial-Parietal Network and Reference Frame Coordination for Spatial Navigation. Behav Neurosci 132 (5):416–429. https://doi.org/10.1037/bne0000260

De Cothi W, Nyberg N, Griesbauer E-M, et al. (2022) Predictive Maps in Rats and Humans for Spatial Navigation. Curr Biol 32 (17):3676–3689. https://doi.org/10.1016/j.cub.2022.06.090

Dobzhansky T (1973) Nothing in Biology Makes Sense except in the Light of Evolution. Am Biol Teach 35 (3):125–129. https://doi.org/10.2307/4444260

Doeller CF, Barry C, and Burgess N (2010) Evidence for Grid Cells in a Human Memory Network. Nature 463 (7281):657–661. https://doi.org/10.1038/nature08704

Dollé L, Sheynikhovich D, Girard B, Chavarriaga R, and Guillot A (2010) Path Planning versus Cue Responding: A Bio-Inspired Model of Switching between Navigation Strategies. Biol Cybern 103 (4):299–317. https://doi.org/10.1007/s00422-010-0400-z

Edvardsen V (2015) A Passive Mechanism for Goal-Directed Navigation Using Grid Cells. In: ECAL 2015: The 13th European Conference on Artificial Life. MIT Press, Boston, pp 191–198

——— (2017) Long-Range Navigation by Path Integration and Decoding of Grid Cells in a Neural Network. In: 2017 International Joint Conference on Neural Networks (IJCNN). New York, IEEE, pp 4348–4355

——— (2018) Navigating with Distorted Grid Cells. In: ALIFE 2018: The 2018 Conference on Artificial Life. MIT Press, Boston, pp 260–267

——— (2019) Goal-Directed Navigation Based on Path Integration and Decoding of Grid Cells in an Artificial Neural Network. Nat Comput 18:13–27. https://doi.org/10.1007/s11047-016-9575-0

Edvardsen V, Bicanski A, and Burgess N (2020) Navigating with Grid and Place Cells in Cluttered Environments. Hippocampus 30 (3):220–232. https://doi.org/10.1002/hipo.23147

Eichenbaum H, and Cohen NJ (2014) Can We Reconcile the Declarative Memory and Spatial Navigation Views on Hippocampal Function? Neuron 83 (4):764–770. https://doi.org/10.1016/j.neuron.2014.07.032

Ekstrom AD, Caplan JB, Fields TA, et al. (2003) Cellular Networks Underlying Human Spatial Navigation. Nature 425:184–188. https://doi.org/10.1038/nature01964

Ekstrom AD, Huffman DJ, and Starrett M (2017) Interacting Networks of Brain Regions Underlie Human Spatial Navigation: A Review and Novel Synthesis of the Literature. J Neurophysiol 118 (6):3328–3344. https://doi.org/10.1152/jn.00531.2017

Ekstrom AD, and Isham EA (2017) Human Spatial Navigation: Representations across Dimensions and Scales. Curr Opin Behav Sci 17:84–89. https://doi.org/10.1016/j.cobeha.2017.06.005

Eliasmith C (2013) How to Build a Brain: A Neural Architecture for Biological Cognition. Oxford Univ Press, New York

Epstein RA, Patai EZ, Julian JB, and Spiers HJ (2017) The Cognitive Map in Humans: Spatial Navigation and Beyond. Nat Neurosci 20 (11):1504–1513. https://doi.org/10.1038/nn.4656

Erdem UM, and Hasselmo M (2012) A Goal-Directed Spatial Navigation Model Using Forward Trajectory Planning Based on Grid Cells. Eur J Neurosci 35 (6):916–931. https://doi.org/10.1111/j.1460-9568.2012.08015.x

Fiete I, Burak Y, and Brookings T (2008) What Grid Cells Convey about Rat Location. J Neurosci 28 (27):6858–6871. https://doi.org/10.1523/jneurosci.5684-07.2008

Finkelstein A, Derdikman D, Rubin A, et al. (2015) Three-Dimensional Head-Direction Coding in the Bat Brain. Nature 517 (7533):159–164. https://doi.org/10.1038/nature14031

Gardner RJ, Hermansen E, Pachitariu M, et al. (2022) Toroidal Topology of Population Activity in Grid Cells. Nature 602 (7895):123–128. https://doi.org/10.1038/s41586-021-04268-7

Ginosar G, Aljadeff J, Las L, Derdikman D, and Ulanovsky N (2023) Are Grid Cells Used for Navigation? On Local Metrics, Subjective Spaces, and Black Holes. Neuron 111 (12):1858–1875. https://doi.org/10.1016/j.neuron.2023.03.027

Giraldo YM, Leitch KJ, Ros IG, et al. (2018) Sun Navigation Requires Compass Neurons in Drosophila. Curr Biol 28 (17):2845–2852. https://doi.org/10.1016/j.cub.2018.07.002
Gofman X, Tocker G, Weiss S, et al. (2019) Dissociation between Postrhinal Cortex and Downstream Parahippocampal Regions in the Representation of Egocentric Boundaries. Curr Biol 29 (16):2751–2757.e2754. https://doi.org/10.1016/j.cub.2019.07.007
Goyal A, and Bengio Y (2022) Inductive Biases for Deep Learning of Higher-Level Cognition. Proc Math Phys Eng Sci 478 (2266):20210068–20210068. https://doi.org/10.1098/rspa.2021.0068
Green J, Adachi A, Shah KK, et al. (2017) A Neural Circuit Architecture for Angular Integration in *Drosophila*. Nature 546 (7656):101–106. https://doi.org/10.1038/nature22343
Green J, Vijayan V, Mussells Pires P, Adachi A, and Maimon G (2019) A Neural Heading Estimate Is Compared with an Internal Goal to Guide Oriented Navigation. Nat Neurosci 22 (9):1460–1468. https://doi.org/10.1038/s41593-019-0444-x
Grillner S, and Wallen P (1985) Central Pattern Generators for Locomotion, with Special Reference to Vertebrates. Annu Rev Neurosci 8 (1):233–261. https://doi.org/10.1146/annurev.ne.08.030185.001313
Hafting T, Fyhn M, Molden S, Moser M-B, and Moser EI (2005) Microstructure of a Spatial Map in the Entorhinal Cortex. Nature 436 (7052):801–806. https://doi.org/10.1038/nature03721
Hartley T, Burgess N, Lever C, Cacucci F, and O'Keefe J (2000) Modeling Place Fields in Terms of the Cortical Inputs to the Hippocampus. Hippocampus 10 (4):369–379. https://doi.org/10.1002/1098-1063(2000)10:4%3C369::aid-hipo3%3E3.0.co;2-0
Hinman JR, Chapman GW, and Hasselmo ME (2019) Neuronal Representation of Environmental Boundaries in Egocentric Coordinates. Nat Commun 10 (1):1–8. https://doi.org/10.1038/s41467-019-10722-y
Horner AJ, Bisby JA, Zotow E, Bush D, and Burgess N (2016) Grid-Like Processing of Imagined Navigation. Curr Biol 26 (6):842–847. https://doi.org/10.1016/j.cub.2016.01.042
Howett D, Castegnaro A, Krzywicka K, et al. (2019) Differentiation of Mild Cognitive Impairment Using an Entorhinal Cortex-Based Test of Virtual Reality Navigation. Brain 142 (6):1751–1766. https://doi.org/10.1093/brain/awz116
Høydal ØA, Skytøen ER, Andersson SO, Moser MB, and Moser EI (2019) Object-Vector Coding in the Medial Entorhinal Cortex. Nature 568:400–404. https://doi.org/10.1038/s41586-019-1077-7
Huang Y, and Rao RPN (2011) Predictive Coding. Wiley Interdiscip Rev Cogn Sci 2 (5):580–593. https://doi.org/10.1002/wcs.142
Jacob P-Y, Casali G, Spieser L, et al. (2017) An Independent, Landmark-Dominated Head-Direction Signal in Dysgranular Retrosplenial Cortex. Nat Neurosci 20 (2):173–175. https://doi.org/10.1038/nn.4465
Jacobs J, Kahana MJ, Ekstrom AD, Mollison MV, and Fried I (2010) A Sense of Direction in Human Entorhinal Cortex. PNAS 107 (14):6487–6492. https://doi.org/10.1073/pnas.0911213107
Jacobs J, Weidemann CT, Miller JF, et al. (2013) Direct Recordings of Grid-Like Neuronal Activity in Human Spatial Navigation. Nat Neurosci 16 (9):1188–1190. https://doi.org/10.1038/nn.3466
Jarvis ED, Güntürkün O, Bruce L, et al. (2005) Avian Brains and a New Understanding of Vertebrate Brain Evolution. Nat Rev Neurosci 6 (2):151–159. https://doi.org/10.1038/nrn1606
Killian NJ, Jutras MJ, and Buffalo EA (2012) A Map of Visual Space in the Primate Entorhinal Cortex. Nature 491 (7426):761–764. https://doi.org/10.1038/nature11587
Kim SS, Rouault H, Druckmann S, and Jayaraman V (2017) Ring Attractor Dynamics in the Drosophila Central Brain. Science 356 (6340):849–853. https://doi.org/10.1126/science.aal4835

Koene RA, Gorchetchnikov A, Cannon RC, and Hasselmo ME (2003) Modeling Goal-Directed Spatial Navigation in the Rat Based on Physiological Data from the Hippocampal Formation. Neural Netw 16 (5-6):577–584. https://doi.org/10.1016/s0893-6080(03)00106-0

Kunz L, Brandt A, Reinacher PC, et al. (2021) A Neural Code for Egocentric Spatial Maps in the Human Medial Temporal Lobe. Neuron 109 (17):2781–2796.e2710. https://doi.org/10.1016/j.neuron.2021.06.019

Lever C, Burton S, Jeewajee A, O'Keefe J, and Burgess N (2009) Boundary Vector Cells in the Subiculum of the Hippocampal Formation. J Neurosci 29 (31):9771–9777. https://doi.org/10.1523/jneurosci.1319-09.2009

Locke J (1860) An Essay Concerning Human Understanding: And a Treatise on the Conduct of the Understanding. Hayes & Zell, Philadelphia

Lyu C, Abbott LF, and Maimon G (2022) Building an Allocentric Travelling Direction Signal via Vector Computation. Nature 601 (7891):92–97. https://doi.org/10.1038/s41586-021-04067-0

Mao D, Avila E, Caziot B, et al. (2021) Spatial Modulation of Hippocampal Activity in Freely Moving Macaques. Neuron 109 (21):3521–3534.e3526. https://doi.org/10.1016/j.neuron.2021.09.032

Marr DC, and Poggio T (1976) From Understanding Computation to Understanding Neural Circuitry. Neurosci Res Program Bull May 1976:1–22. https://dspace.mit.edu/bitstream/handle/1721.1/5782/AIM-357.pdf

McNaughton BL, Battaglia FP, Jensen O, Moser EI, and Moser M-B (2006) Path Integration and the Neural Basis of the 'Cognitive Map'. Nat Rev Neurosci 7 (8):663–678. https://doi.org/10.1038/nrn1932

Milford MJ, Wyeth GF, and Prasser D (2004) RatSLAM: A Hippocampal Model for Simultaneous Localization and Mapping. In: IEEE International Conference on Robotics and Automation. IEEE, New York, pp 403–408

Morris R (1984) Developments of a Water-Maze Procedure for Studying Spatial Learning in the Rat. J Neurosci Methods 11 (1):47–60. https://doi.org/10.1016/0165-0270(84)90007-4

Nau M, Navarro Schröder T, Bellmund JLS, and Doeller CF (2018) Hexadirectional Coding of Visual Space in Human Entorhinal Cortex. Nat Neurosci 21 (2):188–190. https://doi.org/10.1038/s41593-017-0050-8

Nitz DA (2006) Tracking Route Progression in the Posterior Parietal Cortex. Neuron 49 (5):747–756. https://doi.org/10.1016/j.neuron.2006.01.037

O'Keefe J (1976) Place Units in the Hippocampus of the Freely Moving Rat. Exp Neurol 51 (1):78–109. https://doi.org/10.1016/0014-4886(76)90055-8

O'Keefe J, and Burgess N (1996) Geometric Determinants of the Place Fields of Hippocampal Neurons. Nature 381 (6581):425–428. https://doi.org/10.1038/381425a0

Peyrache A, Duszkiewicz AJ, Viejo G, and Angeles-Duran S (2019) Thalamocortical Processing of the Head-Direction Sense. Prog Neurobiol 183:101693. https://doi.org/10.1016/j.pneurobio.2019.101693

Pisokas I, Heinze S, and Webb B (2020) The Head Direction Circuit of Two Insect Species. eLife 9:1–49. https://doi.org/10.7554/elife.53985

Redish AD, Elga AN, and Touretzky DS (1996) A Coupled Attractor Model of the Rodent Head Direction System. Network 7 (4):671–671. https://www.princeton.edu/~adame/papers/coupled-attractor/coupled-attractor.pdf

Reznik D, Majka P, Rosa MGP, Witter MP, and Doeller CF (2024) Phylogeny of Neocortical-Hippocampal Projections Provides Insight in the Nature of Human Memory. eLife 13:RP99203. https://doi.org/10.7554/eLife.99203.1

Rolls ET, Robertson RG, and Georges-François P (1997) Spatial View Cells in the Primate Hippocampus. Eur J Neurosci 9 (8):1789–1794. https://doi.org/10.1111/j.1460-9568.1997.tb01538.x

Salis F (2021) The New Fiction View of Models. Br J Philos Sci 72 (3):717–742. https://doi.org/10.1093/bjps/axz015

Schonhaut DR, Aghajan ZM, Kahana MJ, and Fried I (2023) A Neural Code for Time and Space in the Human Brain. Cell Rep 42 (11):113238. https://doi.org/10.1016/j.celrep.2023.113238

Sheynikhovich D, Chavarriaga R, Strösslin T, Arleo A, and Gerstner W (2009) Is There a Geometric Module for Spatial Orientation? Insights from a Rodent Navigation Model. Psychol Rev 116 (3):540–540. https://doi.org/10.1037/a0016170

Shine JP, Valdés-Herrera JP, Hegarty M, and Wolbers T (2016) The Human Retrosplenial Cortex and Thalamus Code Head Direction in a Global Reference Frame. J Neurosci 36 (24):6371–6381. https://doi.org/10.1523/JNEUROSCI.1268-15.2016

Shine JP, Valdés-Herrera JP, Tempelmann C, and Wolbers T (2019) Evidence for Allocentric Boundary and Goal Direction Information in the Human Entorhinal Cortex and Subiculum. Nat Commun 10 (1):4004. https://doi.org/10.1038/s41467-019-11802-9

Song P, and Wang X-J (2005) Angular Path Integration by Moving "Hill of Activity": A Spiking Neuron Model without Recurrent Excitation of the Head-Direction System. J Neurosci 25 (4):1002–1014. https://doi.org/10.1523/jneurosci.4172-04.2005

Stachenfeld KL, Botvinick MM, and Gershman SJ (2017) The Hippocampus as a Predictive Map. Nat Neurosci 20 (11):1643–1653. https://doi.org/10.1038/nn.4650

Staudigl T, Leszczynski M, Jacobs J, et al. (2018) Hexadirectional Modulation of High-Frequency Electrophysiological Activity in the Human Anterior Medial Temporal Lobe Maps Visual Space. Curr Biol 28 (20):3325–3329.e3324. https://doi.org/10.1016/j.cub.2018.09.035

Stemmler M, Mathis A, and Herz AVM (2015) Connecting Multiple Spatial Scales to Decode the Population Activity of Grid Cells. Sci Adv 1 (11):e1500816–e1500816. https://doi.org/10.1126/science.1500816

Stoianov IP, Pennartz CMA, Lansink CS, and Pezzulo G (2018) Model-Based Spatial Navigation in the Hippocampus-Ventral Striatum Circuit: A Computational Analysis. PLoS Comput Biol 14 (9):e1006316–e1006316. https://doi.org/10.1371/journal.pcbi.1006316

Stringer SM, Trappenberg TP, Rolls ET, and Araujo I (2002) Self-Organizing Continuous Attractor Networks and Path Integration: One-Dimensional Models of Head Direction Cells. Network 13 (2):217–242. https://pubmed.ncbi.nlm.nih.gov/12061421/

Tan HM, Bassett JP, O'Keefe J, Cacucci F, and Wills TJ (2015) The Development of the Head Direction System before Eye Opening in the Rat. Curr Biol 25 (4):479–483. https://doi.org/10.1016/j.cub.2014.12.030

Taube JS (2007) The Head Direction Signal: Origins and Sensory-Motor Integration. Annu Rev Neurosci 30:181–207. https://doi.org/10.1146/annurev.neuro.29.051605.112854

Taube JS, Muller RU, and Ranck JB (1990) Head-Direction Cells Recorded from the Postsubiculum in Freely Moving Rats. II. Effects of Environmental Manipulations. J Neurosci 10 (2):436–447. https://doi.org/10.1523/jneurosci.10-02-00436.1990

Vafidis P, Owald D, D'Albis T, and Kempter R (2022) Learning Accurate Path Integration in Ring Attractor Models of the Head Direction System. eLife 11:e69841. https://doi.org/10.7554/eLife.69841

Vinepinsky E, Cohen L, Perchik S, et al. (2020) Representation of Edges, Head Direction, and Swimming Kinematics in the Brain of Freely-Navigating Fish. Sci Rep 10 (1):14762–14762. https://doi.org/10.1038/s41598-020-71217-1

Wilber AA, Clark BJ, Forster TC, Tatsuno M, and McNaughton BL (2014) Interaction of Egocentric and World-Centered Reference Frames in the Rat Posterior Parietal Cortex. J Neurosci 34 (16):5431–5446. https://doi.org/10.1523/JNEUROSCI.0511-14.2014

Xie X, Hahnloser RHR, and Seung HS (2002) Double-Ring Network Model of the Head-Direction System. Phys Rev E 66 (4):41902–41902. https://doi.org/10.1103/physreve.66.041902

Yan Y, Burgess N, and Bicanski A (2021) A Model of Head Direction and Landmark Coding in Complex Environments. PLoS Comput Biol 17 (9):e1009434–e1009434. https://doi.org/10.1371/journal.pcbi.1009434

Ying J, Keinath AT, Lavoie R, et al. (2022) Disruption of the Grid Cell Network in a Mouse Model of Early Alzheimer's Disease. Nat Commun 13 (1):886–886. httmaps://doi.org/10.1038/s41467-022-28551-x

Zador AM (2019) A Critique of Pure Learning and What Artificial Neural Networks Can Learn from Animal Brains. Nat Commun 10:3770. https://doi.org/10.1038/s41467-019-11786-6

Zhang K (1996) Representation of Spatial Orientation by the Intrinsic Dynamics of the Head-Direction Cell Ensemble: A Theory. J Neurosci 16 (6):2112–2126. https://doi.org/10.1523/jneurosci.16-06-02112.1996

Zhang R, Pitkow X, and Angelaki DE (2024) Inductive Biases of Neural Network Modularity in Spatial Navigation. Sci Adv 10 (29):eadk1256. https://doi.org/10.1126/sciadv.adk1256

6

Do Data from Intracranial Techniques Align (or Not) with What We Have Learned about Human Navigation from fMRI?

Philipp K. Büchel and Lukas Kunz

Abstract Spatial navigation is a fundamental behavior in humans and many other species. Our understanding of the neural mechanisms underlying spatial navigation stems primarily from electrophysiological studies in animals, which have revealed diverse spatial representations at the level of individual neurons, including those of places, directions, boundaries, and landmarks. In humans, research into the neural basis of navigation has largely relied on noninvasive functional magnetic resonance imaging (fMRI) and, to a lesser extent, invasive intracranial recordings, which capture local field potentials and single-neuron activity. This chapter addresses how neural correlates of human navigation, observed through noninvasive fMRI, align with those obtained via invasive recordings. Discussion focuses on grid and scene representations in entorhinal and parahippocampal cortices, respectively, which show consistency across invasive and noninvasive techniques. In contrast, boundary and landmark representations in the entorhinal, parahippocampal, and retrosplenial cortices have been studied using fMRI but rarely with invasive recordings. Finally, several invasive studies have observed place representations in the hippocampus, but it remains uncertain whether noninvasive methods can reliably capture such representations, given their sparse and nontopographical nature. Future research that integrates simultaneous noninvasive and invasive recordings in animal models, sequential noninvasive and invasive recordings in humans, and computational modeling is needed to advance our understanding of how spatial neural representations observed through different neuroscientific methods relate to each other.

Lukas Kunz (✉)
Department of Epileptology, University Hospital Bonn, 53127 Bonn, Germany
Email: Lukas.Kunz@ukbonn.de

Philipp K. Büchel
Department of Epileptology, University Hospital Bonn, 53127 Bonn, Germany

N. S. Newcombe and K. Cheng (eds.), *Challenges in Navigation Research*, Strüngmann Forum Reports, https://doi.org/10.1007/978-3-032-20563-6_6

Keywords Human single-neuron recording, intracranial EEG, functional magnetic resonance imaging, place cell, grid cell, spatial representation

1 Introduction

In humans and many other species, spatial navigation is a vital and complex behavior that requires a highly advanced interplay of numerous brain mechanisms across various brain regions (Ekstrom et al. 2018). Electrophysiological recordings in rodents were instrumental in identifying important neural mechanisms of spatial navigation, including the discoveries of place cells (O'Keefe and Dostrovsky 1971), grid cells (Hafting et al. 2005), and head-direction cells (Taube et al. 1990) in different regions of the hippocampal formation. These cell types provide information about the animal's position and orientation in the environment (Moser et al. 2017) and have inspired influential models on the neural basis of spatial navigation and associated behaviors (e.g., Bicanski and Burgess 2018).

In humans, progress in understanding the neural correlates of spatial navigation has mostly relied on noninvasive fMRI in healthy adults. Guided by the observation of specialized cell types in rodent navigation studies, these fMRI studies have led to major discoveries such as the parahippocampal place area (Epstein and Kanwisher 1998), a macroscopic correlate of grid-cell activity in the entorhinal cortex (Doeller et al. 2010), and neural representations of future goal locations in the hippocampus (Brown et al. 2016). Given its excellent coverage of the human brain, fMRI has enabled insights into the broad distribution of spatial representations across a variety of brain regions, suggesting, for example, that grid representations are present in entorhinal as well as prefrontal cortices (Doeller et al. 2010). Yet fMRI also has disadvantages, including its relatively low spatiotemporal resolution and indirect way of measuring neural activity through the blood-oxygen-level-dependent (BOLD) response (Huettel et al. 2014). Moreover, each voxel typically averages signals from many thousands of neurons, which can mask functionally distinct cell populations, particularly in regions where neurons with similar properties do not cluster but are more randomly interspersed, such as place cells in the hippocampus (Redish et al. 2001). Hence, it often remains uncertain how navigation-related fMRI findings correlate with neural processes at the level of small neuronal networks and individual neurons.

Intracranial recordings of local field potentials (LFPs) and single neurons may help ground fMRI observations in directly measured neural activity from within the human brain. While these direct human brain recordings are only possible in patients with epilepsy (and some other patient groups) in a limited number of brain regions defined clinically, the resulting intracranial recordings provide excellent temporal and spatial resolution, up to the identification of action potentials of individual neurons. These recordings thus operate at a temporal resolution of milliseconds and a spatial scale of milli- to micrometers—several magnitudes more precise than the spatiotemporal resolution of fMRI, where voxels are usually between 1–27 mm^3 in size, and BOLD activity is typically recorded at acquisition windows of 1–3 sec (Huettel et al. 2014). Human intracranial recordings thus provide an opportunity to bridge the gap between human fMRI studies and electrophysiological studies in rodents, potentially explaining the nature and underlying mechanisms of navigation-related neural phenomena observed with fMRI (for further discussion, see Section 3.2).

Here, we discuss to what extent recent findings from intracranial recordings align with what we have learned about human navigation from fMRI research. In general, observing such alignment is highly useful because cross-methodological replication increases our confidence in proposed functions of brain regions, provides insights into the mechanistic basis of observed fMRI phenomena, and enables a broader investigation of neural representations across diverse populations and tasks. Conversely, nonalignment may prompt new investigations into the robustness of results obtained with only one method. It is important to note, however, that whenever intracranial and fMRI recordings do not align, we must consider how fundamental differences in the nature and sensitivity of these approaches may have limited one measure's ability to detect the effect. For example, not being able to detect place representations in the hippocampus would not call into question the discovery of place cells in the rodent hippocampus but rather point to distributions of hippocampal place cells that are nontopographical, and thus undetectable at the macroscopic level of fMRI. Furthermore, nonalignment may result because different methods capture neural activity related to the input or output of a brain region. More specifically, spiking activity reflects primarily a region's output, whereas LFPs and fMRI activity are primarily related to a region's input and local processing (Buzsáki 2004; Buzsáki et al. 2012; Einevoll et al. 2013; Logothetis et al. 2001). Detecting specific types of neural representations using only one technique could thus provide important insights into their dependence on input and output signals. Overall, the alignment and nonalignment of discoveries from intracranial techniques versus fMRI will yield important knowledge about the working principles of the (human) brain and the relationships between different neural recording techniques.

Due to the vast amount of available fMRI literature (Ekstrom et al. 2018), our focus here is on a handful of specialized neural representations in specific brain regions that are considered fundamental for spatial navigation (Figure 6.1). Specifically, we discuss place representations in the hippocampus and grid representations in the entorhinal cortex, as they are among the best-studied neural phenomena and are considered key components of the neural basis of cognitive maps that enable flexible navigation (Behrens et al. 2018; Epstein et al. 2017; Moser et al. 2017). In addition, we address landmark and boundary representations in retrosplenial, parahippocampal, and entorhinal cortices, as they contribute input to place and grid representations (O'Keefe and Burgess 1996; Pérez-Escobar et al. 2016) and may support important spatial behaviors, such as landmark-based navigation. We also consider parahippocampal scene representations, which are one of the most robust neural signals observed with fMRI and may help individuals recognize the geometric layout of environments. Our approach is selective, but we believe that it enables direct comparisons between the different methods and can be useful in suggesting trajectories for future research to pursue. We propose that noninvasive fMRI and invasive neural recordings offer unique yet complementary perspectives on the brain mechanisms that underpin human navigation and suggest that a deeper understanding of their interrelation could advance our understanding of human brain function.

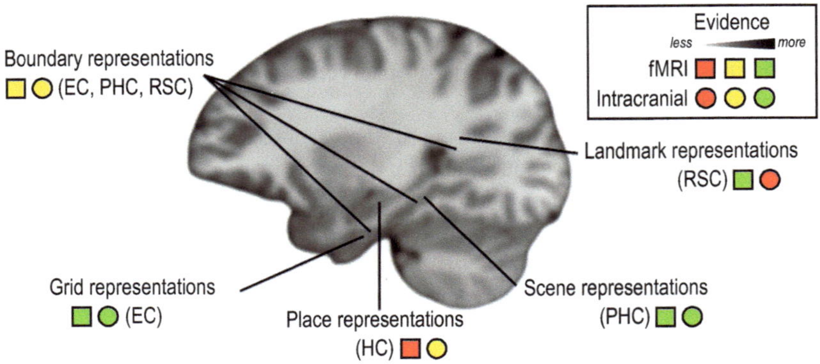

Figure 6.1 Five specialized types of neural representations that are fundamental for spatial navigation: grid representations in the entorhinal cortex (EC), place representations in the hippocampus (HC), scene representations in the parahippocampal cortex (PHC), landmark representations in the retrosplenial cortex (RSC), and boundary representations in the entorhinal, parahippocampal, and retrosplenial cortices. Rectangles indicate evidence derived from fMRI studies; circles represent evidence from intracranial recordings. Based on our subjective assessment, evidentiary strength has been color coded: green denotes robust, yellow indicates moderate, and red signifies little or no evidence.

2 Fundamental Neural Representations for Spatial Navigation

2.1 Grid Representations in the Entorhinal Cortex

Originally discovered in the entorhinal cortex of rodents, grid cells are neurons that activate at multiple locations in space and are arranged in a hexagonal grid pattern (Hafting et al. 2005). Together with place cells (O'Keefe and Dostrovsky 1971), grid cells may be important for self-localization and may support a variety of spatial behaviors, including goal-directed navigation, simulation of future paths, and spatial memory (Moser et al. 2015). In addition, grid cells are believed to support vector computations in the service of path integration and abstract reasoning (Behrens et al. 2018; Bush et al. 2015; Stemmler et al. 2015), but their exact functional roles still remain unclear.

Grid cells are commonly characterized by three major properties: orientation of their grid axes relative to the external environment, spacing between neighboring grid fields, and phase offset of the grid relative to a reference point. Other interesting properties include their modulation by head direction (Sargolini et al. 2006), distortion of their grid patterns due to environmental geometry (Krupic et al. 2015), and shifts in their firing fields toward spatial goals (Boccara et al. 2019). Furthermore, grid cells are organized in several modules that span the anterior-posterior axis of the entorhinal cortex, with grid cells in posterior parts showing increasingly smaller distances between neighboring firing fields as well as smaller field sizes (Stensola et al. 2012). The unique grid pattern of grid cells may result from their organization in recurrently connected continuous attractor networks (Gardner et al. 2022).

Multiple fMRI studies have observed a potential correlate of grid-cell activity in human participants as they navigate through computer-based virtual environments and conceptual spaces (Behrens et al. 2018; Bellmund et al. 2018; Bin Khalid et al. 2024; Kunz et al. 2019). These virtual environments consist, for example, of a grassy plain surrounded by a circular boundary, with distal cues for allocentric orientation (Doeller et al. 2010; Doeller et al. 2008; Kunz et al. 2015). Conceptual spaces can be created by combining any sort of continuous, orthogonal dimensions, such as two independent features of birds (Constantinescu et al. 2016). Typically, participants learn the locations of objects within these virtual environments and conceptual spaces and navigate through them using game controllers.

During data analysis, the putative grid-cell correlate manifests as a hexadirectional modulation of fMRI activity as a function of participants' movement direction through space (Bin Khalid et al. 2024; Doeller et al. 2010). The observation of this hexadirectional modulation at the macroscopic level of fMRI voxels is both fascinating and surprising, given that fMRI voxels aggregate signals from large populations of neurons. Prior research in rodents demonstrated that neighboring grid cells tend to have different grid-phase offsets (Hafting et al. 2005; see also Heys et al. 2014 and Gu et al. 2018), resulting in a similar number of grid fields being traversed along any movement direction. More specifically, when the firing maps of multiple adjacent grid cells—all sharing similar spacing, field size, and orientation but differing in their grid-phase offset—are superimposed, their combined activity uniformly covers the entire environment. In principle, this uniform coverage should preclude any hexadirectional modulation of fMRI activity based on participants' movement direction in space.

Two major hypotheses have been put forward to explain how the activity of grid cells translates into this hexadirectional fMRI signal (Bin Khalid et al. 2024; Doeller et al. 2010; Kunz et al. 2019). The *conjunctive hypothesis* proposes that conjunctive grid-by-head-direction cells exhibit increased firing rates when participants move along the grid axes, resulting in stronger fMRI signals as participants move along the grid axes. The *repetition suppression hypothesis* posits that grid cells exhibit repetition suppression, resulting in weaker fMRI signals when participants move along the grid axes and stronger fMRI signals when participants move between the grid axes. This occurs because fewer individual grid cells are activated more frequently when moving along the grid axes. Whereas the conjunctive hypothesis is unaffected by how participants navigate through space, the repetition suppression hypothesis stipulates that significant hexadirectional modulation of fMRI activity emerges only when participants move along paths that are sufficiently straight for repetition suppression to take effect (Bin Khalid et al. 2024).

The detection of grid representations in fMRI is supported by results from several intracranial studies, including both single-neuron and LFP recordings, although the exact relationship remains unclear. Using human single-neuron recordings, two studies directly observed individual grid cells in the human entorhinal cortex (Jacobs et al. 2013; Nadasdy et al. 2017). This shows that individual grid cells do exist in humans and that they can be examined in epilepsy patients engaged in virtual navigation tasks. Interestingly, these studies found grid cells not only in the entorhinal cortex but also in the cingulate cortex (Jacobs et al. 2013). The observation of grid cells outside the entorhinal cortex aligns with fMRI studies showing hexadirectional modulations in various brain regions (Kunz et al. 2019). Furthermore, following the analysis rationale of the fMRI studies, several intracranial studies observed a hexadirectional modulation of theta (or gamma) power of LFPs recorded from the entorhinal cortex (Chen et al. 2021; Chen et al. 2018; Kunz et al. 2019;

Maidenbaum et al. 2018; Staudigl et al. 2018). As in the fMRI studies, this hexadirectional modulation of LFP power may result from the population activity of grid cells, given that conjunctive grid-by-head-direction cells increase their firing when moving along the grid axes or in case that repetition suppression exists in grid cells.

To make progress, it would be useful to have simultaneous recordings of single grid cells and hexadirectional modulations of LFP power and fMRI activity in animal models. Such recordings could investigate whether the hexadirectional modulation is strongest in voxels with a high density of grid cells. They could also show whether the grid axes of individual grid cells are aligned with the peak directions of the hexadirectional modulation of LFP power and/or fMRI activity. This would speak in favor of the conjunctive hypothesis, whereas a consistent misalignment would support the repetition suppression hypothesis (Bin Khalid et al. 2024; Kunz et al. 2019). In humans, such recordings are not feasible due to safety restrictions when performing fMRI. To establish similar effects, human studies involving epilepsy patients may employ consecutive recordings, using fMRI recordings followed by invasive single-neuron and/or LFP recordings. Building on recent work using computer simulations (Bin Khalid et al. 2024), we propose that tasks for noninvasive and invasive studies in humans be designed that systematically vary the straightness of navigation paths to differentiate between the conjunctive hypothesis and the repetition suppression hypothesis. Furthermore, human single-neuron studies are needed to provide a more detailed characterization of grid-cell properties in humans, including an exact quantification of conjunctive grid-by-head-direction cells and repetition suppression effects in grid cells. Based on these quantitative insights, we recommend that computational studies reassess the likelihood that hexadirectional signals result from the population activity of grid cells. This, in turn, should enable us to evaluate which hypothesis is most likely correct (Figure 6.2).

2.2 Scene Representations in the Parahippocampal Cortex

Similar to the entorhinal cortex, the parahippocampal cortex appears to be a region where evidence converges on navigation-related neural signals from noninvasive fMRI and invasive neural recordings. Specifically, early fMRI studies discovered that the parahippocampal place area within the parahippocampal cortex is a brain region that shows robust fMRI activity when participants view pictures of local environments (Epstein and Kanwisher 1998). Since then, multiple fMRI studies have replicated this finding (e.g., Weiner et al. 2018). Follow-up studies found that the parahippocampal cortex does not exhibit repetition suppression when participants view the same local environment from different angles. Thus, it has been proposed that the parahippocampal place area represents the relationship between the observer and the surfaces that define local space (Epstein et al. 2003).

Several intracranial studies support this fMRI-based finding of the parahippocampal place area. Following the experimental procedure of the fMRI studies, LFP recordings in epilepsy patients identified increased broadband gamma activity in the parahippocampal cortex in response to scene images, along with similar responses in the lateral occipital complex, the occipital place area, and the retrosplenial complex

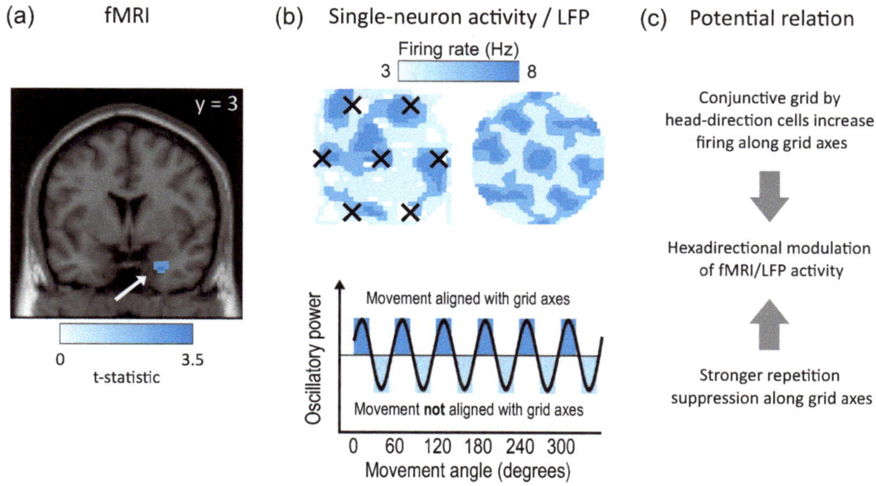

Figure 6.2 Grid representations in the entorhinal cortex. (a) Observation of entorhinal grid representations using fMRI, consisting of a hexadirectional modulation of BOLD activity as a function of participants' movement direction through a computer-based virtual environment (Doeller et al. 2010, reprinted with permission from SNCSC). (b) Top: Firing-rate map (left) and spatial autocorrelogram (right) of an individual grid cell identified using single-neuron recordings in an epilepsy patient (Jacobs et al. 2013, reprinted with permission from SNCSC). Bottom: Entorhinal grid representations observed using intracranial EEG, based on findings from Maidenbaum et al. (2018). Z-scored theta power shows a sixfold sinusoidal modulation by movement direction. (c) Two potential mechanisms showing how the population activity of grid cells may lead to grid representations in fMRI activity and the power of LFPs. The conjunctive hypothesis suggests that conjunctive grid-by-head-direction cells increase their firing when individuals move along the grid axes. The repetition suppression hypothesis suggests that the grid-cell population exhibits stronger repetition suppression when participants move along the grid axes as compared to when they move between the grid axes.

(Bastin et al. 2013; Vicek et al. 2020). Similarly, using single-neuron recordings in patients as they viewed static pictures of spatial scenes (i.e., landscapes), a considerable number of parahippocampal neurons were found to respond strongly to these spatial scenes (Mormann et al. 2017). For both outdoor and indoor pictures, pictures with a spatial layout (scenes) elicited stronger responses in parahippocampal neurons than those without spatial layout (non-scenes). Importantly, scene-responsive neurons had an increased likelihood of being close to other scene-responsive neurons, and stimuli that contained spatial layouts elicited stronger event-related potentials in the LFP than those without a spatial layout. This suggests that scene-selective neurons as well as their input signals (LFPs) are spatially clustered within individuals. This clustering provides a possible explanation for why fMRI voxels show a preference for spatial scenes as the combined activity of many scene-selective neurons may suffice to elicit macroscopically visible fMRI activity. Together, these intracranial studies suggest that the scene-selective activity observed in the parahippocampal place area via fMRI has a direct counterpart at the mesoscopic level of LFPs and the microscopic level of individual neurons (Figure 6.3).

The exact type of spatial information that is represented by neurons in the parahippocampal cortex—and specifically the parahippocampal place area—remains, however, uncertain. One possibility is that it is driven by directional information

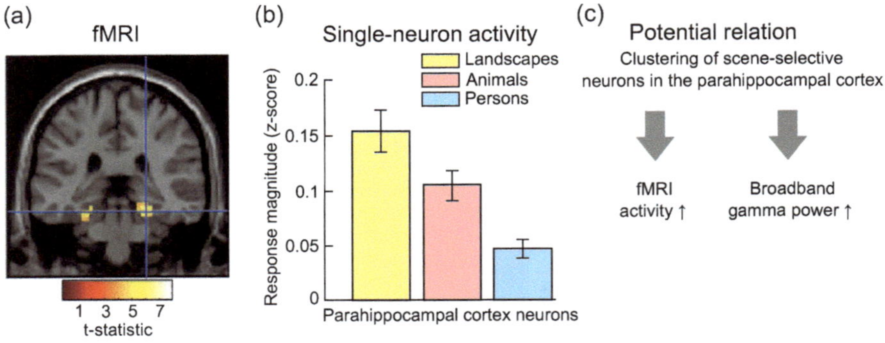

Figure 6.3 Scene representations in the parahippocampal cortex. (a) Observation of the parahippocampal place area using fMRI (Park and Chun 2009, reprinted with permission from Elsevier). (b) Single-neuron activity related to spatial scenes (landscapes) in the parahippocampal cortex, based on findings by Mormann et al. (2017). (c) Potential mechanism of how single-neuron activity related to spatial scenes translates to the observation of the parahippocampal place area in fMRI and intracranial recordings.

that encodes one's first-person perspective relative to the spatial scenes. This idea is based on recent single-neuron studies in rodents and humans, which found such self-centered, egocentric direction signals in several regions, including the rodent postrhinal cortex (LaChance et al. 2019) and the human parahippocampal cortex (Kunz et al. 2021). Specifically, using single-neuron activity in epilepsy patients as they performed a virtual reality task in which they navigated through a computer-based virtual environment, we observed neurons in the parahippocampal cortex that were often tuned to the participant's egocentric direction and distance toward local reference points (Kunz et al. 2021). These so-called egocentric-bearing cells may support human navigation by constituting the neural basis of egocentric cognitive maps, and they may provide a link between sensory inputs and allocentric place, grid, and head-direction cells (Bicanski and Burgess 2018). For example, an individual egocentric-bearing cell showed the strongest spiking activity whenever the environment center—an important geometrical aspect of any environment—was ahead of the participant. We speculate that egocentric-bearing cells contribute to the increased activity of the parahippocampal place area in response to images of spatial scenes, because participants in these studies typically view the scenes from a first-person, egocentric perspective. This idea awaits empirical testing.

Overall, fMRI and intracranial studies consistently indicate that the parahippocampal cortex plays a crucial role in neural computations related to scenes—and potentially egocentric directional information—across different modalities. To make further progress, we propose using sequential noninvasive and invasive recordings in human epilepsy patients to help us better understand the relationship between fMRI activity, LFP power, and single-neuron activity in response to spatial scenes. Such studies could test whether the fMRI-defined parahippocampal place area predicts the loci of strongest responses of intracranial power and neuronal firing. By combining virtual navigation tasks with presentations of static spatial scenes, invasive recordings could probe whether egocentric-bearing cells are indeed the functional cell type that contributes to the responses of individual parahippocampal neurons to static spatial scenes. These future studies could also help delineate the broader set of brain regions involved in representing and processing directional information during human navigation (Nau et al. 2020). We believe that such direction representations are observable in various brain regions, including the retrosplenial cortex (Baumann and Mattingley 2021; Epstein 2008; Marchette et al. 2014) and thalamus (Clark et al. 2012), although finding them with fMRI will depend on whether these regions show similar clustering of spatial responses as in the parahippocampal cortex (Mormann et al. 2017).

2.3 Place Representations in the Hippocampus and Other Medial Temporal Lobe Regions

Place cells are neurons that fire when an animal is at a specific location within the environment (O'Keefe and Dostrovsky 1971). Predominantly found in the hippocampus, but may exist in other regions as well (Long and Zhang 2021; Sharp 2006), place cells have been studied primarily in rodents, with increasing evidence in other species such as birds and bats (Payne et al. 2021; Yartsev and Ulanovsky 2013). Each place cell has one or more place fields where it is most likely to increase its spiking activity. Place cells are considered key cellular substrates for representing allocentric locations in the hippocampus, forming cognitive maps that organisms can flexibly use to navigate adaptively (Behrens et al. 2018). Since their initial discovery, research on rodent place cells has moved at an unprecedented speed to characterize their exact role in spatial navigation (Moser et al. 2017). Research on place cells in humans, however, is relatively sparse. Only a few intracranial studies have aimed at direct recordings of place cells, and even fewer fMRI studies have investigated place representations at the voxel level.

Similar to rodent studies (Muller et al. 1994), the few existing single-neuron recordings in epilepsy patients have identified place cells on virtual linear tracks and in virtual open environments, where place cells display mostly unidirectional and omnidirectional tuning, respectively. In contrast to rodent research, however, human place cells appear to be more difficult to record; they show less reliable place fields and are distributed across several brain regions rather than being exclusive to the hippocampus. For example, in a study by Jacobs et al. (2010), patients navigated to different target locations clockwise or counterclockwise in a narrow computer-based virtual environment (a square-shaped road with buildings on either side) using a game controller. This study observed unidirectionally tuned place cells in the hippocampus as well as in the entorhinal cortex, parahippocampal gyrus, amygdala, and frontal cortex. These findings were corroborated in another study where patients passively moved along a virtual linear track to encode and remember objects at specific locations (Qasim et al. 2019). Again, significant numbers of unidirectionally tuned place cells were identified in the entorhinal cortex, hippocampus, and cingulate. Other work has shown that place cells exhibit unidirectional tuning, as patients navigate across streets in a virtual city, and that the cells reactivated during memory recall outside of the virtual city, directly implicating human place cells in memory formation and retrieval (Miller et al. 2013).

In open environments, place cells typically show omnidirectional tuning, firing at specific locations regardless of varying viewpoints or changing visual information, which makes it more likely that they indeed encode spatial locations rather than visual cues. Such omnidirectionally tuned place cells were observed primarily in the human hippocampus while patients navigated to different shops in a virtual open environment (Ekstrom et al. 2003). Subsequent work supports these findings as significant numbers of omnidirectionally tuned place cells were found in the hippocampus, parahippocampal gyrus, and cingulate cortex of patients navigating through a different type of open environment (Jacobs et al. 2013). Furthermore, we recently obtained evidence that place-cell activity is involved in the formation and recall of associative object-location memories (Kunz et al. 2024). Again, we observed omnidirectionally tuned place-selective cells in the hippocampus, entorhinal cortex, and parahippocampal cortex, with place fields broadly covering the virtual environment.

We note that not every intracranial spatial navigation study using computer-based virtual environments observed a significant number of place cells (e.g., Tsitsiklis et al. 2020). Such null findings may have several reasons, including task designs that do not evoke activity in place cells or unfavorable electrode placement that may miss place cells by pure chance. This also might mirror a general view that place cells in humans are less reliable (i.e., temporally stable), less prevalent, and less distinct (i.e., spatially selective) compared to place cells in rodent research (Han et al. 2023).

Several studies have aimed at investigating place representations with neuroimaging, including fMRI and positron emission tomography (PET). For example, in comparing neural responses between navigation tasks and baseline periods or other cognitive tasks, early neuroimaging studies using PET found increased hippocampal activity during virtual navigation (Maguire et al. 1998), whereas other studies observed navigation-related fMRI activity mostly in other regions, such as the parahippocampus (Aguirre and D'Esposito 1997; Aguirre et al. 1996; Rosenbaum et al. 2004). Subsequent research pursued a more detailed approach by aiming at isolating neural representations for specific locations within the same environment (in the hippocampus and other brain regions). This approach rests on the assumption that place-cell activity patterns are dissociable at the voxel level, which is not immediately intuitive given that empirical and modeling evidence indicates that hippocampal place cells are uniformly distributed and do not cluster, and that nearby place cells have different place fields (Dombeck et al. 2010; Humphreys et al. 1998; Redish et al. 2001; cf. Wirtshafter and Disterhoft 2023). This makes it unlikely for the activations of fMRI voxels to be biased toward particular locations in an environment.

Nevertheless, some fMRI studies showed successful decoding of participants' locations in computer-based virtual environments based on multivariate pattern analyses (Hassabis et al. 2009; Kim et al. 2017; Rodriguez 2010; Sulpizio et al. 2014). A general drawback of these studies, however, may be that target locations were associated with unique visual information (Nolan et al. 2018), such as salient landmarks while navigating to the target location (Hassabis et al. 2009), unique visual panoramas at the target location (Kim et al. 2017), or landmarks visible from the target location (Rodriguez 2010; Sulpizio et al. 2014). Multivariate fMRI activity patterns may thus represent visual inputs rather than spatial locations. Accordingly, using a paradigm that controlled for potential visual confounds and employing multiple analysis pipelines, a more recent fMRI study did not find evidence for hippocampal place representations using multivariate pattern analysis (Nolan et al. 2018). Similarly, another study observed fMRI representations of abstract locations in the presubiculum, retrosplenial complex, and parietal-occipital sulcus, but not in the hippocampus (Vass and Epstein 2013).

In summary, although intracranial recordings show some evidence of place cells, results on place representations from neuroimaging research are inconclusive. Based on our current knowledge of place-cell properties, it remains unclear how populations of (hippocampal) place cells may give rise to meso- and macroscopically visible place representations (Figure 6.4).

2.4 Landmark Representations in the Retrosplenial Cortex

Landmark-based navigation is one of the most important strategies underlying human spatial navigation (Epstein and Vass 2014). Landmarks are useful in various situations as they enable individuals to localize and orient themselves by referencing prominent environmental features. By navigating toward landmarks, one can easily reach goal destinations that are close to these landmarks. A multitude of human fMRI studies, human lesion studies, and electrophysiological studies in animals have thus examined the neural basis of landmarks and landmark-based navigation. Here, we focus on landmark representations in the retrosplenial cortex (Figure 6.5), a highly interconnected brain region that may play various important roles in spatial navigation (Alexander et al. 2023; Mitchell et al. 2018; Vann et al. 2009).

Many human fMRI studies have found evidence for the involvement of the retrosplenial cortex in representing and processing landmark-related information (e.g., Morgan et al. 2011). One line of research specifically suggests that the retrosplenial cortex processes information about landmarks that show high degrees of permanence, thus constituting useful reference points for localization and (re)orientation. In one study, participants viewed pictures of individual landmarks (e.g., a bus, a traffic cone, and a lighthouse) during fMRI scanning (Auger et al. 2012). While BOLD activity in the parahippocampal cortex increased for a wide range of landmark attributes, the retrosplenial cortex was engaged by only the most permanent landmarks. Interestingly, good navigators that more reliably identified the most stable landmarks compared to poor navigators showed heightened BOLD activity in the retrosplenial cortex in response to the most stable landmarks. A follow-up study tested whether retrosplenial cortex activity closely follows the number of permanent landmarks within the visual scene (Auger and Maguire 2013). Using multivariate decoding, Auger and Maguire found that the number of permanent objects could be decoded from BOLD activity in the retrosplenial cortex, suggesting that the retrosplenial cortex can track multiple permanent landmarks simultaneously. The observations were again linked to navigation ability, with good navigators showing significantly higher decoding accuracies compared to poor navigators.

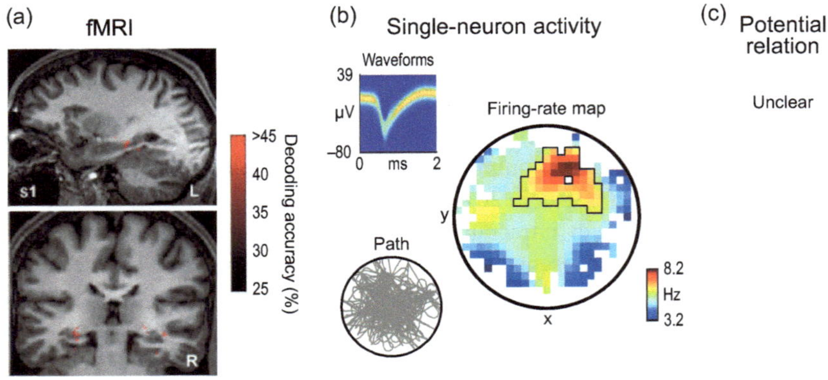

Figure 6.4 Place representations in the hippocampus and other medial temporal lobe regions. (a) Decoding of four different places in a computer-based virtual environment using multivariate fMRI activity in the hippocampus. Each red dot represents the center of a searchlight sphere showing significant four-way decoding of place (Hassabis et al. 2009; https://creativecommons.org/licenses/by/3.0/). (b) Example of a place-selective neuron from the entorhinal cortex of a patient navigating through a circular computer-based virtual environment (Kunz et al. 2024; https://creativecommons.org/licenses/by/4.0/). This neuron exhibited increased spiking activity in the northern part of the environment. Warm colors indicate high firing rate, and cool colors represent low firing rate. (c) It remains unclear whether fMRI can be used to detect macroscopic place representations and how place cells may give rise to such representations.

Subsequent work ruled out that retrosplenial cortex activity was mostly driven by the relevance of these landmarks for orientation, suggesting that the retrosplenial cortex is mainly involved in coding landmark permanence, irrespective of the landmarks' direct behavioral relevance (Auger and Maguire 2018). Furthermore, individuals with better navigation performance appeared to learn landmark permanence better than those with poor navigation performance, and this relationship was mediated by retrosplenial cortex activity (Auger et al. 2017). Interestingly, a recent study that employed neurophysiological recordings in mice also established a link between learning and retrosplenial cortex activity (Fischer et al. 2024). Here, the authors found that learning to use landmarks for navigation amplified their representation in the retrosplenial cortex, with neuronal activity shifting from self-referenced to landmark-referenced coding. Landmark representations in the retrosplenial cortex may also help rescue deficient computations in other brain regions related to different navigation strategies (Bierbrauer et al. 2020): Individuals at increased genetic risk for Alzheimer disease showed impaired path-integration performance in a very sparse virtual environment (without any landmarks or boundaries) but did not show path-integration deficits in a virtual environment with a local landmark. In an accompanying fMRI experiment, increased retrosplenial cortex activity was observed when participants performed path integration in the environment with the landmark, suggesting that retrosplenial landmark representations might compensate for impaired path-integration computations in the entorhinal cortex (and other areas) (Bierbrauer et al. 2020).

In contrast to the considerable number of fMRI studies on the retrosplenial cortex, intracranial investigations of landmark representations in the retrosplenial cortex are basically lacking, as recordings from this region are rare due to its infrequent involvement in patients' epileptic foci. Consequently, to the best of our knowledge,

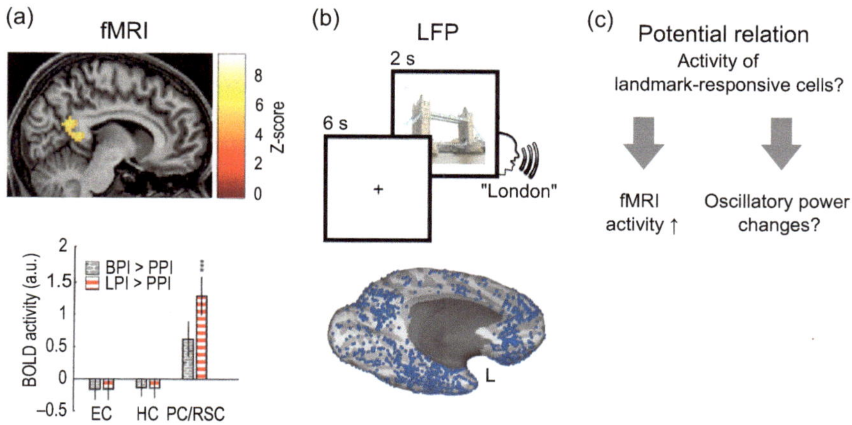

Figure 6.5 Landmark representations in the retrosplenial cortex. (a) Increased fMRI activity related to spatial landmarks in the retrosplenial cortex (top: Auger et al. 2012; https://creativecommons.org/licenses/by/4.0/; bottom: Bierbrauer et al. 2020; https://creativecommons.org/licenses/by/4.0/). (b) An open dataset with retrosplenial recordings during a landmark-viewing task that may be used to identify landmark representations in LFPs (Woolnough et al. 2022; https://creativecommons.org/licenses/by/4.0/). (c) Increased fMRI activity and oscillatory power changes in response to landmarks may result from increased activity of landmark-responsive cells, such as allocentric and egocentric landmark-vector cells. Abbreviations: a.u., arbitrary units; BPI, boundary-supported path integration; EC, entorhinal cortex; HC, hippocampus; LPI, landmark-supported path integration; PC/RSC, posterior cingulate/retrosplenial cortex; PPI, pure path integration.

there are no detailed reports of landmark representations using intracranial EEG or single-neuron recordings in the human retrosplenial cortex. This represents a significant knowledge gap that future studies may wish to address. Such studies could help us understand, for example, how macroscopically visible landmark representations can emerge in the retrosplenial cortex, and why this activity may be mostly related to the permanence of landmarks. They could also show what types of input are needed in the retrosplenial cortex to give rise to landmark representations. Building on recent evidence from mice that visual, motor, and spatial information contributes to landmark coding in the retrosplenial cortex (Fischer et al. 2020), intracranial recordings in the human retrosplenial cortex could help translate these findings to the human domain. Interestingly, a large intracranial EEG dataset on landmark processing was recently published (Woolnough et al. 2022) and included data from 50 epilepsy patients during which they viewed images of famous landmarks. This dataset, which encompassed >6,500 electrodes with some coverage of the retrosplenial cortex, may be useful for exploring landmark processing in the retrosplenial cortex using intracranial recordings.

We note that the retrosplenial cortex is a brain area best characterized as a central processing hub due to its high interconnectivity and comprehensive functionality (Alexander et al. 2023; Ekstrom et al. 2017). Accordingly, in addition to representing landmarks, the retrosplenial cortex most likely plays an important role in various aspects of spatial behavior and other cognitive domains (Alexander et al. 2023; Stacho and Manahan-Vaughan 2022). For example, the retrosplenial cortex may be key in using landmark information to determine the subject's current position and heading (Epstein and Vass 2014). More specifically, it has been suggested that the retrosplenial cortex helps integrate idiothetic cues and spatial relations with environmental features, such as landmarks, in the service of orientation and localization (Stacho and Manahan-Vaughan 2022). Some fMRI studies support this view by demonstrating location-specific fMRI activity in the retrosplenial cortex, independent of head direction (Chen et al. 2024; Vass and Epstein 2013). More broadly, the retrosplenial cortex may constitute the center of the brain's allocentric navigation system, in combination with various other brain regions (Ekstrom et al. 2017). In addition to identifying landmark representations at the cellular level, intracranial studies from the retrosplenial cortex could also investigate these exciting broader hypotheses.

2.5 Boundary Representations in Entorhinal, Parahippocampal, and Retrosplenial Cortices

The final area that we raise for consideration focuses on neural representations of boundaries in entorhinal, parahippocampal, and retrosplenial cortices. Boundary representations are highly significant for spatial navigation, as boundaries define the extent and geometry of the spatial environment and often constitute important obstacles. Electrophysiological studies in rodents identified single-cell responses to boundaries in various brain regions. In early studies (e.g., Lever et al. 2009; Solstad et al. 2008), allocentric border and boundary-vector cells in the entorhinal cortex and subiculum were found to activate whenever a boundary is at a particular allocentric direction and distance from the animal (e.g., 20 cm toward north). Later studies obtained evidence for egocentrically tuned single-cell responses to boundaries in the entorhinal, postrhinal (human parahippocampal), and retrosplenial cortices (Alexander et al. 2020; Gofman et al. 2019; Wang et al. 2018). These egocentric boundary-vector cells fire whenever a boundary is at a particular egocentric direction and distance from the animal (e.g., 20 cm to the left). Computational models suggest that during navigation, the information of egocentric boundary-vector cells is converted into the tuning of allocentric boundary-vector cells by means of a transformation circuit in the retrosplenial cortex (Bicanski and Burgess 2018), and the combined information from several allocentric boundary-vector cells may produce the firing of place cells (Barry et al. 2006; O'Keefe and Burgess 1996).

In human fMRI studies, activity related to environmental boundaries has been described in the parahippocampal place area and retrosplenial cortex. In one study (Ferrara and Park 2016), participants viewed artificial scene images that varied according to their vertical boundary structure (e.g., a flat two-dimensional boundary, a slight addition of three-dimensional boundaries, and full walls). The parahippocampal place area appeared to indicate the presence of a grounded three-dimensional vertical structure (activating already in response to a minimal boundary), whereas the retrosplenial cortex represented whether a boundary introduced a significant impediment to the viewer's potential navigation in space (i.e., only showing activity when boundaries were high enough to constitute an impediment to navigation). In another study (Shine et al. 2019), fMRI activity related to boundaries was observed in the entorhinal cortex and subiculum. Specifically, multivariate decoding analyses of fMRI activity provided evidence for allocentric boundary directions in the entorhinal cortex and posterior subiculum, and control analyses indicated that the effects were not driven by visual information (Shine et al. 2019). These fMRI results provide evidence for boundary-related activity in human brain regions that are consistent with cellular responses in animal models.

Thus far, intracranial studies in epilepsy patients have provided less direct evidence for boundary coding. For example, using intracranial EEG recordings in epilepsy patients engaged in real-world navigation through a rectangular room, theta power increased whenever participants were close to the boundary as compared to when they were in the environment center (Stangl et al. 2020). The same result emerged when participants observed the experimenter walking through the room: theta power was again higher whenever the experimenter was close to a wall. Relatedly, increased theta power was observed when participants encoded target locations near boundaries as compared to target locations near the center of the environment (Lee et al. 2018). These findings indicate that spatial boundaries have a major effect on theta-frequency power. Still, the exact kind of boundary information that theta power represents remains unclear, and future intracranial studies are needed to determine whether theta power is related to a particular type of boundary information (e.g., distance or direction toward the boundary, simply the presence of boundaries, or another type of information). These investigations may also elucidate how such boundary-related changes in theta power relate to the well-established effects of movement on theta power and frequency (Kennedy et al. 2022). To investigate boundary representations at the single-cell level, future intracranial studies are needed to probe whether allocentric and egocentric boundary-vector cells exist not only in rodents but also in humans.

It seems plausible to us that fMRI studies which modulate the prevalence of boundaries in an environment (Ferrara and Park 2016) may find differential responses in regions with large numbers of boundary-sensitive neurons described in rodents. This is because environments with many boundaries will trigger activity in more boundary-tuned neurons, which may directly translate into stronger BOLD signals. Currently, however, it is unclear how fMRI activity related to particular allocentric and egocentric directions of boundaries may emerge because, to the best of our knowledge, a topographical clustering of specific boundary directions has not been observed in single-neuron studies. Similarly, the mechanism that drives increases of theta power close to environmental boundaries has not yet been elucidated. Although place cells and other spatially modulated neurons show theta-associated phenomena such as theta-phase locking and theta-phase precession, it is unclear why this should lead to increases in theta power regarding the presence of boundaries. Hence, more detailed studies with specific manipulations of the boundaries (e.g., using various heights and varying behavioral relevance) are needed to shed light on the boundary properties that drive theta-power increases. By combining LFP and single-neuron recordings, intracranial studies may establish whether boundary-related theta-power effects are most prominent on electrodes also showing allocentric and/or egocentric boundary-vector cells. Furthermore, consecutive noninvasive–invasive studies may reveal whether voxels with boundary-related fMRI activity have the strongest theta-power increases in relation to boundaries (Figure 6.6).

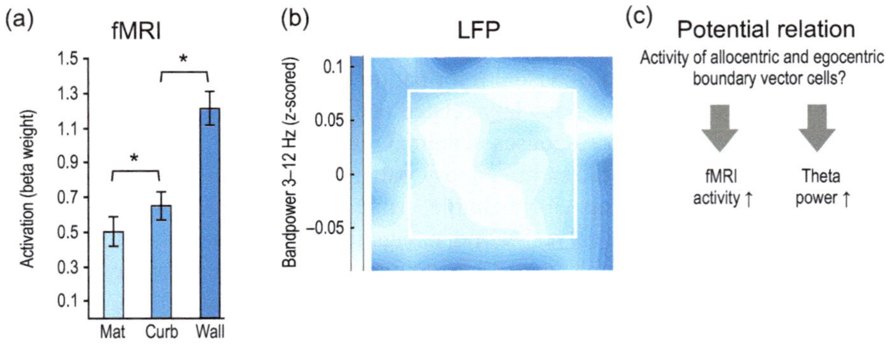

Figure 6.6 (a) Observation of increased fMRI activity in the parahippocampal cortex in response to the presence of boundaries with minimal vertical height (Ferrara and Park 2016, reprinted with permission from Elsevier). (b) Increased theta power when participants moved through the periphery of a physical environment close to the boundary as compared to when moving in the center (Stangl et al. 2020, reprinted with permission from SNCSC). (c) Potential relationship between invasive and noninvasive observations. Increased activity of allocentric and egocentric boundary-vector cells closer to spatial boundaries may lead to increased theta power and fMRI activity. The exact mechanisms underlying the meso- and macroscopic effects remain unclear.

3 Discussion

Thus far, we have focused on whether data from intracranial techniques align (or not) with what we have learned about human navigation from fMRI. Here, we raise three questions related to our approach in this article, the general relationship between fMRI and neural activity, and the fast timescale of brain mechanisms.

3.1 Why Focus on Particular Types of Spatial Representations?

In the previous sections, our discussion focused on five specific types of spatial representations in particular brain regions:

1. Grid representations in the entorhinal cortex
2. Scene representations in the parahippocampal cortex
3. Place representations in the hippocampus
4. Landmark representations in the retrosplenial cortex
5. Boundary representations in entorhinal, parahippocampal, and retrosplenial cortices

We opted for this focused approach because it enabled the most direct comparisons between findings from invasive and noninvasive methods. We believe that these specialized neural representations function together to support a variety of different navigation strategies in humans. For instance, grid, direction, and landmark representations may flexibly interact during path integration in landmark-rich environments.

We acknowledge, however, that other accounts have questioned the usefulness of differentiating between elemental cognitive processes underlying spatial navigation that can be ascribed to unique brain regions. Ekstrom et al. (2017) suggests, for example, that human navigation relies on a network of distributed brain areas—centered on the retrosplenial cortex for allocentric strategies and on the posterior parietal cortex for egocentric strategies—whose relative importance flexibly changes as a function of task demands. To test these types of network models, researchers need to pursue other types of analysis approaches than the ones we discussed here. Specifically, network analyses such as interregional correlations of BOLD, LFPs, or single-neuron activity need to be performed to test whether these networks change between different types of navigation in a similar manner. This may provide complementary insights into the question of how well data from intracranial techniques align with what we have learned about human navigation from fMRI.

3.2 What Is the Basic Relationship between fMRI and Intracranial Signals?

fMRI measures the BOLD response that increases whenever neuronal activity in a region increases (Huettel et al. 2014). In short, neuronal activity leads to increased blood flow, resulting in a higher concentration of oxygenated blood in the local area, which, in turn, leads to a higher BOLD signal in that region. The BOLD response is much slower than neuronal activity, leading to a highly smoothed version of the underlying neuronal activity that peaks about five seconds later. Two major advantages of fMRI are the opportunity of performing fMRI recordings in healthy adults and across the entire brain. Disadvantages comprise the low temporal resolution (typically between 1 and 3 sec), medium to low spatial resolution (typical voxel sizes range between 1 and 27 mm^3), and the indirect measurement of neuronal activity through blood oxygenation.

In contrast, intracranial LFP and single-neuron recordings measure neuronal activity directly from within the human brain. LFPs arise mainly from synaptic currents based on synchronized inputs into the dendrites of neurons in a particular region (Buzsáki et al. 2012), and single-neuron recordings detect extracellular action potentials of individual (single units) or small combinations (multi-units) of neurons. Their advantages are an excellent temporal and spatial resolution up to the recording of single neurons. In addition, intracranial recordings can be conducted in patients as they physically navigate through real-world environments rather than performing virtual navigation (Stangl et al. 2023; Stangl et al. 2020). This may enable discoveries that arise uniquely during real-world navigation and are missed in studies using virtual reality tasks, such as fMRI research where participants navigate virtually while lying still inside the scanner. The three major disadvantages of intracranial recordings are that (a) they can only be done in epilepsy patients (and some other groups of patients), (b) the experimenter has no control over the exact locations of the electrodes (as they are solely determined by clinical needs), and (c) coverage is typically limited to a few brain regions, such as the hippocampus and neighboring medial temporal lobe regions.

How BOLD responses relate to the activity of individual neurons and different aspects of LFPs (including power in different frequency bands) remains largely unclear, because simultaneous recordings are difficult in animal models and impossible in humans. In cortical regions, there may be positive linear relationships between spiking activity and BOLD as well as between the LFP and BOLD (Logothetis et al. 2001), but particularly in subcortical regions, these relationships are probably more complex. This discussion, however, is beyond the scope of this chapter but has been reviewed elsewhere (Ekstrom 2010; Hill et al. 2021; Kunz et al. 2019). Here, we thus focused on the issue of whether noninvasive fMRI and invasive LFP and single-neuron recordings have led to the identification of similar functional representations related to human navigation in a select number of brain regions. Nevertheless, we hold that it is of fundamental importance to establish which types of electrophysiological phenomena in intracranial research relate to which kind of BOLD signal changes in fMRI.

3.3 What About Fast Brain Mechanisms Such as Theta Sequences?

Because of its low temporal resolution and the sluggishness of the BOLD response, fMRI is limited regarding the detection of fast temporal dynamics of neuronal activity (Huettel et al. 2014). These dynamics include phenomena such as theta sequences, theta-phase precession, and replay during sharp-wave ripples. Recording fast dynamics is essential, however, as they may constitute mechanisms for important navigational functions, such as planning of future routes (theta sequences), formation of memories for sequences (theta-phase precession), and memory retrieval and consolidation (replay during sharp-wave ripples). Intracranial recordings provide a major advantage as they are able to investigate these fast temporal dynamics.

Currently it is unclear whether fMRI may be able to detect indirect measures of theta sequences, theta-phase precession, and replay. Interestingly, using multivariate decoding techniques that can disentangle subtle and spatially overlapping activity patterns, it has been possible to detect fast replay events as ordered activations of sequential fMRI patterns (Schapiro et al. 2018; Schuck and Niv 2019; Wittkuhn and Schuck 2021). Theta sequences and theta-phase precession may be approachable by means of similar methods, but its empirical feasibility has not yet been tested. Another way forward might be to combine surface EEG with fMRI to detect, for example, replay events using temporally delayed linear modeling (Liu et al. 2021). Researchers could then investigate BOLD responses aligned with these replay events and examine their brain-wide effects (Huang et al. 2024).

4 Conclusion and Outlook

Here, we discussed whether noninvasive fMRI in healthy adults and invasive neural recordings in epilepsy patients have obtained similar or divergent insights into the neural correlates of human navigation. We focused on five important components of spatial navigation in a limited set of brain regions and observed that grid and scene representations in the entorhinal and parahippocampal cortices, respectively, show the strongest alignment between techniques. In contrast, although boundary and landmark representations in the entorhinal, parahippocampal, and retrosplenial cortices have been studied in various fMRI studies, invasive studies have been used to a lesser extent. Several invasive studies have observed place representations in the hippocampus, but it remains unclear whether noninvasive methods can reliably capture such representations.

For grid representations, there are at least two interesting hypotheses that address the direct relationship between grid-cell population activity and hexadirectional fMRI activity, which can be directly tested in future studies (Bin Khalid et al. 2024). For most of these representations, however, the exact relationship between cellular activity, LFP changes, and BOLD responses in fMRI remains vague. This is mostly because anatomically nearby neurons can show very different spatial tuning properties, making systematic variations at the level of fMRI voxels unlikely. Hence, to make further progress, systematic characterizations of spatial representations at the cellular level will be useful, particularly when combined with LFP recordings (Mormann et al. 2017) and computational modeling (Bin Khalid et al. 2024) that can assess the likelihood and make predictions regarding the large-scale fMRI effects of specific cellular tuning properties.

Simultaneous recordings of noninvasive fMRI and invasive neural recordings in animal models may be another useful approach for advancing our knowledge on the relationship between fMRI phenomena and intracranial neural activity. Such recordings have achieved groundbreaking insights into the general relationship between BOLD, LFPs, and spiking activity in cortical regions (Logothetis et al. 2001) but have, to the best of our knowledge, never been performed with regard to particular neural representations of space. Furthermore, consecutive recordings with epilepsy patients performing the same spatial task twice—before electrode implantation, while inside the fMRI scanner, and after implantation of intracranial electrodes, outside the fMRI scanner—may lead to novel, cross-methodological insights. Such consecutive recordings have been performed several times (e.g., Hill et al. 2021) and have shown, for example, that BOLD signal changes correlated positively with theta power in the parahippocampal cortex but not in the hippocampus, and that BOLD activity was unrelated to neuronal spiking activity in either region (Ekstrom et al. 2009). Although one must carefully account for caveats unique to this recording procedure (e.g., potential brain shifts due to neurosurgery), further studies using this methodological approach (potentially including ultra-high-field fMRI at 7T) will advance our knowledge of whether data from intracranial techniques align with what we can learn about human navigation from fMRI.

Acknowledgments We acknowledge funding support through the return program of the Ministry of Culture and Science of the State of North Rhine-Westphalia.

References

Aguirre GK, and D'Esposito M (1997) Environmental Knowledge Is Subserved by Separable Dorsal/Ventral Neural Areas. J Neurosci 17 (7):2512–2518. https://doi.org/10.1523/JNEUROSCI.17-07-02512.1997

Aguirre GK, Detre JA, Alsop DC, and D'Esposito M (1996) The Parahippocampus Subserves Topographical Learning in Man. Cereb Cortex 6 (6):823–829. https://doi.org/10.1093/cercor/6.6.823

Alexander AS, Carstensen LC, Hinman JR, et al. (2020) Egocentric Boundary Vector Tuning of the Retrosplenial Cortex. Sci Adv 6 (8):eaaz2322. https://doi.org/10.1126/sciadv.aaz2322

Alexander AS, Place R, Starrett MJ, Chrastil ER, and Nitz DA (2023) Rethinking Retrosplenial Cortex: Perspectives and Predictions. Neuron 111 (2):150–175. https://doi.org/10.1016/j.neuron.2022.11.006

Auger SD, and Maguire EA (2013) Assessing the Mechanism of Response in the Retrosplenial Cortex of Good and Poor Navigators. Cortex 49 (10):2904–2913. https://doi.org/10.1016/j.cortex.2013.08.002

——— (2018) Retrosplenial Cortex Indexes Stability Beyond the Spatial Domain. J Neurosci 38 (6):1472–1481. https://doi.org/10.1523/JNEUROSCI.2602-17.2017

Auger SD, Mullally SL, and Maguire EA (2012) Retrosplenial Cortex Codes for Permanent Landmarks. PLoS ONE 7 (8):e43620. https://doi.org/10.1371/journal.pone.0043620

Auger SD, Zeidman P, and Maguire EA (2017) Efficacy of Navigation May Be Influenced by Retrosplenial Cortex-Mediated Learning of Landmark Stability. Neuropsychologia 104:102–112. https://doi.org/10.1016/j.neuropsychologia.2017.08.012

Barry C, Lever C, Hayman R, et al. (2006) The Boundary Vector Cell Model of Place Cell Firing and Spatial Memory. Rev Neurosci 17 (1–2):71–97. DOI:10.1515/REVNEURO.2006.17.1-2.71

Bastin J, Committeri G, Kahane P, et al. (2013) Timing of Posterior Parahippocampal Gyrus Activity Reveals Multiple Scene Processing Stages. Hum Brain Mapp 34 (6):1357–1370. https://doi.org/10.1002/hbm.21515

Baumann O, and Mattingley JB (2021) Extrahippocampal Contributions to Spatial Navigation in Humans: A Review of the Neuroimaging Evidence. Hippocampus 31 (7):640–657. https://doi.org/10.1002/hipo.23313

Behrens TEJ, Muller TH, Whittington JCR, et al. (2018) What Is a Cognitive Map? Organizing Knowledge for Flexible Behavior. Neuron 100 (2):490–509. https://doi.org/10.1016/j.neuron.2018.10.002

Bellmund JLS, Gärdenfors P, Moser EI, and Doeller CF (2018) Navigating Cognition: Spatial Codes for Human Thinking. Science 362 (6415):eaat6766. https://doi.org/10.1126/science.aat6766

Bicanski A, and Burgess N (2018) A Neural-Level Model of Spatial Memory and Imagery. eLife 7 (7052):1–3. https://doi.org/10.7554/elife.33752

Bierbrauer A, Kunz L, Gomes CA, et al. (2020) Unmasking Selective Path Integration Deficits in Alzheimer's Disease Risk Carriers. Sci Adv 6:eaba1394. https://doi.org/10.1126/SCIADV.ABA1394

Bin Khalid I, Reifenstein ET, Auer N, Kunz L, and Kempter R (2024) Quantitative Modeling of the Emergence of Macroscopic Grid-Like Representations. eLife 13:e85742. https://doi.org/10.7554/eLife.85742

Boccara CN, Nardin M, Stella F, O'Neill J, and Csicsvari J (2019) The Entorhinal Cognitive Map Is Attracted to Goals. Science 363 (6434):1443–1447. https://doi.org/10.1126/science.aav4837

Brown TI, Carr VA, LaRocque KF, et al. (2016) Prospective Representation of Navigational Goals in the Human Hippocampus. Science 352 (6291):1323–1326. https://doi.org/10.1126/science.aaf0784

Bush D, Barry C, Burgess N, et al. (2015) Using Grid Cells for Navigation. Neuron 87 (3):507–520. https://doi.org/10.1016/j.neuron.2015.07.006

Buzsáki G (2004) Large-Scale Recording of Neuronal Ensembles. Nat Neurosci 7 (5):446–451. https://doi.org/10.1038/nn1233

Buzsáki G, Anastassiou CA, and Koch C (2012) The Origin of Extracellular Fields and Currents: EEG, ECoG, LFP and Spikes. Nat Rev Neurosci 13 (6):407–420. https://doi.org/10.1038/nrn3241

Chen D, Kunz L, Lv P, et al. (2021) Theta Oscillations Coordinate Grid-Like Representations between Ventromedial Prefrontal and Entorhinal Cortex. Sci Adv 7 (44):eabj0200. https://doi.org/10.1126/sciadv.abj0200

Chen D, Kunz L, Wang W, et al. (2018) Hexadirectional Modulation of Theta Power in Human Entorhinal Cortex during Spatial Navigation. Curr Biol 28 (20):3310–3315.e3314. https://doi.org/10.1016/j.cub.2018.08.029

Chen X, Wei Z, and Wolbers T (2024) Repetition Suppression Reveals Cue-Specific Spatial Representations for Landmarks and Self-Motion Cues in the Human Retrosplenial Cortex. eNeuro 11 (4):ENEURO.0294–0223.2024. https://doi.org/10.1523/ENEURO.0294-23.2024

Clark BJ, Harris MJ, and Taube JS (2012) Control of Anterodorsal Thalamic Head Direction Cells by Environmental Boundaries: Comparison with Conflicting Distal Landmarks. Hippocampus 22 (2):172–187. https://doi.org/10.1002/hipo.20880

Constantinescu AO, O'Reilly JX, and Behrens TEJ (2016) Organizing Conceptual Knowledge in Humans with a Gridlike Code. Science 352 (6292):1464–1468. https://doi.org/10.1126/science.aaf0941

Doeller CF, Barry C, and Burgess N (2010) Evidence for Grid Cells in a Human Memory Network. Nature 463 (7281):657–661. https://doi.org/10.1038/nature08704

Doeller CF, King JA, and Burgess N (2008) Parallel Striatal and Hippocampal Systems for Landmarks and Boundaries in Spatial Memory. PNAS 105 (15):5915–5920. https://doi.org/10.1073/pnas.0801489105

Dombeck DA, Harvey CD, Tian L, Looger LL, and Tank DW (2010) Functional Imaging of Hippocampal Place Cells at Cellular Resolution during Virtual Navigation. Nat Neurosci 13 (11):1433–1440. https://doi.org/10.1038/nn.2648

Einevoll GT, Kayser C, Logothetis NK, and Panzeri S (2013) Modelling and Analysis of Local Field Potentials for Studying the Function of Cortical Circuits. Nat Rev Neurosci 14 (11):770–785. https://doi.org/10.1038/nrn3599

Ekstrom AD (2010) How and When the fMRI BOLD Signal Relates to Underlying Neural Activity: The Danger in Dissociation. Brain Res Rev 62 (2):233–244. https://doi.org/10.1016/j.brainresrev.2009.12.004

Ekstrom AD, Caplan JB, Fields TA, et al. (2003) Cellular Networks Underlying Human Spatial Navigation. Nature 425:184–188. https://doi.org/10.1038/nature01964

Ekstrom AD, Huffman DJ, and Starrett M (2017) Interacting Networks of Brain Regions Underlie Human Spatial Navigation: A Review and Novel Synthesis of the Literature. J Neurophysiol 118 (6):3328–3344. https://doi.org/10.1152/jn.00531.2017

Ekstrom AD, Spiers HJ, Bohbot VD, and Rosenbaum RS (2018) Human Spatial Navigation. Princeton Univ Press, Princeton

Ekstrom AD, Suthana N, Millett D, Fried I, and Bookheimer S (2009) Correlation between BOLD fMRI and Theta-Band Local Field Potentials in the Human Hippocampal Area. J Neurophysiol 101 (5):2668–2678. https://doi.org/10.1152/jn.91252.2008

Epstein RA (2008) Parahippocampal and Retrosplenial Contributions to Human Spatial Navigation. Trends Cogn Sci 12 (10):388–396. https://doi.org/10.1016/j.tics.2008.07.004

Epstein RA, Graham KS, and Downing PE (2003) Viewpoint-Specific Scene Representations in Human Parahippocampal Cortex. Neuron 37 (5):865–876. https://doi.org/10.1016/S0896-6273(03)00117-X

Epstein RA, and Kanwisher N (1998) A Cortical Representation of the Local Visual Environment. Nature 392 (6676):598–601. https://doi.org/10.1038/33402

Epstein RA, Patai EZ, Julian JB, and Spiers HJ (2017) The Cognitive Map in Humans: Spatial Navigation and Beyond. Nat Neurosci 20 (11):1504–1513. https://doi.org/10.1038/nn.4656

Epstein RA, and Vass LK (2014) Neural Systems for Landmark-Based Wayfinding in Humans. Philos Trans R Soc Lond B Biol Sci 369 (1635):20120533. https://doi.org/10.1098/rstb.2012.0533

Ferrara K, and Park S (2016) Neural Representation of Scene Boundaries. Neuropsychologia 89:180–190. https://doi.org/10.1016/j.neuropsychologia.2016.05.012

Fischer LF, Mojica Soto-Albors R, Buck F, and Harnett MT (2020) Representation of Visual Landmarks in Retrosplenial Cortex. eLife 9:e51458. https://doi.org/10.7554/eLife.51458

Fischer LF, Xu L, Murray KT, and Harnett MT (2024) Learning to Use Landmarks for Navigation Amplifies Their Representation in Retrosplenial Cortex. bioRxiv Preprint. https://doi.org/10.1101/2024.08.18.607457

Gardner RJ, Hermansen E, Pachitariu M, et al. (2022) Toroidal Topology of Population Activity in Grid Cells. Nature 602 (7895):123–128. https://doi.org/10.1038/s41586-021-04268-7

Gofman X, Tocker G, Weiss S, et al. (2019) Dissociation between Postrhinal Cortex and Downstream Parahippocampal Regions in the Representation of Egocentric Boundaries. Curr Biol 29 (16):2751–2757.e2754. https://doi.org/10.1016/j.cub.2019.07.007

Gu Y, Lewallen S, Kinkhabwala AA, et al. (2018) A Map-Like Micro-Organization of Grid Cells in the Medial Entorhinal Cortex. Cell 175 (3):736–750.e730. https://doi.org/10.1016/j.cell.2018.08.066

Hafting T, Fyhn M, Molden S, Moser M-B, and Moser EI (2005) Microstructure of a Spatial Map in the Entorhinal Cortex. Nature 436 (7052):801–806. https://doi.org/10.1038/nature03721

Han CZ, Donoghue T, Cao R, et al. (2023) Using Multi-Task Experiments to Test Principles of Hippocampal Function. Hippocampus 33 (5):646–657. https://doi.org/10.1002/hipo.23540

Hassabis D, Chu C, Rees G, et al. (2009) Decoding Neuronal Ensembles in the Human Hippocampus. Curr Biol 19 (7):546–554. https://doi.org/10.1016/j.cub.2009.02.033

Heys James G, Rangarajan Krsna V, and Dombeck Daniel A (2014) The Functional Micro-Organization of Grid Cells Revealed by Cellular-Resolution Imaging. Neuron 84 (5):1079–1090. https://doi.org/10.1016/j.neuron.2014.10.048

Hill PF, Seger SE, Yoo HB, et al. (2021) Distinct Neurophysiological Correlates of the fMRI BOLD Signal in the Hippocampus and Neocortex. J Neurosci 41 (29):6343–6352. https://doi.org/10.1523/JNEUROSCI.0278-21.2021

Huang Q, Xiao Z, Yu Q, et al. (2024) Replay-Triggered Brain-Wide Activation in Humans. Nat Commun 15 (1):7185. https://doi.org/10.1038/s41467-024-51582-5

Huettel SA, Song AW, and McCarthy G (2014) Functional Magnetic Resonance Imaging. Sinauer Assoc, Sunderland, MA

Humphreys GW, Duncan J, Treisman A, et al. (1998) Place Cells, Navigational Accuracy, and the Human Hippocampus. Philos Trans R Soc Lond B Biol Sci 353 (1373):1333–1340. https://doi.org/10.1098/rstb.1998.0287

Jacobs J, Kahana MJ, Ekstrom AD, Mollison MV, and Fried I (2010) A Sense of Direction in Human Entorhinal Cortex. PNAS 107 (14):6487–6492. https://doi.org/10.1073/pnas.0911213107

Jacobs J, Weidemann CT, Miller JF, et al. (2013) Direct Recordings of Grid-Like Neuronal Activity in Human Spatial Navigation. Nat Neurosci 16 (9):1188–1190. https://doi.org/10.1038/nn.3466

Kennedy JP, Zhou Y, Qin Y, et al. (2022) A Direct Comparison of Theta Power and Frequency to Speed and Acceleration. J Neurosci 42 (21):4326–4341. https://doi.org/10.1523/JNEUROSCI.0987-21.2022

Kim M, Jeffery KJ, and Maguire EA (2017) Multivoxel Pattern Analysis Reveals 3D Place Information in the Human Hippocampus. J Neurosci 37 (16):4270–4279. https://doi.org/10.1523/JNEUROSCI.2703-16.2017

Krupic J, Bauza M, Burton S, Barry C, and O'Keefe J (2015) Grid Cell Symmetry Is Shaped by Environmental Geometry. Nature 518 (232):235. https://doi.org/10.1038/nature14153

Kunz L, Brandt A, Reinacher PC, et al. (2021) A Neural Code for Egocentric Spatial Maps in the Human Medial Temporal Lobe. Neuron 109 (17):2781–2796.e2710. https://doi.org/10.1016/j.neuron.2021.06.019

Kunz L, Maidenbaum S, Chen D, et al. (2019) Mesoscopic Neural Representations in Spatial Navigation. Trends Cogn Sci 23 (7):615–630. https://doi.org/10.1016/j.tics.2019.04.011

Kunz L, Schröder TN, Lee H, et al. (2015) Reduced Grid-Cell-Like Representations in Adults at Genetic Risk for Alzheimer's Disease. Science 350 (6259):430–433. https://doi.org/10.1126/science.aac8128

Kunz L, Staresina BP, Reinacher PC, et al. (2024) Ripple-Locked Coactivity of Stimulus-Specific Neurons and Human Associative Memory. Nat Neurosci 27 (3):587–599. https://doi.org/10.1038/s41593-023-01550-x

LaChance PA, Todd TP, and Taube JS (2019) A Sense of Space in Postrhinal Cortex. Science 365 (6449):eaax4192. https://doi.org/10.1126/science.aax4192

Lee SA, Miller JF, Watrous AJ, et al. (2018) Electrophysiological Signatures of Spatial Boundaries in the Human Subiculum. J Neurosci 38:3265–3272. https://doi.org/10.1523/jneurosci.3216-17.2018

Lever C, Burton S, Jeewajee A, O'Keefe J, and Burgess N (2009) Boundary Vector Cells in the Subiculum of the Hippocampal Formation. J Neurosci 29 (31):9771–9777. https://doi.org/10.1523/jneurosci.1319-09.2009

Liu Y, Dolan RJ, Higgins C, et al. (2021) Temporally Delayed Linear Modelling (TDLM) Measures Replay in Both Animals and Humans. eLife 10:e66917. https://doi.org/10.7554/eLife.66917

Logothetis NK, Pauls J, Augath M, Trinath T, and Oeltermann A (2001) Neurophysiological Investigation of the Basis of the fMRI Signal. Nature 412 (6843):150–157. https://doi.org/10.1038/35084005

Long X, and Zhang S-J (2021) A Novel Somatosensory Spatial Navigation System Outside the Hippocampal Formation. Cell Res 31 (6):649–663. https://doi.org/10.1038/s41422-020-00448-8

Maguire EA, Burgess N, Donnett JG, et al. (1998) Knowing Where and Getting There: A Human Navigation Network. Science 280 (5365):921–924. https://doi.org/10.1126/science.280.5365.921

Maidenbaum S, Miller J, Stein JM, and Jacobs J (2018) Grid-Like Hexadirectional Modulation of Human Entorhinal Theta Oscillations. PNAS 115 (42):10798–10803. https://doi.org/10.1073/pnas.1805007115

Marchette SA, Vass LK, Ryan J, and Epstein RA (2014) Anchoring the Neural Compass: Coding of Local Spatial Reference Frames in Human Medial Parietal Lobe. Nat Neurosci 17 (11):1598–1606. https://doi.org/10.1038/nn.3834

Miller JF, Neufang M, Solway A, et al. (2013) Neural Activity in Human Hippocampal Formation Reveals the Spatial Context of Retrieved Memories. Science 342:1111–1114. https://doi.org/10.1126/science.1244056

Mitchell AS, Czajkowski R, Zhang N, Jeffery K, and Nelson AJD (2018) Retrosplenial Cortex and Its Role in Spatial Cognition. Brain Neurosci Adv 2:2398212818757098. https://doi.org/10.1177/2398212818757098

Morgan LK, MacEvoy SP, Aguirre GK, and Epstein RA (2011) Distances between Real-World Locations Are Represented in the Human Hippocampus. J Neurosci 31 (4):1238–1245. https://doi.org/10.1523/JNEUROSCI.4667-10.2011

Mormann F, Kornblith S, Cerf M, et al. (2017) Scene-Selective Coding by Single Neurons in the Human Parahippocampal Cortex. PNAS 114 (5):1153–1158. https://doi.org/10.1073/pnas.1608159113

Moser EI, Moser M-B, and McNaughton BL (2017) Spatial Representation in the Hippocampal Formation: A History. Nat Neurosci 20 (11):1448–1464. https://doi.org/10.1038/nn.4653

Moser M-B, Rowland DC, and Moser EI (2015) Place Cells, Grid Cells, and Memory. Cold Spring Harb Perspect Biol 7 (2):a021808. https://doi.org/10.1101/cshperspect.a021808

Muller RU, Bostock E, Taube JS, and Kubie JL (1994) On the Directional Firing Properties of Hippocampal Place Cells. J Neurosci 14 (12):7235–7251. https://doi.org/10.1523/JNEUROSCI.14-12-07235.1994

Nadasdy Z, Nguyen TP, Török Á, et al. (2017) Context-Dependent Spatially Periodic Activity in the Human Entorhinal Cortex. PNAS 114 (17):E3516–E3525. https://doi.org/10.1073/pnas.1701352114

Nau M, Navarro Schröder T, Frey M, and Doeller CF (2020) Behavior-Dependent Directional Tuning in the Human Visual-Navigation Network. Nat Commun 11 (1):3247. https://doi.org/10.1038/s41467-020-17000-2

Nolan CR, Vromen JMG, Cheung A, and Baumann O (2018) Evidence against the Detectability of a Hippocampal Place Code Using Functional Magnetic Resonance Imaging. eNeuro 5 (4):ENEURO.0177–0118.2018. https://doi.org/10.1523/ENEURO.0177-18.2018

O'Keefe J, and Burgess N (1996) Geometric Determinants of the Place Fields of Hippocampal Neurons. Nature 381 (6581):425–428. https://doi.org/10.1038/381425a0

O'Keefe J, and Dostrovsky J (1971) The Hippocampus as a Spatial Map: Preliminary Evidence from Unit Activity in the Freely-Moving Rat. Brain Res 34 (1):171–175. https://doi.org/10.1016/0006-8993(71)90358-1

Park S, and Chun MM (2009) Different Roles of the Parahippocampal Place Area (PPA) and Retrosplenial Cortex (RSC) in Panoramic Scene Perception. Neuroimage 47 (4):1747–1756. https://doi.org/10.1016/j.neuroimage.2009.04.058

Payne HL, Lynch GF, and Aronov D (2021) Neural Representations of Space in the Hippocampus of a Food-Caching Bird. Science 373 (6552):343–348. https://doi.org/10.1126/science.abg2009

Pérez-Escobar JA, Kornienko O, Latuske P, Kohler L, and Allen K (2016) Visual Landmarks Sharpen Grid Cell Metric and Confer Context Specificity to Neurons of the Medial Entorhinal Cortex. eLife 5:e16937. https://doi.org/10.7554/eLife.16937

Qasim SE, Miller J, Inman CS, et al. (2019) Memory Retrieval Modulates Spatial Tuning of Single Neurons in the Human Entorhinal Cortex. Nat Neurosci 22 (12):2078–2086. https://doi.org/10.1038/s41593-019-0523-z

Redish AD, Battaglia FP, Chawla MK, et al. (2001) Independence of Firing Correlates of Anatomically Proximate Hippocampal Pyramidal Cells. J Neurosci 21 (5):RC134. https://doi.org/10.1523/JNEUROSCI.21-05-j0004.2001

Rodriguez PF (2010) Neural Decoding of Goal Locations in Spatial Navigation in Humans with fMRI. Hum Brain Mapp 31 (3):391–397. https://doi.org/10.1002/hbm.20873

Rosenbaum RS, Ziegler M, Winocur G, Grady CL, and Moscovitch M (2004) "I Have Often Walked Down This Street Before": fMRI Studies on the Hippocampus and Other Structures during Mental Navigation of an Old Environment. Hippocampus 14 (7):826–835. https://doi.org/10.1002/hipo.10218

Sargolini F, Fyhn M, Hafting T, et al. (2006) Conjunctive Representation of Position, Direction, and Velocity in Entorhinal Cortex. Science 312 (5774):758–762. https://doi.org/10.1126/science.1125572

Schapiro AC, McDevitt EA, Rogers TT, Mednick SC, and Norman KA (2018) Human Hippocampal Replay during Rest Prioritizes Weakly Learned Information and Predicts Memory Performance. Nat Commun 9 (1):3920. https://doi.org/10.1038/s41467-018-06213-1

Schuck NW, and Niv Y (2019) Sequential Replay of Nonspatial Task States in the Human Hippocampus. Science 364 (6447):eaaw5181. https://doi.org/10.1126/science.aaw5181

Sharp P (2006) Subicular Place Cells Generate the Same "Map" for Different Environments: Comparison with Hippocampal Cells. Behav Brain Res 174 (2):206–214. https://doi.org/10.1016/j.bbr.2006.05.034

Shine JP, Valdés-Herrera JP, Tempelmann C, and Wolbers T (2019) Evidence for Allocentric Boundary and Goal Direction Information in the Human Entorhinal Cortex and Subiculum. Nat Commun 10 (1):4004. https://doi.org/10.1038/s41467-019-11802-9

Solstad T, Boccara CN, Kropff E, Moser M-B, and Moser EI (2008) Representation of Geometric Borders in the Entorhinal Cortex. Science 322 (5909):1865–1868. https://doi.org/10.1126/science.1166466

Stacho M, and Manahan-Vaughan D (2022) Mechanistic Flexibility of the Retrosplenial Cortex Enables Its Contribution to Spatial Cognition. Trends Neurosci 45 (4):284–296. https://doi.org/10.1016/j.tins.2022.01.007

Stangl M, Maoz SL, and Suthana N (2023) Mobile Cognition: Imaging the Human Brain in the "Real World." Nat Rev Neurosci 24 (6):347–362. https://doi.org/10.1038/s41583-023-00692-y

Stangl M, Topalovic U, Inman CS, et al. (2020) Boundary-Anchored Neural Mechanisms of Location-Encoding for Self and Others. Nature:1–6. https://doi.org/10.1038/s41586-020-03073-y

Staudigl T, Leszczynski M, Jacobs J, et al. (2018) Hexadirectional Modulation of High-Frequency Electrophysiological Activity in the Human Anterior Medial Temporal Lobe Maps Visual Space. Curr Biol 28 (20):3325–3329.e3324. https://doi.org/10.1016/j.cub.2018.09.035

Stemmler M, Mathis A, and Herz AVM (2015) Connecting Multiple Spatial Scales to Decode the Population Activity of Grid Cells. Sci Adv 1 (11):e1500816–e1500816. https://doi.org/10.1126/science.1500816

Stensola H, Stensola T, Solstad T, et al. (2012) The Entorhinal Grid Map Is Discretized. Nature 492 (7427):72–78. https://doi.org/10.1038/nature11649

Sulpizio V, Committeri G, and Galati G (2014) Distributed Cognitive Maps Reflecting Real Distances between Places and Views in the Human Brain. Front Hum Neurosci 8:00716. https://doi.org/10.3389/fnhum.2014.00716

Taube JS, Muller R, and Ranck J (1990) Head-Direction Cells Recorded from the Postsubiculum in Freely Moving Rats. I. Description and Quantitative Analysis. J Neurosci 10 (2):420–435. https://doi.org/10.1523/jneurosci.10-02-00420.1990

Tsitsiklis M, Miller J, Qasim SE, et al. (2020) Single-Neuron Representations of Spatial Targets in Humans. Curr Biol 30 (2):245–253.e244. https://doi.org/10.1016/j.cub.2019.11.048

Vann SD, Aggleton JP, and Maguire EA (2009) What Does the Retrosplenial Cortex Do? Nat Rev Neurosci 10 (11):792–802. https://doi.org/10.1038/nrn2733

Vass LK, and Epstein RA (2013) Abstract Representations of Location and Facing Direction in the Human Brain. J Neurosci 33 (14):6133–6142. https://doi.org/10.1523/JNEUROSCI.3873-12.2013

Vicek K, Fajnerova I, Nekovarova T, et al. (2020) Mapping the Scene and Object Processing Networks by Intracranial EEG. Front Hum Neurosci 14:561399. https://doi.org/10.3389/fnhum.2020.561399

Wang C, Chen X, Lee H, et al. (2018) Egocentric Coding of External Items in the Lateral Entorhinal Cortex. Science 362 (6417):945–949. https://doi.org/10.1126/science.aau4940

Weiner KS, Barnett MA, Witthoft N, et al. (2018) Defining the Most Probable Location of the Parahippocampal Place Area Using Cortex-Based Alignment and Cross-Validation. Neuroimage 170:373–384. https://doi.org/10.1016/j.neuroimage.2017.04.040

Wirtshafter HS, and Disterhoft JF (2023) Place Cells Are Nonrandomly Clustered by Field Location in CA1 Hippocampus. Hippocampus 33 (2):65–84. https://doi.org/10.1002/hipo.23489

Wittkuhn L, and Schuck NW (2021) Dynamics of fMRI Patterns Reflect Sub-Second Activation Sequences and Reveal Replay in Human Visual Cortex. Nat Commun 12 (1):1795. https://doi.org/10.1038/s41467-021-21970-2

Woolnough O, Kadipasaoglu CM, Conner CR, et al. (2022) Dataset of Human Intracranial Recordings during Famous Landmark Identification. Sci Data 9 (1):28. https://doi.org/10.1038/s41597-022-01125-8

Yartsev MM, and Ulanovsky N (2013) Representation of Three-Dimensional Space in the Hippocampus of Flying Bats. Science 340 (6130):367–372. https://doi.org/10.1126/science.1235338

7

What Do We Know about Navigation Development from the Perspective of Cognitive Neuroscience in Humans?

Daniel D. Dilks, Rebecca J. Rennert, and Yaelan Jung

Abstract Our ability to navigate through the environment forms the bedrock of many of our essential, everyday behaviors. Given this ecological importance, it is perhaps not surprising that we have dedicated neural tissue supporting navigation, including the hippocampus and entorhinal cortex in the medial temporal lobe (MTL) as well as the retrosplenial complex (RSComp) (also known as the medial place area), occipital place area (OPA), and superior parietal lobule (SPL) in neocortex. Recently, it has been proposed that each of these regions may support a distinct role in adult human navigation, with MTL and RSComp involved in "map-based navigation" and OPA and SPL involved in "visually guided navigation." Despite advances in understanding how these brain regions support navigation in adulthood, surprisingly little is known about how they develop. To address this gap, the limited body of literature on the development of MTL and RSComp, and their involvement in map-based navigation, is reviewed in this chapter. In addition, a novel, and perhaps counterintuitive, hypothesis is proposed; namely, that map-based navigation (including MTL and RSComp) emerges early, by at least five years of age. Thereafter, the few studies available on the development of OPA and its involvement in visually guided navigation are reviewed, and a further hypothesis is advanced that OPA and its involvement in visually navigation develops late, possibly in a discontinuous manner, and does not emerge until around eight years of age.

Keywords Spatial development, brain imaging, higher level vision, scene processing, pediatric fMRI

Daniel D. Dilks (✉)
Dept. of Psychology, Emory University, Atlanta, GA 30322, USA
Email: dilks@emory.edu

Rebecca J. Rennert
Dept. of Psychology, Emory University, Atlanta, GA 30322, USA

Yaelan Jung
Dept. of Psychology, Emory University, Atlanta, GA 30322, USA

N. S. Newcombe and K. Cheng (eds.), *Challenges in Navigation Research*, Strüngmann Forum Reports, https://doi.org/10.1007/978-3-032-20563-6_7

1 Introduction

Over the past three decades, cognitive neuroscience research has uncovered several key brain regions that support navigation in adult humans, including the hippocampus and entorhinal cortex in the MTL, as well as three cortical regions: the RSComp (Maguire 2001), the OPA (Dilks et al. 2013), and the SPL (Yoon et al. 2025). Beyond establishing the general involvement of these structures in navigation, however, recent functional magnetic resonance imaging (fMRI) research suggests that each of these regions may play a distinct role within the broader domain of navigation (Dilks et al. 2022).

More specifically, MTL and RSComp—also referred to as the medial place area (Silson et al. 2016)—have been implicated in "map-based navigation," which is the ability to travel from a specific location to distant place (e.g., from your house to a favorite restaurant across town). These brain regions represent information necessary for map-based navigation, including location, orientation, distance, borders and boundaries, landmarks, and spatial layout (for reviews, see Dilks et al. 2022 and Chapter 9). By contrast, it has been proposed that OPA and SPL are involved in a distinct form of navigation, referred to as "visually guided navigation"—the ability to move about the immediately visible environment while avoiding boundaries and obstacles, such walking around a kitchen without bumping into the cabinets or the kitchen table (Dilks et al. 2022). Indeed, both OPA and SPL represent information necessary for visually guided navigation, including egocentric "sense" information (i.e., left/right orientation) (Dilks et al. 2011; Yoon et al. 2025) and first-person perspective motion information through scenes that mimic the visual experience of navigating through immediately visible environments (Kamps et al. 2016; Yoon et al. 2025). Moreover, OPA and SPL are preferentially functionally connected to each other (Yoon et al. 2025). Critically, however, OPA does not represent information necessary for map-based navigation (i.e., location and orientation) (Persichetti and Dilks 2019), and SPL has not yet been explicitly tested on this information. Finally, studies investigating the "function" of these regions, not just the information represented, found that OPA and SPL show greater activation when participants are engaged in a visually guided navigation task compared to two control tasks, a pattern not seen in MTL and RSComp (Persichetti and Dilks 2018; Yoon et al. 2025). Yet, despite this growing understanding about how MTL, RSComp, OPA, and SPL support navigation in adult humans, we know shockingly little about how these systems develop. So, how do these brain regions develop?

After reviewing what little is known about the development of MTL and RSComp and their involvement in map-based navigation, we offer the novel, perhaps counterintuitive, hypothesis that map-based navigation (including MTL and RSComp) emerges early in life—by at least five years of age, if not earlier. We then discuss the few available studies on the development of OPA, which suggest that OPA and its involvement in visually guided navigation develop surprisingly late, not until around eight years of age. Further, we propose that OPA and its involvement in visually guided navigation develops discontinuously. Finally, since studies have not yet investigated the development of SPL and its involvement in visually guided navigation, future research is needed.

2 The Early Development of MTL and Its Involvement in Map-Based Navigation

Despite the well-established developmental models of MTL in animals (for a review, see Chapter 8), how MTL begins to support map-based navigation in humans remains largely unknown. Instead, the developmental trajectory of hippocampus has been studied in terms of its structure and connectivity to the rest of the brain, and findings suggest a prolonged development. For example, the volume of the hippocampus continues to change until approximately 15 years of age (Daugherty et al. 2016; Gogtay et al. 2006; Krogsrud et al. 2014). Likewise, functional connectivity between the hippocampus and other cortical regions continues to mature throughout late childhood (DeMaster et al. 2014). In line with these anatomical and connectivity findings, numerous behavioral studies have suggested prolonged development in map-based navigation and related abilities presumably supported by MTL (Burles et al. 2020; Cousins et al. 1983; Lehnung et al. 2003; Nazareth et al. 2018; Overman et al. 1996). A child's ability to encode and utilize landmarks, for example, improves between the ages of 6–10 years (Allen et al. 1979; Cohen and Schuepfer 1980; Jansen-Osmann and Wiedenbauer 2004; Nys et al. 2015), and it is still not adultlike even at age 10 (Allen et al. 1979; Cohen and Schuepfer 1980). Furthermore, route learning (i.e., the acquisition of sequential or metric information between landmarks) and path-integration skills in seven- to ten-year-olds are less developed than those of adults (Allen et al. 1979; Burles et al. 2020; Cohen and Schuepfer 1980; Cousins et al. 1983; Nazareth et al. 2018). Together, these behavioral studies provide indirect evidence supporting the hypothesis that the MTL may develop quite late in childhood.

More direct evidence in support of this hypothesis comes from a recent behavioral study that adapted a virtual environment paradigm from an adult fMRI study that was aimed at investigating the role of hippocampus in map-based navigation (Doeller et al. 2008). In that study, which used a virtual large-scale space with unique geometry and landmark objects, Doeller et al. (2008) found that the hippocampus was involved in geometry-based (or boundary-based) learning of the environment, while the dorsal striatum was involved in landmark- or object-based learning. Using this same paradigm, Julian et al. (2019) behaviorally tested six- to ten-year-old children on geometry- versus landmark-based learning. They found that children, compared to adults, were worse at geometry-based learning but not landmark-based learning, which suggests that geometry-based learning (supported by the hippocampus) matures later than landmark-based learning (supported by the dorsal striatum). Although these behavioral measures serve only as proxies for neural development, they nevertheless provide more direct evidence for the hypothesis that map-based navigation supported by MTL undergoes prolonged development.

However, the prolonged development of the hippocampal structure, its connectivity, and even map-based navigation abilities themselves does *not* necessarily mean that the hippocampus was not involved in map-based navigation earlier in life. Indeed, in a recent fMRI study with five-year-old children (Jung and Dilks 2025), we directly examined the "function" of the developing hippocampus and demonstrated that the hippocampus is, in fact, involved in map-based navigation *early* in childhood. Specifically, we created a virtual town, which children learned by navigating through it via button presses. The town had three different "locations": a mountain, a lake, and trees (see Figure 7.1), and each location contained two buildings, for a total of six buildings in the town. Each building was a specific type; there were two ice cream stores, two playgrounds, and two fire stations. Importantly, the location of any two buildings belonging to the same category was completely dissociable from the category information. For instance, if one ice cream store was in the mountain location, the other ice cream store was in the lake location. This design allowed us to characterize the neural response to buildings located in the same location (e.g., an ice cream store and a fire station both at the mountain location) separately from the neural response to buildings in the same category (e.g., ice cream stores regardless of location). We then asked whether the hippocampus in five-year-olds represents location information critical for map-based navigation and found that indeed the hippocampus (along with RSComp, see below) represents the location information of buildings. Thus, by at least five years of age, the hippocampus is able to support map-based navigation; this is consistent with animal models that have suggested the early development of the hippocampus (e.g., Wills et al. 2014). Interestingly, the hippocampus also represented category information of buildings, which is not particularly relevant for map-based navigation and may reflect its broader role in declarative memory and category learning (e.g., Eichenbaum 2000). Taken together, these findings provide the first direct neural evidence that the hippocampus in humans is involved in map-based navigation by at least age five, suggesting the early emergence of the hippocampal navigation system.

How then are we to reconcile this finding with the large body of behavioral literature (discussed above) that highlights the late development of map-based navigation skills? It is important to note that most, if not all, of these studies focused on when children's navigation abilities reach full maturity (i.e., when they become adult-like) rather than on when these abilities first emerge. In fact, studies investigating younger children have shown that certain foundational skills, such as landmark recognition, begin to emerge relatively early, around five years of age (Acredolo 1977; Herman and Siegel 1978), consistent with our finding that the hippocampus represents location information at this age. Therefore, we hypothesize that map-based navigation (including the hippocampus) emerges relatively early in life, by around five years of age, if not earlier, and undergoes subsequent refinement as children

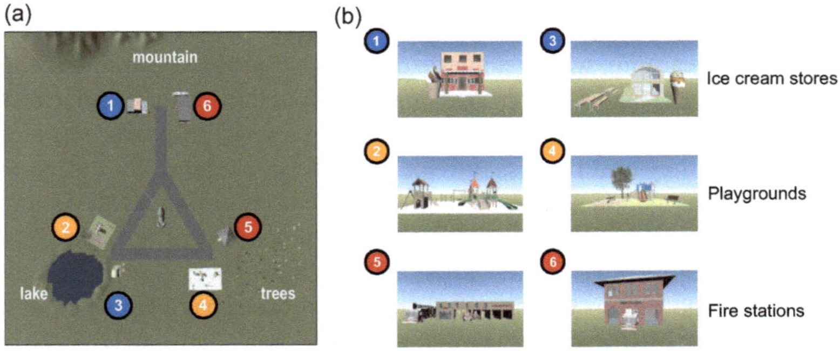

Figure 7.1 (a) An aerial view of the virtual town. Six buildings are placed in either the mountain, lake, or tree locations. (b) Each building belongs to one of three categories: ice cream store (1 and 3), playground (2 and 4), and fire station (5 and 6). Note the children never saw the aerial view of the town. Figure from Jung and Dilks (2025), used with permission.

gain more experience and acquire more advanced map-based navigational skills (e.g., route learning, integration of landmark and route). Future research, however, is needed to clarify how continued development of the hippocampus supports such refinement in children's map-based navigational abilities.

3 The Early Development of RSComp and Its Involvement in Map-Based Navigation

To date, no studies have directly examined how RSComp begins to support map-based navigation, although many have assumed that, like the hippocampus, it undergoes prolonged development (Burles et al. 2020; Cousins et al. 1983; Lehnung et al. 2003; Nazareth et al. 2018; Overman et al. 1996). Our recent research challenges this idea, revealing that RSComp, like hippocampus, develops much earlier than previously assumed (Jung and Dilks 2025).

In this work (Jung and Dilks 2025), again using fMRI and a virtual town paradigm (see Figure 7.1), we tested whether RSComp, like hippocampus, represents the location of specific places (i.e., information necessary for map-based navigation). Similar to the findings in adults (Persichetti and Dilks 2019), where RSComp represents location information but not category information (i.e., information *not* necessary for map-based navigation), we found that in five-year-old children, RSComp also represents location information, but not category information. By contrast, the parahippocampal place area—another cortical region that is scene selective (like RSComp), not involved in map-based navigation, but involved in scene categorization—showed the exact opposite pattern, representing category information but not location information. These findings suggest that the map-based navigation supported by RSComp develops early, emerging by at least age five, if not earlier.

4 The Late Development of OPA and Its Involvement in Visually Guided Navigation

Children begin to move about their immediately visible environments relatively early in life. At around 7–12 months they begin to crawl and progress to walking around 9–18 months (Newcombe 2019). This early behavioral emergence of locomotion skills often leads to the intuition that visually guided navigation would mature relatively early in childhood. Recent fMRI work in children (Jung et al. 2024; Kamps et al. 2020), however, suggests that OPA develops surprisingly late in childhood. The information necessary for visually guided navigation is not even represented at five years of age; it only emerges around eight years of age.

One of the first studies to support the hypothesis that OPA is late developing comes from an fMRI study in adults. In this study, we asked whether OPA represents visual information about crawling and walking—the two types of locomotion by which humans move about their immediately visible environment (Jones et al. 2023). Interestingly, we found that OPA does *not* represent visual information about crawling at all, showing similar response to videos from a crawling perspective and to two other videos depicting perspectives by which humans do not or cannot move about the immediately visible environment (i.e., "flying" and "scrambled" videos). Rather, OPA represents visual information from a walking perspective only, showing a greater response to the videos from a walking perspective than to any of the other videos, including crawling. These results suggest that OPA may never have supported crawling and develops only after children begin to walk. Following up on this finding, a recent fMRI study conducted on two groups of children who know how to walk (i.e., five- and eight-year-olds) revealed that OPA only shows "walking selectivity" (i.e., the greater response to the walking videos than to any of the others, including crawling) in eight-year-olds, but not in five-year-olds. This finding further supports the hypothesis that OPA is late developing and suggests that OPA develops only after a child is walking like an adult (Jung et al. 2024). Similarly, another fMRI study conducted on children in this same age range demonstrated that OPA does not represent first-person perspective motion information through scenes in five-year-olds, and only does so in 8-year-olds (Kamps et al. 2020). Taken together, these results suggest that OPA undergoes protracted development; walking selectivity and first-person perspective motion selectivity (information necessary for visually guided navigation) are not even present until around five years of age and are only detectable at around eight years of age.

The findings that OPA undergoes protracted development dovetails with several other behavioral findings. For example, in a recent behavioral study, Kamps et al. (2023) tested the development of visually guided navigation and scene categorization abilities in children between four and nine years of age and found that visually guided navigation is slower to develop (adultlike performance is not reached until around eight years of age) compared to scene categorization (adultlike performance is reached around five years of age). Finally, obstacle avoidance and efficient path taking—two key features of visually guided navigation—develop slowly, again not reaching adultlike levels until around age eight (Berard and Vallis 2006; Choi et al. 2016; Michel et al. 2010; Pryde et al. 1997).

At the same time, and as alluded to above, humans begin moving around their environment early in life (e.g., crawling at approximately nine months) and show remarkably sophisticated navigational ability within the first few years (Newcombe 2019). For example, young children can use boundaries to recover their orientation after becoming disoriented (Hermer and Spelke 1994), and infants understand whether it is safe to locomote over a "visual cliff" (Gibson and Walk 1960). How, then, can we reconcile these observations with the hypothesis that visually guided navigation undergoes protracted development? One possibility is that these tasks rely on qualitatively different systems. For example, the reorientation task may involve spatial memory systems in the hippocampus, whereas the visual cliff task may depend on basic depth perception in early visual cortex. Another possibility is that these tasks rely on the same visually guided navigation system (i.e., OPA or perhaps even SPI), and that despite development extending well into childhood, the foundations of this system are nevertheless intact early and sufficient to support these early navigational behaviors (discussed further below).

What then drives such late development of OPA? One possibility is that the cortical connectivity of OPA to primary visual cortex (V1), which enables OPA to receive visual information in the first place, undergoes protracted development. More specifically, given that peripheral vision is important for visually guided navigation (Cao and Händel 2019; Franchak and Adolph 2010; Stoffregen et al. 1987), the connectivity between OPA and the region of V1, which represents the peripheral visual field, may develop late in childhood. Using both behavior and neural data, our group is currently investigating this exact question, asking (a) when children begin to use peripheral vision for visually guided navigation and (b) when their OPA shows connectivity to "peripheral" V1. Preliminary data reveal the following:

- Eight-year-olds, like adults, use peripheral vision to perform a visually guided navigation task. Five-year-olds, however, do *not*; instead, they use central vision.
- The functional connectivity between OPA and peripheral V1 is present only in eight-year-olds (like adults), not in five-year-olds.

These findings support the hypothesis that the late development of cortical connectivity between OPA and peripheral V1 may drive the late development of OPA. They also strongly support the hypothesis that OPA (and visually guided navigation) is late developing.

Does this mean then that OPA is not involved in visually guided navigation *at all* until around eight years of the age? Since, at age five, OPA does not represent at least two kinds of navigationally relevant information (i.e., visual information from a walking perspective and first-person perspective motion information), we are left with the fascinating possibility that OPA develops abruptly (or discontinuously) only at around age eight, and not beforehand. Future research is required to address this possibility, as it could still be the case that OPA supports visually guided navigation before age eight (at five years of age even) and develops continuously with earlier emerging representations of more "primitive" kinds of navigationally relevant information (e.g., sense and egocentric distance information). Indeed, studies have shown that young toddlers (Hermer and Spelke 1994) and even nonhuman species, including fish (Sovrano et al. 2002) and insects (Wehner et al. 1996), can perceive and utilize sense and egocentric distance information for navigation. Thus, the question remains: How does OPA develop—continuously or discontinuously? To approach this question, a recent study by our lab used fMRI adaptation to analyze whether OPA represents sense information at age five (Rennert and Dilks, in press). This data reveal that sense information is not at all represented in OPA in five-year-olds but is represented in eight-year-olds, consistent with the hypothesis that OPA develops discontinuously. Interestingly, preliminary data also reveal that sense information is represented in RSComp in five-year-olds (as well as eight-year-olds), consistent with the hypothesis that RSComp develops early.

5 Conclusion

Despite decades of research on navigation in the adult human brain, how these systems develop remains largely unknown. Here, we examined behavioral and fMRI studies on the development of the map-based navigation system, including MTL and RSComp, as well as recent behavioral and fMRI studies on the development of the visually guided navigation system, focusing on OPA. Despite the anatomical and connectivity findings and behavioral work, which suggests that the map-based navigation system develops late (and is really revealed when the system is fully mature), a recent fMRI study found that MTL and RSComp are both involved in map-based navigation much earlier in life, by at least age five, if not before. By contrast, findings for OPA reveal a surprisingly late developmental trajectory, with no evidence of visually guided information processing until around age eight. This raises the intriguing possibility of discontinuous development. Taken together, these findings lead to the novel (and perhaps counterintuitive) developmental hypothesis that the visually guided navigation system (including OPA and perhaps SPL) develops later than the map-based navigation system (including MTL and RSComp).

Several important questions, however, remain. First, although human fMRI has illuminated the roles of RSComp, OPA, and SPL in navigation, very little is known about these systems in animal models. Future research is thus needed to examine how these cortical navigation systems function and develop in nonhuman species (if they even exist), which would provide critical mechanistic insight into the evolution and function of navigation systems across species. Second, while the development of MTL has been extensively studied in animals, relatively little is known about how these regions develop to support map-based navigation in humans. Understanding how the MTL emerges and potentially interacts with the cortical navigation regions (e.g., RSComp, and perhaps OPA and SPL) during development remains a central challenge for the field—one that will be essential for a comprehensive theory of navigation across both species and development.

Acknowledgments We would like to thank the Child Study Center at Emory for supporting recruitment efforts and the Facility for Education and Research in Neuroscience (FERN) at Emory University for providing neuroimaging resources. Finally, we would like to thank all of the children and their families who participated in these studies.

References

Acredolo LP (1977) Developmental Changes in the Ability to Coordinate Perspectives of a Large-Scale Space. Dev Psychol 13 (1):1–8. https://doi.org/10.1037/0012-1649.13.1.1

Allen GL, Kirasic KC, Siegel AW, and Herman JF (1979) Developmental Issues in Cognitive Mapping: The Selection and Utilization of Environmental Landmarks. Child Dev 50 (4):1062–1070. https://doi.org/10.2307/1129332

Berard JR, and Vallis LA (2006) Characteristics of Single and Double Obstacle Avoidance Strategies: A Comparison between Adults and Children. Exp Brain Res 175 (1):21–31. https://doi.org/10.1007/s00221-006-0529-0

Burles F, Liu I, Hart C, et al. (2020) The Emergence of Cognitive Maps for Spatial Navigation in 7– to 10-Year-Old Children. Child Dev 91 (3):e733–e744. https://doi.org/10.1111/cdev.13285

Cao L, and Händel B (2019) Walking Enhances Peripheral Visual Processing in Humans. PLoS Biol 17 (10):e3000511. https://doi.org/10.1371/journal.pbio.3000511

Choi JT, Jensen P, and Nielsen JB (2016) Locomotor Sequence Learning in Visually Guided Walking. J Neurophysiol 115 (4):2014–2020. https://doi.org/10.1152/jn.00938.2015
Cohen R, and Schuepfer T (1980) The Representation of Landmarks and Routes. Child Dev 51 (4):1065–1071. https://psycnet.apa.org/doi/10.2307/1129545
Cousins JH, Siegel AW, and Maxwell SE (1983) Way Finding and Cognitive Mapping in Large-Scale Environments: A Test of a Developmental Model. J Exp Child Psychol 35 (1):1–20. https://doi.org/10.1016/0022-0965(83)90066-8
Daugherty AM, Bender AR, Raz N, and Ofen N (2016) Age Differences in Hippocampal Subfield Volumes from Childhood to Late Adulthood. Hippocampus 26 (2):220–228. https://doi.org/10.1002/hipo.22517
DeMaster D, Pathman T, Lee JK, and Ghetti S (2014) Structural Development of the Hippocampus and Episodic Memory: Developmental Differences Along the Anterior/Posterior Axis. Cereb Cortex 24 (11):3036–3045. https://doi.org/10.1093/cercor/bht160
Dilks DD, Julian JB, Kubilius J, Spelke ES, and Kanwisher N (2011) Mirror-Image Sensitivity and Invariance in Object and Scene Processing Pathways. J Neurosci 31 (31):11305–11312. https://doi.org/10.1523/JNEUROSCI.1935-11.2011
Dilks DD, Julian JB, Paunov AM, and Kanwisher N (2013) The Occipital Place Area Is Causally and Selectively Involved in Scene Perception. J Neurosci 33 (4):1331–1336. https://doi.org/10.1523/JNEUROSCI.4081-12.2013
Dilks DD, Kamps FS, and Persichetti AS (2022) Three Cortical Scene Systems and Their Development. Trends Cogn Sci 26 (2):117–127. https://doi.org/1di0.1016/j.tics.2021.11.002
Doeller CF, King JA, and Burgess N (2008) Parallel Striatal and Hippocampal Systems for Landmarks and Boundaries in Spatial Memory. PNAS 105 (15):5915–5920. https://doi.org/10.1073/pnas.0801489105
Eichenbaum H (2000) A Cortical-Hippocampal System for Declarative Memory. Nat Rev Neurosci 1:41–50. https://doi.org/10.1038/35036213
Franchak JM, and Adolph KE (2010) Visually Guided Navigation: Head-Mounted Eye-Tracking of Natural Locomotion in Children and Adults. Vision Res 50 (24):2766–2774. https://doi.org/10.1016/j.visres.2010.09.024
Gibson EJ, and Walk RD (1960) The "Visual Cliff." Sci Am 202 (4):64–71. https://psycnet.apa.org/doi/10.1038/scientificamerican0460-64
Gogtay N, Nugent III TF, Herman DH, et al. (2006) Dynamic Mapping of Normal Human Hippocampal Development. Hippocampus 16 (8):664–672. https://doi.org/10.1002/hipo.20193
Herman JF, and Siegel AW (1978) The Development of Cognitive Mapping of the Large-Scale Environment. J Exp Child Psychol 26 (3):389–406. https://doi.org/10.1016/0022-0965(78)90120-0
Hermer L, and Spelke ES (1994) A Geometric Process for Spatial Reorientation in Young Children. Nature 370 (6484):57–59. https://doi.org/10.1038/370057a0
Jansen-Osmann P, and Wiedenbauer G (2004) The Representation of Landmarks and Routes in Children and Adults: A Study in a Virtual Environment. J Environ Psychol 24 (3):347–357. https://doi.org/10.1016/j.jenvp.2004.08.003
Jones CM, Byland J, and Dilks DD (2023) The Occipital Place Area Represents Visual Information about Walking, Not Crawling. Cereb Cortex 33 (12):7500–7505. https://doi.org/10.1093/cercor/bhad055
Julian JB, Kamps FS, Epstein RA, and Dilks DD (2019) Dissociable Spatial Memory Systems Revealed by Typical and Atypical Human Development. Dev Sci 22 (2):e12737. https://doi.org/10.1111/desc.12737
Jung Y, and Dilks DD (2025) Early Development of Navigationally Relevant Location Information in the Retrosplenial Complex. PNAS 122 (19):e2503569122. https://doi.org/10.1073/pnas.2503569122

Jung Y, Hsu D, and Dilks DD (2024) "Walking Selectivity" in the Occipital Place Area in 8-Year-Olds, Not 5-Year-Olds. Cereb Cortex 34 (3):bhae101. https://doi.org/10.1093/cercor/bhae101

Kamps FS, Lall V, and Dilks DD (2016) The Occipital Place Area Represents First-Person Perspective Motion in Formation through Scenes. Cortex 83:17–26. https://doi.org/10.1016/j.cortex.2016.06.022

Kamps FS, Pincus JE, Radwan SF, Wahab S, and Dilks DD (2020) Late Development of Navigationally Relevant Motion Processing in the Occipital Place Area. Curr Biol 30 (3):544–550.e543. https://doi.org/10.1016/j.cub.2019.12.008

Kamps FS, Rennert RJ, Radwan SF, et al. (2023) Dissociable Cognitive Systems for Recognizing Places and Navigating through Them: Developmental and Neuropsychological Evidence. J Neurosci 43 (36):6320–6329. https://doi.org/10.1523/JNEUROSCI.0153-23.2023

Krogsrud SK, Tamnes CK, Fjell AM, et al. (2014) Development of Hippocampal Subfield Volumes from 4 to 22 Years. Hum Brain Mapp 35 (11):5646–5657. https://doi.org/10.1002/hbm.22576

Lehnung M, Leplow B, Ekroll V, et al. (2003) The Role of Locomotion in the Acquisition and Transfer of Spatial Knowledge in Children. Scand J Psychol 44 (2):79–86. https://doi.org/10.1111/1467-9450.00324

Maguire EA (2001) The Retrosplenial Contribution to Human Navigation: A Review of Lesion and Neuroimaging Findings. Scand J Psychol 42 (3):225–238. https://doi.org/10.1111/1467-9450.00233

Michel J, Grobet C, Dietz V, and Hedel HJA (2010) Obstacle Stepping in Children: Task Acquisition and Performance. Gait Posture 31 (3):341–346. https://doi.org/10.1016/j.gaitpost.2009.12.001

Nazareth A, Weisberg SM, Margulis K, and Newcombe NS (2018) Charting the Development of Cognitive Mapping. J Exp Child Psychol 170:86–106. https://doi.org/10.1016/j.jecp.2018.01.009

Newcombe NS (2019) Navigation and the Developing Brain. J Exp Biol 222 (Suppl_1):jeb186460. https://doi.org/10.1242/jeb.186460

Nys M, Gyselinck V, Orriols E, and Hickmann M (2015) Landmark and Route Knowledge in Children's Spatial Representation of a Virtual Environment. Front Psychol 5:fpsyg.2014.01522. https://doi.org/10.3389/fpsyg.2014.01522

Overman WH, Pate BJ, Moore K, and Peuster A (1996) Ontogeny of Place Learning in Children as Measured in the Radial Arm Maze, Morris Search Task, and Open Field Task. Behav Neurosci 110 (6):1205–1228. https://doi.org/10.1037//0735-7044.110.6.1205

Persichetti AS, and Dilks DD (2018) Dissociable Neural Systems for Recognizing Places and Navigating through Them. J Neurosci 38 (48):10295–10304. https://doi.org/10.1523/JNEUROSCI.1200-18.2018

——— (2019) Distinct Representations of Spatial and Categorical Relationships across Human Scene-Selective Cortex. PNAS 116 (42):21312–21317. https://doi.org/10.1073/pnas.1903057116

Pryde KM, Roy EA, and Patla AE (1997) Age-Related Trends in Locomotor Ability and Obstacle Avoidance. Hum Mov Sci 16 (4):507–516. https://doi.org/10.1016/S0167-9457(96)00064-4

Rennert RJ and Dilks DD (in press). Two Cortical Navigation Systems and Their Development. Cerebral Cortex, in press. https://doi.org/10.1093/cercor/bhaf326

Silson E, Steel A, and Baker C (2016) Scene-Selectivity and Retinotopy in Medial Parietal Cortex. Front Hum Neurosci 10:412. https://doi.org/10.3389/fnhum.2016.00412

Sovrano VA, Bisazza A, and Vallortigara G (2002) Modularity and Spatial Reorientation in a Simple Mind: Encoding of Geometric and Nongeometric Properties of a Spatial Environment by Fish. Cognition 85 (2):B51–B59. https://doi.org/10.1016/S0010-0277(02)00110-5

Stoffregen TA, Schmuckler MA, and Gibson EJ (1987) Use of Central and Peripheral Optical Flow in Stance and Locomotion in Young Walkers. Perception 16 (1):113–119. https://doi.org/10.1068/p160113

Wehner R, Michel B, and Antonsen P (1996) Visual Navigation in Insects: Coupling of Egocentric and Geocentric Information. J Exp Biol 199 (Pt 1):129–140. https://doi.org/10.1242/jeb.199.1.129

Wills TJ, Muessig L, and Cacucci F (2014) The Development of Spatial Behaviour and the Hippocampal Neural Representation of Space. Philos Trans R Soc Lond B Biol Sci 369 (1635):20130409. https://doi.org/10.1098/rstb.2013.0409

Yoon HK, Jung Y, Persichetti AS, and Dilks DD (2025) A Scene-Selective Region in the Superior Parietal Lobule for Visually Guided Navigation. Cereb Cortex 35 (4):bhaf082. https://doi.org/10.1093/cercor/bhaf082

8

How Can Development Inform Us about the Neural Mechanisms of Navigation in Rats?

Thomas J. Wills and Francesca Cacucci

Abstract Studying the ontogeny of spatial memory and neural maps has improved our understanding of navigation in adulthood. Specifically, development affords the opportunity to uncover the cognitive building blocks that contribute to navigation and pinpoint which specific neural circuits support these. Focusing on the rat as a model organism, this chapter reviews current knowledge regarding the development of navigational behavior and the emergence of spatially modulated neural activity. Conclusions regarding links between behavioral and neural development are highlighted: map-based navigation, involving the ability to construct a vectorial path to an unmarked goal, emerges at around three weeks of age concurrently with grid cells in entorhinal cortex. However, several elements of spatial cognition, including context and object recognition and neural representations of direction and boundaries, are present before this age. The gain in function that occurs around three weeks of age may, therefore, be relatively specific (e.g., calculation of a path vector to a goal). Although three weeks marks the first emergence of map-based navigation, fully adultlike spatial memory continues to undergo a protracted development. Potential neural substrates for this prolonged development include the spatial stability and ordered temporal firing patterns of hippocampal place cells.

Keywords Memory, navigation, development, hippocampus

Thomas J. Wills (✉)
Dept. of Cell and Development Biology, University College London, London WC1E 6BT, UK
Email: t.wills@ucl.ac.uk

Francesca Cacucci
Dept. of Neuroscience, Physiology and Pharmacology, University College London,
London WC1E 6BT, UK

N. S. Newcombe and K. Cheng (eds.), *Challenges in Navigation Research*,
Strüngmann Forum Reports, https://doi.org/10.1007/978-3-032-20563-6_8

1 Introduction

1.1 The Value of Studying Development

In the broadest sense, navigation may be thought of as any organized movement in space that increases survival and evolutionary fitness. This broad definition, however, captures a variety of different strategies, based on different elemental cognitive building blocks and/or neural representations (Jeffery et al. 2024). Dissociating these cognitive processes through the manipulation of the activity in specific neural circuits in the adult is possible but remains technically challenging.

To address this complexity, one promising approach is to study how neural circuits for navigation are *assembled* during postnatal development. Here, our focus is on the laboratory rat, *Rattus norvegicus*, an altricial species that displays (similarly to humans) a relatively late ontogenetic emergence of navigation. This protracted development affords the opportunity to study whether different navigational strategies, and the cognitive processes and representations upon which they rely, emerge at different times and, if so, in what order. Knowing the timeline of development may offer insights as to which cognitive processes and neural representations depend on the input from others, and which may instead be able to exist independently. Development also allows us to look at the relative developmental timelines of behavioral markers of spatial memory and neural activity, allowing inference of causal links between these two. Finally, once the timeline of development is established, interventional experiments, aimed at manipulating developing circuits at crucial temporal junctures, may help provide causal and mechanistic evidence on how navigation is achieved.

1.2 Navigation and Neural Representations of Space in Rodents

Two well-understood navigation strategies described in rodents are those based on routes and those based on maps (equivalent to "place" and "taxon" strategies) (O'Keefe and Nadel 1978). Both can be observed in the behaving animal, depending on task demands and training schedule (Packard and McGaugh 1996). Route-based navigation involves learning a sequence of *egocentric* (body-based) responses to an ordered sequence of environmental cues (e.g., continue straight until you pass the park, then turn right). Route-based navigation is efficient but highly inflexible: one can only follow previously learned routes, and just one error can render the route useless. Map-based navigation, by contrast, involves constructing a mental representation of the environment, independent of one's body (hence, *allocentric*), into which the locations of landmarks and/or goals can be stored. A defining characteristic of map-based navigation is the ability to use this information to flexibly construct novel routes to goal locations, which may include previously unvisited locations (most easily highlighted behaviorally as the ability to take novel shortcuts). Here, we focus on the emergence of map-based navigation, due to the relative paucity of evidence regarding route navigation development.

A helpful approach to understanding navigation in more detail was recently published by Jeffery et al. (2024). To aid cross-species comparisons, they proposed a hierarchy of spatial cognition processes ordered from simple spatial responses, driven by unimodal sensory cues through to full navigation (e.g., based on maps or routes). Of particular interest to development is an intermediate category of cognitive processes and representations, which they call "spatial primitives," involving multimodal abstractions of environmental input that can aid (but do not constitute, in themselves) navigation. Pertinent examples include recognition of landmarks and spatial contexts as well as representations of boundaries, heading direction, and distance traveled. Below, we discuss each of these elements of spatial cognition and will argue that many of them emerge sequentially during development before the emergence of bona fide map-based navigation.

There is a long tradition, dating back over five decades, of studying the underlying neural circuitry of map-based navigation. Here, we focus on the development of these neural circuits; namely, those neurons that respond to an animal's position or orientation in space (Poulter et al. 2018). This includes position, defined both in an absolute reference frame as well as in relation to salient features of the environment, such as boundaries and objects. These neurons can be thought of as belonging to an integrated circuit and are likely to support map-based navigation: for instance, lesions or inactivation of brain structures containing spatially tuned neurons disrupt navigation (Morris et al. 1982). Furthermore, interindividual variability in navigational performance, or between correct and incorrect trials, can be leveraged to track correlations between spatial neuron activity and behavioral spatial responses (Lenck-Santini et al. 2001; Valerio and Taube 2012).

We begin with a review of what is known about the ontogenetic development of navigation and spatially tuned neurons in the brain. Thereafter, we discuss what has been learned about the relationship, during development, between neural firing and behavior, and highlight important, outstanding questions that remain to be addressed.

2 Development of Navigation

2.1 Sensory and Motor Development

Sensory development in the rat follows the general mammalian blueprint: the vestibular and olfactory functions are the first to emerge (around birth), followed by tactile, auditory, and visual function, respectively (Alberts 1984). All modalities, except vision, are largely mature before the emergence of navigation, with hearing the last to mature by postnatal day 12 (P12). Vision continues to develop as navigation emerges: eyelids open around P14 but electrophysiological recordings from primary visual cortex reveal immature responses at P19 (Fagiolini et al. 1994).

Within the first two weeks of life, movement is predominantly seen in the context of the "huddle," a loose aggregation of pups through which individuals move in a convection-like manner, most likely in service of thermoregulation (Alberts 1978). More relevant to navigation are the motor capabilities pups display when away from the nest: of these, the first distinctive pattern is "pivoting" (i.e., turning on the spot) that is driven by the forelimbs only, which peaks at P7 and reduces thereafter. Full quadrupedal walking emerges from P14 onward, and distances moved in an open field increase abruptly as of P15. By P21, pups are capable of a large range of adultlike motor skills (Altman and Sudarshan 1975).

Rats, therefore, become capable of sustained, independent movement around two weeks of age. This is the earliest point at which the cognitive capacity to navigate may plausibly become useful for survival, and the earliest point at which navigation and spatial memory can easily be tested. This age, however, may not reflect the earliest spontaneous emergence of egression from the nest, which is temperature dependent (Gerrish and Alberts 1996) and may occur as late as four to five weeks in naturalistic conditions (Thiels et al. 1990).

2.2 Earliest Spatial Behaviors at the Onset of Movement

Around two weeks of age marks the emergence of behaviors which may represent the earliest components of the adult navigation system: homing to the nest (initially with sensory guidance, but later without) and recognition memory of recently experienced objects.

Pups removed from the nest by the experimenter can learn to turn consistently left or right to return to a home cage between P10 and P15 (Bulut and Altman 1974; Green and Stanton 1989). We note, however, that auditory and olfactory cues emanating from the nest do not appear to have been controlled for in these studies. This suggests that homing behavior is likely based on sensory taxis toward the home cage. An alternative possibility, however, is that these cues represent a rudimentary form of position habit (a consistent and predictable series of body-based responses), of the type that is dependent on striatal system in adults (Packard and McGaugh 1996).

Very shortly afterward (P16), true homing behavior emerges in the rat, defined as the ability to take a shorter route to return to the nest compared with the outbound foray. In the absence of external sensory cues, homing requires path integration, the calculation of position on the basis of internally generated distance and direction information. Loewen et al. (2005) tracked the spontaneous outbound and inbound journeys of rats placed with a huddle of littermates in an open field arena (Figure 8.1a). Even at the earliest age tested (P16), inbound journeys were shorter than outbound, indicative of homing behavior. Importantly, this was true even for pups tested in the dark as well as with a heat pad in the place of littermates, making path integration, rather than straightforward sensory taxis, likely to be the mechanism underlying this behavior.

The ability to remember a previously seen landmark can be tested using the novel object recognition (NOR) experimental paradigm, where a prominent object is inserted into an open field arena as a proxy for a landmark (Figure 8.1b). This method tests spontaneous recognition, measured by rats' propensity to explore novel objects more than familiar ones. Object recognition emerges early in the second week of life: two different research groups reported that recognition memory for objects placed in an arena was present as early as tested, in cohorts aged either P14 or P17 pups. For a comprehensive review of NOR development, see Cruz-Sanchez et al. (2021). Notably, the duration of the retention span of recognition memory was short: 10 or 5 mins, respectively.

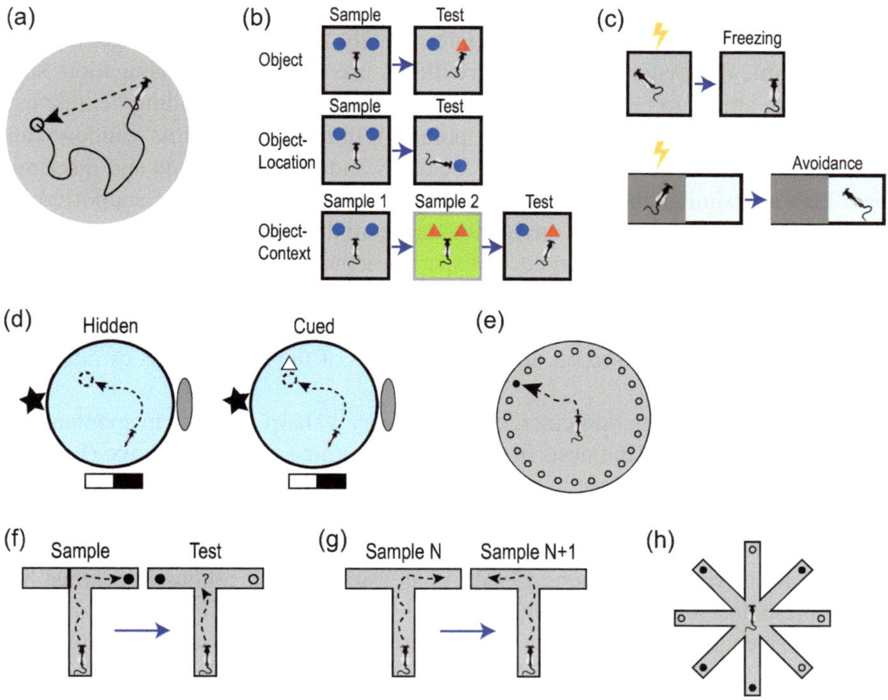

Figure 8.1 Tasks used in the assessment of navigation development. (a) Path integration in an open field. Animals make spontaneous, tortuous excursions from a home base (solid line and black circle, respectively). Return trips (dashed line) are more direct than outward trips (even in the absence of direct sensory information from the home base), indicating the presence of path integration of position relative to the home base. (b) Novel object recognition tasks. Top: Novel object recognition, where memory is tested by enhanced exploration of a novel object. Middle: Object-location recognition, where memory is tested by enhanced exploration of a familiar object in a novel location. Bottom: Object-context recognition, where memory is tested by enhanced exploration of an object not previously encountered in the current context. (c) Aversive tests of context recognition. Top: Context fear conditioning. A shock is delivered in a learning trial. Memory is assessed by the amount of freezing (an innate response to painful or fearful stimuli) on return to the same context after a time delay. Bottom: Inhibitory avoidance. The environment is divided in light and dark zones, the latter of which is innately preferred by rodents. Shocks are delivered upon entry to the dark zone, and learning is assessed by the tendency to avoid the dark zone following shock delivery. (d) Water maze. In the "hidden" platform version, animals search for a platform hidden under the surface of an opaque pool of water. No cues directly mark the platform site, which must be inferred on the basis of distal visual and boundary information. In the "cued" version, a prominent visual beacon marks the platform site. (e) Barnes maze. An array of holes line the periphery of a circular open field, only one of which is rewarded. In a developmental version of this maze, the reward takes the form of an escape route to a home cage. (f) Forced choice T-maze alternation. In a sample trial, the nonreward arm is blocked such that the animal is forced to visit the rewarded arm. In a following free choice trial (after a variable delay), the animal must choose the previously non-rewarded arm. (g) Spontaneous T-maze alternation. In the absence of reward, the natural "win-shift" tendency of rats to alternate visited arms is tested. (h) 8-arm radial maze (reference/working memory protocol). In every trial, the same four arms are baited. Rats must learn which arms are baited (reference memory) as well as which rewards have already been collected on a single trial (working memory).

2.3 The Rapid Emergence of Navigation and Spatial Memory at Weaning

Over a critical age range that spans approximately 2.5–3.5 weeks of age, rats become capable of navigation consistent with the use of map-based navigation, specifically the ability to recall and navigate to unmarked goals, including flexible use of different routes from multiple start points. Within the same time window, rats become able to associate objects with locations and spatial contexts and discriminate between similar but different contexts. These conclusions are supported by evidence from a wide range of navigation and spatial memory tasks, and we will attempt to lay out this rapid emergence of spatial memory behavior in a chronological sequence. It is important to note, however, that individual studies almost always use only one behavioral task; therefore, one important caveat to the timeline described below is that it is not based on direct comparison of the same cohort of rats from the same lab.

Shortly before navigation emerges, rats display a sharp increase in spontaneous exploration away from the nest. In the study of homing described above (Loewen et al. 2005), a steep increase in the number and length of spontaneous excursions from the huddle occurred between P17 and P19. Similar results are seen in studies where both mother and pups have access to a larger environment connected to the home cage (Gerrish and Alberts 1996). As an aside, there is a notable contrast between these data and the behavior of isolated pups in an open field, which show an anxiety-driven peak in movement at P16 followed by a decline in movement (for a review, see Wills et al. 2014). This highlights the importance of using ethologically valid scenarios for behavioral testing.

By the end of the second week, associative memory for spatial contexts emerges, including recognition memory of spatial contexts per se, as well as the associations of discrete objects with both the general context and specific location in which they appear. The ability to recognize a spatial context is most commonly tested using aversive association tests such as context fear conditioning or inhibitory avoidance (Figure 8.1c). Using these tests, context recognition is present from P17, tested over short time intervals (Travaglia et al. 2016). The ability to recognize associations between objects and contexts or locations can be tested using a variant of NOR in which the object-context or object-location pairing is novel (Figure 8.1b): object-context associations are present at P17, and associations between an object and a specific location in an environment from P16 (Cruz-Sanchez et al. 2021; cf. Westbrook et al. 2014, where object-location associations were present at P21, but not at P17.) Finally, some studies have tested the specificity of context recognition during fear conditioning (i.e., whether animals freeze only in an aversive context but not in another context that is only partially different). This ability emerges slightly later but is present by the end of the rapid emergence of navigation at P24 (Ramsaran et al. 2023).

The Morris water maze (Figure 8.1d) is often considered a gold-standard test for map-based navigation. Its advantages include the absence of a local cue marking the goal location, which must instead be triangulated from an array of distal cues, and the necessity to generate a movement vector from any intramaze location to the goal to solve the task efficiently. The overall consensus of several studies (for a review, see Wills et al. 2014) is that learning (as measured by reduced escape latencies) is definitely present at P21, though some studies find evidence of this as early as P19. Notably, those studies which adapted the apparatus to suit pup size and physiology by warming the water and reducing the size of the maze, report the earliest evidence of learning (Carman and Mactutus 2001). Most studies find that the visually cued version of the task can be solved 1–2 days earlier than the hidden platform version, with the earliest evidence of learning at P17. This demonstrates the early emergence of beacon homing as a navigation method and suggests that sensorimotor development is unlikely to be the limiting step in the emergence of map-based navigation. It should be emphasized, however, that although the first evidence of learning emerges around P19, fully adultlike performance emerges in a gradual and protracted manner over the next weeks of life (discussed further below). In relation to the water maze, it is worth noting that a prolonged search in the platform quadrant following platform removal emerges a few days later than the first reduction in search latencies, and that learning rate and memory retention remain less than adultlike until at least P50 (Schenk 1985).

Similar to the water maze, the Barnes maze tests navigation in an open arena (Figure 8.1e), with two key differences. First, it is a dry-land task, and thus there are lower motor demands on young pups. Second, all goals are arranged at the arena edge, meaning that theoretically, only a directional fix is needed to solve the task. Consistent with this latter point evidence of learning appears earlier than in the water maze, with reduced latencies to goal at P18. However, persistent searching at the goal location, following removal, is not seen until P23 (McHail et al. 2018).

A different category of mazes consists of those in which subjects are confined to constrained paths, with spatial memory assayed by the animals' behavior at discrete choice points. The most commonly used during development is the T-maze, usually run with a forced alternation protocol. On the "sample" run, the subject makes a forced run into arm A and finds a reward; the "choice" run shortly afterward then gives a free choice between rewarded arm B and non-rewarded arm A (Figure 8.1f). A recent study (Bevandić et al. 2024) found a gradual emergence of alternation P20–P26 at the population level. Interestingly, this gradual pattern of development is based on the abrupt emergence of alternation in individual rats at varying times (modal emergence P22). This is consistent with a previous finding of *spontaneous* alternation in T-maze (Figure 8.1g), which emerged abruptly within each animal across the age range P20–P30. Thus, the emergence of the alternation habit appears to be all-or-none within each animal.

Two differences in spatial memory required for the T-maze and water maze are noteworthy. First, T-maze alternation is a test of spatial working memory (using rapidly changing goals and short-term memory load), as opposed to the reference memory tested in the water maze and Barnes maze (which have fixed goals and long-term memory load). Despite this difference, the age at which spatial memory emerges in these tasks is similar, though it may be slightly later for spatial working memory. Reference memory can also be tested in discrete choice mazes, such as the 8-arm maze (Figure 8.1h): here, reference memory appears first within the same time window, at around P21 (Rauch and Raskin 1984). The second important difference is that discrete choice mazes do not necessarily require a vector- or map-based navigation strategy but can be solved using a route-based strategy of sequences of egocentric responses (O'Keefe and Nadel 1978). There are well-established protocols for distinguishing these strategies in adults (Packard and McGaugh 1996). To our knowledge, however, none has been explicitly applied to developing rats, meaning that the developmental timing of route-based navigation remains unknown.

2.4 The Protracted Development of Adultlike Spatial Memory

As described above, rats begin to show some evidence of learning on most spatial memory tasks, in a fairly narrow developmental window, between 2.5 and 3.5 weeks of age. It is important to emphasize, however, that this age does not mark the emergence of adultlike navigation or memory, which is not in place until roughly 7–8 weeks of age.

Reduced retention intervals of goal locations, contexts, or location-object associations are characteristic of postweaning spatial memory. Contextual fear conditioning, for example, is seen at P18 when the conditioning and test trials occur within the same day, but a long retention interval (42 days), typical of adult memory, is only present at P42 (Campbell and Campbell 1962). The earliest instances of object-location recognition memory (P16–19; see above) were observed when testing across delays of one, five, or ten minutes. When delays of two to four hours were used instead, preference for the novel object location was not observed until much later, with ages ranging from P24 to P31 (Cruz-Sanchez et al. 2021). Water maze reference memory tested across a three- or seven-day delay is equally impaired in both P20 and P34 rats, whereas these delays have no effect on adult function (Carman et al. 2002). In all cases, information can be demonstrated to have been acquired during the initial learning episode, but the interval over which this information is available for recall (or recognition) gradually increases over the juvenile and adolescent period.

Another late-developing element of spatial memory is the ability to recognize more complex associations between objects, locations, and contexts. For example, the variants of the NOR task in which different objects are experienced in the same location emerges at P30 (Ainge and Langston 2012), and specific three-way associations of locations, contexts, and objects emerges only from seven weeks (Asiminas et al. 2019). These results suggest that higher-level control of spatial memory by context also emerges considerably later in development than map-based navigation.

3 Development of Spatially Tuned Neurons

Over the past 15 years, a series of studies has delineated the timeline of spatial neuron development. Below, in our review of this work, we examine head-direction (HD) and grid cells, as both types are thought to reflect a characteristic network architecture in adults (the "continuous attractor"), and both share similar developmental trajectory profiles (if not absolute timelines). In addition, we look at the protracted development of CA1 place cells, for which several important functional milestones have been established, beyond their characteristic location-tuned firing.

Finally, we review the development of neurons whose primary spatial firing correlates are environmental boundaries.

3.1 HD and Grid Cells

HD cells, which fire when rats orient their heads in a specific direction in the azimuth plane, are the earliest developing spatially modulated neuron, appearing in an immature form as early as P12 and becoming adultlike on the day of eye opening, P14–P15 (Bjerknes et al. 2015; Tan et al. 2015). This age coincides with the first emergence of sustained efficient movement outside the huddle and well before spontaneous egression from the nest into the outside world. At this age, HD cells have an adultlike degree of directional tuning as well as an adultlike stability of their preferred firing direction (measured across different exposures to the recording arena). The preferred firing direction of adult HD cells is principally controlled by distal visual landmarks: in pups, this control is already present by one day after eye opening. HD cells have been found to emerge early in several of the network of regions where they are located in adults, including anterodorsal thalamic nucleus, dorsal presubiculum, parasubiculum, and medial entorhinal cortex (Langston et al. 2010; Tan et al. 2015). This suggests a global and sudden onset of adultlike HD signaling.

In an adult brain, the HD cell population is thought to be connected following a "continuous attractor" architecture (Zhang 1996). The key experimental finding supporting this theory is that within an animal, the relative angular offsets of all HD cells are always locked together, even when absolute directional preferences of HD cells change (e.g., following disorientation). This is indicative of a network topology in which the population activity is constrained to a low-dimensional surface (in this case, a 1D ring), producing the functional advantage of a unitary, low-noise estimate of direction across the HD cell population. During development, activity consistent with a continuous attractor develops *before* stable HD signaling in single neurons, which suggests that this architecture may be "pre-wired" without reference to environmental landmarks (Bassett et al. 2018). In this view, following the opening of the eyelids, mature HD cells emerge very rapidly after the continuous attractor network first receives access to distal visual landmark information. Indeed, a detailed investigation of sensory control of HD attractor activity showed that vestibular inputs are present as early as tested and that enhanced olfactory and tactile cues can stabilize HD cells at P13 (Bassett et al. 2018), even before eye opening, which then allows the expression of mature HD signaling by P15.

Grid-cell firing is modulated by an animal's position, with each grid cell displaying a series of firing fields spread out across the entire environment in a repeating, regular, hexagonal grid. The age at which grid cells emerge is markedly later than HD cells: the earliest age at which clear grid patterns have been observed in medial entorhinal cortex (MEC) is P20 (Wills et al. 2010). Similar to HD cells, grid-cell activity is organized into a continuous attractor network: every pair of grid cells have a fixed (2D) spatial offset between their hexagonally distributed firing rate fields (Burak and Fiete 2009). Population activity is thus constrained to a low-dimensional space, in the case of grid cells this being a two-dimensional torus. The developmental profile of grid-cell emergence is also similar to that of HD cells: adultlike grid cells emerge rather abruptly and appear to have adultlike characteristics already at this age (Wills et al. 2012). A tempting speculation is that grid cells, like HD cells, develop as a preformed attractor, anchored to the environment by external sensory stimuli at the appropriate time during development. The emergence of stable grid cells does require environmental experience (specifically, exploration of an open field with boundaries in the presence of visual cues) (Ulsaker-Janke et al. 2023), but this does not rule out a preconfigured attractor which develops independently of experience and is then stabilized with respect to the outside world.

A further important detail is that the grid-cell network consists of not just one, but multiple continuous attractors or "modules" (Stensola et al. 2012), each of which is composed of grid cells with different grid spacings between their fields. Most data on grid-cell development relate to one module only. Thus, when and how other modules emerge remains unknown.

3.2 Place-Cell Development

The first studies of CA1 place-cell development appeared to set out a simple story: the modulation of place cells by an animal's current location emerged early in an immature form and then underwent a gradual and protracted development, reaching maturity only after approximately six weeks (Langston et al. 2010; Scott et al. 2010; Wills et al. 2010). Since then, further studies have elucidated how specific aspects of place-cell firing develop (and modulate by location), allowing a more detailed timeline of place-cell development to be constructed.

In the adult rat, the large majority of active CA1 pyramidal cells (ca. 90%) are place cells (Wills et al. 2010). The activity of these cells is significantly modulated by an animal's current location, with the spatial tuning function (place field) normally approximating a single 2D Gaussian. In rats aged P14–16, the majority of pyramidal cells have poor spatial tuning, with multiple, large place fields or indeed no tuning at all (Langston et al. 2010; Scott et al. 2010; Wills et al. 2010). However, CA1 pyramidal cells exhibit very heterogeneous developmental profiles, and a small minority of cells show clear place fields even early in development (Muessig et al. 2015). On average, the degree of spatial tuning as well as the spatial stability of the place field within the environment (both within and across exposures to the recording arena) continue to improve in a gradual, monotonic fashion. These reach maturity some time from P35 onward, though a large degree of individual variation is still apparent, with the place cells in the slowest developing animals still immature at P45 (Scott et al. 2010).

Contrary to the simple picture above, more recent studies have outlined how early place cells do show some specific, mature functional properties as early as P16. Bjerknes et al. (2018) tested whether place-cell firing was primarily controlled by the distance traveled or by fixed environmental cues in a task in which these were dissociated by moving a start box relative to the running track in both light and dark conditions. At all ages tested (including as young as P17), place fields recorded in darkness were primarily controlled by distance traveled from the start box, indicating that even early place cells can calculate position by path integration. Muessig et al. (2016) investigated whether and how early place cells encode maps for different spatial contexts. As in adults, they found that place cells encode multiple maps for multiple environments. Furthermore, the appropriate map can be recalled for each context, even when rats are presented only with a partial set of the cues associated with the originally learned context. This is consistent with the CA1 acting as an associative memory encoder of spatial context, as observed in adults, even in animals as young as P16.

Other functional characteristics of place cells are known to develop later, approximately across the pre- to post-weanling divide. In adult rats, place fields are distributed approximately evenly across an open field environment and are equally spatially stable throughout this space. In pre-weanling rats, however, place fields cluster close to the boundaries, and there exists a functional gradient across space, whereby place cells further from boundaries are less stable. Around weaning (P20), place fields abruptly switch to a mature pattern of equal stability throughout the environment. Thus, P20 appears to mark a point at which place cells are no longer solely reliant on boundaries for their spatial stability (Muessig et al. 2015).

On a short timescale—from tens to hundreds of milliseconds—adult place-cell activity is organized into temporally ordered sequences, including in different behavioral states such as sleep and exploration. During slow-wave sleep and awake immobility, population-level bursts of place-cell activity that last hundreds of milliseconds occur, during which the temporal order of place-cell activation recapitulates recently experienced trajectories through the environment ("place-cell replay"). By contrast, during movement, place-cell firing is modulated by the hippocampal theta oscillation (7–10 Hz in adults), and within each theta cycle, place cells are activated in an order that reflects the animal's current trajectory. Place cells with fields just behind the animal fire first in the theta cycle, whereas those just ahead of the current position fire later during the cycle ("theta sequences"). Both of these forms of temporal organization emerge over the fourth week of life (Farooq and Dragoi 2019; Muessig et al. 2019). Early replay events (tested at P17) represent only single locations in the environment, whereas replays that reflect trajectories through the environment emerge gradually between approximately P19 and P35. In a striking example of coincident development, theta sequences emerge over the same time period, and the relative emergence of both replay and theta sequences is tightly correlated (even after controlling for chronological age). This strongly suggests that the two phenomena share a common underlying mechanism (Muessig et al. 2019). It is noteworthy that the "stationary" replays seen at P17 are evidence of Hebbian plasticity between place cells at this age: cells which fire together during exploration appear to have wired together, as evidenced by their later enhanced co-firing during sleep. What develops after approximately three weeks is the mechanism by which this plasticity becomes directional and can produce ordered sequences of firing, enabling them to be a significant mechanism for spatial memory.

In adults, place-cell firing can be modulated not only by location but also by the trajectory along which an animal is moving as well as latent variables, such as task phase. One specific example of this is differential firing (or "remapping") on the T-maze apparatus between "sample" and "choice" runs during the forced choice alternation protocol. During development, remapping between sample and choice emerges concurrently with successful alternation behavior, suggesting that this may be one mechanism underlying spatial working memory (Bevandić et al. 2024).

3.3 Boundary-Modulated Neurons

A further distinctive type of spatially modulated neuron is one whose firing correlates with an animal's position with respect to environmental boundaries. These include border cells in the MEC (Solstad et al. 2008), which fire only in close proximity to boundaries, and boundary vector cells (BVCs) in the subiculum (Lever et al. 2009), which fire instead at a range of vectorial (distance and direction) offsets from boundaries. The functional role of boundary-driven neurons may be to anchor the neural map of space following contact with boundaries, and/or allow for a vectorial map of space constructed in a boundary-linked reference frame (Bicanski and Burgess 2018).

During development, boundary-driven firing proceeds on markedly different trajectories, depending on the cell type and anatomical region. MEC border cells are present in adultlike form as early as tested (P17) (Bjerknes et al. 2014), whereas the development of subiculum BVCs is most similar to that of place cells (Muessig et al. 2024). Cells matching the formal definition of a BVC are present from as early as tested (P16), but with a greatly reduced spatial specificity and stability compared to adults. Their development proceeds gradually and remains far from mature at P25 (the oldest age at which animals were tested in this study). Interestingly, one specific aspect of BVC firing is present at the earliest ages tested: BVC directional preferences are distorted by boundary geometry, such that BVC fields line up with walls in square environments but are evenly distributed in circular arenas. The early development of BVC distortion by geometry suggests this may be an inherent property of BVCs. Recent work has shown, however, that place-cell firing can be altered by changing the experience of boundaries during rearing; specifically, rearing in a sphere that removes discrete boundaries altogether (Farooq and Dragoi 2024). An important future direction would be to test explicitly how neural representations of boundaries are affected by rearing conditions, and how this relates to the later use of boundary geometry as a navigational cue (Twyman et al. 2013).

4 Conclusions

The above reviews provide a concise timeline of the functional development of navigation, both at the behavioral and neural levels. For a graphical summary of these findings, see Figure 8.2. Here, we aim to highlight conclusions that can be drawn regarding how the development of the cognitive ability of navigation unfolds, and which neural mechanisms support it.

Multiple lines of evidence suggest that in the rat, map-based navigation emerges across the approximate period of P19–P23. However, during behavioral and neural

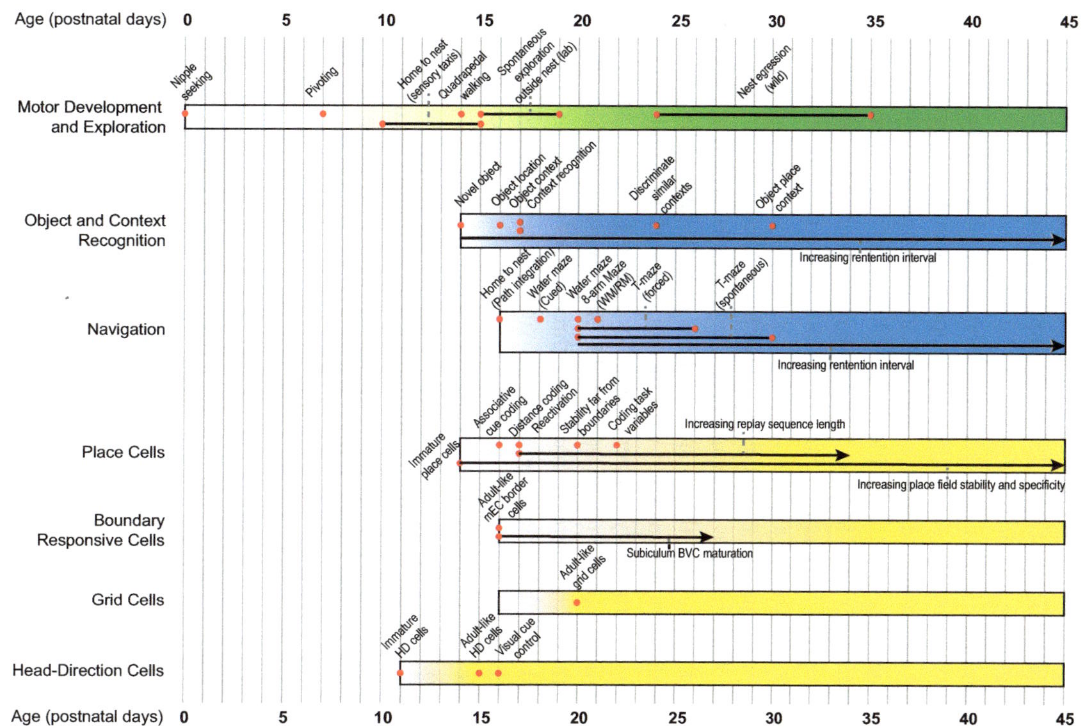

Figure 8.2 Summary timeline of the development of movement, spatial memory, and navigation in rats and mice (evidence is based primarily on rats). Green bar shows development of movement and exploration, blue bars show development of spatial memory and navigation, and yellow bars show development of spatially modulated neuron firing. Color intensity shows an estimate of overall maturity within each domain. Bars start at the earliest age of relevant evidence within each domain. Red dots annotated with text indicate developmental emergence of specific characteristics. Unless otherwise specified, these ages denote the first emergence of a characteristic rather than its full maturity. Black lines with red dots at both ends denote the beginning and end of a maturational process. Black lines with an arrowhead indicate a maturational process whose endpoint is not well established.

development, "spatial primitives" (i.e., those simpler cognitive processes and neural constructs that contribute to navigation) are generally present a few days before map-based navigation emerges (Jeffery et al. 2024). Context and object recognition are present before P19, as well as a form of Hebbian plasticity in CA1 place cells. This indicates that neural mechanisms that allow the association of reinforcers with locations may already be present at this age. Adultlike representations of head direction and boundaries can be observed from P15 and P17, respectively, and homing by path integration appears to be present very early, from P16 onward. Despite the presence of all these ingredients of spatial memory, however, map-based navigation is absent: it only emerges at around three weeks.

The developmental dissociation between path integration and map-based navigation is particularly noteworthy and at odds with proposals which suggest that path integration grid cells constitute the primary mechanism by which the hippocampal formation supports navigation (McNaughton et al. 2006). It is also consistent with neuropsychological findings showing successful path integration in human subjects with hippocampal and entorhinal cortex damage, as long as there is no memory load in the navigation task (Shrager et al. 2008).

What could drive the relatively rapid emergence of map-based navigation around three weeks of age? The most obvious candidate are grid cells, which are present in a mature form and anchored to the outside world as of P20. An enabling role for grid cells in map navigation is consistent with theoretical work that highlights the capacity of grid cells to form highly efficient and unique codes for global space and, in particular, to calculate vectors (Bush et al. 2015; Sreenivasan and Fiete 2011). One attractive hypothesis, therefore, is that before three weeks of age, animals can maintain a consistent directional heading, sense proximity to a boundary, track the distance traveled, recognize spatial contexts and locations within those contexts, and may even be able to associate rewards with locations using place-cell plasticity. What they cannot do, however, is to calculate a vector between their current position and the known goal, due to a lack of spatially stable grid-cell signaling. One important caveat to this conclusion is that calculating vectors (or mapping large-scale space) with grid cells requires multiple "modules" containing grid cells of different field spacings. Currently, when multiple grid-cell modules emerge during development is unknown.

One alternative mechanism for the abrupt development of navigation is the emergence of place-cell replay, from around P20. Replay inherently represents the linking together of spatially separated locations in the environment and has been linked to planning or imagination of possible paths through the environment (Foster 2017). The first emergence of replay is approximately concurrent with the first ability to navigate. Furthermore, the gradual maturation of replay (with replayed trajectories becoming longer over the fourth to fifth weeks of life) offers a natural explanation for the continued maturation of spatial memory. Moreover, the "grid-cell" and "replay" explanations for the emergence of navigation may be mechanistically linked. Recent evidence from adult rats has found that MEC inactivation leads to a loss of replay in CA1, thus suggesting a possible role for grid cells in replay generation (Liu et al. 2023).

Navigation in discrete choice mazes, such as the T-maze, also emerges at approximately three weeks of age, concurrently with map-based navigation in the water maze. This may indicate that developing rats are using map-based strategies to solve the T-maze, though as noted above, we are not aware of an explicit test of whether simpler, egocentric, route-based strategies are used to solve these tasks during development. A specific requirement of tasks, such as the forced choice alternation T-maze, is spatial working memory: this also appears to emerge at around three weeks of age, and recent results suggest an intriguing candidate neural mechanism (i.e., coding of task phase as a latent variable by place cells) (Bevandić et al. 2024).

While the abrupt emergence of spatially stable grid cells and/or trajectory-specific place-cell replay may underlie the first emergence of navigation, it is still unclear what might explain the protracted maturation of spatial memory. One likely candidate is the continued instability of place-cell firing, which lasts until 7–8 weeks of age and prevents stable encoding and recall of long-term spatial reference memories. This instability is a separate phenomenon to "representational drift," the long-term, experience-dependent turnover in the adult place-cell population representing an environment (Ziv et al. 2013). Representational drift is caused by changing firing rates within place fields that are nonetheless stable in space. Immature place fields, by contrast, are unstable in space, even across short (inter-trial) intervals. Long-term stable spatial reference memory could, therefore, depend on the protracted emergence of stability in place-cell firing. A second possibility, as mentioned above, is the slow development of place-cell replay. A third possibility lies with the protracted immaturity of subiculum BVCs. If BVCs create a vectorial map of space based on the boundaries of the environment (Bicanski and Burgess 2018), the continued imprecision and instability of this map contribute to immature spatial memory.

In conclusion, studying the ontogeny of navigation has produced important breakthroughs in our understanding of this cognitive system and the neural circuits which underpin it. These include, for example, the ontological primacy of the HD cell system, whose adultlike form precedes any other "spatial primitive," either behavioral or neural. Future work should investigate whether HD cells, and other early-developing neural representations such as border cells, can support the early spatial behaviors observed before full map-based navigation. By contrast, grid cells develop late, and their emergence coincides with map-based navigation. Some of the most powerful studies of spatial neuron development have been those that have leveraged individual developmental differences to find correlations between either neural representations and behavior (Bevandić et al. 2024) or different neural mechanisms for memory, previously not shown to be directly related (Muessig et al. 2019). An important future direction would be to apply the same methodology to grid-cell and map-based navigation to demonstrate a more direct link between the emergence of these two phenomena. Finally, an important goal for developmental research will be to design behavioral protocols that will allow direct comparison of different navigational strategies, or different cognitive components of spatial behavior, so that how their relative emergence relates to the development of neural circuits can be tested. One specific example would be to test the emergence of map-based versus route navigation strategies and the neural circuits supporting them.

References

Ainge JA, and Langston RF (2012) Ontogeny of Neural Circuits Underlying Spatial Memory in the Rat. Front Neural Circuits. Front Neural Circuits 6:8. https://doi.org/10.3389/fncir.2012.00008

Alberts JR (1978) Huddling by Rat Pups: Group Behavioral Mechanisms of Temperature Regulation and Energy Conservation. J Comp Physiol Psychol 92:231–245. https://doi.org/10.1037/h0077459

——— (1984) Sensory-Perceptual Studies in the Norway Rat: A View toward Comparative Studies. In: Kail R, Spear NS (eds) Comparative Perspectives on Memory Development. Erlbaum, Hillsdale, pp 65–101

Altman J, and Sudarshan K (1975) Postnatal Development of Locomotion in the Laboratory Rat. Anim Behav 23:896–920. https://doi.org/10.1016/0003-3472(75)90114-1

Asiminas A, Jackson AD, Louros SR, et al. (2019) Sustained Correction of Associative Learning Deficits after Brief, Early Treatment in a Rat Model of Fragile X Syndrome. Sci Transl Med 11:eaao0498. https://doi.org/10.1126/scitranslmed.aao0498

Bassett JP, Wills TJ, and Cacucci F (2018) Self-Organized Attractor Dynamics in the Developing Head Direction Circuit. Curr Biol 28 (4):609–615.e603. https://doi.org/10.1016/j.cub.2018.01.010

Bevandić J, Stella F, and Ólafsdóttir HF (2024) Parallel Maturation of Rodent Hippocampal Memory and CA1 Task Representations. Curr Biol 34 (21):5062–5072.e5065. https://doi.org/10.1016/j.cub.2024.08.048

Bicanski A, and Burgess N (2018) A Neural-Level Model of Spatial Memory and Imagery. eLife 7 (7052):1–3. https://doi.org/10.7554/elife.33752

Bjerknes TL, Daglsott NC, Moser EI, and Moser MB (2018) Path Integration in Place Cells of Developing Rats. PNAS 115 (7):E1637–E1646. https://doi.org/10.1073/pnas.1719054115

Bjerknes TL, Langston RF, Kruge IU, Moser EI, and Moser MB (2015) Coherence among Head Direction Cells before Eye Opening in Rat Pups. Curr Biol 25:103–108. https://doi.org/10.1016/j.cub.2014.11.009

Bjerknes TL, Moser EI, and Moser MB (2014) Representation of Geometric Borders in the Developing Rat. Neuron 82:71–78. https://doi.org/10.1016/j.neuron.2014.02.014

Bulut F, and Altman J (1974) Spatial and Tactile Discrimination Learning in Infant Rats Motivated by Homing. Dev Psychobiol 7:465–473. https://doi.org/10.1002/dev.420070510

Burak Y, and Fiete IR (2009) Accurate Path Integration in Continuous Attractor Network Models of Grid Cells. PLoS Comput Biol 5:e1000291. https://doi.org/10.1371/journal.pcbi.1000291

Bush D, Barry C, Burgess N, et al. (2015) Using Grid Cells for Navigation. Neuron 87 (3):507–520. https://doi.org/10.1016/j.neuron.2015.07.006

Campbell BA, and Campbell EH (1962) Retention and Extinction of Learned Fear in Infant and Adult Rats. J Comp Physiol Psychol 55:1–8. https://doi.org/10.1037/h0049182

Carman HM, Booze RM, and Mactutus CF (2002) Long-Term Retention of Spatial Navigation by Preweanling Rats. Dev Psychobiol 40:68–77. https://doi.org/10.1002/dev.10014

Carman HM, and Mactutus CF (2001) Ontogeny of Spatial Navigation in Rats: A Role for Response Requirements? Behav Neurosci 115:870–879. https://pubmed.ncbi.nlm.nih.gov/11508726/

Cruz-Sanchez A, Wilkin J, and Arruda-Carvalho M (2021) Ontogeny of Spontaneous Recognition Memory in Rodents. Neurobiol Learn Mem 177:107361. https://doi.org/10.1016/j.nlm.2020.107361

Fagiolini M, Pizzorusso T, Berardi N, Domenici L, and Maffei L (1994) Functional Postnatal Development of the Rat Primary Visual Cortex and the Role of Visual Experience: Dark Rearing and Monocular Deprivation. Vision Res 34:709–720. https://doi.org/10.1016/0042-6989(94)90210-0

Farooq U, and Dragoi G (2019) Emergence of Preconfigured and Plastic Time-Compressed Sequences in Early Postnatal Development. Science 363:168–173. https://doi.org/10.1126/science.aav0502

——— (2024) Experience of Euclidean Geometry Sculpts the Development and Dynamics of Rodent Hippocampal Sequential Cell Assemblies. Nat Commun 15:8417. https://doi.org/10.1038/s41467-024-52758-9

Foster DJ (2017) Replay Comes of Age. Annu Rev Neurosci 40:581–602. https://doi.org/10.1146/annurev-neuro-072116-031538

Gerrish CJ, and Alberts JR (1996) Environmental Temperature Modulates Onset of Independent Feeding: Warmer Is Sooner. Dev Psychobiol 29:483–495. https://doi.org/10.1002/(sici)1098-2302(199609)29:6%3C483::aid-dev1%3E3.0.co;2-i

Green RJ, and Stanton ME (1989) Differential Ontogeny of Working Memory and Reference Memory in the Rat. Behav Neurosci 103:98–105. https://doi.org/10.1037//0735-7044.103.1.98

Jeffery KJ, Cheng K, Newcombe NS, Bingman VP, and Menzel R (2024) Unpacking the Navigation Toolbox: Insights from Comparative Cognition. Proc R Soc Lond B Biol Sci 291 (2016):20231304. https://doi.org/10.1098/rspb.2023.1304

Langston RF, Ainge JA, Couey JJ, et al. (2010) Development of the Spatial Representation System in the Rat. Science (1979) 328:1576–1580. https://doi.org/10.1126/science.1188210

Lenck-Santini PP, Save E, and Poucet B (2001) Evidence for a Relationship between Place-Cell Spatial Firing and Spatial Memory Performance. Hippocampus 11:377–390. https://doi.org/10.1002/hipo.1052

Lever C, Burton S, Jeewajee A, O'Keefe J, and Burgess N (2009) Boundary Vector Cells in the Subiculum of the Hippocampal Formation. J Neurosci 29 (31):9771–9777. https://doi.org/10.1523/jneurosci.1319-09.2009

Liu C, Todorova R, Tang W, Oliva A, and Fernandez-Ruiz A (2023) Associative and Predictive Hippocampal Codes Support Memory-Guided Behaviors. Science (1979) 382:eadi8237. https://doi.org/10.1126/science.adi8237

Loewen I, Wallace DG, and Whishaw IQ (2005) The Development of Spatial Capacity in Piloting and Dead Reckoning by Infant Rats: Use of the Huddle as a Home Base for Spatial Navigation. Dev Psychobiol 46:350–361. https://doi.org/10.1002/dev.20063

McHail D, Valibeigi N, and Dumas TA (2018) Barnes Maze for Juvenile Rats Delineates the Emergence of Spatial Navigation Ability. Learn Mem 25:138–146. https://doi.org/10.1101/lm.046300.117

McNaughton BL, Battaglia FP, Jensen O, Moser EI, and Moser M-B (2006) Path Integration and the Neural Basis of the 'Cognitive Map'. Nat Rev Neurosci 7 (8):663–678. https://doi.org/10.1038/nrn1932

Morris RG, Garrud P, Rawlins JN, and O'Keefe J (1982) Place Navigation Impaired in Rats with Hippocampal Lesions. Nature 297 (5868):681–683. https://doi.org/10.1038/297681a0

Muessig L, Hauser J, Wills TJ, and Cacucci F (2015) A Developmental Switch in Place Cell Accuracy Coincides with Grid Cell Maturation. Neuron 86 (5):1167–1173. https://doi.org/10.1016/j.neuron.2015.05.011

——— (2016) Place Cell Networks in Pre-Weanling Rats Show Associative Memory Properties from the Onset of Exploratory Behavior. Cereb Cortex 26:3627–3636. https://doi.org/10.1093/cercor/bhw174

Muessig L, Lasek M, Varsavsky I, Cacucci F, and Wills TJ (2019) Coordinated Emergence of Hippocampal Replay and Theta Sequences during Post-Natal Development. Curr Biol 29:834–840.e834. https://doi.org/10.1016/j.cub.2019.01.005

Muessig L, Rodrigues FR, Bjerknes TL, et al. (2024) Environment Geometry Alters Subiculum Boundary Vector Cell Receptive Fields in Adulthood and Early Development. Nat Commun 15 (1):982. https://doi.org/10.1038/s41467-024-45098-1

O'Keefe J, and Nadel L (1978) The Hippocampus as a Cognitive Map. Oxford Univ Press, London

Packard MG, and McGaugh JL (1996) Inactivation of Hippocampus or Caudate Nucleus with Lidocaine Differentially Affects Expression of Place and Response Learning. Neurobiol Learn Mem 65(1):65–72. https://doi.org/10.1006/nlme.1996.0007

Poulter S, Hartley T, and Lever C (2018) The Neurobiology of Mammalian Navigation. Curr Biol 28:R1023–R1042. https://doi.org/10.1016/j.cub.2018.05.050

Ramsaran AI, Wang Y, Golbabaei A, et al. (2023) A Shift in the Mechanisms Controlling Hippocampal Engram Formation during Brain Maturation. Science (1979) 380:543–551. https://doi.org/10.1126/science.ade6530

Rauch SL, and Raskin LA (1984) Cholinergic Mediation of Spatial Memory in the Preweanling Rat: Application of the Radial Arm Maze Paradigm. Behav Neurosci 98:35–43. https://doi.org/10.1037//0735-7044.98.1.35

Schenk F (1985) Development of Place Navigation in Rats from Weaning to Puberty. Behav Neural Biol 43:69–85. https://doi.org/10.1016/s0163-1047(85)91510-9

Scott RC, Richard GR, Holmes GL, and Lenck-Santini PP (2010) Maturational Dynamics of Hippocampal Place Cells in Immature Rats. Hippocampus 4:347–353. https://doi.org/10.1002/hipo.20789

Shrager Y, Kirwan CB, and Squire LR (2008) Neural Basis of the Cognitive Map: Path Integration Does Not Require Hippocampus or Entorhinal Cortex. PNAS 105:12034–12038. https://doi.org/10.1073/pnas.0805414105

Solstad T, Boccara CN, Kropff E, Moser M-B, and Moser EI (2008) Representation of Geometric Borders in the Entorhinal Cortex. Science 322 (5909):1865–1868. https://doi.org/10.1126/science.1166466

Sreenivasan S, and Fiete I (2011) Grid Cells Generate an Analog Error-Correcting Code for Singularly Precise Neural Computation. Nat Neurosci 14:1330–1337. https://doi.org/10.1038/nn.2901

Stensola H, Stensola T, Solstad T, et al. (2012) The Entorhinal Grid Map Is Discretized. Nature 492 (7427):72–78. https://doi.org/10.1038/nature11649

Tan HM, Bassett JP, O'Keefe J, Cacucci F, and Wills TJ (2015) The Development of the Head Direction System before Eye Opening in the Rat. Curr Biol 25 (4):479–483. https://doi.org/10.1016/j.cub.2014.12.030

Thiels E, Alberts JR, and Cramer CP (1990) Weaning in Rats: II. Pup Behavior Patterns. Dev Psychobiol 23:495–510. https://doi.org/10.1002/dev.420230605

Travaglia A, Bisaz R, Sweet ES, Blitzer RD, and Alberini CM (2016) Infantile Amnesia Reflects a Developmental Critical Period for Hippocampal Learning. Nat Neurosci 19:1225–1233. https://doi.org/10.1038/nn.4348

Twyman AD, Newcombe NS, and Gould TJ (2013) Malleability in the Development of Spatial Reorientation. Dev Psychobiol 55:243–255. https://doi.org/10.1002/dev.21017

Ulsaker-Janke I, Waaga T, Waaga T, Moser EI, and Moser M-B (2023) Grid Cells in Rats Deprived of Geometric Experience during Development. PNAS 120:e2310820120. https://doi.org/10.1073/pnas.2310820120

Valerio S, and Taube JS (2012) Path Integration: How the Head Direction Signal Maintains and Corrects Spatial Orientation. Nat Neurosci 15:1445–1453. https://doi.org/10.1038/nn.3215

Westbrook SR, Brennan LE, and Stanton ME (2014) Ontogeny of Object versus Location Recognition in the Rat: Acquisition and Retention Effects. Dev Psychobiol 56:1492–1506. https://doi.org/10.1002/dev.21232

Wills TJ, Barry C, and Cacucci F (2012) The Abrupt Development of Adult-Like Grid Cell Firing in the Medial Entorhinal Cortex. Front Neural Circuits 6:21. https://doi.org/10.3389/fncir.2012.00021

Wills TJ, Cacucci F, Burgess N, and O'Keefe J (2010) Development of the Hippocampal Cognitive Map in Preweanling Rats. Science 328 (5985):1573–1576. https://doi.org/10.1126/science.1188224

Wills TJ, Muessig L, and Cacucci F (2014) The Development of Spatial Behaviour and the Hippocampal Neural Representation of Space. Philos Trans R Soc Lond B Biol Sci 369 (1635):20130409. https://doi.org/10.1098/rstb.2013.0409

Zhang K (1996) Representation of Spatial Orientation by the Intrinsic Dynamics of the Head-Direction Cell Ensemble: A Theory. J Neurosci 16:2112–2126. https://doi.org/10.1523/jneurosci.16-06-02112.1996

Ziv Y, Burns LD, Cocker ED, et al. (2013) Long-Term Dynamics of CA1 Hippocampal Place Codes. Nat Neurosci 16:264–266. https://doi.org/10.1038/nn.3329

9

Spatial Navigation in Humans and Rodents

Bridging Levels of Analyses

H. Freyja Ólafsdóttir, Andrej Bicanski, Sarah H. Creem-Regehr, Daniel D. Dilks, Flavio Donato, Russell A. Epstein, Emily K. Farran, Joshua Jacobs, Lukas Kunz, Thomas J. Wills, and Nora S. Newcombe

Abstract This chapter reviews and synthesizes current understanding of the neural and cognitive mechanisms that support spatial navigation, with the goal of linking neuronal activity to complex spatial behavior across species. Single-neuron recordings in rodent studies have identified key neuronal representations such as place cells, grid cells, and head-direction cells that form the basis of cognitive maps. Human neuroimaging and intracranial studies have identified indirect homologues of the neural spatial signals and have provided insights into the neural basis of higher-order spatial processes, including path planning and vector computation. Despite this progress, the generation of a unified account that integrates across different processing levels has remained challenging due to methodological differences across species and techniques. To bridge these gaps, three strategies are proposed: (a) methodological bridges, in which cellular-level and macroscopic measures, obtained from rodent and human studies respectively, are aligned through computational modeling and signal alignment; (b) behavioral bridges, in which comparable navigation tasks are employed across species in naturalistic or large-scale environments; and (c) developmental bridges, in which rodent neurodevelopment is used to identify milestones in the emergence of navigational abilities across species. By integrating findings from neurophysiology, neuroimaging, computational modeling, and developmental studies, a comprehensive framework linking neuronal mechanisms with spatial behavior can be achieved.

Keywords Navigation, rodents, humans, cognitive map, spatial cells

---

H. Freyja Ólafsdóttir (✉)
Donders Institute for Brain, Cognition and Behaviour, Radboud University, 6525 EN Nijmegen, The Netherlands
Email: freyja.olafsdottir@donders.ru.nl

Coauthor affiliations are available in the List of Contributors.

N. S. Newcombe and K. Cheng (eds.), *Challenges in Navigation Research*, Strüngmann Forum Reports, https://doi.org/10.1007/978-3-032-20563-6_9

Group photos (top left to bottom right) H. Freyja Ólafsdóttir, Andrej Bicanski, Nora Newcombe, Daniel Dilks, Sarah Creem-Regehr, Flavio Donato, Russell Epstein, Joshua Jacobs, H. Freyja Ólafsdóttir, Lukas Kunz, Emily Farran, Daniel Dilks, Nora Newcombe, Thomas Wills, Andrej Bicanski, Emily Farran, Russell Epstein, Sarah Creem-Regehr

1 Introduction

Understanding how humans and other animals navigate in search of food, shelter, and mates is a central goal in biology, psychology, and neuroscience. Over the past decades, research efforts have focused on elucidating the neuronal networks and cellular representations that underpin spatial navigation. Human neuroimaging studies have provided a comprehensive overview of the network of brain regions that underlie spatial navigation and the specific functional contributions of individual regions (for reviews see Dilks et al. 2022; Epstein 2008; Epstein et al. 2017; Julian et al. 2018b). In addition, human intracranial recording studies have yielded important insights about how spatial information is encoded at the neuronal level (e.g., Ekstrom et al. 2003; Jacobs et al. 2013; Tsitsiklis et al. 2020). However, as intracranial studies are only possible in small clinical populations (e.g., individuals with drug-resistant epilepsy), and the recording locations and techniques are determined by clinicians rather than researchers, it is essential for insights gained from human research to be complemented by data acquired from animal models, where neuronal circuit mechanisms and representations can be studied systematically.

The most studied animal model in the field is the rodent (mice, rats). Although rodents and humans differ in important ways (see Section 2.2), they share a similar ecological niche as central-place foragers (Metcalfe and Barlow 1992), rely on similar strategies during navigation and some of the anatomical structures that support navigation are conserved across the two taxa. Indeed, nonhuman, tree-dwelling primates that are evolutionarily closer to humans have radically different foraging patterns. Rodent studies have provided detailed insight into the neuronal representations and mechanisms that support mammalian spatial behavior. These studies have, in turn, inspired detailed computational models of diverse spatial cognitive processes. Thus, the field of spatial navigation is in a promising position to integrate the different processing levels—from single-cell and cell population activity to complex spatial behaviors—to create a comprehensive framework that links brain and behavior.

Connecting different processing levels, however, has been impeded by fragmentation in the research field. Systems neuroscientists studying the spatial responses of single neurons in the rodent brain, for example, interact only occasionally with cognitive neuroscientists studying navigation-related responses in humans via noninvasive brain imaging techniques. To create a comprehensive, cross-level description of navigation, these separate areas of research must be connected. Here, our aim is to link these areas by creating three bridges. We begin by reviewing and comparing side-by-side research derived from human and rodent studies that employ different neuronal recording methods to enable a unified account of the neuronal representations and population mechanisms that underlie our ability to orient ourselves in space and navigate to spatial goals. We then discuss areas where methodological differences between studies hinder the creation of a cross-level description of navigation and propose solutions to bridge these gaps (methodological bridge). Thereafter we discuss aspects of spatial behavior that lack a cross-level explanation and offer concrete suggestions for tasks and approaches to fill these knowledge gaps (behavioral bridge). Finally, we explore how recent studies on rodent neurodevelopment provide an alternative approach for creating a cross-level description of navigation, offering insights into the neurodevelopmental milestones that underlie navigation development (developmental bridge) as well as suggestions for future research.

2 The Neural Basis of Spatial Navigation

2.1 Cognitive Maps in Rodents and Humans

To navigate efficiently from place to place, humans and other animals rely on internal representation of spaces, which are sometimes referred to as a "cognitive map" (Tolman 1948). The reference frame of a cognitive map is world-(allo)centric: it emphasizes the independence of the cognitive map representation from the current position or orientation of the organism and enables flexible navigation, such as taking shortcuts or detours. In the classical view, the reference frame of the cognitive map is Euclidean (O'Keefe and Nadel 1978), but alternative views suggest that maps may take a more graph-like format and violate Euclidean properties of distances between places (Chrastil and Warren 2014; Peer et al. 2024; Warren 2019).

A wealth of human and rodent studies have identified the central involvement of the medial temporal lobe (MTL) and, in particular, the hippocampus in mediating cognitive maps (Hartley et al. 2003; O'Keefe and Conway 1978; O'Keefe and Dostrovsky 1971; Vargha-Khadem et al. 1997). Single-neuron electrophysiological recording studies in freely moving rodents (Figure 9.1a, i) have shown that hippocampal principal cells (recorded from the CA1–CA3 subregions) encode a unique position in space (Figure 9.1a, ii) and, as a population, keep track of an animal's spatial location (O'Keefe and Dostrovsky 1971; O'Keefe and Nadel 1978). Importantly, the population of place cells that are active in different environments as well as the position in which they fire in these environments are not correlated; that is, place cells form a new map for each environment, a process termed *remapping* (Muller and Kubie 1987). Place cells, however, seem to be metric free, lacking an important property of maps. Aside from place cells with neighboring place fields that overlap partially, place cells alone cannot be used to infer relative distances. Grid cells likely provide a metric for distance metric and path-integration abilities (Hafting et al. 2005; Mittelstaedt and Mittelstaedt 1982 and discussed further below).

Cognitive maps are anchored to the environment via allothetic cues such as landmarks (e.g., tall buildings, statues, or other prominent visual cues) and environmental geometry (Cheng 1986; Epstein and Vass 2014). For example, when walking out of a London underground station in search of a restaurant located on the river Thames, one may look for the Big Ben tower to determine in which direction and how far to walk. One could also rely on geometric cues about the shape of the environment, particularly in small, confined spaces such as rooms. In the classic reorientation task, if organisms learn that a target object can be found in one corner of a rectangular environment and undergo disorientation before testing, they are equally likely to search in a location that is diagonally opposite to the correct location, an area that is geometrically equivalent to the location where they had found the object previously (Cheng 1986).

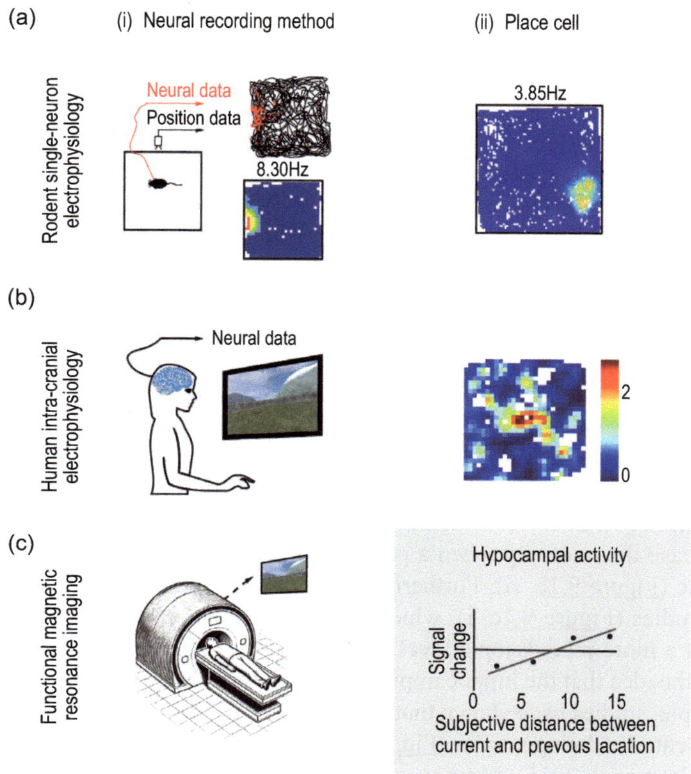

Figure 9.1 Studying the hippocampal cognitive map in humans and rodents. (a) Rodent neuronal hippocampal recording and spatial representation. (i) Schematic drawing of a setup to record spatial cell types in the rodent brain. While single-neuron activity is recorded (left), the position of the animal is recorded on a moment-to-moment basis with an overhead camera. Spikes recorded from a single hippocampal place cell (top right, red dots) are plotted on top of the animal's path (black lines). Spatial rate map of a cell's activity in the square environment (bottom right); hotter colors indicate higher activity. (ii) A rate map of an example place cell, showing peak firing rate of cell. (b) Single-neuron hippocampal representations recorded in the human brain using intracranial electrophysiological techniques. (i) Schematic of a standard setup to record spatial representations. As subjects engage in virtual reality (VR) spatial navigation tasks, electrophysiological signals are recorded from the brain. (ii) Example rate map of a human hippocampal place cell recorded during a VR navigation task (Jacobs et al. 2013, used with permission from Springer Nature). (c) Evidence for hippocampal spatial representations in the human brain obtained via fMRI experiments. (i) Schematic of subject undergoing fMRI while conducting VR navigation tasks. (ii) Change in hippocampal fMRI BOLD signal (y-axis) in relation to distance between landmarks shown on successive trials (after Morgan et al. 2011, used with permission).

Akin to the cues known to guide orientation behavior, visual landmarks and environmental geometry are known to influence place-cell activity. For example, if a prominent distal visual cue rotates around the environment, the spatial firing fields (place fields) of place cells rotate by a proportional amount (Muller and Kubie 1987; O'Keefe and Conway 1978). Similarly, morphing a square-shaped environment into a circular-shaped environment has been found to induce remapping (Wills et al. 2005); even a mere stretch of one axis of an environment can lead to a concomitant "stretching" of place fields (Burgess and O'Keefe 1996). However, place cells are not exclusively influenced by allothetic cues. Place cells can maintain their place fields in the dark (Save et al. 1998; Save et al. 2000), and a wealth of studies have shown that self-motion (idiothetic) cues are critical for place-cell activity (Terrazas et al. 2005). This demonstrates sensitivity to internal (idiothetic) cues (e.g., vestibular and proprioceptive cues) as well as signals relayed from the motor system (motor efference copy) and optic flow, all of which are known to guide navigation through a process called *path integration* (Darwin 1873). Navigation seldom relies exclusively on path integration as it is inherently error prone: it relies on the addition of multiple incremental movement steps—errors which could be catastrophic in larger environments. Thus, accurate navigation is thought to rely on the integration of both allo- and idiothetic cues (e.g., Cheung et al. 2012; Parra-Barrero et al. 2023), with different cues being weighted according to the reliability of sensory information (Chen et al. 2013; Evans et al. 2016; Gothard et al. 1996). Hippocampal place cells are sensitive to both types of inputs.

Neurons with similar activity properties have also been identified in humans implanted with intracranial electrodes as they navigate through virtual reality (VR) environments (see Figure 9.1b, i) (Ekstrom et al. 2003; Kunz et al. 2021; Kunz et al. 2024; Miller et al. 2013; Tsitsiklis et al. 2020). Similar to rodent place cells, these cells increase their activity when a participant occupies a specific location in (virtual) space (Figure 9.1b, ii). Furthermore, functional magnetic resonance imaging (fMRI) studies (Figure 9.1c, i), which examine hemodynamic correlates of brain activity at a more macroscopic level (~1–27 mm^3), have reported findings consistent with the idea that the hippocampus encodes a map-like representation of space. For example, studies have shown that positions located near each other in VR space are represented as being similar (Figure 9.1c, ii) (Deuker et al. 2016; Morgan et al. 2011; Nielson et al. 2015). Moreover, fMRI studies have shown that the hippocampus is preferentially engaged when participants participate in map-based navigation tasks (Chadwick et al. 2015; Hartley et al. 2003; Howard et al. 2014).

As noted above, although the human hippocampus may represent space in a manner similar to the rodent, neurons in the human hippocampus may display additional encoding capabilities. For example, in certain tasks, the activity of human hippocampal neurons can couple to the position of a spatial target (i.e., spatial target cells; Tsitsiklis et al. 2020). Nonhuman primate (NHP) research has reported cells that encode remote places, such as where a NHP is looking regardless of their own position (Mao et al. 2021; Rolls and O'Mara 1995; Wirth et al. 2017). In addition, the encoding of abstract representations of objects (concept cells) has been reported in human intracranial studies (Quiroga et al. 2005). These cells are thought to contribute to the hippocampus's role in episodic memory. Whether these findings point to divergent- or convergent-encoding mechanisms across species is still a matter of debate, a topic we explore next.

2.2 Alternative Model Organism for Bridging Analysis Levels in Navigation Research

Evolutionarily closer to humans than rodents, NHPs offer the possibility of bridging brain activity and navigation behavior and allow intracranial recordings to be conducted. A prominent early study of NHP hippocampal responses emphasized their primary function as "spatial view" cells; that is, hippocampal neurons fired not according to where an animal was, but rather to where it was currently looking (Rolls and O'Mara 1995). This finding diverged notably from the established place-cell coding observed in rodents, and subsequent controversy focused on whether the information processed in primate and rodent hippocampal systems was fundamentally different (Mao et al. 2021; Rolls and Wirth 2018; Zhu et al. 2023). A key point of contention was whether the immobility of the experimental primate subjects (fixed in a chair) altered the activity of hippocampal neurons in a fundamental way such that freely moving subjects would have demonstrated place-tuned activity in their hippocampus, akin to that observed in rodents. Resolution of this controversy came through experiments that had NHPs actively navigate in VR (Furuya et al. 2014; Gulli et al. 2020; Wirth et al. 2017) or that freely moved in a manner analogous to rodent experiments (Mao et al. 2021; Xu et al. 2024) as well as a recent avian study that recorded the activity of place cells in relation to spatial and gaze behavior (Payne and Aronov 2025). Key consensus points emerged from these studies:

- The majority of NHP hippocampal neurons appear to encode complex multivariate information, including combinations of position, facing or gaze direction, linear and angular speed, and head tilt.
- Both the current position and the position being looked at or faced are among the variables consistently coded. Indeed, of all gaze-tuned hippocampal neurons in the black-capped chickadee, 95% exhibited place-specific responses when the bird occupied the location corresponding to their preferred gaze target (Payne and Aronov 2025).

Comparable findings have been reported in human intracranial studies, where cells that encode the position of a spatial target have been observed (Tsitsiklis et al. 2020). Thus, although differences may exist in terms of the inputs that underpin spatial representations in different species, and view-based coding may dominate, it appears that the "place-coding" function of the hippocampus is preserved accross species and that an important "remote" place-coding function emerged in the primate order.

Human intracranial studies have also reported neurons in the MTL that encode abstract representations of a person (e.g., Jennifer Aniston) or building (e.g., Sydney Opera House) (Quiroga et al. 2005). These abstract representations, termed *concept cells*, have been argued to be evidence for encoding divergence between rodent and humans and are hypothesized to give rise to episodic memory capabilities, which may be unique to humans (but see Bicanski 2025; Buzsaki and Tingley 2018). At this point, however, we caution against such an interpretation because the encoding of concepts has not been investigated explicitly in rodents. Thus, with the right paradigm, comparable neuronal representations may be found in rodents. Furthermore, rodent place-cell activity has been linked to memory processes such as recall (e.g., during replay when nonlocal place-cell representations are activated, sometimes reflecting upcoming paths) (Pfeiffer and Foster 2013; Wilson and McNaughton 1994). Even more tantalizing evidence comes from a recent brain-machine interface study, which showed that rodents can volitionally activate remote place representations (Lai et al. 2023). Mounting evidence suggests that hippocampal neuronal representations show a considerable degree of convergence across species. Based on common anatomical features of the hippocampal formation and some convergence in neural responses, we suggest that, to a first approximation, the hippocampal formation performs similar computations across species, albeit operating on different kinds of inputs: rodents predominantly explore with their body, whereas primates also use eye movements to explore space. We encourage researchers interested in species-specific differences to employ paradigms suited for cross-species testing so that such differences can be identified without bias.

2.3 Coding of Orientation, Distance, and Boundaries

To generate a map-like representation, hippocampal place cells need to integrate orientation and distance information. This information is thought to derive from brain regions afferent to the hippocampus and is summarized here, along with the spatial representations they encode.

Orientation information is thought to be relayed to place cells by head-direction (HD) cells (Taube et al. 1990). These cells encode an animal's allocentric orientation in space irrespective of the animal's spatial position (Figure 9.2a, ii). HD cells have been extensively recorded in rats. First discovered in the postsubiculum (PoS) by Ranck (1984), they have since been found in multiple brain regions, with the main loci being the medial entorhinal cortex (MEC) (Sargolini et al. 2006), retrosplenial cortex (RSC) (Cho and Sharp 2001), the anterior dorsal thalamic nucleus (ADN), and the lateral mammillary nucleus (LMN) (Stackman and Taube 1998). HD cell activity is driven by vestibular input from the brain stem that projects to the HD circuit via the LMN and the dorsal tegmental nucleus (DTN) (Blair et al. 1998; Sharp et al. 2001). External cues such as landmarks, however, also exert strong influence over HD cells in all loci of the circuit (Taube 1995; Taube et al. 1990). The RSC and PoS are thought to be critical for registering HD tuning to landmark cues (Goodridge and Taube 1997; Vann and Aggleton 2005) and serve as a key gateway for visual information (Taube 2007). Modeling studies have shown the RSC is needed to correct for drift of the heading estimate coded by HD cells (Bicanski and Burgess 2016; Page et al. 2018; Yan et al. 2021). In addition, RSC has been suggested to play a role in egocentric-to-allocentric transformation of visually derived object and boundary representations (Bicanski and Burgess 2018; Byrne et al. 2007).

HD cells have not been studied directly in humans, due to the nonexistence of human single-neuron data during real-world navigation studies. Moreover, in human neuroimaging studies, subjects navigate VR environments while lying still in a scanner, thus observed neural correlates of head direction relate necessarily to *virtual*, or even imagined, orientation rather than actual head direction. Nonetheless, a few single-neuron studies have identified neurons in the human MTL (hippocampus, parahippocampal cortex, EC), where activity correlates with a person's direction in virtual space (Figure 9.2b, ii) (Jacobs et al. 2010; Kunz et al. 2021). In addition,

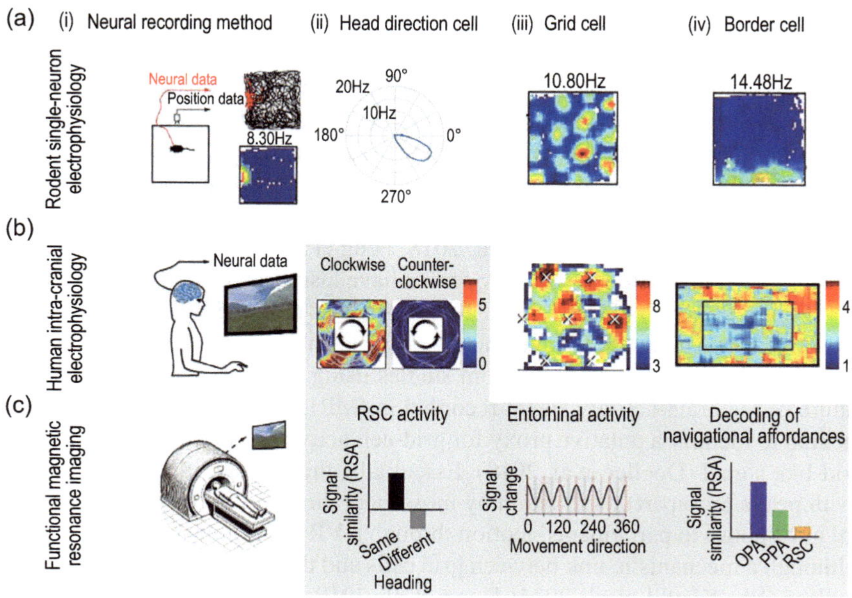

Figure 9.2 Neuronal representations for head direction, distance and boundaries in rodents and humans. (a) Rodent neuronal spatial representations. (i) Same as Figure 9.1. (ii) Polar plot showing directionally modulated activity of a HD cell in the anterior dorsal thalamic nucleus. Rate map of sample grid (iii) and border (iv) cells. (b) Spatial representations recorded in the human brain using intracranial electrophysiological techniques. (i) Same as Figure 9.1. (ii) Rate maps of a cell recorded in the entorhinal cortex while subjects navigated a virtual square-shaped corridor in a clockwise (left) or counterclockwise (right) manner (Jacobs et al. 2010, used with permission). (iii) Rate map of an entorhinal grid cell recorded during VR navigation (from Jacobs et al. 2013, used with permission). (iv) Rate map of local field potential theta power recorded from the subiculum while subjects learned the locations of objects in a rectangular-shaped environment (Lee et al. 2018, used with permission). (c) Evidence for spatial representations in the human brain obtained via fMRI experiments where subjects engage in VR navigation tasks. (i) Same as Figure 9.1. (ii) Similarity between BOLD signal recorded from the retrosplenial complex for the same and different head directions (Vass and Epstein 2016, used with permission from Oxford University Press). (iii) BOLD signal change, displaying sixfold symmetry, from the entorhinal cortex in relation to angular difference between sampled paths (after Bin Khalid et al. 2024). (iv) BOLD signal similarity from the parahippocampal place area (PPA), occipital place area (OPA), and RSComp in relation to the similarity of the navigational affordances of different scenes (after Bonner and Epstein 2017). Shaded panels indicate indirect evidence for a particular spatial representation.

fMRI studies have shown orientation-related signals in the EC and retrosplenial complex (RSComp, including the RSC) during VR spatial navigation (Figure 9.2c, ii), suggesting an orientation code may also exist in humans (Bellmund et al. 2016; Chadwick et al. 2015; Lu et al. 2025; Marchette et al. 2014; Persichetti and Dilks 2019; Vass and Epstein 2017). Still, it remains unclear whether this code is driven by inputs that are similar to rodent HD cell activity, which are generated by the same neural circuit (see Bicanski, Chapter 5).

Information about distance traversed is thought to be mediated by a class of MEC neurons known as grid cells (Hafting et al. 2005). Grid cells display periodic spatial firing fields that tessellate entire environments in a triangular grid-like array

(Figure 9.2a, iii) at different scales. Through their periodic activity, grid cells are thought to provide a continuous update regarding an animal's translational movements (i.e., odometry). In addition, a subset of MEC grid cells display a conjunctive code, responding only when an animal's head is oriented in a particular allocentric direction (Sargolini et al. 2006). In contrast to place cells, grid cells are active in all environments. Given the ubiquity of the grid code and its encoding of distance and orientation, these cells have been hypothesized to be the source of the path-integration signal in the hippocampal cognitive map, a hypothesis that has received some experimental support (Gil et al. 2018; Peng et al. 2025; Tennant et al. 2018).

Single neurons with grid-like properties have also been described in the human EC (Figure 9.2b, iii) (Jacobs et al. 2013; Miller et al. 2015). Yet because single-neuron recordings in the EC are rarely possible in humans, the vast majority of human grid-cell research comes from studies using fMRI. Given the macroscopic nature of the brain-activity signal recorded in fMRI, researchers have developed a method to identify a putative proxy for grid-cell activity, sometimes referred to as a grid-like signal (Doeller et al. 2010). Essentially, this method identifies six-peaked (with peaks 60° apart) modulation by movement direction of the fMRI BOLD signal in response to participants' motion through a VR environment (Figure 9.2c, iii). Although a mechanistic link between grid cells and the grid-like signal still remains unclear (Bin Khalid et al. 2024; Kunz et al. 2019; see also Chapter 6), multiple groups have independently found grid-like activity in the human EC during motion in VR environments (Doeller et al. 2010; Kunz et al. 2015) and mental navigation (Bellmund et al. 2016; Horner et al. 2016), as well as in abstract two-dimensional spaces (Bellmund et al. 2018; Constantinescu et al. 2016), suggesting a grid-cell code akin to that seen in rodents may also exist in humans.

Information about the geometry of an environment's boundaries (e.g., walls and drop edges) is thought to be integrated in rodents by boundary vector cells (BVCs) found in the subiculum (Lever et al. 2009) and border cells in the MEC (Solstad et al. 2008). BVCs fire at fixed distances from environmental boundaries that are located at a particular allocentric direction from the animal (e.g., west wall, see Figure 9.2a, iv). Border cells in MEC similarly code for boundary-related information, but they fire exclusively when the animal is located at or near an environmental border. No intracranial recording studies have investigated boundary-related single-neuron representations in humans, and only a few studies have investigated changes in intracranial EEG signals in the MTL in response to humans moving freely in space. These latter studies observed an increase in theta power when subjects navigated or learned items close to environmental boundaries (Figure 9.2b, iv) (Lee et al. 2018; Stangl et al. 2020). Similarly, a small number of studies have investigated boundary-related modulation of the BOLD signal. These studies identified a number of brain regions, such as the parahippocampal place area (PPA) and occipital place area (OPA), which respond when subjects are presented with an environmental boundary or an object that impedes their navigation through space (Figure 9.2c, iv) (for recent reviews, see Dilks et al. 2022; Epstein 2008; Julian et al. 2018b). Thus, there is some evidence for the representation of geometric information in the navigation network of the human brain, although the cortical regions implicated in geometric processing in the human brain may use boundary or object information for different purposes (Dilks et al. 2022).

Studies in humans have identified several neocortical regions that are proposed to provide the described spatial cell types with allothetic input. These include the PPA, the OPA, the RSComp, and the superior parietal lobe (Epstein and Baker 2019; Kennedy et al. 2024; Yoon et al. 2024). These regions are often activated in fMRI studies of navigation. They also respond strongly during mere passive viewing of spatial scenes (e.g., landscapes, cityscapes, or individual rooms) and are hence often referred to as the "scene network." The OPA is specifically implicated in the processing of the local spatial layout of scenes (e.g., the location of boundaries and navigational affordances), information necessary to move about one's immediately visible environment (often referred to as "visually guided navigation") (Bonner and Epstein 2017; Dilks et al. 2011, 2013; Kamps et al. 2016a, 2016b; Persichetti and Dilks 2016, Dilks 2018). The PPA is thought to encode visual and spatial features that allow for the identification and categorization of scenes (Epstein and Kanwisher 1998; Persichetti and Dilks 2018). The RSComp, on the other hand, is purported to mediate between the local scene information encoded by the PPA and OPA and the map-based representation of the hippocampal cognitive map (Epstein 2008). Indeed, lesions to the RSComp are known to result in heading disorientation in humans (Ino et al. 2007); that is, the inability to understand the relationship between the local scene and distal locations. Recent rodent studies have shown that the RSC encodes head direction in terms of a local and a global reference frame (Jacob et al. 2017). RSComp may also be a cortical locus for spatial knowledge of familiar environments that has become consolidated over time (Wolbers and Buchel 2005). The superior parietal lobe has been less studied but exhibits similarities to RSComp (Dutriaux et al. 2024; Marchette et al. 2014) and OPA (Yoon et al. 2025).

It is important to note that the neuroanatomical organization of the scene network, as well as its rodent homologues, continues to be investigated. Scene regions can be reliably localized based on their response to scenes (Weiner et al. 2017, 2018), but these functionally defined regions do not correspond neatly to anatomically defined counterparts. For example, PPA is not the same as parahippocampal cortex, and RSComp is not the same as RSC. The mapping to the corresponding regions in rodents (e.g., postrhinal and retrosplenial cortices) is even more tentative, and for some regions (e.g., OPA), it is unclear whether there is a rodent equivalent. Recent work suggests a superseding organization that cuts across the scene regions, consisting of an anterior subnetwork (including anterior PPA and most of RSComp) that is more involved in mnemonic aspects of scene processing and a posterior network (including OPA and posterior PPA) that is more involved in perceptual aspects (Baldassano et al. 2013, 2016; Steel et al. 2021).

An alternative approach to mapping anatomical homologues between species is to draw on recent advances in molecular techniques, which have enabled the generation of spatially resolved atlases of gene expression and anatomical connectivity at single-cell resolution (e.g., Yao et al. 2023). These atlases hold transformative potential for comparative neuroanatomy. Just as classical neuronal staining allowed anatomists to identify parallels in cellular architectures across species, contemporary molecular approaches may now enable the discovery of fundamental genetic correspondences between rodent and human brains (Joglekar et al. 2024; Zemke et al. 2023). Critically, these genetic signatures could go beyond identifying anatomical homologues and reveal meaningful evolutionary relationships that might otherwise remain undetected by anatomical or functional analyses alone.

2.4 Neuronal Mechanisms of Goal-Directed Navigation

Above, we described the neuronal representations that underlie our ability to orient ourselves in space. Navigation, however, requires more than just self-localization. Once we establish our spatial position and heading, we then need to estimate the distance and direction (i.e., a vector) between our current location and that of our goal. Furthermore, since most environments through which we navigate are complex (i.e., they contain multiple barriers or objects that obstruct a path), computing a vector that depicts the straight line distance to a goal is rarely sufficient. A route that respects the physical constraints of an environment must also be constructed. Here, we synthesize available knowledge on the proposed neuronal mechanisms that underlie these critical navigational processes.

Theoretical and experimental work has made progress in elucidating the neuronal computations that underlie vector-based navigation. Convergent findings suggest the EC and, more specifically, the grid cells of the MEC are the candidate neural substrates for computing a navigational vector (Bierbrauer et al. 2020; Bush et al. 2015; Erdem and Hasselmo 2012; Howard et al. 2014; Kubie and Fenton 2012). For example, in modeling the activity of grid cells across distinct anatomical modules (defined in terms of scale and orientation of the grid), Bush et al. (2015) showed that the activity of grid cells belonging to distinct modules was sufficient to resolve unambiguously the position (distance and direction) of a goal location, even in large environments where distance to the goal exceeds the largest grid scale (see also Edvardsen 2019; Mathis et al. 2012; Sreenivasan and Fiete 2011). Moreover, in studying the vectorial codes in the human brain while participants engaged in a VR navigation task inside an fMRI scanner, Howard et al. (2014) designed a task to decorrelate Euclidean and path distances to learned goals and found that entorhinal activity correlated positively with the Euclidean distance to a goal, and not the path distance. Subsequent work showed that entorhinal activity also tracks the (imagined) heading of a navigational goal (Chadwick et al. 2015) and exhibits grid-like responses that correspond to the goal vector (Bellmund et al. 2016). Relatedly, Bierbrauer et al. (2020) showed that fMRI grid-like activity in the EC predicted performance on a path-integration task, and that performance was impaired in carriers of the ε4 allele of the *APOE* gene, which carries an increased risk of developing Alzheimer disease (Corder et al. 1993) and is associated with neural degeneration in the EC in young adults.

The hippocampus, on the other hand, has been hypothesized to support the ability to plan navigational paths. For example, fMRI studies have shown that the hippocampus preferentially tracks path distance over Euclidean distance to a goal (Howard et al. 2014) and represents intermediate steps along the route when participants plan a route to a goal (Brown et al. 2016). Furthermore, MTL regions are thought to work in concert with neocortical regions, such as RSComp (Patai et al. 2019) and prefrontal cortex (Patai and Spiers 2021), during path and vector planning. The neuronal mechanism for path planning may be time-compressed reactivation of rodent place-cell sequences, which encode paths an animal has already traversed (Jackson et al. 2006; Lee and Wilson 2002; Wilson and McNaughton 1994). These sequences occur either during oscillatory states associated with locomotion (theta sequences) or sleep/immobility periods (replay). Goal-directed paths are often preceded by replay that is centered at locations and paths near the animal; in some instances, replay even predicts future behavior (Olafsdottir et al. 2017; Pfeiffer and Foster 2013; Singer et al. 2013). At decision points, theta sequences have been found to depict upcoming paths (Johnson and Redish 2007; Wikenheiser and Redish 2015). When animals are unsure of their path in a complex maze, theta sequences continuously cycle between alternative paths (Kay et al. 2020), reminiscent of deliberative processes. In a task where animals alternated between random foraging and goal-directed navigation, replay only predicted upcoming goal-directed paths (Pfeiffer and Foster 2013). Similarly, the spatial extent of theta sequences has been found to be modulated by navigational goals, extending further ahead of the animal when they take place during vicarious trial-and-error behavior (Wikenheiser and Redish 2015). Finally, and important for its hypothesized role in path planning, replay is sensitive to an environment's navigational affordances (Widloski and Foster 2022). Such flexible navigation could result from the interplay of map-based planning and vector navigation (Edvardsen et al. 2020). In an environment containing multiple barriers, replay sequences respected the topological structure of the environment, even when the structure changed on every visit (Widloski and Foster 2022).

It is important to note that not all replay studies report results that are consistent with its proposed planning function. For example, Gillespie et al. (2021) recorded replay while animals engaged in a flexible goal-directed navigation task on a multi-arm maze and found no bias for replay of paths leading to an animal's current navigational goal. Instead, they observed a strong bias for past rewarded paths. Replay during sleep has long been hypothesized to support memory consolidation (i.e., the stabilization of new memories) (Wilson and McNaughton 1994), a hypothesis that has considerable experimental support (Ego-Stengel and Wilson 2010; Girardeau et al. 2009; Gridchyn et al. 2020; Maingret et al. 2016). Whether replay recorded during waking periods serves exclusively to support memory consolidation or is additionally involved in planning remains an area of ongoing research (e.g., Mallory et al. 2025).

In summary, research conducted in humans and rodents over the past decades has elucidated a network of brain regions and their neuronal representations that encode spatial variables critical for self-orientation, and an emerging body of research has highlighted candidate neuronal mechanisms that may support goal-directed navigation. There are, however, areas of contention. For example, due to methodological differences in the way neuronal representations are studied in rodents and humans, questions remain regarding the nature of neuronal spatial encoding in different species and even whether brain regions strongly implicated in spatial behavior in rodents encode space in humans. Moreover, the interplay of different navigation strategies (e.g., landmark-based, map-based, vector-based) is also understudied. We turn to these issues next and discuss possible solutions.

3 Bridging Methodologies in Navigation Research

A key challenge to creating a cross-level description of navigation is to bridge the varied methodologies employed to study the neural basis of navigation, particularly when methods vary between species-specific researchers. Findings reported in rodent research are often derived from direct measurements of neuronal activity that operate at the timescale of brain activity (< 1 ms) and allow the activity of single cells to be resolved. Human research, on the other hand, often relies on indirect, hemodynamic correlates of brain activity (e.g., via the fMRI BOLD signal) over long timescales (i.e., seconds) with limited spatial resolution (1–27 mm^3 in typical fMRI studies). Thus, obtaining unequivocal evidence for the existence of the distinct spatial cell types in the human brain has been challenging. To address this issue, we propose the use of (a) computational modeling of rodent data to predict how a single-neuron functional representation would be expressed in population activity that fMRI measures and (b) signal alignment of rodent and human data so that comparable representations can be studied in the two species.

3.1 Computational Modeling: Linking Macroscopic and Neural Signals

As described above, grid-like signals have been reported in fMRI research (Doeller et al. 2010), yet how do these signals relate to cellular responses? Computational simulations of single-neuron signals have explored several plausible accounts. One view, the *conjunctive hypothesis*, is that the fMRI grid signal may emerge from population activity of conjunctive grid × HD cells (Figure 9.3a) (Bin Khalid et al. 2024; Kunz et al. 2019). As these conjunctive cells may increase their activity when individuals navigate along the grid axes, this could result in a heightened BOLD response when participants move along the axes of the grid-cell network. An alternative view, the *repetition suppression hypothesis*, is that grid cells and/or related fMRI activations show repetition suppression (Figure 9.3b), leading to reduced fMRI activity when individuals move along the grid axes (because the traversed grid fields repeatedly come from a smaller set of unique neurons). A potential third option, the *clustering hypothesis*, is that grid cells with similar phase offsets relative to environmental boundaries cluster anatomically, resulting in stronger signals in these anatomical subregions when subjects move along the axes of the grid cells belonging to a given cluster. This signal could then potentially modulate the BOLD signal, as long as the size of the grid-cell cluster does not exceed the spatial resolution of fMRI (Figure 9.3c).

Computer simulations of populations of grid cells showed that the conjunctive and repetition suppression hypotheses are good candidates for explaining grid-like signals, whereas the clustering hypothesis is less plausible (Bin Khalid et al. 2024). Furthermore, the repetition suppression hypothesis is dependent on the exact type of navigation paths, triggering new ideas for designing fMRI experiments on grid-like signals. Future studies could examine whether grid-like signals show alterations due to environmental changes, such as the reported elliptic distortions in individual grid cells (Stensola et al. 2015), which would provide useful additional information

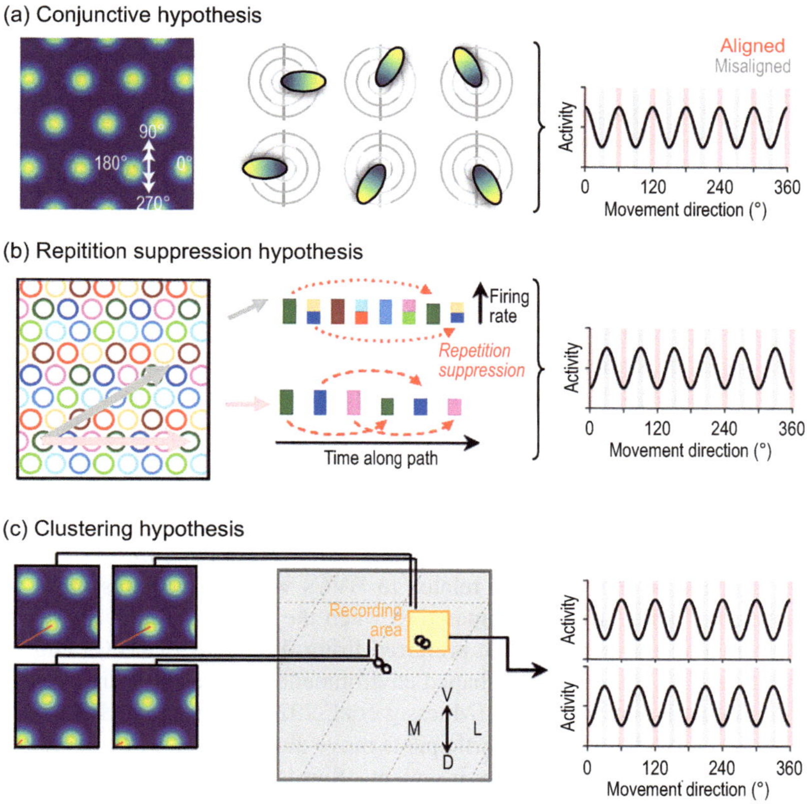

Figure 9.3 Computational modeling of macroscopic grid-like signals. Reused with permission from Bin Khalid et al. (2024) http://creativecommons.org/licenses/by/4.0/. (a) Conjunctive hypothesis: Macroscopic grid-like representations (right) may emerge from the firing of conjunctive grid by HD cells (left and middle panels) that exhibit increased firing when the subject moves aligned as compared to misaligned with the grid axes (right) (Doeller et al. 2010). This relies on directional preference across the grid-cell population being biased toward on-axis directions (middle, which shows preferred HD tuning for grid cells akin to the left panel). (b) Repetition suppression hypothesis: In a grid-cell population with similar grid orientations and grid spacings but with distributed phases (left: colored circles represent firing fields of different grid cells), aligned movements (horizontal pink arrow) lead to more frequent activation of a smaller number of different grid cells, whereas misaligned movements (diagonal gray arrow) lead to less frequent activation of a higher number of different grid cells (Doeller et al. 2010). Thus, aligned movements may lead to more pronounced repetition suppression as compared to misaligned movements (middle), resulting in a hexadirectional modulation of population spiking activity (right). (c) Clustering hypothesis: Because anatomically adjacent grid cells exhibit similar grid phase offsets, in addition to similar grid orientations and grid spacings (Gu et al. 2018), recordings from a limited number of grid cells with a nonrandom distribution of phase offsets may lead to macroscopic grid-like representations. Left: grid phase offset of four different grid cells, whose anatomical locations are illustrated in the middle panel. Depending on the subject's starting location relative to the phase offset of the grid fields, movements aligned or misaligned with the grid axes lead to higher total grid-cell activity compared to misaligned or aligned movements; upper and lower portions of the right panel illustrate hexadirectional modulation for two different anatomical clusters. Moreover, the orientation of hexadirectional modulation may shift when recording from neighboring voxels in anatomical space due to a shift in the clustered phase offsets.

for computational studies and enhance confidence in the fMRI grid-like signal. For a discussion of how fMRI grid-like signals are modulated by the shape and orientation of the environment, see Julian et al. (2018a).

A comparable approach could be adopted to investigate other spatial representations (e.g., boundary and HD cells) in the human brain. By considering the properties of these cell types in rodents and simulating populations of them, researchers could make predictions regarding the functional properties of these cells and the likelihood of observing population activity of boundary and HD cells with fMRI and intracranial EEG. For example, if a population of HD cells preferred a particular heading direction (e.g., toward a major distal cue), fMRI studies might record increased BOLD signals when participants face that direction. In contrast, if the cells' preferred heading directions were distributed randomly, a macroscopically visible population signal would become unlikely as the same magnitude of BOLD signal would be recorded along any heading direction. In relation to boundary representations, BVCs in the subiculum tend to fire near environmental boundaries (Muessig et al. 2024). Interestingly, recent intracranial studies have observed increased theta power in epilepsy patients when they were close to the boundary of a virtual rectangular room (Lee et al. 2018; Stangl et al. 2020), which may stem from the overrepresentation of short distances in BVCs. Accordingly, fMRI studies may also find increased fMRI signals related to BVCs when participants move along environmental boundaries as compared to when they move through the center of an environment. Furthermore, the preferred firing directions of BVCs align with environmental walls in a square-shaped environment, whereas they are evenly distributed in circular environments (Muessig et al. 2024). Hence, fMRI studies may find fourfold modulation while participants move along the edges of a rectangular environment but not in circular environments. Similar to models of the emergence of sixfold symmetry in the BOLD signal, computational modeling could help us to understand how and under which conditions a boundary signal could be measured in the fMRI BOLD signal by simulating populations of BVC cells with their known properties. Additionally, these models may elucidate why representations of boundaries and heading directions are most often observed in the theta band, by considering how strongly the spiking of individual neurons is modulated by theta rhythms (Lever et al. 2009).

3.2 Signal Alignment

An alternative approach is to try to match the neuronal signals recorded in rodent and human studies. For example, many human intracranial studies investigate representations in the local field potential (LFP), such as of environmental boundaries, rather than in single-neuron activity as is standard in rodent studies. Thus, a straightforward approach to determine whether results like these truly reflect a common functional representation is to assess whether rodent LFP shows similar spatial tuning. For example, by applying high-density, multisite electrophysiological probes, this approach might show whether regions with a higher density of place cells (e.g., dorsal vs. ventral CA1) are also more likely to contain LFP signals modulated by specific locations (Agarwal et al. 2014). Conversely, human intracranial studies could use single-neuron recordings through Behnke-Fried electrodes (Fried et al. 1999) to detect the activity of individual neurons in addition to LFPs. Using this technique, studies in epilepsy patients have obtained evidence for grid and place cells in the human MTL (Jacobs et al. 2013; Kunz et al. 2021, 2024; Miller et al. 2013; Nadasdy et al. 2017; Qasim et al. 2019, 2021). More broadly, this line of research may show whether information encoded at the level of individual neurons and LFPs is redundant or complementary. For example, studies might show whether LFP signals are more likely to represent allocentric information if individual neurons in that brain region are also more specialized in processing allothetic instead of idiothetic cues. Consistent with this view, several studies in humans have demonstrated theta oscillations—an oscillatory signal implicated in core spatial and mnemonic processes—in the MTL related to allocentric distance coding (Bush et al. 2017; Buzsaki and Moser 2013; Solomon et al. 2019; Vass et al. 2016).

Matching signals acquired through indirect hemodynamic correlates of neural activity (e.g., the fMRI BOLD signal) with single-unit electrophysiological signals is more challenging. Using fMRI scanners, neuroimaging is possible in rodents. However, these experiments require the rodent to be at least lightly anesthetized, precluding the study of spatial representations. An alternative is to image hemodynamics in rodents via an optical approach (e.g., 1- and 2-photon microscopy, functional ultrasound imaging) that is compatible with *in vivo* experiments (both in head-fixed and freely moving animals). An added advantage of this approach is that neural activity could be simultaneously recorded via calcium imaging, allowing researchers to unequivocally ascertain whether, for example, the sixfold modulation of the hemodynamic signal truly reflects grid-cell activity. This approach has yet to be attempted in the context of navigation, yet it has been successfully implemented in experiments studying resting-state activity (Kozberg and Hillman 2016; Ma et al. 2016). Thus, it represents a potentially viable and exciting approach to bridging the diverse methodologies employed in navigation research.

4 Bridging Neuronal Activity to Navigation Behavior

Rodent neurophysiology research has provided crucial insights into the single-neuron code that supports spatial behaviors. However, the majority of rodent studies examine neuronal representations in the context of simple spatial behaviors in controlled experimental apparatuses (e.g., foraging in open field environments or traversals on linear tracks) in oriented subjects or when disoriented animals are not required to express any meaningful spatial behavior. As a result, it is difficult to link neural activity to navigation because it is unclear which processes and cues are guiding behavior. Furthermore, most rodent studies investigate spatial representations in vista environments (i.e., environments viewed from a single vantage point such as a single chamber) as opposed to large-scale and complex environments (environmental spaces) that better mimic real-world settings (but see Eliav et al. 2025; Ray et al. 2025). This point is particularly important as processes which guide spatial behavior are strongly affected by whether navigation occurs in a vista or environmental space (Ekstrom and Isham 2017; Meilinger et al. 2016; for further discussion, see Appendix 9.1). Finally, future navigation research should strive to make their studies more naturalistic. Highly controlled and constrained experimental environments limit the range of spatial behaviors animals can exhibit and may cause researchers to overlook behaviors that occur only in the wild. Such naturalistic approaches have already been successfully applied in studies of flying bats and could readily be extended to research on other species (for further discussion, see Ulanovsky 2025).

We are only beginning to understand the neuronal processes that underlie higher-order navigational processes, such as planning and vector computation. Some initial insight has been gained from studies on replay (Pfeiffer and Foster 2013; Singer et al. 2013) and theta sequences (Johnson and Redish 2007; Kay et al. 2020; Wikenheiser and Redish 2015; Yu et al. 2025) as well as a small number of human fMRI studies that have mapped the neural networks engaged at different stages of navigation (e.g., Brown et al. 2016; Chadwick et al. 2015; Howard et al. 2014; Patai et al. 2019). Given the lack of consensus regarding the role of replay in planning and limited experimental evidence for the involvement of the MEC and grid cells in vector computation, more systematic studies are needed in these areas.

The reason for these knowledge gaps likely reflects (a) the difficulty with studying single-neuronal representations in experimental contexts which do not allow for extensive and repetitive sampling of the spatial firing fields of individual neurons, as is often the case in studies that employ spatial navigation tasks, and (b) technical challenges associated with carrying out electrophysiological experiments in complex environments. The advent of high-density single-neuron recording techniques like Neuropixels probes (Jun et al. 2017) may help address the first challenge. These probes enable researchers to record from hundreds of neurons simultaneously, allowing them to analyze a population code rather than relying solely on a single-neuron code, which requires less extensive spatial sampling. To address the second, technical challenge, emerging wireless electrophysiology technologies (e.g., neurologgers) offer a potential solution (for an extensive discussion, see Ulanovsky 2025). In recent years, neurologgers have rapidly improved: some models are capable of recording from 300–400 channels simultaneously, making them comparable to wired systems. Additionally, advances in lightweight battery technology now allow for long-duration wireless recordings for low channel-count systems (e.g., 32–64 channels), lasting several hours. Continued technological advances in wireless power transfer will hopefully soon enable wireless recordings to become on par with state-of-the-art wired solutions.

Importantly, when studying navigation with the view of developing a cross-level analysis, efforts should be made to ensure that behavioral paradigms used are comparable between species-specific studies (e.g., test similar navigational processes), while respecting the ecological niches of different organisms. There are several tests in the field that meet this criterion. For example, the Morris water maze (Morris et al. 1982)—where subjects need to find an underwater escape platform located in a murky pool (Figure 9.4a) and for which VR alternatives exist (Astur et al. 1998; Cornwell et al. 2008; Doeller and Burgess 2008; Doeller et al. 2008)—allows researchers to assess the role of different orientation cues (landmarks, boundaries). Classic reorientation paradigms (Cheng 1986; Cheng et al. 2013) provide another

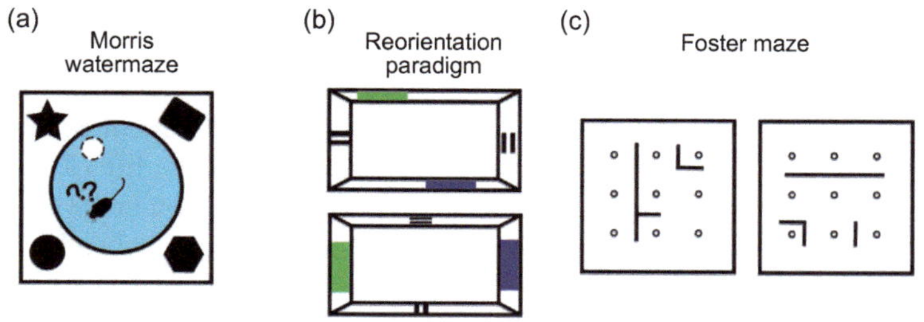

Figure 9.4 Navigational tasks suitable for cross-species investigation. (a) The Morris water maze: Organisms learn to locate a hidden escape platform in a circular pool. To solve the task, participants must learn the location of the escape platform in relation to a set of distal cues. (b) Reorientation paradigm: Participants learn the location of a target object in a rectangular environment with a set of visual cues (top). At testing (bottom), memory for the learned location is assessed after disorientation (e.g., visual cues rotated). (c) The Foster maze: Rodents learn the location of a reward in a square environment with configurable barriers. On each visit, the barrier may be reconfigured to change the topology of the environment.

example: subjects need to recover a learned location after disorientation (Figure 9.4b), and comparative data has been extensively collected. One recently developed alternative test is the Foster maze (Widloski and Foster 2022): its reconfigurable barriers (Figure 9.4c) allow researchers to examine how an environment's topology influences spatial behavior and resulting neural responses, and to disentangle the different types of goal-directed navigational processes (e.g., vector computation and path planning). These paradigms can be adapted to VR testing, making them suitable to test via a wide range of brain-recording methodologies. Efforts should also be made to further develop navigational paradigms that more closely match navigation in the real world, such as navigation in complex environments that contain multiple enclosures, passages and levels, and large-scale spaces (e.g., Eliav et al. 2021; Sarel et al. 2022). Similarly, it is important to determine whether the cognitive and neuronal processes that underpin navigation are similar in outdoor environments, where spatial cues often differ significantly from indoor spaces (e.g., no ceiling, no wall borders).

5 Studying Development to Bridge Neural Activity and Navigation Behavior

Another promising approach to bridging the gap between analysis levels is to look at the assembly of neural systems for navigation in development. The advantages of this approach lie in stereotyped developmental trajectories of navigation ability. Navigation does not emerge in an all-or-none fashion but through a series of incremental steps. Place learning (the ability to locate a target object relative to the location of distal cues) is known to develop relatively early, in both humans (~2 years) (Newcombe et al. 1998) and rodents (~3 weeks) (Brown and Whishaw 2000), but its full development is protracted (lasting years in humans and weeks in rodents). Moreover, human research has shown that children often rely on geometric cues over visual cues for orientation in small, enclosed spaces (Hermer and

Spelke 1994), although visual cues are used in larger environments (Learmonth et al. 2008), which are arguably closer to environmental space. Cross-species research has also shown that the ability to integrate and resolve conflicting cues develops over an extended period, continuing over the first decade in human development (Nardini et al. 2008) and over the first 2–3 months in a rodent's life (Rossier and Schenk 2003). Thus, studying the emergence of neuronal function in the early postnatal period allows a functional alternative to a lesion approach, whereby the system can be studied during a time period when certain components of it are naturally not active yet (see Chapter 7 and Dilks et al. 2023).

Some notable successes of this approach have been seen in rodents (see Chapter 8). A series of descriptive studies have set out the core timeline for the maturation of neural networks for navigation in rodents, including some clear temporal dissociations between different functional spatial neuron types (Figure 9.5c) (Wills et al. 2014). HD cells emerge initially: by just over two weeks of age, HD cells display adultlike tuning (Bjerknes et al. 2015; Langston et al. 2010; Tan et al. 2015; Wills et al. 2010). Also, although less reliable and stable, HD signals can even be recorded before eye opening (Tan et al. 2015). Boundary-responsive cells show different developmental trajectories in different anatomical regions, with EC border cells developing early (adultlike from P17; Bjerknes et al. 2014). By contrast, subiculum BVCs show a protracted development, even into the fourth week of life (Muessig et al. 2024). Place cells have also been recorded from P16 (Langston et al. 2010; Wills et al. 2010). At this age, however, they are less stable and display more diffuse spatial firing fields relative to adults, particularly in locations far away from environmental boundaries (Muessig et al. 2015). They also are not yet able to encode for nonspatial variables (e.g., task phases; Bevandić et al. 2024) as they do in adults. Place-cell development is thought to continue until at least five weeks of age, and in slow-developing individuals, full maturity may take several more weeks (Scott et al. 2010). Grid cells displaying clear, sixfold symmetry, on the other hand, have not been recorded before P20 (Wills et al. 2010). In contrast to place cells, however, grid-cell maturation is abrupt, with adultlike activity characteristics displayed at the earliest age at which they have been recorded. Interestingly, at the age when grid cells emerge, place-cell tuning improves, particularly for locations far away from environmental boundaries (Muessig et al. 2015), suggesting that place-cell tuning early in life may be preferentially influenced by the location of boundaries. Furthermore, hippocampal replay (Figure 9.5a) and theta sequences (Figure 9.5b) have been reliably identified in the fourth postnatal week (Farooq and Dragoi 2019; Muessig et al. 2019), and their emergence is tightly correlated, suggesting that a shared mechanism underlies both (Muessig et al. 2019).

In addition to bridging analysis levels, research on rodent development provides crucial insights into the neurodevelopmental milestones that underpin the maturation of human navigational capabilities. In school-aged children, fMRI studies have become increasingly common in recent years (e.g., Jung and Dilks 2025; Jung et al. 2024; Kamps et al. 2020; Mooney et al. 2021; Prabhakar et al. 2018). However, functional brain imaging remains rare in infants (but see, e.g., Deen et al. 2017; Yates et al. 2025) and is nearly impossible in young toddlers, thereby precluding direct examination of the neural development of navigation during the period in

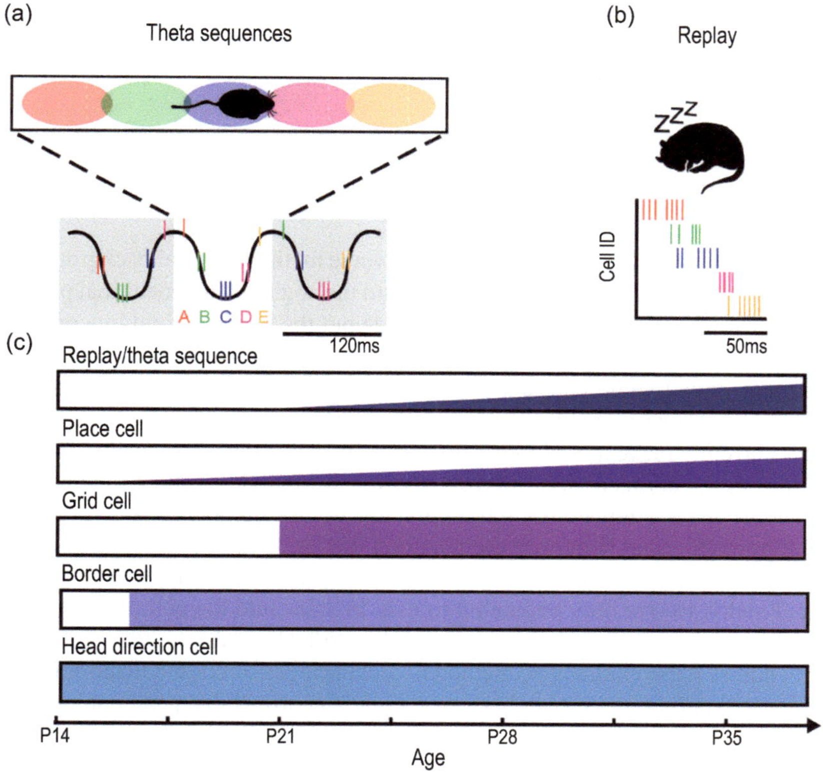

Figure 9.5 The ontogeny of neuronal spatial representations and mechanisms. (a) Traversals on a linear track are encoded by sequential activity of a group of place cells. During individual cycles of theta oscillations, place-cell sequences that reflect path segments are activated (theta sequences). (b) During sleep and rest periods, place-cell sequences that reflect paths taken are spontaneously replayed at a sped-up pace (~10–20 times faster than during behavior). (c) Developmental timeline for the emergence of different spatial cell types and replay and theta sequences in rats. P: postnatal day.

which the foundational milestones for this ability emerge. If we apply the findings from rodent development, however, some insights may be gained. For example, the gradual development of place learning may reflect the gradual development of place-cell tuning. Furthermore, the presumed reliance on geometric cues over landmark cues early in life may reflect the early development of neurons with boundary-related tuning and place cells with stable spatial firing fields near environment walls. However, many knowledge gaps remain. For example, the development of allo- and idiothetic cue integration in the different spatial cell types has not been systematically investigated, and a number of nodes in the brain's navigation network have not been studied early in life (e.g., different head direction loci). Finally, most rodent developmental research does not focus on the maturation of neuronal function in the context of the development of spatial capabilities. Studying neuronal activity and behavior in tandem is critical, as significant individual variability exists in terms of age of inflection for spatial capabilities (Bevandić et al. 2024; Douglas et al. 1973; Nadel et al. 1992). In addition, we lack a consensus on the age equivalence between rodents and humans. Examining populations with unique neurodevelopmental profiles (e.g., Williams syndrome) may offer a complementary approach to dissecting the navigational system (see Appendix 9.2) Hopefully, future research will address these knowledge gaps.

6 Conclusions

Over the past few decades, scientists from diverse methodological backgrounds and disciplines have provided detailed insights into the cognitive and neuronal processes that underlie spatial navigation. This work has put the research field in a promising position to create a cross-level description of navigation that links cellular activity to complex cognitive processes. Nonetheless, significant gaps in our understanding remain. The divergent brain-recording methodologies used in human and rodent studies currently make it hard for scientists to determine the consistency of neuronal spatial representations across different species. Moreover, as most rodent navigation studies do not study neural activity in tandem with complex spatial behavior or use environments that better represent natural spaces, linking neuronal mechanisms with specific navigational abilities applicable to real-world settings remains a challenge. Finally, we are only beginning to gain insights into the neurodevelopmental milestones that underpin the emergence of this critical capability early in life.

To address these challenges, methodological bridges through collaboration need to connect experimental researchers, who collect neuronal data, and computational scientists, who model how one type of neuronal signal translates into another, and their data should be aligned to match the signals recorded across techniques. Behavioral bridges could be advanced by recording from the brain's navigation network during naturalistic rodent navigation using state-of-the-art electrophysiology to overcome challenges of complex behavior (Ulanovsky 2025). To elucidate the neuronal basis of navigation development, we propose studying developing rodents, whose staggered developmental timeline for navigational capabilities closely resembles that of humans and for which tools for recording from the developing brain have been established in recent years. Connecting these areas of research will enable the field to generate a complete description of navigation, spanning multiple levels of analyses and linking single-neuron activity to complex cognition.

Appendix 9.1: Vista versus Environmental Spaces

Spatial environments can be differentiated into vista and environmental spaces by the visibility of the relevant surroundings (Montello 1993; Parra-Barrero et al. 2023; Peer et al. 2024). Vista spaces are environments that can be perceived from a single vantage point, such as in the Morris water maze, an open field, or a single room. Environmental spaces require traversal and multiple views to gather information and to integrate these views over different points of time, such as in a complex maze with multiple turns, a building, or a city. In studies with humans, vista and environmental spaces often correlate with the scale of the space (e.g., a vista-scale room is typically smaller than an environmental-scale building or city). However, large-scale open environments whose features and layout can be perceived from a distance are also considered vista spaces.

Rodent navigation research typically employs small-scale vista spaces. In human neuroimaging and behavioral studies, large-scale environmental spaces—sometimes implemented as a VR setting—have become increasingly common (Wolbers and Wiener 2014). This difference poses a challenge for designing parallel tasks and environments suitable for cross-species comparisons of the neural mechanisms underlying navigation. There are important differences between vista and environmental spaces, both in terms of the timescale of navigation and the way information is acquired. Navigators are capable of locomoting through and forming spatial memories of both types of spaces, but human research shows that the types of spatial representations formed in these spaces are likely different (Ekstrom and Isham 2017; Meilinger et al. 2016). For example, borders of vista space influence the ease of spatial updating, with self-to-object locations more readily updated within the border than outside of the border (Avraamides and Kelly 2010). In a direct comparison of environmental and vista space (i.e., corridors present or removed from a VR building), memory for environmental space (but not vista) was influenced by distance traveled and number of objects encountered (Peer et al. 2024). Studies suggest that allocentric representations in environmental spaces are formed in an experience-dependent and compartmentalized way, building on representations of smaller interconnected units (Meilinger et al. 2016).

Appendix 9.2: What Can Special Populations Tell Us about the Neural Basis of Navigation?

Navigation can be studied in the context of neurodevelopmental conditions such as Williams syndrome, a genetic syndrome associated with moderate learning difficulties. Individuals with Williams syndrome exhibit an uneven cognitive profile, characterized by poorer spatial than verbal abilities (Farran and Karmiloff-Smith 2012), including significant impairments in navigation (e.g., Banta Lavenex and Lavenex 2021; Farran et al. 2010, 2015; Kamps et al. 2023). Notably, Williams syndrome is characterized by structural and functional abnormalities in brain regions implicated in navigation in typical development, including the hippocampus and parietal cortex (for a review, see Thom et al. 2023), raising the intriguing possibility of a causal link between these behavioral and neural navigation systems. However, individuals with Williams syndrome may rely on alternative strategies or brain regions (Karmiloff-Smith 1998), complicating attempts to draw definitive conclusions about the causal relationship between behavioral and neural systems underpinning navigation across development.

Indeed, several studies have demonstrated that individuals with Williams syndrome are impaired on several putative hippocampal abilities. For example, allocentric spatial coding (a known hippocampal function) has been investigated through tasks requiring participants to learn two routes (A to B and A to C) and then to find the shortest route between two places (B to C). In a task using a large-scale

VR environment, Farran et al. (2015) reported that only 35% of individuals with Williams syndrome walked the shortest route compared to 59% of neurotypical 5- to 11-year-old children (see also Broadbent et al. 2014a; Farran et al. 2010). In addition, individuals with Williams syndrome also exhibit impairments in other putative hippocampal abilities:

1. Geometric reorientation in a traditional version of the reorientation task (Ferrara and Landau 2015; Lakusta et al. 2010)
2. Boundary-based memory, but *not* landmark-based memory, in a "human version" of the water maze task (Julian et al. 2019)
3. Visuospatial memory in both a real-world radial arm maze task and the open field task with multiple rewards, as well deficits in understanding spatial layout (Foti et al. 2011; Mandolesi et al. 2009)

Landmark manipulation and view-matching tasks further suggest that individuals with Williams syndrome rely on a view-matching strategy for learning routes, thus providing an example of the alternative strategies discussed above (Broadbent et al. 2014b, 2015; Purser et al. 2015). In their studies of vista space, Banta Lavenex and Lavenex (2021) attribute the use of a view-matching strategy in Williams syndrome to activation of the parahippocampal and retrosplenial cortices and medial parietal regions.

Next, cortical systems, beyond the hippocampus, in the parietal cortex are hypothesized to be involved in navigation in typical development, albeit a different type of navigation. For example, the OPA has been hypothesized to support "visually guided navigation" (for a review, see Dilks et al. 2022). What, then, do we know about visually guided navigation in Williams syndrome—in which, curiously, cortical thinning in and around the OPA has been shown (Meyer-Lindenberg et al. 2004)? A recent behavioral study found that adults with Williams syndrome exhibit impaired visually guided navigation abilities, yet relatively proficient scene categorization abilities (i.e., the ability to recognize a place as a particular type of environment), relative to neurotypical mental age-matched controls. These findings suggest that OPA is casually involved in visually guided navigation.

While evidence indicates that individuals with Williams syndrome exhibit impaired navigation abilities, as well as structural and functional abnormalities in the hippocampus and parietal cortex, future research is needed to test whether these behavioral impairments result from these neural atypicalities. It is possible that individuals with Williams syndrome rely on alternative strategies or brain regions during navigation, complicating definitive conclusions about the behavioral and neural systems underlying navigation across typical development.

References

Agarwal G, Stevenson IH, Berenyi A, et al. (2014) Spatially Distributed Local Fields in the Hippocampus Encode Rat Position. Science 344 (6184):626–630. https://doi.org/10.1126/science.1250444

Astur RS, Ortiz ML, and Sutherland RJ (1998) A Characterization of Performance by Men and Women in a Virtual Morris Water Task: A Large and Reliable Sex Difference. Behav Brain Res 93 (1-2):185–190. https://doi.org/10.1016/s0166-4328(98)00019-9

Avraamides MN, and Kelly JW (2010) Multiple Systems of Spatial Memory: Evidence from Described Scenes. J Exp Psychol Learn Mem Cogn 36 (3):635–645. https://doi.org/10.1037/a0017040

Baldassano C, Beck DM, and Fei-Fei L (2013) Differential Connectivity within the Parahippocampal Place Area. Neuroimage 75:228–237. https://doi.org/10.1016/j.neuroimage.2013.02.073

Baldassano C, Esteva A, Fei-Fei L, and Beck DM (2016) Two Distinct Scene-Processing Networks Connecting Vision and Memory. eNeuro 3 (5):0178–0116. https://doi.org/10.1523/ENEURO.0178-16.2016

Banta Lavenex P, and Lavenex P (2021) A Critical Review of Spatial Abilities in Down and Williams Syndromes: Not All Space Is Created Equal. Front Psychiatry 12:669320. https://doi.org/10.3389/fpsyt.2021.669320

Bellmund JL, Deuker L, Navarro Schroder T, and Doeller CF (2016) Grid-Cell Representations in Mental Simulation. eLife 5:17089. https://doi.org/10.7554/eLife.17089

Bellmund JLS, Gärdenfors P, Moser EI, and Doeller CF (2018) Navigating Cognition: Spatial Codes for Human Thinking. Science 362 (6415):eaat6766. https://doi.org/10.1126/science.aat6766

Bevandić J, Stella F, and Ólafsdóttir HF (2024) Parallel Maturation of Rodent Hippocampal Memory and CA1 Task Representations. Curr Biol 34 (21):5062–5072.e5065. https://doi.org/10.1016/j.cub.2024.08.048

Bicanski A (2025) Breaking the Chains: Toward a Neural-Level Account of Episodic Memory. Psychol Rev Sept. 8. https://doi.org/10.1037/rev0000571

Bicanski A, and Burgess N (2016) Environmental Anchoring of Head Direction in a Computational Model of Retrosplenial Cortex. J Neurosci 36 (46):11601–11618. https://doi.org/10.1523/jneurosci.0516-16.2016

——— (2018) A Neural-Level Model of Spatial Memory and Imagery. eLife 7 (7052):1–3. https://doi.org/10.7554/elife.33752

Bierbrauer A, Kunz L, Gomes CA, et al. (2020) Unmasking Selective Path Integration Deficits in Alzheimer's Disease Risk Carriers. Sci Adv 6:eaba1394. https://doi.org/10.1126/SCIADV.ABA1394

Bin Khalid I, Reifenstein ET, Auer N, Kunz L, and Kempter R (2024) Quantitative Modeling of the Emergence of Macroscopic Grid-Like Representations. eLife 13:e85742. https://doi.org/10.7554/eLife.85742

Bjerknes TL, Langston RF, Kruge IU, Moser EI, and Moser MB (2015) Coherence among Head Direction Cells before Eye Opening in Rat Pups. Curr Biol 25:103–108. https://doi.org/10.1016/j.cub.2014.11.009

Bjerknes TL, Moser EI, and Moser MB (2014) Representation of Geometric Borders in the Developing Rat. Neuron 82:71–78. https://doi.org/10.1016/j.neuron.2014.02.014

Blair HT, Cho J, and Sharp PE (1998) Role of the Lateral Mammillary Nucleus in the Rat Head Direction Circuit: A Combined Single Unit Recording and Lesion Study. Neuron 21 (6):1387–1397. https://doi.org/10.1016/s0896-6273(00)80657-1

Bonner MF, and Epstein RA (2017) Coding of Navigational Affordances in the Human Visual System. PNAS 114 (18):4793–4798. https://doi.org/10.1073/pnas.1618228114

Broadbent HJ, Farran EK, and Tolmie A (2014a) Egocentric and Allocentric Navigation Strategies in Williams Syndrome and Typical Development. Dev Sci 17 (6):920–934. https://doi.org/10.1111/desc.12176

——— (2014b) Object-Based Mental Rotation and Visual Perspective-Taking in Typical Development and Williams Syndrome. Dev Neuropsychol 39 (3):205–225. https://doi.org/10.1080/87565641.2013.876027

——— (2015) Sequential Egocentric Navigation and Reliance on Landmarks in Williams Syndrome and Typical Development. Front Psychol 6:216. https://doi.org/10.3389/fpsyg.2015.00216

Brown RW, and Whishaw IQ (2000) Similarities in the Development of Place and Cue Navigation by Rats in a Swimming Pool. Dev Psychobiol 37 (4):238–245. https://doi.org/10.1002/1098-2302(2000)37:4%3C238::aid-dev4%3E3.0.co;2-j

Brown TI, Carr VA, LaRocque KF, et al. (2016) Prospective Representation of Navigational Goals in the Human Hippocampus. Science 352 (6291):1323–1326. https://doi.org/10.1126/science.aaf0784

Burgess N, and O'Keefe J (1996) Neuronal Computations Underlying the Firing of Place Cells and Their Role in Navigation. Hippocampus 6 (6):749–762. https://doi.org/10.1002/(SICI)1098-1063(1996)6:6<749::AID-HIPO16>3.0.CO;2-0

Bush D, Barry C, Burgess N, et al. (2015) Using Grid Cells for Navigation. Neuron 87 (3):507–520. https://doi.org/10.1016/j.neuron.2015.07.006

Bush D, Bisby JA, Bird CM, et al. (2017) Human Hippocampal Theta Power Indicates Movement Onset and Distance Travelled. Proc Natl Acad Sci U S A 114 (46):12297–12302. https://doi.org/10.1073/pnas.1708716114

Buzsaki G, and Moser EI (2013) Memory, Navigation and Theta Rhythm in the Hippocampal-Entorhinal System. Nat Neurosci 16 (2):130–138. https://doi.org/10.1038/nn.3304

Buzsaki G, and Tingley D (2018) Space and Time: The Hippocampus as a Sequence Generator. Trends Cogn Sci 22 (10):853–869. https://doi.org/10.1016/j.tics.2018.07.006

Byrne P, Becker S, and Burgess N (2007) Remembering the Past and Imagining the Future: A Neural Model of Spatial Memory and Imagery. Psychol Rev 114 (2):340–375. https://doi.org/10.1037/0033-295X.114.2.340

Chadwick MJ, Jolly AE, Amos DP, Hassabis D, and Spiers HJ (2015) A Goal Direction Signal in the Human Entorhinal/Subicular Region. Curr Biol 25 (1):87–92. https://doi.org/10.1016/j.cub.2014.11.001

Chen G, King JA, Burgess N, and O'Keefe J (2013) How Vision and Movement Combine in the Hippocampal Place Code. PNAS 110 (1):378–383. https://doi.org/10.1073/pnas.1215834110

Cheng K (1986) A Purely Geometric Module in the Rat's Spatial Representation. Cognition 23 (2):149–178. https://doi.org/10.1016/0010-0277(86)90041-7

Cheng K, Huttenlocher J, and Newcombe NS (2013) 25 Years of Research on the Use of Geometry in Spatial Reorientation: A Current Theoretical Perspective. Psychon Bull Rev 20 (6):1033–1054. https://doi.org/10.3758/s13423-013-0416-1

Cheung A, Ball D, Milford M, Wyeth G, and Wiles J (2012) Maintaining a Cognitive Map in Darkness: The Need to Fuse Boundary Knowledge with Path Integration. PLoS Comput Biol 8 (8):e1002651. https://doi.org/10.1371/journal.pcbi.1002651

Cho J, and Sharp PE (2001) Head Direction, Place, and Movement Correlates for Cells in the Rat Retrosplenial Cortex. Behav Neurosci 115 (1):3–25. https://doi.org/10.1037/0735-7044.115.1.3

Chrastil ER, and Warren WH (2014) From Cognitive Maps to Cognitive Graphs. PLoS ONE 9 (11):e112544. https://doi.org/10.1371/journal.pone.0112544

Constantinescu AO, O'Reilly JX, and Behrens TEJ (2016) Organizing Conceptual Knowledge in Humans with a Gridlike Code. Science 352 (6292):1464–1468. https://doi.org/10.1126/science.aaf0941

Corder EH, Saunders AM, Strittmatter WJ, et al. (1993) Gene Dose of Apolipoprotein E Type 4 Allele and the Risk of Alzheimer's Disease in Late Onset Families. Science 261 (5123):921–923. https://doi.org/10.1126/science.8346443

Cornwell BR, Johnson LL, Holroyd T, Carver FW, and Grillon C (2008) Human Hippocampal and Parahippocampal Theta during Goal-Directed Spatial Navigation Predicts Performance on a Virtual Morris Water Maze. J Neurosci 28 (23):5983–5990. https://doi.org/10.1523/JNEUROSCI.5001-07.2008

Darwin C (1873) Origin of Certain Instincts. Nature 7:417–418. https://doi.org/10.1038/007417a0

Deen B, Richardson H, Dilks DD, et al. (2017) Organization of High-Level Visual Cortex in Human Infants. Nat Commun 8:13995. https://doi.org/10.1038/ncomms13995

Deuker L, Bellmund JL, Navarro Schröder T, and Doeller CF (2016) An Event Map of Memory Space in the Hippocampus. eLife 5:e16534. https://doi.org/10.7554/eLife.16534

Dilks DD, Julian JB, Kubilius J, Spelke ES, and Kanwisher N (2011) Mirror-Image Sensitivity and Invariance in Object and Scene Processing Pathways. J Neurosci 31 (31):11305–11312. https://doi.org/10.1523/JNEUROSCI.1935-11.2011

Dilks DD, Julian JB, Paunov AM, and Kanwisher N (2013) The Occipital Place Area Is Causally and Selectively Involved in Scene Perception. J Neurosci 33 (4):1331–1336. https://doi.org/10.1523/JNEUROSCI.4081-12.2013

Dilks DD, Jung Y, and Kamps FS (2023) The Development of Human Cortical Scene Processing. Curr Dir Psychol Sci 32 (6):479–486. https://doi.org/10.1177/09637214231191772

Dilks DD, Kamps FS, and Persichetti AS (2022) Three Cortical Scene Systems and Their Development. Trends Cogn Sci 26 (2):117–127. https://doi.org/1di0.1016/j.tics.2021.11.002

Doeller CF, Barry C, and Burgess N (2010) Evidence for Grid Cells in a Human Memory Network. Nature 463 (7281):657–661. https://doi.org/10.1038/nature08704

Doeller CF, and Burgess N (2008) Distinct Error-Correcting and Incidental Learning of Location Relative to Landmarks and Boundaries. Proc Natl Acad Sci U S A 105 (15):5909–5914. https://doi.org/10.1073/pnas.0711433105

Doeller CF, King JA, and Burgess N (2008) Parallel Striatal and Hippocampal Systems for Landmarks and Boundaries in Spatial Memory. PNAS 105 (15):5915–5920. https://doi.org/10.1073/pnas.0801489105

Douglas RJ, Peterson JJ, and Douglas DP (1973) The Ontogeny of a Hippocampus-Dependent Response in Two Rodent Species. Behavioral biology 8 (1):27–37. https://doi.org/10.1016/s0091-6773(73)80003-3

Dutriaux L, Xu Y, Sartorato N, Lhuillier S, and Bottini R (2024) Disentangling Reference Frames in the Neural Compass. Imaging Neurosci (Camb) 2:1–18. https://doi.org/10.1162/imag_a_00149

Edvardsen V (2019) Goal-Directed Navigation Based on Path Integration and Decoding of Grid Cells in an Artificial Neural Network. Nat Comput 18:13–27. https://doi.org/10.1007/s11047-016-9575-0

Edvardsen V, Bicanski A, and Burgess N (2020) Navigating with Grid and Place Cells in Cluttered Environments. Hippocampus 30 (3):220–232. https://doi.org/10.1002/hipo.23147

Ego-Stengel V, and Wilson MA (2010) Disruption of Ripple-Associated Hippocampal Activity during Rest Impairs Spatial Learning in the Rat. Hippocampus 20 (1):1–10. https://doi.org/10.1002/hipo.20707

Ekstrom AD, Caplan JB, Fields TA, et al. (2003) Cellular Networks Underlying Human Spatial Navigation. Nature 425:184–188. https://doi.org/10.1038/nature01964

Ekstrom AD, and Isham EA (2017) Human Spatial Navigation: Representations across Dimensions and Scales. Curr Opin Behav Sci 17:84–89. https://doi.org/10.1016/j.cobeha.2017.06.005

Eliav T, Maimon SR, Aljadeff J, et al. (2021) Multiscale Representation of Very Large Environments in the Hippocampus of Flying Bats. Science 372 (6545):eabg4020. https://doi.org/10.1126/science.abg4020

Eliav T, Maimon SR, Sarel A, et al. (2025) Fragmented Replay of Very Large Environments in the Hippocampus of Bats. Cell 188 (15):4091–4105.e4016. https://doi.org/10.1016/j.cell.2025.05.024

Epstein RA (2008) Parahippocampal and Retrosplenial Contributions to Human Spatial Navigation. Trends Cogn Sci 12 (10):388–396. https://doi.org/10.1016/j.tics.2008.07.004

Epstein RA, and Baker CI (2019) Scene Perception in the Human Brain. Annu Rev Vis Sci 5:373–397. https://doi.org/10.1146/annurev-vision-091718-014809
Epstein RA, and Kanwisher N (1998) A Cortical Representation of the Local Visual Environment. Nature 392 (6676):598–601. https://doi.org/10.1038/33402
Epstein RA, Patai EZ, Julian JB, and Spiers HJ (2017) The Cognitive Map in Humans: Spatial Navigation and Beyond. Nat Neurosci 20 (11):1504–1513. https://doi.org/10.1038/nn.4656
Epstein RA, and Vass LK (2014) Neural Systems for Landmark-Based Wayfinding in Humans. Philos Trans R Soc Lond B Biol Sci 369 (1635):20120533. https://doi.org/10.1098/rstb.2012.0533
Erdem UM, and Hasselmo M (2012) A Goal-Directed Spatial Navigation Model Using Forward Trajectory Planning Based on Grid Cells. Eur J Neurosci 35 (6):916–931. https://doi.org/10.1111/j.1460-9568.2012.08015.x
Evans T, Bicanski A, Bush D, and Burgess N (2016) How Environment and Self-Motion Combine in Neural Representations of Space. J Physiol 594 (22):6535–6546. https://doi.org/10.1113/JP270666
Farooq U, and Dragoi G (2019) Emergence of Preconfigured and Plastic Time-Compressed Sequences in Early Postnatal Development. Science 363:168–173. https://doi.org/10.1126/science.aav0502
Farran EK, Blades M, Boucher J, and Tranter LJ (2010) How Do Individuals with Williams Syndrome Learn a Route in a Real-World Environment? Dev Sci 13 (3):454–468. https://doi.org/10.1111/j.1467-7687.2009.00894.x
Farran EK, and Karmiloff-Smith A (2012) Neurodevelopmental Disorders across the Lifespan: A Neuroconstructivist Approach. Oxford Univ Press, Oxford
Farran EK, Purser HR, Courbois Y, et al. (2015) Route Knowledge and Configural Knowledge in Typical and Atypical Development: A Comparison of Sparse and Rich Environments. J Neurodev Disord 7:37. https://doi.org/10.1186/s11689-015-9133-6
Ferrara K, and Landau B (2015) Geometric and Featural Systems, Separable and Combined: Evidence from Reorientation in People with Williams Syndrome. Cognition 144:123–133. https://doi.org/10.1016/j.cognition.2015.07.010
Foti F, Petrosini L, Cutuli D, et al. (2011) Explorative Function in Williams Syndrome Analyzed through a Large-Scale Task with Multiple Rewards. Res Dev Disabil 32 (3):972–985. https://doi.org/10.1016/j.ridd.2011.02.001
Fried I, Wilson CL, Maidment NT, et al. (1999) Cerebral Microdialysis Combined with Single-Neuron and Electroencephalographic Recording in Neurosurgical Patients. Technical Note. J Neurosurg 91 (4):697–705. https://doi.org/10.3171/jns.1999.91.4.0697
Furuya Y, Matsumoto J, Hori E, et al. (2014) Place-Related Neuronal Activity in the Monkey Parahippocampal Gyrus and Hippocampal Formation during Virtual Navigation. Hippocampus 24 (1):113–130. https://doi.org/10.1002/hipo.22209
Georges-Francois P (1999) Spatial View Cells in the Primate Hippocampus: Allocentric View Not Head Direction or Eye Position or Place. Cereb Cortex 9 (3):197–212. https://doi.org/10.1093/cercor/9.3.197
Gil M, Ancau M, Schlesiger MI, et al. (2018) Impaired Path Integration in Mice with Disrupted Grid Cell Firing. Nat Neurosci 21 (1):81–91. https://doi.org/10.1038/s41593-017-0039-3
Gillespie AK, Astudillo Maya DA, Denovellis EL, et al. (2021) Hippocampal Replay Reflects Specific Past Experiences Rather Than a Plan for Subsequent Choice. Neuron 109 (19):3149–3163.e3146. https://doi.org/10.1016/j.neuron.2021.07.029
Girardeau G, Benchenane K, Wiener SI, Buzsaki G, and Zugaro MB (2009) Selective Suppression of Hippocampal Ripples Impairs Spatial Memory. Nat Neurosci 12 (10):1222–1223. https://doi.org/10.1038/nn.2384
Goodridge JP, and Taube JS (1997) Interaction between the Postsubiculum and Anterior Thalamus in the Generation of Head Direction Cell Activity. J Neurosci 17 (23):9315–9330. https://doi.org/10.1523/JNEUROSCI.17-23-09315.1997

Gothard KM, Skaggs WE, and McNaughton BL (1996) Dynamics of Mismatch Correction in the Hippocampal Ensemble Code for Space: Interaction between Path Integration and Environmental Cues. J Neurosci 16 (24):8027–8040. https://doi.org/10.1523/JNEUROSCI.16-24-08027.1996

Gridchyn I, Schoenenberger P, O'Neill J, and Csicsvari J (2020) Assembly-Specific Disruption of Hippocampal Replay Leads to Selective Memory Deficit. Neuron 106 (2):291–300 e296. https://doi.org/10.1016/j.neuron.2020.01.021

Gu Y, Lewallen S, Kinkhabwala AA, et al. (2018) A Map-Like Micro-Organization of Grid Cells in the Medial Entorhinal Cortex. Cell 175 (3):736–750.e730. https://doi.org/10.1016/j.cell.2018.08.066

Gulli RA, Duong LR, Corrigan BW, et al. (2020) Context-Dependent Representations of Objects and Space in the Primate Hippocampus during Virtual Navigation. Nat Neurosci 23 (1):103–112. https://doi.org/10.1038/s41593-019-0548-3

Hafting T, Fyhn M, Molden S, Moser M-B, and Moser EI (2005) Microstructure of a Spatial Map in the Entorhinal Cortex. Nature 436 (7052):801–806. https://doi.org/10.1038/nature03721

Hartley T, Maguire EA, Spiers HJ, and Burgess N (2003) The Well-Worn Route and the Path Less Traveled: Distinct Neural Bases of Route Following and Wayfinding in Humans. Neuron 37 (5):877–888. https://doi.org/10.1016/s0896-6273(03)00095-3

Hermer L, and Spelke ES (1994) A Geometric Process for Spatial Reorientation in Young Children. Nature 370 (6484):57–59. https://doi.org/10.1038/370057a0

Horner AJ, Bisby JA, Zotow E, Bush D, and Burgess N (2016) Grid-Like Processing of Imagined Navigation. Curr Biol 26 (6):842–847. https://doi.org/10.1016/j.cub.2016.01.042

Howard LR, Javadi AH, Yu Y, et al. (2014) The Hippocampus and Entorhinal Cortex Encode the Path and Euclidean Distances to Goals during Navigation. Curr Biol 24 (12):1331–1340. https://doi.org/10.1016/j.cub.2014.05.001

Ino T, Doi T, Hirose S, et al. (2007) Directional Disorientation Following Left Retrosplenial Hemorrhage: A Case Report with fMRI Studies. Cortex 43 (2):248–254. https://doi.org/10.1016/s0010-9452(08)70479-9

Jackson JC, Johnson A, and Redish AD (2006) Hippocampal Sharp Waves and Reactivation during Awake States Depend on Repeated Sequential Experience. J Neurosci 26 (48):12415–12426. https://doi.org/10.1523/JNEUROSCI.4118-06.2006

Jacob P-Y, Casali G, Spieser L, et al. (2017) An Independent, Landmark-Dominated Head-Direction Signal in Dysgranular Retrosplenial Cortex. Nat Neurosci 20 (2):173–175. https://doi.org/10.1038/nn.4465

Jacobs J, Kahana MJ, Ekstrom AD, Mollison MV, and Fried I (2010) A Sense of Direction in Human Entorhinal Cortex. PNAS 107 (14):6487–6492. https://doi.org/10.1073/pnas.0911213107

Jacobs J, Weidemann CT, Miller JF, et al. (2013) Direct Recordings of Grid-Like Neuronal Activity in Human Spatial Navigation. Nat Neurosci 16 (9):1188–1190. https://doi.org/10.1038/nn.3466

Joglekar A, Hu W, Zhang B, et al. (2024) Single-Cell Long-Read Sequencing-Based Mapping Reveals Specialized Splicing Patterns in Developing and Adult Mouse and Human Brain. Nat Neurosci 27 (6):1051–1063. https://doi.org/10.1038/s41593-024-01616-4

Johnson A, and Redish AD (2007) Neural Ensembles in CA3 Transiently Encode Paths Forward of the Animal at a Decision Point. J Neurosci 27 (45):12176–12189. https://doi.org/10.1523/jneurosci.3761-07.2007

Julian JB, Kamps FS, Epstein RA, and Dilks DD (2019) Dissociable Spatial Memory Systems Revealed by Typical and Atypical Human Development. Dev Sci 22 (2):e12737. https://doi.org/10.1111/desc.12737

Julian JB, Keinath AT, Frazzetta G, and Epstein RA (2018a) Human Entorhinal Cortex Represents Visual Space Using a Boundary-Anchored Grid. Nat Neurosci 21 (2):191–194. https://doi.org/10.1038/s41593-017-0049-1

Julian JB, Keinath AT, Marchette SA, and Epstein RA (2018b) The Neurocognitive Basis of Spatial Reorientation. Curr Biol 28 (17):R1059–R1073. https://doi.org/10.1016/j.cub.2018.04.057

Jun JJ, Steinmetz NA, Siegle JH, et al. (2017) Fully Integrated Silicon Probes for High-Density Recording of Neural Activity. Nature 551 (7679):232–236. https://doi.org/10.1038/nature24636

Jung Y, and Dilks DD (2025) Early Development of Navigationally Relevant Location Information in the Retrosplenial Complex. PNAS 122 (19):e2503569122. https://doi.org/10.1073/pnas.2503569122

Jung Y, Hsu D, and Dilks DD (2024) "Walking Selectivity" in the Occipital Place Area in 8-Year-Olds, Not 5-Year-Olds. Cereb Cortex 34 (3):bhae101. https://doi.org/10.1093/cercor/bhae101

Kamps FS, Julian JB, Kubillius J, Kanwisher N, and Dilks DD (2016a) The Occipital Place Area Represents the Local Elements of Scenes. Neuroimage 132:417–424. https://doi.org/10.1016/j.neuroimage.2016.02.062

Kamps FS, Lall V, and Dilks DD (2016b) The Occipital Place Area Represents First-Person Perspective Motion in Formation through Scenes. Cortex 83:17–26. https://doi.org/10.1016/j.cortex.2016.06.022

Kamps FS, Pincus JE, Radwan SF, Wahab S, and Dilks DD (2020) Late Development of Navigationally Relevant Motion Processing in the Occipital Place Area. Curr Biol 30 (3):544–550.e543. https://doi.org/10.1016/j.cub.2019.12.008

Kamps FS, Rennert RJ, Radwan SF, et al. (2023) Dissociable Cognitive Systems for Recognizing Places and Navigating through Them: Developmental and Neuropsychological Evidence. J Neurosci 43 (36):6320–6329. https://doi.org/10.1523/JNEUROSCI.0153-23.2023

Karmiloff-Smith A (1998) Development Itself Is the Key to Understanding Developmental Disorders. Trends Cogn Sci 2 (10):389–398. https://doi.org/10.1016/S1364-6613(98)01230-3

Kay K, Chung JE, Sosa M, et al. (2020) Constant Sub-Second Cycling between Representations of Possible Futures in the Hippocampus. Cell 180 (3):552–567 e525. https://doi.org/10.1016/j.cell.2020.01.014

Kennedy B, Malladi SN, Tootell RBH, and Nasr S (2024) A Previously Undescribed Scene-Selective Site Is the Key to Encoding Ego-Motion in Naturalistic Environments. eLife 13:91601. https://doi.org/10.7554/eLife.91601

Kozberg MG, and Hillman EM (2016) Neurovascular Coupling Develops Alongside Neural Circuits in the Postnatal Brain. Neurogenesis (Austin) 3 (1):e1244439. https://doi.org/10.1080/23262133.2016.1244439

Kubie JL, and Fenton AA (2012) Linear Look-Ahead in Conjunctive Cells: an Entorhinal Mechanism for Vector-Based Navigation. Front Neural Circuits 6:20. https://doi.org/10.3389/fncir.2012.00020

Kunz L, Brandt A, Reinacher PC, et al. (2021) A Neural Code for Egocentric Spatial Maps in the Human Medial Temporal Lobe. Neuron 109 (17):2781–2796.e2710. https://doi.org/10.1016/j.neuron.2021.06.019

Kunz L, Maidenbaum S, Chen D, et al. (2019) Mesoscopic Neural Representations in Spatial Navigation. Trends Cogn Sci 23 (7):615–630. https://doi.org/10.1016/j.tics.2019.04.011

Kunz L, Schröder TN, Lee H, et al. (2015) Reduced Grid-Cell-Like Representations in Adults at Genetic Risk for Alzheimer's Disease. Science 350 (6259):430–433. https://doi.org/10.1126/science.aac8128

Kunz L, Staresina BP, Reinacher PC, et al. (2024) Ripple-Locked Coactivity of Stimulus-Specific Neurons and Human Associative Memory. Nat Neurosci 27 (3):587–599. https://doi.org/10.1038/s41593-023-01550-x

Lai C, Tanaka S, Harris TD, and Lee AK (2023) Volitional Activation of Remote Place Representations with a Hippocampal Brain-Machine Interface. Science 382 (6670):566–573. https://doi.org/10.1126/science.adh5206

Lakusta L, Dessalegn B, and Landau B (2010) Impaired Geometric Reorientation Caused by Genetic Defect. PNAS 107 (7):2813–2817. https://doi.org/10.1073/pnas.0909155107
Langston RF, Ainge JA, Couey JJ, et al. (2010) Development of the Spatial Representation System in the Rat. Science (1979) 328:1576–1580. https://doi.org/10.1126/science.1188210
Learmonth AE, Newcombe NS, Sheridan N, and Jones M (2008) Why Size Counts: Children's Spatial Reorientation in Large and Small Enclosures. Dev Sci 11 (3):414–426. https://doi.org/10.1111/j.1467-7687.2008.00686.x
Lee AK, and Wilson MA (2002) Memory of Sequential Experience in the Hippocampus during Slow Wave Sleep. Neuron 36 (6):1183–1194. https://doi.org/10.1016/s0896-6273(02)01096-6
Lee SA, Miller JF, Watrous AJ, et al. (2018) Electrophysiological Signatures of Spatial Boundaries in the Human Subiculum. J Neurosci 38:3265–3272. https://doi.org/10.1523/jneurosci.3216-17.2018
Lever C, Burton S, Jeewajee A, O'Keefe J, and Burgess N (2009) Boundary Vector Cells in the Subiculum of the Hippocampal Formation. J Neurosci 29 (31):9771–9777. https://doi.org/10.1523/jneurosci.1319-09.2009
Lu Z, Julian JB, Aguirre GK, and Epstein RA (2025) A Neural Compass in the Human Brain during Naturalistic Virtual Navigation. journal of Neuroscience. J Neurosci 45 (34):e1765242025. https://doi.org/10.1523/JNEUROSCI.1765-24.2025
Ma Y, Shaik MA, Kozberg MG, et al. (2016) Resting-State Hemodynamics Are Spatiotemporally Coupled to Synchronized and Symmetric Neural Activity in Excitatory Neurons. PNAS 113 (52):E8463–E8471. https://doi.org/10.1073/pnas.1525369113
Maingret N, Girardeau G, Todorova R, Goutierre M, and Zugaro M (2016) Hippocampo-Cortical Coupling Mediates Memory Consolidation during Sleep. Nat Neurosci 19 (7):959–964. https://doi.org/10.1038/nn.4304
Mallory CS, Widloski J, and Foster DJ (2025) The Time Course and Organization of Hippocampal Replay. Science 387 (6733):541–548. https://doi.org/10.1126/science.ads4760
Mandolesi L, Addona F, Foti F, et al. (2009) Spatial Competences in Williams Syndrome: A Radial Arm Maze Study. Int J Dev Neurosci 27 (3):205–213. https://doi.org/10.1016/j.ijdevneu.2009.01.004
Mao D, Avila E, Caziot B, et al. (2021) Spatial Modulation of Hippocampal Activity in Freely Moving Macaques. Neuron 109 (21):3521–3534.e3526. https://doi.org/10.1016/j.neuron.2021.09.032
Marchette SA, Vass LK, Ryan J, and Epstein RA (2014) Anchoring the Neural Compass: Coding of Local Spatial Reference Frames in Human Medial Parietal Lobe. Nat Neurosci 17 (11):1598–1606. https://doi.org/10.1038/nn.3834
Mathis A, Herz AV, and Stemmler M (2012) Optimal Population Codes for Space: Grid Cells Outperform Place Cells. Neural Comput 24 (9):2280–2317. https://doi.org/10.1162/NECO_a_00319
Meilinger T, Strickrodt M, and Bülthoff HH (2016) Qualitative Differences in Memory for Vista and Environmental Spaces Are Caused by Opaque Borders, Not Movement or Successive Presentation. Cognition 155:77–95. https://doi.org/10.1016/j.cognition.2016.06.003
Metcalfe D, and Barlow KR (1992) A Model for Exploring the Optimal Trade-Off between Field Processing and Transport. Am Anthropol 94 (2):340–356. https://doi.org/10.1525/aa.1992.94.2.02a00040
Meyer-Lindenberg A, Kohn P, Mervis CB, et al. (2004) Neural Basis of Genetically Determined Visuospatial Construction Deficit in Williams Syndrome. Neuron 43 (5):623–631. https://doi.org/10.1016/j.neuron.2004.08.014
Miller J, Fried I, Suthana N, and Jacobs J (2015) Repeating Spatial Activations in Human Entorhinal Cortex. Curr Biol 25 (8):1080–1085. http://dx.doi.org/10.1016/j.cub.2015.02.045

Miller JF, Neufang M, Solway A, et al. (2013) Neural Activity in Human Hippocampal Formation Reveals the Spatial Context of Retrieved Memories. Science 342:1111–1114. https://doi.org/10.1126/science.1244056

Mittelstaedt H, and Mittelstaedt M-L (1982) Homing by Path Integration. In: Papi F, Wallraff HG (eds) Avian Navigation. Springer, New York, pp 290–297

Montello DR (1993) Scale and Multiple Psychologies of Space. In: Frank AU, Campari I (eds). Spatial Information Theory a Theoretical Basis for GIS. Springer, Berlin, Heidelberg, pp 312–321

Mooney LN, Johnson EG, Prabhakar J, and Ghetti S (2021) Memory-Related Hippocampal Activation during Sleep and Temporal Memory in Toddlers. Dev Cogn Neurosci 47:100908. https://doi.org/10.1016/j.dcn.2020.100908

Morgan LK, MacEvoy SP, Aguirre GK, and Epstein RA (2011) Distances between Real-World Locations Are Represented in the Human Hippocampus. J Neurosci 31 (4):1238–1245. https://doi.org/10.1523/JNEUROSCI.4667-10.2011

Morris RG, Garrud P, Rawlins JN, and O'Keefe J (1982) Place Navigation Impaired in Rats with Hippocampal Lesions. Nature 297 (5868):681–683. https://doi.org/10.1038/297681a0

Muessig L, Hauser J, Wills TJ, and Cacucci F (2015) A Developmental Switch in Place Cell Accuracy Coincides with Grid Cell Maturation. Neuron 86 (5):1167–1173. https://doi.org/10.1016/j.neuron.2015.05.011

Muessig L, Lasek M, Varsavsky I, Cacucci F, and Wills TJ (2019) Coordinated Emergence of Hippocampal Replay and Theta Sequences during Post-Natal Development. Curr Biol 29:834–840.e834. https://doi.org/10.1016/j.cub.2019.01.005

Muessig L, Rodrigues FR, Bjerknes TL, et al. (2024) Environment Geometry Alters Subiculum Boundary Vector Cell Receptive Fields in Adulthood and Early Development. Nat Commun 15 (1):982. https://doi.org/10.1038/s41467-024-45098-1

Muller R, and Kubie J (1987) The Effects of Changes in the Environment on the Spatial Firing of Hippocampal Complex-Spike Cells. J Neurosci 7 (7):1951–1968. https://doi.org/10.1523/JNEUROSCI.07-07-01951.1987

Nadasdy Z, Nguyen TP, Torok A, et al. (2017) Context-Dependent Spatially Periodic Activity in the Human Entorhinal Cortex. PNAS 114 (17):E3516–E3525. https://doi.org/10.1073/pnas.1701352114

Nadel L, Wilson L, and Kurz EM (1992) Hippocampus: Effects of Alterations in Timing of Development. In: Turkewitz G, Devenny D (eds) Developmental Time and Timing. Erlbaum, Hillsdale, pp 233–247

Nardini M, Jones P, Bedford R, and Braddick O (2008) Development of Cue Integration in Human Navigation. Curr Biol 18 (9):689–693. https://doi.org/10.1016/j.cub.2008.04.021

Newcombe N, Huttenlocher J, Drummey AB, and Wiley JG (1998) The Development of Spatial Location Coding: Place Learning and Dead Reckoning in the Second and Third Years. Cogn Dev 13 (2):185–200. https://doi.org/10.1016/S0885-2014(98)90038-7

Nielson DM, Smith TA, Sreekumar V, Dennis S, and Sederberg PB (2015) Human Hippocampus Represents Space and Time during Retrieval of Real-World Memories. PNAS 112 (35):11078–11083. https://doi.org/10.1073/pnas.1507104112

O'Keefe J, and Conway DH (1978) Hippocampal Place Units in the Freely Moving Rat: Why They Fire Where They Fire. Exp Brain Res 31 (4):573–590. https://doi.org/10.1007/bf00239813

O'Keefe J, and Dostrovsky J (1971) The Hippocampus as a Spatial Map: Preliminary Evidence from Unit Activity in the Freely-Moving Rat. Brain Res 34 (1):171–175. https://doi.org/10.1016/0006-8993(71)90358-1

O'Keefe J, and Nadel L (1978) The Hippocampus as a Cognitive Map. Oxford Univ Press, London

Olafsdottir HF, Carpenter F, and Barry C (2017) Task Demands Predict a Dynamic Switch in the Content of Awake Hippocampal Replay. Neuron 96 (4):925–935 e926. https://doi.org/10.1016/j.neuron.2017.09.035

Page HJI, Wilson JJ, and Jeffery KJ (2018) A Dual-Axis Rotation Rule for Updating the Head Direction Cell Reference Frame during Movement in Three Dimensions. J Neurophysiol 119:192–208. https://doi.org/10.1152/jn.00501.2017

Parra-Barrero E, Vijayabaskaran S, Seabrook E, Wiskott L, and Cheng S (2023) A Map of Spatial Navigation for Neuroscience. Neurosci Biobehav Rev 152:105200. https://doi.org/10.1016/j.neubiorev.2023.105200

Patai EZ, Javadi AH, Ozubko JD, et al. (2019) Hippocampal and Retrosplenial Goal Distance Coding after Long-Term Consolidation of a Real-World Environment. Cereb Cortex 29 (6):2748–2758. https://doi.org/10.1093/cercor/bhz044

Patai EZ, and Spiers HJ (2021) The Versatile Wayfinder: Prefrontal Contributions to Spatial Navigation. Trends Cogn Sci 25 (6):520–533. https://doi.org/10.1016/j.tics.2021.02.010

Payne HL, and Aronov D (2025) Remote Activation of Place Codes by Gaze in a Highly Visual Animal. Nature 643 (8073):1037–1043. https://doi.org/10.1038/s41586-025-09101-z

Peer M, Nadar C, and Epstein RA (2024) The Format of the Cognitive Map Depends on the Structure of the Environment. J Exp Psychol Gen 153 (1):224–240. https://doi.org/10.1037/xge0001498

Peng J-J, Throm B, Najafian Jazi M, et al. (2025) Grid Cells Accurately Track Movement during Path Integration-Based Navigation Despite Switching Reference Frames. Nat Neurosci 28 (10):2092–2105. https://doi.org/10.1038/s41593-025-02054-6

Persichetti AS, and Dilks DD (2016) Perceived Egocentric Distance Sensitivity and Invariance across Scene-Selective Cortex. Cortex 77:155–163. https://doi.org/10.1016/j.cortex.2016.02.006

——— (2018) Dissociable Neural Systems for Recognizing Places and Navigating through Them. J Neurosci 38 (48):10295–10304. https://doi.org/10.1523/JNEUROSCI.1200-18.2018

——— (2019) Distinct Representations of Spatial and Categorical Relationships across Human Scene-Selective Cortex. PNAS 116 (42):21312–21317. https://doi.org/10.1073/pnas.1903057116

Pfeiffer BE, and Foster DJ (2013) Hippocampal Place-Cell Sequences Depict Future Paths to Remembered Goals. Nature 497 (7447):74–79. https://doi.org/10.1038/nature12112

Prabhakar J, Johnson EG, Nordahl CW, and Ghetti S (2018) Memory-Related Hippocampal Activation in the Sleeping Toddler. PNAS 115 (25):6500–6505. https://doi.org/10.1073/pnas.1805572115

Purser HR, Farran EK, Courbois Y, et al. (2015) The Development of Route Learning in Down Syndrome, Williams Syndrome and Typical Development: Investigations with Virtual Environments. Dev Sci 18 (4):599–613. https://doi.org/10.1111/desc.12236

Qasim SE, Fried I, and Jacobs J (2021) Phase Precession in the Human Hippocampus and Entorhinal Cortex. Cell 184 (12):3242–3255 e3210. https://doi.org/10.1016/j.cell.2021.04.017

Qasim SE, Miller J, Inman CS, et al. (2019) Memory Retrieval Modulates Spatial Tuning of Single Neurons in the Human Entorhinal Cortex. Nat Neurosci 22 (12):2078–2086. https://doi.org/10.1038/s41593-019-0523-z

Quiroga RQ, Reddy L, Kreiman G, Koch C, and Fried I (2005) Invariant Visual Representation by Single Neurons in the Human Brain. Nature 435 (7045):1102–1107. https://doi.org/10.1038/nature03687

Ranck JB (1984) Head Direction Cells in the Deep Cell Layer of Dorsal Presubiculum in Freely Moving Rats. Akademiai Kiado, Budapest

Ray S, Yona I, Elami N, et al. (2025) Hippocampal Coding of Identity, Sex, Hierarchy, and Affiliation in a Social Group of Wild Fruit Bats. Science 387 (6733):eadk9385. https://doi.org/10.1126/science.adk9385

Rolls ET, and O'Mara SM (1995) View-Responsive Neurons in the Primate Hippocampal Complex. Hippocampus 5 (5):409–424. https://doi.org/10.1002/hipo.450050504

Rolls ET, and Wirth S (2018) Spatial Representations in the Primate Hippocampus, and Their Functions in Memory and Navigation. Prog Neurobiol 171:90–113. https://doi.org/10.1016/j.pneurobio.2018.09.004

Rossier J, and Schenk F (2003) Olfactory and/or Visual Cues for Spatial Navigation through Ontogeny: Olfactory Cues Enable the Use of Visual Cues. Behav Neurosci 117 (3):412–425. https://doi.org/10.1037/0735-7044.117.3.412

Sarel A, Palgi S, Blum D, et al. (2022) Natural Switches in Behaviour Rapidly Modulate Hippocampal Coding. Nature 609 (7925):119–127. https://doi.org/10.1038/s41586-022-05112-2

Sargolini F, Fyhn M, Hafting T, et al. (2006) Conjunctive Representation of Position, Direction, and Velocity in Entorhinal Cortex. Science 312 (5774):758–762. https://doi.org/10.1126/science.1125572

Save E, Cressant A, Thinus-Blanc C, and Poucet B (1998) Spatial Firing of Hippocampal Place Cells in Blind Rats. J Neurosci 18 (5):1818–1826. https://doi.org/10.1523/JNEUROSCI.18-05-01818.1998

Save E, Nerad L, and Poucet B (2000) Contribution of Multiple Sensory Information to Place Field Stability in Hippocampal Place Cells. Hippocampus 10 (1):64–76. https://doi.org/10.1002/(SICI)1098-1063(2000)10:1<64::AID-HIPO7>3.0.CO;2-Y

Scott RC, Richard GR, Holmes GL, and Lenck-Santini PP (2010) Maturational Dynamics of Hippocampal Place Cells in Immature Rats. Hippocampus 4:347–353. https://doi.org/10.1002/hipo.20789

Sharp PE, Tinkelman A, and Cho J (2001) Angular Velocity and Head Direction Signals Recorded from the Dorsal Tegmental Nucleus of Gudden in the Rat: Implications for Path Integration in the Head Direction Cell Circuit. Behav Neurosci 115 (3):571–588. https://doi.org/10.1037/0735-7044.115.3.571

Singer AC, Carr MF, Karlsson MP, and Frank LM (2013) Hippocampal SWR Activity Predicts Correct Decisions during the Initial Learning of an Alternation Task. Neuron 77 (6):1163–1173. https://doi.org/10.1016/j.neuron.2013.01.027

Solomon EA, Lega BC, Sperling MR, and Kahana MJ (2019) Hippocampal Theta Codes for Distances in Semantic and Temporal Spaces. PNAS 116 (48):24343–24352. https://doi.org/10.1073/pnas.1906729116

Sreenivasan S, and Fiete I (2011) Grid Cells Generate an Analog Error-Correcting Code for Singularly Precise Neural Computation. Nat Neurosci 14:1330–1337. https://doi.org/10.1038/nn.2901

Stackman RW, and Taube JS (1998) Firing Properties of Rat Lateral Mammillary Single Units: Head Direction, Head Pitch, and Angular Head Velocity. J Neurosci 18 (21):9020–9037. https://doi.org/10.1523/jneurosci.18-21-09020.1998

Stangl M, Topalovic U, Inman CS, et al. (2020) Boundary-Anchored Neural Mechanisms of Location-Encoding for Self and Others. Nature:1–6. https://doi.org/10.1038/s41586-020-03073-y

Steel A, Billings MM, Silson EH, and Robertson CE (2021) A Network Linking Scene Perception and Spatial Memory Systems in Posterior Cerebral Cortex. Nat Commun 12 (1):2632. https://doi.org/10.1038/s41467-021-22848-z

Stensola T, Stensola H, Moser MB, and Moser EI (2015) Shearing-Induced Asymmetry in Entorhinal Grid Cells. Nature 518 (7538):207–212. https://doi.org/10.1038/nature14151

Tan HM, Bassett JP, O'Keefe J, Cacucci F, and Wills TJ (2015) The Development of the Head Direction System before Eye Opening in the Rat. Curr Biol 25 (4):479–483. https://doi.org/10.1016/j.cub.2014.12.030

Taube JS (1995) Head Direction Cells Recorded in the Anterior Thalamic Nuclei of Freely Moving Rats. J Neurosci 15 (1):70–86. https://doi.org/10.1523/jneurosci.15-01-00070.1995

Taube JS (2007) The Head Direction Signal: Origins and Sensory-Motor Integration. Annu Rev Neurosci 30:181–207. https://doi.org/10.1146/annurev.neuro.29.051605.112854

Taube JS, Muller RU, and Ranck JB (1990) Head-Direction Cells Recorded from the Postsubiculum in Freely Moving Rats. II. Effects of Environmental Manipulations. J Neurosci 10 (2):436–447. https://doi.org/10.1523/jneurosci.10-02-00436.1990

Tennant SA, Fischer L, Garden DLF, et al. (2018) Stellate Cells in the Medial Entorhinal Cortex Are Required for Spatial Learning. Cell Rep 22 (5):1313–1324. https://doi.org/10.1016/j.celrep.2018.01.005

Terrazas A, Krause M, Lipa P, et al. (2005) Self-Motion and the Hippocampal Spatial Metric. J Neurosci 25 (35):8085–8096. https://doi.org/10.1523/jneurosci.0693-05.2005

Thom RP, Canales C, Tresvalles M, et al. (2023) Neuroimaging Research in Williams Syndrome: Beginning to Bridge the Gap with Clinical Care. Neurosci Biobehav Rev 153:105364. https://doi.org/10.1016/j.neubiorev.2023.105364

Tolman EC (1948) Cognitive Maps in Rats and Men. Psychol Rev 55 (4):189–208. https://doi.org/10.1037/h0061626

Tsitsiklis M, Miller J, Qasim SE, et al. (2020) Single-Neuron Representations of Spatial Targets in Humans. Curr Biol 30 (2):245–253.e244. https://doi.org/10.1016/j.cub.2019.11.048

Ulanovsky, U (205) Natural Neuroscience: Toward a Systems Neuroscience of Natural Behaviors. MIT Press, Cambridge, MA

Vann SD, and Aggleton JP (2005) Selective Dysgranular Retrosplenial Cortex Lesions in Rats Disrupt Allocentric Performance of the Radial-Arm Maze Task. Behav Neurosci 119 (6):1682–1686. https://doi.org/10.1037/0735-7044.119.6.1682

Vargha-Khadem F, Gadian DG, Watkins KE, et al. (1997) Differential Effects of Early Hippocampal Pathology on Episodic and Semantic Memory. Science 277 (5324):376–380. https://doi.org/10.1126/science.277.5324.376

Vass LK, Copara MS, Seyal M, et al. (2016) Oscillations Go the Distance: Low-Frequency Human Hippocampal Oscillations Code Spatial Distance in the Absence of Sensory Cues during Teleportation. Neuron 89 (6):1180–1186. https://doi.org/10.1016/j.neuron.2016.01.045

Vass LK, and Epstein RA (2017) Common Neural Representations for Visually Guided Reorientation and Spatial Imagery. Cereb Cortex 27 (2):1457–1471. https://doi.org/10.1093/cercor/bhv343

Warren WH (2019) Non-Euclidean Navigation. J Exp Biol 222:jeb187971. https://doi.org/10.1242/jeb.187971

Weiner KS, Barnett MA, Lorenz S, et al. (2017) The Cytoarchitecture of Domain-Specific Regions in Human High-Level Visual Cortex. Cereb Cortex 27 (1):146–161. https://doi.org/10.1093/cercor/bhw361

Weiner KS, Barnett MA, Witthoft N, et al. (2018) Defining the Most Probable Location of the Parahippocampal Place Area Using Cortex-Based Alignment and Cross-Validation. Neuroimage 170:373–384. https://doi.org/10.1016/j.neuroimage.2017.04.040

Widloski J, and Foster DJ (2022) Flexible Rerouting of Hippocampal Replay Sequences around Changing Barriers in the Absence of Global Place Field Remapping. Neuron 110 (9):1547–1558 e1548. https://doi.org/10.1016/j.neuron.2022.02.002

Wikenheiser AM, and Redish AD (2015) Hippocampal Theta Sequences Reflect Current Goals. Nat Neurosci 18 (2):289–294. https://doi.org/10.1038/nn.3909

Wills TJ, Cacucci F, Burgess N, and O'Keefe J (2010) Development of the Hippocampal Cognitive Map in Preweanling Rats. Science 328 (5985):1573–1576. https://doi.org/10.1126/science.1188224

Wills TJ, Lever C, Cacucci F, Burgess N, and O'Keefe J (2005) Attractor Dynamics in the Hippocampal Representation of the Local Environment. Science 308 (5723):873–876. https://doi.org/10.1126/science.1108905

Wills TJ, Muessig L, and Cacucci F (2014) The Development of Spatial Behaviour and the Hippocampal Neural Representation of Space. Philos Trans R Soc Lond B Biol Sci 369 (1635):20130409. https://doi.org/10.1098/rstb.2013.0409

Wilson MA, and McNaughton BL (1994) Reactivation of Hippocampal Ensemble Memories during Sleep. Science 265 (5172):676–679. https://doi.org/10.1126/science.8036517

Wirth S, Baraduc P, Planté A, Pinède S, and Duhamel J-R (2017) Gaze-Informed, Task-Situated Representation of Space in Primate Hippocampus during Virtual Navigation. PLoS Biol 15 (2):e2001045. https://doi.org/10.1371/journal.pbio.2001045

Wolbers T, and Buchel C (2005) Dissociable Retrosplenial and Hippocampal Contributions to Successful Formation of Survey Representations. J Neurosci 25 (13):3333–3340. https://doi.org/10.1523/JNEUROSCI.4705-04.2005

Wolbers T, and Wiener JM (2014) Challenges for Identifying the Neural Mechanisms That Support Spatial Navigation: The Impact of Spatial Scale. Front Hum Neurosci 8:00571. https://doi.org/10.3389/fnhum.2014.00571

Xu X, Du K, and Mao D (2024) Spatial Dissociation between Recognition and Navigation in the Primate Hippocampus. Sci Adv 10 (38):eado7392. https://doi.org/10.1126/sciadv.ado7392

Yan Y, Burgess N, and Bicanski A (2021) A Model of Head Direction and Landmark Coding in Complex Environments. PLoS Comput Biol 17 (9):e1009434–e1009434. https://doi.org/10.1371/journal.pcbi.1009434

Yao Z, van Velthoven CTJ, Kunst M, et al. (2023) A High-Resolution Transcriptomic and Spatial Atlas of Cell Types in the Whole Mouse Brain. Nature 624 (7991):317–332. https://doi.org/10.1038/s41586-023-06812-z

Yates TS, Fel J, Choi D, et al. (2025) Hippocampal Encoding of Memories in Human Infants. Science 387 (6740):1316–1320. https://doi.org/10.1126/science.adt7570

Yoon HK, Jung Y, and Dilks D (2024) A New Scene-Selective Region in the Superior Parietal Lobule and Its Potential Involvement in Visually-Guided Navigation. J Vis 24 (10):1160. https://doi.org/10.1167/jov.24.10.1160

Yoon HK, Jung Y, Persichetti AS, and Dilks DD (2025) A Scene-Selective Region in the Superior Parietal Lobule for Visually Guided Navigation. Cereb Cortex 35 (4):bhaf082. https://doi.org/10.1093/cercor/bhaf082

Yu C, Ji Z, Ormond J, O'Keefe J, and Burgess N (2025) Hippocampal Theta Sweeps Indicate Goal Direction. bioRxiv Preprint:71551 https://doi.org/10.1101/2025.08.21.671551

Zemke NR, Armand EJ, Wang W, et al. (2023) Conserved and Divergent Gene Regulatory Programs of the Mammalian Neocortex. Nature 624 (7991):390–402. https://doi.org/10.1038/s41586-023-06819-6

Zhu SL, Lakshminarasimhan KJ, and Angelaki DE (2023) Computational Cross-Species Views of the Hippocampal Formation. Hippocampus 33 (5):586–599. https://doi.org/10.1002/hipo.23535

Navigation in a Wider Perspective

10

Planning and Imagination in the Navigational Domain

The Role of the Hippocampus and Prefrontal Cortex

Shantanu P. Jadhav

Abstract Planning is a fundamental part of navigation. The notion of a cognitive map of space in the brain is intertwined with the idea that organisms can plan, imagine, and mentally navigate this cognitive map to choose which routes to follow, and to execute trajectories to reach a desired goal. Memory systems in the mammalian brain that function to store information and remember the past also have a role in imagining the future and planning actions; memories of the past contribute to planning the future. This link between the past and future aligns with the role of a default mode network in the brain, which is active both for memories and imagination. The primary drive behind navigational planning is linked to goal representations. The hippocampus and prefrontal cortex—core of the episodic memory and executive functioning systems, respectively, in the mammalian brain—are key for using past experiences to guide the future and support planning, to deliberate options leading to goals, to execute decisions to choose a path, and to make online changes to the navigation plan, such as rerouting. It is therefore critical to understand (a) navigational representations in the hippocampus and prefrontal cortical regions at multiple timescales (e.g., spatial maps, reward signals, goal coding, context and rule representations) and (b) how interactions between the two regions support planning and execution of navigational strategies. This chapter examines hippocampal-prefrontal spatial representations and their interactions, especially in rodents, for navigational planning.

Keywords Cognitive maps, goals, timescales, replay, sequences, prospection, deliberation

Shantanu P. Jadhav (✉)
Neuroscience Program, Dept. of Psychology, and Volen National Center for Complex Systems, Brandeis University, Waltham, MA 02453, USA
Email: shantanu@brandeis.edu

N. S. Newcombe and K. Cheng (eds.), *Challenges in Navigation Research*, Strüngmann Forum Reports, https://doi.org/10.1007/978-3-032-20563-6_10

1 Introduction

Animals must navigate the environment in search of food and water to find mates, to seek shelter, and to avoid predators. The neural circuits that support the ability to learn from past experiences are critical in navigating both familiar environments and in adapting to new environments; remembering the past is linked to imagining and planning for the future (Addis and Schacter 2011; Hassabis et al. 2007; Buckner 2010). These systems are also active in the idling brain and during internal thought, known as the default mode network, which includes the medial prefrontal cortex, the hippocampus, and the cingulate cortex (Buckner et al. 2008).

The neural basis of planning and imagination in the navigational domain relies on spatial cognitive maps in the hippocampus and associated areas in the medial temporal lobe. Tolman's pioneering studies showed that rats form spatial cognitive maps that support flexible expression in navigation through short cuts and indirect routes (Tolman 1948), and cognitive maps of spatial representations mediated by place cells are well studied in the hippocampus (Cohen and Eichenbaum 1993; O'Keefe and Nadel 1978). Place codes in the hippocampus and grid codes in the entorhinal cortex (EC) (Moser et al. 2017; Hafting et al. 2005; Epstein et al. 2017) underlie spatial representations that are central for cognitive map-based navigation. Navigation can be achieved through path integration using these maps, but errors accumulate over time. Cognitive maps thus need to be anchored to landmarks, with a corresponding important role of inputs, such as boundary cells, and other networks in the transformation from allocentric to egocentric reference frames required for navigation (Epstein et al. 2017; Moser et al. 2017). Map-based navigation is also used in artificial systems implemented by algorithms such as SLAM (simultaneous localization and mapping) (see also Chapter 3). However, planning is required in a model-based framework to enable flexible behavior in more complex environments (e.g., dynamic contexts that require evaluation of several contingencies and progress tracking, and rerouting) and to adapt to novel environments that have similarities to past experiences. This planning process involves learning the structure of the world, enabling mental imagery and prospection to choose a route of action. It also relies on the prefrontal cortex (PFC) and on interactions between hippocampal cognitive maps and prefrontal systems for evaluation and goal representations.

The PFC has complementary and overlapping memory functions to the hippocampus, including memory consolidation, retrieval, and working memory (Preston and Eichenbaum 2013; Eichenbaum 2017; Shin and Jadhav 2016). Further, as the central executive region, PFC has a key role in planning, decision making, and goal representations. In the process of imagining routes based on the cognitive map and planning trajectories toward goals, PFC is expected to have a primary role and has been a focus of recent studies in both rodents and humans (Patai and Spiers 2021; Tang et al. 2021). PFC plays a role in selecting and representing goals that guide behavior, in visualizing routes through the cognitive map, in planning a course of action, and in overseeing the execution of the decision. Similar to its role in deliberative decision making in the perceptual domain (Shadlen and Kiani 2013), PFC networks are required for deliberating between choices in memory-guided navigation (Ratcliff et al. 2004), evaluating options based on the balance between cost effort and reward before committing to a decision for execution, rerouting as needed during navigation, and inferring novel routes and shortcuts. These deliberative processes require the integration of cognitive maps, goal representations, value coding, retrospection, and prospection.

Continuous interactions between hippocampal and prefrontal networks are required for these processes (Shin and Jadhav 2016). Some elements of planning and evaluation are consciously accessible, akin to a "thinking aloud" protocol in problem solving (Wolcott and Lobczowski 2021). This episodic future thinking aligns with mental imagery, occurs in real time, and is deliberate. In rodents, overt behavioral cues such as "vicarious trial and error" are thought to indicate uncertainty underlying such a deliberative process (Redish 2016). However, moment-by-moment interactions between regions can also occur on faster timescales, with internal computations occurring without deliberative effort that are not accessible to consciousness, corresponding to sudden or instantaneous emergence of plans, akin to insight. Consider the case of a soccer player navigating through a maze of defenders toward the goal. The player can deliberate strategies for dribbling through space between defenders to pass the ball to an anticipated location of a teammate or toward the goal. When under pressure with a split second available, "instinct" developed through years of practice kicks in, and a perfectly placed pass is executed in a flash through an internal plan that executes the motor memory. This spatial instinct has similarities to insight in problem solving in nonspatial cases. Imagination in the navigational domain, therefore, has parallels with creativity. Such improvisation is also important for cognitive flexibility.

The cellular and circuit mechanisms underlying planning and imagination in the navigational domain are most accessible in rodent models, where navigational systems driven by spatial representations have been well studied. Thus, I will focus here on neurophysiological correlates of hippocampal-prefrontal interactions in planning during navigation in rodents. In particular, I emphasize the need to investigate multiple timescale computations in those networks that can subserve different aspects of planning and imagination. I examine features of cognitive maps in the hippocampus and PFC, goal representations, as well as the nature of representations for navigational planning and conclude with recommendations for further research.

2 Cognitive Maps

Cognitive maps of spatial representations are mediated by hippocampal place cells and entorhinal grid cells that underlie a spatial coordinate system. Map-based navigation has been heavily studied in rodents, with several aspects of spatial representations confirmed in human studies (Maguire et al. 1998; Epstein et al. 2017). Place cells in the hippocampus (primarily in regions CA3 and CA1) are active when an animal is in a particular location of an environment, and place-cell ensembles form unique population codes for a given spatial environment or context (O'Keefe and Nadel 1978), with remapping across contexts. Sequences of place cells represent trajectories in an environment. When animals explore an environment, grid cells in superficial layers of the EC fire at regular spacing intervals, forming triangular grids (Moser et al. 2017; Hafting et al. 2005). Grid cells are thought to underlie a spatial coordinate system across the environment, thus forming an essential part of the brain's "GPS" navigation system. This system allows animals to understand position in space by storing and integrating information about location, distance, and direction. Head-direction (HD) cells, a primary input to the grid network in EC, are located in retrosplenial cortex, anterior dorsal thalamic nucleus, and postsubiculum. EC also consists of HD cells and mixed HD-grid cells as well as boundary cells and object vector cells that can anchor the spatial maps to landmarks and boundaries (Moser et al. 2017). Further, networks in parahippocampal regions, retrosplenial cortex, and posterior parietal cortex play a role in transforming allocentric signals to egocentric reference frames to aid navigation (Patai and Spiers 2021).

Initial models of navigation used hippocampal place maps in a simple goal-cell implementation, with reward or goal signals located in other networks, such as PFC (Burgess et al. 1997), and extended this framework using a model-free reinforcement learning approach to select appropriate navigational actions (Foster et al. 2000). Relatively rigid strategies, such as route-based navigation (Nyberg et al. 2022) via a set of egocentric responses, can also be implemented by learning the value of cue-based actions through reinforcement learning in a model-free framework, likely involving striatal regions. This is, however, a less flexible approach to navigation that might fail when facing complex contingencies and novelty.

Several models utilize place-cell and grid-cell maps in a topological or vector-based approach for navigation toward goal locations. For example, successor representation (SR) models generate predicted states/locations toward goals, using place-cell and grid-cell maps, and is a form of value-based decision making that uses a learning approach (Stachenfeld et al. 2017). Vector-based approaches need to determine heading vectors between an animal's current location and remembered goals, and there are several implementations using place-cell and grid-cell maps, such as a hierarchical look-ahead (HiLAM) model (Erdem et al. 2015), which uses a prefrontal reward/goal signal and entorhinal mechanisms for vector-based navigation (Kubie and Fenton 2012).

Does the PFC have any role in this cognitive mapping? Several recent studies have established spatial responses in rodent PFC, including trajectory-specific firing (Tang et al. 2021; Zielinski et al. 2019; Ito et al. 2015). This PFC spatial representation is impaired by hippocampal lesions (Esteves et al. 2021), and there is strong coordination between hippocampal and prefrontal responses mediated by ongoing theta oscillations (local field potential theta coherence and theta-phase spike locking) in spatial memory tasks (Shin and Jadhav 2016; Jadhav et al. 2016). However, there is increasing consensus that the PFC cognitive map is distinct from the hippocampal spatial map: PFC maps represent task structure in the spatial domain, including task rules, in contrast to the hippocampal topological map. For example, geometric coding and dimensionality-reduction analyses in W/M mazes[1] show that PFC cognitive maps represent the alternation task rule and generalize the rule across contexts, in contrast to the hippocampal spatial map, which remaps across contexts (Tang et al. 2023). Similar results from other studies show that PFC maps

[1] Maze structures in a W or M shape, with three arms and reward wells that animals learn to visit according to rules such as alternation in memory tasks.

are governed by task structure that reflect relationships between locations and trajectories based on abstract rules (Samborska et al. 2022). Thus, PFC maps are likely important in top-down control during navigation toward goals.

These rule-based PFC cognitive maps may be integrated with other signals for planning ahead, tracking progress toward goals, and rerouting and aligning with the PFC functions of goal coding, decision making, and adaptive behavior (Miller and Cohen 2001). As noted above, to enable flexible behavior in both complex and novel environments, a model-based framework is needed to enable learning of the structure of the world and navigational planning. Indeed, some models include PFC in the initial planning process as well as tracking distances to goals using PFC interaction with hippocampal networks (Patai and Spiers 2021).

In rodents, there have been recent suggestions, gathered from comparative anatomy across species, that the entire rodent medial PFC can be classified as cingulate cortex (area 24) based on connectivity patterns (van Heukelum et al. 2020). Substantial evidence from other studies, however, points to distinct roles of rodent medial PFC subnetworks in cognition and behavior, owing to unique connectivity patterns. Primary evidence about distinct roles of the prefrontal subregions in the prelimbic cortex (PrL) and infralimbic cortex (IL) comes from fear and addiction studies: PrL is thought to drive the expression of fear and drug-seeking, whereas IL drives suppression of this behavior after extinction (Peters et al. 2009; Moorman and Aston-Jones 2023). This "PrL-go/IL-stop" dichotomy extends beyond fear and addiction studies to goal-directed behavior and rule switching (Gourley and Taylor 2016; Mukherjee and Caroni 2018), with PrL involved in selection and maintenance of goal-directed response strategies, and IL involved in learning and switching to alternative strategies while suppressing inappropriate actions. In rodents, the anterior cingulate cortex (ACC) plays a crucial role in rule monitoring and alternative strategies, as well as effort-related response strategies (Karlsson et al. 2012; Hillman and Bilkey 2012). In parallel, studies in humans suggest that dorsomedial PFC may track progress to goals, ventral regions may play a key role in goal maintenance during travel, and the dorsal ACC may represent values of alternatives, thus supporting backtracking and strategy switches (Patai and Spiers 2021). Finally, orbitofrontal cortex has a role in value coding, which may influence goal-directed behavior (Basu and Ito 2023).

3 Goal Representations for Navigation in the Hippocampus and Prefrontal Cortex

Inherent in navigation is the concept of a goal. Goal representations and planning actions toward goals are central components of models of navigation and reinforcement learning. Several models incorporate goal representations in hippocampal and EC maps, based on evidence of goal coding during different phases of navigation (for comprehensive reviews, see Nyberg et al. 2022; Sosa and Giocomo 2021).

Goal locations have salient representations in the hippocampal-EC system, with place-cell overrepresentation of goal locations developing with experience (Sosa and Giocomo 2021). It is, however, unclear whether this overrepresentation has any functional role in guiding navigation (Duvelle et al. 2019). A dedicated reward-coding population in the hippocampus has also been suggested (Gauthier and Tank 2018). Similarly, entorhinal grid cells show increased firing rate near rewards and have been reported to shift their fields toward rewards (Boccara et al. 2019; Butler et al. 2019).

Goal signals are also prominent in the hippocampus during navigation and during goal approach or goal arrival (Sosa and Giocomo 2021). This is manifest as selective firing of place cells during travel toward a specific goal. On common trajectories leading toward multiple goals, selective firing leads to trajectory-specific coding, or splitter cells, with preferential firing of cells on common paths when heading toward specific goals versus heading toward other goals (prospective trajectory coding) and heading away from goals (retrospective trajectory coding) (Wood et al. 2000). Trajectory-specific coding can be a potential means of supporting ongoing navigation by maintaining a representation of the current goal or keeping track of progress toward the goal.

Finally, egocentric goal coding has also been reported in the hippocampus, with cells encoding distance and direction toward goals in bats (Sarel et al. 2017) and in rats (Aoki et al. 2019). This form of egocentric goal coding has also been reported in hippocampal–entorhinal regions in humans (Howard et al. 2014).

Although several navigation models consider goal representations only in the hippocampal–entorhinal systems, goal coding is a canonical cognitive function of PFC, and similar goal representations are prevalent in the PFC during navigation. Although the orbitofrontal cortex also represents goals in rodents (Basu and Ito 2023), PFC often overrepresents goals (Hok et al. 2005). Some models, such as HiLAM, consider PFC goal signals to act as a beacon to guide navigation using hippocampal–entorhinal cognitive maps (Erdem et al. 2015).

Trajectory-specific coding of goal locations during navigation is seen in PFC as in hippocampus, resulting in similar unique sequential representation of trajectories in the PFC and hippocampus, with accurate decoding of current location and intended goal possible from both populations (Shin et al. 2019; Tang et al. 2021). In fact, the trajectory-specific coding in PFC and hippocampus has been shown to be interdependent via connections from the thalamic region, nucleus reuniens (Ito et al. 2015). Finally, egocentric goal coding in PFC has also been reported in human studies (Maguire et al. 1998; Patai and Spiers 2021).

Given the substantial evidence for parallel goal representations in the hippocampus and PFC, there is a great need for experiments and models to integrate these goal representations, as well as cognitive maps in all regions, for a comprehensive understanding of planning and execution of navigation toward a goal. Goal representation is considered a canonical function of PFC, so it is possible that hippocampal goal representations are downstream of PFC. It is, however, also likely that these representations arise simultaneously in both regions. Further, hippocampal cognitive maps are considered to represent structure of an environment by representing relationships between components of the environment; PFC cognitive maps likely build on these further by representing rules of linking elements across environments for generalization (Tang et al. 2023).

4 Multiple Timescale Representations for Navigation Planning

Cognitive maps and goal representations focus on phenomena that occur in real time, especially during travel execution. However, similar to sudden or instantaneous emergence of plans ("insight") in the nonspatial domain, internal computations for navigational planning can occur without deliberative effort and are not accessible to consciousness. This ability entails the flexibility to operate based on changes to external cues as well as to internal needs, referencing both real-world and subjective time across temporal scales. Time-compressed neuronal sequences in the hippocampus are prevalent at fast timescales that can support multi-timescale computation, with hypothesized roles in memory as well as navigational processes (Tang and Jadhav 2022; Buzsaki and Tingley 2018) (Figure 10.1).

The hippocampus generates internal sequences for memory and navigation (Buzsaki and Tingley 2018). Sequences of place cells encode trajectories during travel, firing at a series of locations in space at a relatively slow timescale of seconds. Embedded within these slow sequences, the same sequences are temporally organized within single cycles of theta oscillations (6–12 Hz) during navigation (i.e., a temporal code), with a timescale of ~100–200 ms (i.e., 5–10 times faster than real time); these fast sequences are called "theta sequences" (Colgin 2013; Buzsaki and Moser 2013). Interestingly, in non-exploratory states when theta oscillations are most attenuated, place-cell sequences reoccur during brief (~50–200 ms), high-frequency (150–250 Hz) sharp-wave ripples, a phenomenon known as *replay* (Figure 10.1) (Buzsaki 2015; Tang and Jadhav 2019). The fast timescale theta and replay sequences can be organized in the network without reliance on external changes and thus are potentially useful for exploiting internally generated signals that are vital for memory and cognition (Buzsaki and Tingley 2018). Theta sequences occur during movement as well as during brief pauses in locomotion associated with the overt behavioral characteristic of vicarious trial and error (VTE) (Redish 2016). VTEs correspond to active head sweeps in the direction of possible options, often at decision points prior to a choice, and are considered an overt marker of mental planning (i.e., exploration of options) or of uncertainty (Johnson and Redish 2007). Thus, it is possible that theta sequences associated with VTEs correspond to a deliberative decision-making process. During locomotion, theta sequences correspond to look-ahead sequences, extending from slightly behind the animal to ahead of the animal's location, and are modulated by distance to goal locations (Zheng et al. 2021). Intriguingly, theta sequences can cycle between the representation of potential routes during travel, with alternate cycles of theta oscillations corresponding to one route (Kay et al. 2020). This may reflect a route deliberation and maintenance process during travel.

Theta sequences have also been reported in PFC (Tang et al. 2021). When hippocampal theta sequences alternate between representing different route options, PFC theta sequences more strongly predict the upcoming choice (Tang et al., 2021). Thus, theta sequences during locomotion may act as a complement to real-time splitter sequences in the maintenance of goal representations during locomotion, with PFC theta sequences maintaining progress toward the goal. The representation of alternate goals likely corresponds to an online planning process to maintain flexibility in case of need to change routes. PFC inactivation reduces VTE behavior

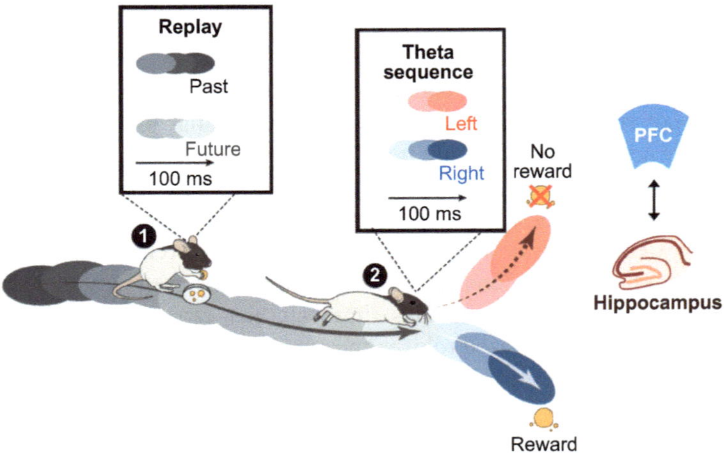

Figure 10.1 Schematic of potential functions of multiple timescale sequences in hippocampus and PFC in rodents, adapted from Tang and Jadhav (2022). During ongoing navigation, successive assemblies of place cells are sequentially activated in accordance with changes of environmental and body-derived cues at the behavioral timescale (i.e., place-field sequences). Similar sequences of PFC cells are seen during travel, and sequences in both regions are modulated by upcoming goals. The hippocampal and prefrontal cognitive maps are linked through theta oscillations. (1) During periods of immobility at the reward well, when environmental cues are relatively constant, the place-cell firing patterns of past and future running experience are replayed in reverse and forward orders at a timescale of 100–200 ms during sharp-wave ripples. These forward-and-reverse replay sequences showed opposing gradients during learning. Hippocampal replay reactivates past and future options and coordinated hippocampal-PFC replay is stronger for taken paths, with a potential role in learning paths to goals as well as planning upcoming paths. (2) During travel, fast theta sequences that sweep into future locations can represent multiple possible future scenarios before a choice point without overtly deliberative behaviors. At decision points, theta sequences are associated with overt vicarious trial-and-error behaviors, sweeping ahead of the animal during head movements. Theta sequences are also coordinated across hippocampal-PFC networks, with a potential role in online planning and maintenance.

and impacts hippocampal theta sequences (Hasz and Redish 2020); therefore, both top-down modulation from PFC to hippocampus and the bottom-up influence of hippocampus on PFC activity are possible during theta sequences.

In contrast to theta sequences, replay sequences in the hippocampus occur during immobility, consummatory behaviors, and non-REM sleep associated with sharp-wave ripples. Replay was first described in non-REM sleep as a potential physiological substrate of memory consolidation, reactivating time-compressed sequences of place cells representing trajectories in recently experienced environments. Temporally compressed replay of place-cell sequences reactivates trajectories in both the reverse and forward direction during awake immobility, especially at goal locations when animals pause between trajectories in maze tasks (Buzsaki 2015). Reverse replay is akin to retrospection, representing paths leading to reward in a reverse order, and can be used as a potential signal to solve the temporal credit assignment problem in reinforcement learning (Shin et al. 2019). In contrast, forward replay represents upcoming trajectories, similar to a prospection process that can evaluate paths in planning. Indeed, reverse replay rates are higher early in learning, and forward replay rate increases as animals switch from exploration to exploitation of learned routes toward reward (Shin et al. 2019). Models with reinforcement learning frameworks also suggest a role for replay in planning (Mattar and Daw 2018), offering solutions to reinforcement learning problems in the real world, where model-free and model-based reinforcement learning can provide optimal solutions to problems with multiple contingencies.

Coordinated replay occurs in hippocampal CA1-prefrontal networks during sharp-wave ripples, with increased coordination for trajectories that lead to reward or those subsequently taken. The strength of this relationship predicts performance on the spatial task. These results suggest a process in which hippocampal replay supports an internal evaluation of route options in the cognitive map, which can be utilized by downstream prefrontal regions to commit to a plan (Shin et al. 2019) (Figure 10.2). During retrospection, a similar process can consolidate memories of paths leading toward goals, which is further continued during post-experience non-REM sleep.

Other studies support the notion of hippocampal replay as a planning process, with replay of future paths seen in other mazes (Ólafsdóttir et al. 2015), 8-arm radial mazes (Xu et al. 2019), two-dimensional environments (Pfeiffer and Foster 2013), and zones to avoid (Wu et al. 2017). Some studies, however, suggest that hippocampal replay reflects reactivation of previously rewarded locations without supporting planning, at least in specific behavioral contexts (Gillespie et al. 2021). Of note, human studies of replay support a role in retrieval and planning (Vaz et al. 2020; Liu et al. 2019). Only a very few studies on entorhinal replay have been conducted, with reports of EC replay occurring independently of hippocampal replay during sleep (O'Neill et al. 2017). Further studies that examine network-wide replay are therefore required to address the role of replay in planning, since it offers the potential of an internal planning mechanism before the onset of travel.

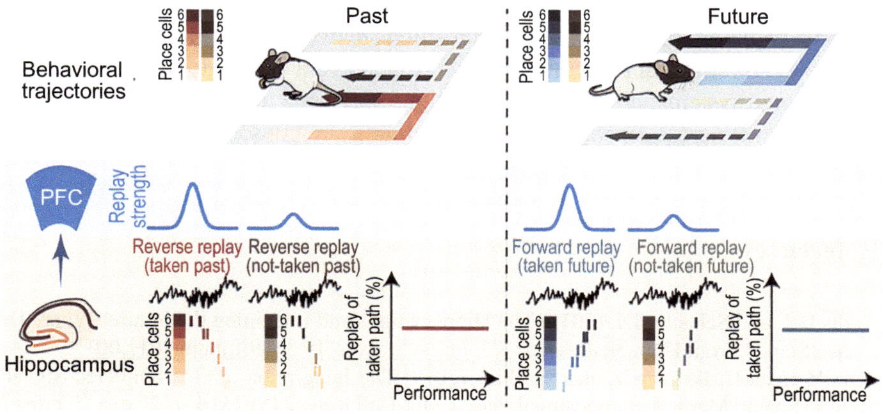

Figure 10.2 Coordinated replay in hippocampal-prefrontal networks for retrospective evaluation and prospective planning. Left: Reverse replay of hippocampal place-cell sequences at goal or reward locations reactivates paths taken to goals. Coordinated hippocampal-PFC replay is stronger for paths taken in the past, which led to reward. Right: Forward replay of hippocampal place-cell sequences reactivates paths toward goals. Coordinated hippocampal-PFC replay is stronger for future paths to take toward reaching rewards. Thus, hippocampal replay can represent internal evaluation of route options in the spatial cognitive map, utilized by downstream PFC networks for learning and planning. Figure adapted from Shin et al. (2019) and used with permission from Elsevier.

5 Conclusions

Extensive research has pinpointed the involvement of brain systems—in particular, the hippocampus, EC, and PFC—in navigational planning. However, to understand the mechanisms that underlie the planning and execution of goal-directed navigation, we need to adopt a more holistic approach—one that integrates evidence about spatial, goal, and sequential representations in these regions at different timescales. Initially, navigational models considered only the hippocampal place-cell cognitive map and goal representations in either hippocampal or cortical networks. With the discovery of grid cells, these models have been updated to incorporate both place-cell and grid-cell cognitive maps in a topological or vector-based code for goal-directed navigation. While the PFC has been acknowledged to play a critical role in navigation—due to its canonical functions as an executive control center, with a role in goal coding, planning, decision making, and memory-guided behavior—bottom-up and top-down interactions between the hippocampus and PFC at different timescales are just beginning to be understood. These interactions, however, may play a vital role in every aspect of navigation, including goal representations, goal selection, initial planning of a route to goals, progress monitoring toward a goal, online maintenance of route plan, and the ability to adjust (e.g., rerouting, inferring novel paths, learning optimal routes).

It is vital that we consider multiple timescales of operations for computing in these networks. Imagination and planning in the nonspatial domain in humans comprise both deliberative planning with conscious effort as well as internal computations leading to inference (bridging across experienced episodes) and insight, an important contributor to creativity and flexibility. Fast timescale sequences that occur during "offline" states, which are distinct from engaged or exploration states, are likely to contribute to this process and need to be considered in models of navigational planning. Investigating the interactions between the hippocampus and PFC is a good starting point for this inquiry. To achieve a complete understanding of navigational systems, however, investigations may need to include broader distributed networks across sensory, cognitive, and motor regions.

References

Addis DR, and Schacter DL (2011) The Hippocampus and Imagining the Future: Where Do We Stand? Front Hum Neurosci 5:173. https://doi.org/10.3389/fnhum.2011.00173

Aoki Y, Igata H, Ikegaya Y, and Sasaki T (2019) The Integration of Goal-Directed Signals onto Spatial Maps of Hippocampal Place Cells. Cell Rep 27 (5):1516–1527 e1515. https://doi.org/10.1016/j.celrep.2019.04.002

Basu R, and Ito HT (2023) A Goal Pointer for a Cognitive Map in the Orbitofrontal Cortex. Curr Opin Neurobiol 83:102803. https://doi.org/10.1016/j.conb.2023.102803

Boccara CN, Nardin M, Stella F, O'Neill J, and Csicsvari J (2019) The Entorh inal Cognitive Map Is Attracted to Goals. Science 363 (6434):1443–1447. https://doi.org/10.1126/science.aav4837

Buckner RL (2010) The Role of the Hippocampus in Prediction and Imagination. Annu Rev Psychol 61:27–48, C21–28. https://doi.org/10.1146/annurev.psych.60.110707.163508

Buckner RL, Andrews-Hanna JR, and Schacter DL (2008) The Brain's Default Network: Anatomy, Function, and Relevance to Disease. Ann N Y Acad Sci 1124:1–38. https://doi.org/10.1196/annals.1440.011
Burgess N, Donnett JG, Jeffery KJ, and O–keefe J (1997) Robotic and Neuronal Simulation of the Hippocampus and Rat Navigation. Philos Trans R Soc Lond B Biol Sci 352 (1360):1535–1543. https://doi.org/10.1098/rstb.1997.0140
Butler WN, Hardcastle K, and Giocomo LM (2019) Remembered Reward Locations Restructure Entorhinal Spatial Maps. Science 363 (6434):1447–1452. https://doi.org/10.1126/science.aav5297
Buzsaki G (2015) Hippocampal Sharp Wave-Ripple: A Cognitive Biomarker for Episodic Memory and Planning. Hippocampus 25 (10):1073–1188. https://doi.org/10.1002/hipo.22488
Buzsaki G, and Moser EI (2013) Memory, Navigation and Theta Rhythm in the Hippocampal-Entorhinal System. Nat Neurosci 16 (2):130–138. https://doi.org/10.1038/nn.3304
Buzsaki G, and Tingley D (2018) Space and Time: The Hippocampus as a Sequence Generator. Trends Cogn Sci 22 (10):853–869. https://doi.org/10.1016/j.tics.2018.07.006
Cohen NJ, and Eichenbaum H (1993) Memory, Amnesia, and the Hippocampal System. MIT Press, Cambridge, MA
Colgin LL (2013) Mechanisms and Functions of Theta Rhythms. Annu Rev Neurosci 36:295–312. https://doi.org/10.1146/annurev-neuro-062012-170330
Duvelle E, Grieves RM, Hok V, et al. (2019) Insensitivity of Place Cells to the Value of Spatial Goals in a Two-Choice Flexible Navigation Task. J Neurosci 39 (13):2522–2541. https://doi.org/10.1523/jneurosci.1578-18.2018
Eichenbaum H (2017) Prefrontal-Hippocampal Interactions in Episodic Memory. Nat Rev Neurosci 18 (9):547–558. https://doi.org/10.1038/nrn.2017.74
Epstein RA, Patai EZ, Julian JB, and Spiers HJ (2017) The Cognitive Map in Humans: Spatial Navigation and Beyond. Nat Neurosci 20 (11):1504–1513. https://doi.org/10.1038/nn.4656
Erdem UM, Milford MJ, and Hasselmo ME (2015) A Hierarchical Model of Goal Directed Navigation Selects Trajectories in a Visual Environment. Neurobiol Learn Mem 117:109–121. https://doi.org/10.1016/j.nlm.2014.07.003
Esteves IM, Chang H, Neumann AR, et al. (2021) Spatial Information Encoding across Multiple Neocortical Regions Depends on an Intact Hippocampus. J Neurosci 41 (2):307–319. https://doi.org/10.1523/jneurosci.1788-20.2020
Foster DJ, Morris RG, and Dayan P (2000) A Model of Hippocampally Dependent Navigation, Using the Temporal Difference Learning Rule. Hippocampus 10 (1):1–16. https://doi.org/10.1002/(sici)1098-1063(2000)10:1%3C1::aid-hipo1%3E3.0.co;2-1
Gauthier JL, and Tank DW (2018) A Dedicated Population for Reward Coding in the Hippocampus. Neuron 99 (1):179–193 e177. https://doi.org/10.1016/j.neuron.2018.06.008
Gillespie AK, Astudillo Maya DA, Denovellis EL, et al. (2021) Hippocampal Replay Reflects Specific Past Experiences Rather Than a Plan for Subsequent Choice. Neuron 109 (19):3149–3163.e3146. https://doi.org/10.1016/j.neuron.2021.07.029
Gourley SL, and Taylor JR (2016) Going and Stopping: Dichotomies in Behavioral Control by the Prefrontal Cortex. Nat Neurosci 19 (6):656–664. https://doi.org/10.1038/nn.4275
Hafting T, Fyhn M, Molden S, Moser M-B, and Moser EI (2005) Microstructure of a Spatial Map in the Entorhinal Cortex. Nature 436 (7052):801–806. https://doi.org/10.1038/nature03721
Hassabis D, Kumaran D, and Maguire EA (2007) Using Imagination to Understand the Neural Basis of Episodic Memory. J Neurosci 27 (52):14365–14374. https://doi.org/10.1523/jneurosci.4549-07.2007
Hasz BM, and Redish AD (2020) Spatial Encoding in Dorsomedial Prefrontal Cortex and Hippocampus Is Related during Deliberation. Hippocampus 30 (11):1194–1208. https://doi.org/10.1002/hipo.23250

Hillman KL, and Bilkey DK (2012) Neural Encoding of Competitive Effort in the Anterior Cingulate Cortex. Nat Neurosci 15 (9):1290–1297. https://doi.org/10.1038/nn.3187

Hok V, Save E, Lenck-Santini PP, and Poucet B (2005) Coding for Spatial Goals in the Prelimbic/Infralimbic Area of the Rat Frontal Cortex. PNAS 102 (12):4602–4607. https://doi.org/10.1073/pnas.0407332102

Howard LR, Javadi AH, Yu Y, et al. (2014) The Hippocampus and Entorhinal Cortex Encode the Path and Euclidean Distances to Goals during Navigation. Curr Biol 24 (12):1331–1340. https://doi.org/10.1016/j.cub.2014.05.001

Ito HT, Zhang SJ, Witter MP, Moser EI, and Moser MB (2015) A Prefrontal-Thalamo-Hippocampal Circuit for Goal-Directed Spatial Navigation. Nature 522 (7554):50–55. https://doi.org/10.1038/nature14396

Jadhav SP, Rothschild G, Roumis DK, and Frank LM (2016) Coordinated Excitation and Inhibition of Prefrontal Ensembles during Awake Hippocampal Sharp-Wave Ripple Events. Neuron 90 (1):113–127. https://doi.org/10.1016/j.neuron.2016.02.010

Johnson A, and Redish AD (2007) Neural Ensembles in CA3 Transiently Encode Paths Forward of the Animal at a Decision Point. J Neurosci 27 (45):12176–12189. https://doi.org/10.1523/jneurosci.3761-07.2007

Karlsson MP, Tervo DG, and Karpova AY (2012) Network Resets in Medial Prefrontal Cortex Mark the Onset of Behavioral Uncertainty. Science 338 (6103):135–139. https://doi.org/10.1126/science.1226518

Kay K, Chung JE, Sosa M, et al. (2020) Constant Sub-Second Cycling between Representations of Possible Futures in the Hippocampus. Cell 180 (3):552–567 e525. https://doi.org/10.1016/j.cell.2020.01.014

Kubie JL, and Fenton AA (2012) Linear Look-Ahead in Conjunctive Cells: an Entorhinal Mechanism for Vector-Based Navigation. Front Neural Circuits 6:20. https://doi.org/10.3389/fncir.2012.00020

Liu Y, Dolan RJ, Kurth-Nelson Z, and Behrens TEJ (2019) Human Replay Spontaneously Reorganizes Experience. Cell 178 (3):640–652.e614. https://doi.org/10.1016/j.cell.2019.06.012

Maguire EA, Burgess N, Donnett JG, et al. (1998) Knowing Where and Getting There: A Human Navigation Network. Science 280 (5365):921–924. https://doi.org/10.1126/science.280.5365.921

Mattar MG, and Daw ND (2018) Prioritized Memory Access Explains Planning and Hippocampal Replay. Nat Neurosci 21 (11):1609–1617. https://doi.org/10.1038/s41593-018-0232-z

Miller EK, and Cohen JD (2001) An Integrative Theory of Prefrontal Cortex Function. Annu Rev Neurosci 24:167–202. https://doi.org/10.1146/annurev.neuro.24.1.167

Moorman DE, and Aston-Jones G (2023) Prelimbic and Infralimbic Medial Prefrontal Cortex Neuron Activity Signals Cocaine Seeking Variables across Multiple Timescales. Psychopharmacology (Berl) 240 (3):575–594. https://doi.org/10.1007/s00213-022-06287-2

Moser EI, Moser M-B, and McNaughton BL (2017) Spatial Representation in the Hippocampal Formation: A History. Nat Neurosci 20 (11):1448–1464. https://doi.org/10.1038/nn.4653

Mukherjee A, and Caroni P (2018) Infralimbic Cortex Is Required for Learning Alternatives to Prelimbic Promoted Associations through Reciprocal Connectivity. Nat Commun 9 (1):2727. https://doi.org/10.1038/s41467-018-05318-x

Nyberg N, Duvelle E, Barry C, and Spiers HJ (2022) Spatial Goal Coding in the Hippocampal Formation. Neuron 110 (3):394–422. https://doi.org/10.1016/j.neuron.2021.12.012

O'Keefe J, and Nadel L (1978) The Hippocampus as a Cognitive Map. Oxford Univ Press, London

O'Neill J, Boccara C, Stella F, Schoenenberger P, and Csicsvari J (2017) Superficial Layers of the Medial Entorhinal Cortex Replay Independently of the Hippocampus. Science 355 (6321):184–188. https://doi.org/10.1126/science.aag2787

Ólafsdóttir HF, Barry C, Saleem AB, Hassabis D, and Spiers HJ (2015) Hippocampal Place Cells Construct Reward Related Sequences through Unexplored Space. eLife 4:e06063. https://doi.org/10.7554/elife.06063

Patai EZ, and Spiers HJ (2021) The Versatile Wayfinder: Prefrontal Contributions to Spatial Navigation. Trends Cogn Sci 25 (6):520–533. https://doi.org/10.1016/j.tics.2021.02.010

Peters J, Kalivas PW, and Quirk GJ (2009) Extinction Circuits for Fear and Addiction Overlap in Prefrontal Cortex. Learn Mem 16 (5):279–288. https://doi.org/10.1101/lm.1041309

Pfeiffer BE, and Foster DJ (2013) Hippocampal Place-Cell Sequences Depict Future Paths to Remembered Goals. Nature 497 (7447):74–79. https://doi.org/10.1038/nature12112

Preston AR, and Eichenbaum H (2013) Interplay of Hippocampus and Prefrontal Cortex in Memory. Curr Biol 23 (17):R764–773. https://doi.org/10.1016/j.cub.2013.05.041

Ratcliff R, Gomez P, and McKoon G (2004) A Diffusion Model Account of the Lexical Decision Task. Psychol Rev 111 (1):159–182. https://doi.org/10.1037/0033-295x.111.1.159

Redish AD (2016) Vicarious Trial and Error. Nat Rev Neurosci 17 (3):147–159. https://doi.org/10.1038/nrn.2015.30

Samborska V, Butler JL, Walton ME, Behrens TEJ, and Akam T (2022) Complementary Task Representations in Hippocampus and Prefrontal Cortex for Generalizing Structure of Problems. Nat Neurosci 25 (10):1314–1326. https://doi.org/10.1038/s41593-022-01149-8

Sarel A, Finkelstein A, Las L, and Ulanovsky N (2017) Vectorial Representation of Spatial Goals in the Hippocampus of Bats. Science 355 (6321):176–180. https://doi.org/10.1126/science.aak9589

Shadlen MN, and Kiani R (2013) Decision Making as a Window on Cognition. Neuron 80 (3):791–806. https://doi.org/10.1016/j.neuron.2013.10.047

Shin JD, and Jadhav SP (2016) Multiple Modes of Hippocampal-Prefrontal Interactions in Memory-Guided Behavior. Curr Opin Neurobiol 40:161–169. https://doi.org/10.1016/j.conb.2016.07.015

Shin JD, Tang W, and Jadhav SP (2019) Dynamics of Awake Hippocampal-Prefrontal Replay for Spatial Learning and Memory-Guided Decision Making. Neuron 104 (6):1110–1125 e1117. https://doi.org/10.1016/j.neuron.2019.09.012

Sosa M, and Giocomo LM (2021) Navigating for Reward. Nat Rev Neurosci 22 (8):472–487. https://doi.org/10.1038/s41583-021-00479-z

Stachenfeld KL, Botvinick MM, and Gershman SJ (2017) The Hippocampus as a Predictive Map. Nat Neurosci 20 (11):1643–1653. https://doi.org/10.1038/nn.4650

Tang W, and Jadhav SP (2019) Sharp-Wave Ripples as a Signature of Hippocampal-Prefrontal Reactivation for Memory during Sleep and Waking States. Neurobiol Learn Mem 160:11–20. https://doi.org/10.1016/j.nlm.2018.01.002

——— (2022) Multiple-Timescale Representations of Space: Linking Memory to Navigation. Annu Rev Neurosci 45 (1):1–21. https://doi.org/10.1146/annurev-neuro-111020-084824

Tang W, Shin JD, and Jadhav SP (2021) Multiple Time-Scales of Decision-Making in the Hippocampus and Prefrontal Cortex. eLife 10:e66227. https://doi.org/10.7554/elife.66227

——— (2023) Geometric Transformation of Cognitive Maps for Generalization across Hippocampal-Prefrontal Circuits. Cell Rep 42 (3):112246. https://doi.org/10.1016/j.celrep.2023.112246

Tolman EC (1948) Cognitive Maps in Rats and Men. Psychol Rev 55 (4):189–208. https://doi.org/10.1037/h0061626

van Heukelum S, Mars RB, Guthrie M, et al. (2020) Where Is Cingulate Cortex? A Cross-Species View. Trends Neurosci 43 (5):285–299. https://doi.org/10.1016/j.tins.2020.03.007

Vaz AP, Wittig JH, Jr., Inati SK, and Zaghloul KA (2020) Replay of Cortical Spiking Sequences during Human Memory Retrieval. Science 367 (6482):1131–1134. https://doi.org/10.1126/science.aba0672

Wolcott MD, and Lobczowski NG (2021) Using Cognitive Interviews and Think-Aloud Protocols to Understand Thought Processes. Curr Pharm Teach Learn 13 (2):181–188. https://doi.org/10.1016/j.cptl.2020.09.005

Wood ER, Dudchenko PA, Robitsek RJ, and Eichenbaum H (2000) Hippocampal Neurons Encode Information about Different Types of Memory Episodes Occurring in the Same Location. Neuron 27 (3):623–633. https://doi.org/10.1016/s0896-6273(00)00071-4

Wu C-T, Haggerty D, Kemere C, and Ji D (2017) Hippocampal Awake Replay in Fear Memory Retrieval. Nat Neurosci 20 (4):571–580. https://doi.org/10.1038/nn.4507

Xu H, Baracskay P, O'Neill J, and Csicsvari J (2019) Assembly Responses of Hippocampal CA1 Place Cells Predict Learned Behavior in Goal-Directed Spatial Tasks on the Radial Eight-Arm Maze. Neuron 101 (1):119–132 e114. https://doi.org/10.1016/j.neuron.2018.11.015

Zheng C, Hwaun E, Loza CA, and Colgin LL (2021) Hippocampal Place Cell Sequences Differ during Correct and Error Trials in a Spatial Memory Task. Nat Commun 12 (1):3373. https://doi.org/10.1038/s41467-021-23765-x

Zielinski MC, Shin JD, and Jadhav SP (2019) Coherent Coding of Spatial Position Mediated by Theta Oscillations in the Hippocampus and Prefrontal Cortex. J Neurosci 39 (23):4550–4565. https://doi.org/10.1523/jneurosci.0106-19.2019

11

The Role of Motivation and Emotion in the Hippocampal Navigation System

Thomas Forro and Stéphane Ciocchi

Abstract Navigation in humans and nonhuman animals is reliant on critical interactions between motivational-emotional systems and the hippocampus. Whereas emotions guide the navigational process to help an organism achieve the most efficient and least dangerous pathway to a reward, the hippocampus integrates sensory information and creates cognitive maps of the environment. To enable this interaction, the hippocampus is integrated into circuits that process emotions and motivation, but the precise mechanisms by which this occurs are poorly understood. One view posits that emotionally enhanced memories are better consolidated, which makes places and events associated with strong emotions easier to recall and more enduring in memory. This chapter considers how emotions and goals are intertwined within the hippocampus, and their influence on spatial maps and memories. It explores the anatomy of hippocampal circuits and their integration with cognitive, emotional, and motivational circuits, as well as hippocampal spatial representations and interactions with cognitive, emotional, and motivational processes. It examines how emotional and motivational conflict impacts the hippocampus and can disrupt hippocampal circuits, leading to emotional disorders. Throughout, future research trajectories are highlighted to fill existing knowledge gaps.

Keywords Hippocampus, cognitive maps, spatial navigation, motivation, emotion, neural computations

Stéphane Ciocchi (✉)
Laboratory of Systems Neuroscience, Dept. of Physiology, University of Bern, CH-3012 Bern, Switzerland
Email: stephane.ciocchi@unibe.ch

Thomas Forro
Laboratory of Systems Neuroscience, Dept. of Physiology, University of Bern, CH-3012 Bern, Switzerland

N. S. Newcombe and K. Cheng (eds.), *Challenges in Navigation Research*, Strüngmann Forum Reports, https://doi.org/10.1007/978-3-032-20563-6_11

1 Introduction

The interaction between motivational-emotional systems and the hippocampal navigational system is a critical aspect of understanding how humans and animals orient themselves in space while responding to emotional and goal-driven needs. The hippocampus, well-known for its role in spatial memory and navigation, integrates sensory information to create cognitive maps of the environment. Simultaneously, motivational and emotional systems, which primarily involve the amygdala and ventral striatum, drive behavior based on goals, rewards, and emotional states. These systems converge to influence navigation by modulating attention, memory encoding, and decision-making processes. For instance, emotional states like anxiety or fear can alter hippocampal function and impact pathfinding, risk assessment, and spatial learning. Conversely, goal-oriented behaviors, motivated by reward-related signals from the dopaminergic system, guide navigation toward desired outcomes. Thus, the interaction between motivational and emotional states and the hippocampal navigational system shapes how organisms navigate their environments, linking external spatial cues with internal drives and emotional responses. To explore this interplay, we examine four key topics: (a) the anatomy of hippocampal circuits and their integration into cognitive and emotional and motivational circuits, (b) hippocampal spatial representations and their interactions with cognitive and emotional and motivational processes, (c) the hippocampus during emotional and motivational conflict, and (d) the disruption of hippocampal circuits leading to emotional disorders.

2 Anatomy of Hippocampal Circuits and Their Integration into Cognitive, Emotional, and Motivational Circuits

The hippocampus is one of the most extensively studied brain regions in neuroscience, with substantial evidence demonstrating its critical role in cognition, as shown by decades of research in both humans and animals (Knierim 2015). While it is best known for its involvement in learning, memory, and spatial navigation, the exact nature of the information processed by the hippocampus, and how this processing occurs, is still a matter of ongoing debate and continuous refinement (Li et al. 2024; Rolls and Treves 2024). Beyond these well-known cognitive functions, the hippocampus has also been found to be involved in the representation and modulation of emotional and motivational processes (Bannerman et al. 2003; Fanselow and Dong 2010; Gray and McNaughton 2003; Tracy et al. 2001). We are only beginning to unravel how these various cognitive, emotional, and motivational processes interact within the hippocampus. A significant part of our evolving understanding comes from anatomical studies of the hippocampus, which have led to new hypotheses regarding its functional roles. The diverse inputs to distinct regions of the hippocampus determine which types of information can be processed, while the hippocampal circuit architecture dictates the computational possibilities for how this information is transformed. Thus, the hippocampus is not an isolated functional unit; it is embedded within a larger network of cognitive systems and is connected to multiple brain regions, thus highlighting its integrative role.

Among cortical and subcortical brain areas, the hippocampus is notable for its unique cytoarchitectural organization, which has evolved and has been largely conserved across different mammalian lineages, including marsupials, placentals, and humans. This organization is characterized by dense neuronal layers of archicortex that fold into each other through invaginations and rotations, originating from the dorsomedial telencephalon. Moreover, the hippocampus acquires its adultlike structure by mid-gestation in both humans and rodents (Kier et al. 1997). These foldings give rise to distinct subfields, including the cornu ammonis (CA) areas CA1, CA2, CA3, and CA4, the dentate gyrus, and the subiculum. Together with adjacent parahippocampal regions (e.g., the perirhinal and entorhinal cortices), these areas form what is often referred to as the *hippocampal formation*. Within the hippocampus is a canonical feedforward circuit known as the *trisynaptic circuit*. In this circuit, projections from the entorhinal cortex travel via the perforant pathway to granule cells in the dentate gyrus, which then project via mossy fibers to CA3 neurons. CA3 neurons, in turn, project through the Schaffer collaterals to the CA1 region, which sends feedback to the entorhinal cortex. This connectivity has been instrumental in forming hypotheses about the hippocampus's role in memory storage and recall (Rolls 1989; Rolls and Treves 1994) as well as sequential memory representation (Buzsaki and Tingley 2018; Rueckemann et al. 2021). The heavy development of this circuit in young children during toddlerhood (ca. 1–3 yrs of age) coincides with the emergence of sequence-learning abilities (Lavenex and Lavenex 2013). Higher-order sensory information from the entorhinal cortex is highly divergent, with about a 1:4 ratio of inputs to granule cells in the dentate gyrus, which supports the process of pattern separation in this region. The dentate gyrus is also one of the few regions in the adult mammalian brain where neurogenesis occurs, further emphasizing its importance in learning processes (Kempermann et al. 2015). In CA3, pyramidal neurons not only project to CA1 but also exhibit dense recurrent collateral projections within CA3 itself, leading to the hypothesis that pattern completion takes place in this region. In CA1, it has been proposed (Vinogradova 2001) that recalled information from CA3 is compared with current sensory input, as CA1 also receives direct projections from the entorhinal cortex, facilitating the integration of past and present information. A key insight is that information flow through the canonical trisynaptic circuit is not homogenous across the subfields of the hippocampus, as there are significant differences in inputs along its dorsoventral axis.

The hippocampus forms a C-shaped structure along its long axis, also known as the *dorsoventral* or *septotemporal axis*, with a dorsal (posterior in humans) and ventral (anterior in humans) portion. Anatomical studies have identified topographically distinct connections along this axis, with the medial entorhinal cortex (MEC) preferentially innervating the dorsal hippocampus and the lateral entorhinal cortex (LEC) innervating the ventral hippocampus (van Groen et al. 2003). Importantly, the MEC has been associated with spatial information processing, whereas the LEC is more connected to nonspatial and sensory processing, a distinction that has been largely confirmed in humans as well (Burwell 2000; Deshmukh and Knierim 2011; Schröder et al. 2015). Moreover, parahippocampal areas are extensively connected to a wide range of brain regions, including the prefrontal cortex, cingulate cortex, insula, septal nuclei, thalamus, nucleus accumbens, and amygdala, suggesting that they process and relay diverse types of information to the hippocampus. Although inputs from the entorhinal cortex and intrahippocampal projections account for the majority of inputs to the hippocampus, recent studies have begun to highlight the functional significance of direct inputs from extrahippocampal regions, such as from the amygdala and dopaminergic and noradrenergic projections from the ventral tegmental area (VTA) and locus coeruleus (LC). Anatomical analyses of inputs to CA1 pyramidal neurons suggest that the differentiation of these direct inputs along the dorsoventral axis may contribute to the functional specialization of the dorsal and ventral hippocampus (Tao et al. 2021). Dorsal CA1 pyramidal neurons primarily receive inputs from the ipsilateral and contralateral CA3 and CA2 areas, whereas ventral CA1 neurons receive substancially more inputs from the olfactory areas and the amygdala. Understanding the precise inputs to the hippocampus and linking them to functional data will be essential for comprehending the roles of the hippocampal subfields, as well as the distinctions between dorsal and ventral domains.

Rather than viewing the hippocampus as an isolated unit responsible for creating and transmitting finalized memory representations, it may be more appropriate to consider it a conserved processing structure that transforms information in close interaction with other brain areas. This view is supported by evidence which shows that hippocampal output projections are highly diverse and differentially organized along the dorsoventral axis (Cenquizca and Swanson 2007). Tracing studies reveal that the hippocampus not only receives input from motivational and emotional systems but also sends direct projections to regions such as the amygdala, nucleus accumbens, lateral septum, and prefrontal cortex. For instance, direct projections from the hippocampus to the amygdala originate predominantly from the ventral CA1, whereas both dorsal and ventral CA1 project to the nucleus accumbens and medial prefrontal cortex. Projections from CA1 to the entorhinal cortex also exhibit dorsoventral differences, with dorsal and ventral regions targeting distinct areas and neurons within the MEC (Ohara et al. 2023). Similarly, while both dorsal and ventral projections innervate the lateral septum, they target different subregions within this nucleus. Additional projections from CA1 to regions such as the cingulate cortex and anterior insular cortex have been described, although their functional significance remains to be fully explored.

Many aspects of the dorsoventral division appear conserved across species (Dalton et al. 2022; Dalton et al. 2019), however. It is important to note that differences have also been observed (Bergmann et al. 2016; Huang et al. 2021). For example, in rodents, the hippocampus constitutes a relatively large portion of the brain (around 5%), whereas in humans, where other cortical areas have expanded significantly, the hippocampus accounts for only about 0.3% of total brain volume. Despite these relative size differences, the hippocampus remains functionally essential in humans, as evidenced by lesion studies that show impairments in learning, memory, and spatial navigation (Burgess et al. 2002; Scoville and Milner 1957; Squire 1992).

Hippocampal circuitry is highly distinctive and is perhaps the most extensively studied circuit in the brain. Nonetheless, many gaps persist in our understanding of the connectivity between various cell types and how they connect to other brain regions, particularly regarding potential differences along the dorsoventral axis. The influence of other key memory-related brain areas—such as the nucleus reuniens of the thalamus or the retrosplenial cortex, which modulate hippocampal encoding and communication (Mao et al. 2018; Totty et al. 2023)—deserves greater attention, particularly in terms of their impact on emotional and motivational representations across the hippocampus.

A crucial question that remains is the level of structural detail required to advance our understanding. Can we uncover fundamental principles of circuit mechanisms in one part of the hippocampus and apply them to other regions? To what extent does the brain use the same tools to process different types of information? How much of our detailed knowledge of the rodent hippocampus can be applied to the human hippocampus, and to what degree should we focus on basic principles? While fundamental connectivity patterns, such as inputs from the entorhinal cortex, appear to be similar in humans, projections from the human hippocampus need to be examined in much greater detail. For example, it has been found that direct projections from the human hippocampus are relatively sparse, raising questions about how communication with decision-making areas, such as the prefrontal cortex, might differ in humans—possibly relying more on indirect pathways—and what functional implications this may have (Dalton et al. 2022).

3 Hippocampal Spatial Representations and Their Interactions with Cognitive, Emotional, and Motivational Processes

Research on the rodent hippocampus has primarily focused on the remarkable finding that neurons in the hippocampus and entorhinal cortex are spatially tuned, leading to the hypothesis that the hippocampus encodes a cognitive map necessary for spatial navigation (Moser et al. 2008; O'Keefe and Nadel 1978). Specifically, hippocampal place cells—pyramidal neurons that fire when an animal is in a specific location—and grid cells in the entorhinal cortex—which fire across a triangular grid of locations—may provide the spatial code required to create this cognitive map. This spatial coding system operates independently of the sensory modality an animal uses to navigate a familiar environment, and different environments are encoded by distinct neural populations, a phenomenon known as *remapping*. Place cells have also been discovered in the human hippocampus, and damage to this region is linked to impaired spatial navigation (Ekstrom et al. 2003). While it is clear that the hippocampal formation is essential for spatial navigation, there is ongoing debate about whether this is its primary function. Some argue that all memory may derive from spatial encoding (e.g., O'Keefe and Krupic 2021), whereas others suggest that the hippocampus encodes memory more broadly, with spatial memory being just one form of it (e.g., Lisman et al. 2017). In contrast to animal studies, human research has historically focused on the hippocampus's role in memory, beginning with the case of Henry Molaison (patient HM), who lost the ability to form new episodic memories after his hippocampus was removed to treat epilepsy in 1953. The link between the hippocampus and memory was further supported by extensive clinical data (Scoville and Milner 1957; Squire 1992), but the question remains: Are memory processes and spatial encoding two sides of the same hippocampal function as Eichenbaum (2017) suggests, or are they distinct cognitive roles performed by the hippocampus? In practice, it is difficult to fully separate spatial processing from memory during natural behavior, although laboratory experiments have attempted to quantify how much nonspatial information is encoded by the hippocampus. Tasks conducted in the same spatial environment, but associated with different events or sequences of nonspatial sensory inputs (e.g., odors), have demonstrated that these inputs can be sequentially encoded in a manner similar to place-cell sequences via time cells or memory event-related cells (MacDonald et al. 2011; Schiller et al. 2015). This suggests that sequence encoding may be a fundamental property of the hippocampus, potentially uniting the concepts of space and memory.

The hippocampus integrates many nonspatial inputs. Experiments have shown that these inputs can be encoded similarly to spatial information. Furthermore, nonspatial sensory inputs can be integrated into spatial maps, modulate spatial activity levels, or even trigger remapping of place encoding (Eichenbaum 2018). Multiple studies have demonstrated that the hippocampus represents a variety of sensory inputs, including odors, tactile sensations, and auditory stimuli, but this representation is modulated by the salience and behavioral relevance of the stimuli. For instance, a novel tone initially elicits hippocampal responses regardless of its behavioral significance, but these responses quickly diminish if the tone lacks further relevance (Billig et al. 2022). This is logical for animals navigating their environments and for humans navigating their lives within society, where choices are purposeful rather than random. Natural behavior is goal driven, aimed at objectives like food, survival, or reproduction. Achieving these goals activates the brain's reward system and produces positive emotions, whereas failure elicits negative emotions. Emotions, in turn, guide decision making and motivate goal-directed behavior. It seems unlikely that the hippocampus could have evolved purely as a memory system or solely for spatial encoding, as these cognitive processes must interact closely with the brain's reward and emotion systems. Indeed, anatomical studies have shown that the hippocampus not only receives spatial and sensory information but also integrates input related to emotion (e.g., from the amygdala) and reward (e.g., from the VTA and LC) and sends feedback to these systems as well.

How are these different inputs represented in the neural code along the dorsoventral axis of the hippocampus? A comparison of the spatial codes along the dorsoventral axis of the hippocampus provides initial insights. Recordings from the dorsal hippocampus reveal small, precise place fields, whereas the ventral hippocampus shows larger, more diffuse place fields (Kjelstrup et al. 2008; Royer et al. 2010), reflecting a functional subdivision: the dorsal hippocampus encodes a fine spatial map, whereas the ventral hippocampus integrates a broader range of behaviorally relevant information to form a contextual representation of the environment. As a whole, the hippocampus encodes spatial location (predominantly dorsal) along with learned associations of sensory (e.g., odors), internally generated, and emotional as well as motivational inputs (primarily ventral). Fear, however, may induce remapping of place-cell activity (Moita et al. 2004), suggesting that the hippocampus differentiates between motivational contexts. While the entire hippocampus participates in contextual encoding, as demonstrated by connectivity and lesion studies, representations of internal states become more pronounced toward the ventral hippocampus and dominate contextual representations in situations involving anxiety (Ciocchi et al. 2015; Jimenez et al. 2018). For example, when rodents explore an elevated plus maze and enter its high, open arms, innate anxiety is triggered, and a significant proportion of neurons are activated by the anxiogenic context rather than by spatial factors alone. The population of neurons activated by anxiety includes not only pyramidal neurons but also a large number of inhibitory interneurons, such as parvalbumin interneurons, which may play a role in shaping anxiety-related experiences and memories (Forro et al. 2022; Li et al. 2024; Volitaki et al. 2024). One of the main sources of anxiogenic input to the hippocampus comes from the basolateral amygdala, which has been shown to alter spatial firing and synchronization, thereby influencing anxiogenic behavior (Jackson et al. 2024; Kim et al. 2015; Kong et al. 2021). Although negative valence is strongly represented in the ventral hippocampus, reward representations are also critical. The introduction or discovery of a reward leads to remapping and an overrepresentation of the rewarded area within the environment.

Furthermore, specific hippocampal pyramidal neurons and population representations encode and even predict future rewards, with these signals varying along the dorsoventral axis (Gauthier and Tank 2018; Jarzebowski et al. 2022). Vasoactive intestinal peptide and a subtype of cholecystokinin interneurons have also been linked to reward processing and may contribute to the integration of reward signals (Forro and Klausberger 2023; Turi et al. 2019). For some time, the extent to which the hippocampus is influenced by dopaminergic inputs was debated. It has now been established, however, that dopamine and norepinephrine from the VTA and LC are vital for long-term potentiation, contextual learning, and memory retention (Tsetsenis et al. 2022). LC and VTA projections to the hippocampus have been shown to directly enhance the spatial representation of reward locations and signal reward expectations, revealing once again the close interaction between valence processing and spatial encoding (Kaufman et al. 2020; Krishnan et al. 2022). Rewards beyond food, such as social rewards, have also been strongly associated with the ventral hippocampus, where they induce remapping and are modulated by LC inputs to form social reward memories (Duarte et al. 2024). Studies of ventral hippocampal representations infused with either negative or positive valence have demonstrated that communicating this information to other brain areas is essential for motivational behavior. Projections to the lateral septum and the nucleus accumbens are particularly important for reward- and goal-related spatial memories (Décarie-Spain et al. 2022; Jung et al. 2024; Trouche et al. 2019).

Importantly, hippocampal information is routed in a "functional" manner, meaning anxiety-related information is preferentially transmitted to the prefrontal cortex and lateral hypothalamic area while goal-directed information is directed to the nucleus accumbens (Ciocchi et al. 2015; Jimenez et al. 2018). Future progress will depend on viewing hippocampal outputs through the lens of how target brain regions decode them (Buzsaki 2010). Computational analysis can help reveal different methods of encoding information by analyzing neural activity in specific areas. Encoding operations do not, however, necessarily reflect how the brain decodes information, making it crucial to determine which neurons inside and outside the hippocampus receive specific types of input.

The wide range of cognitive processes that affect spatial encoding and can be manipulated via optogenetics is striking, suggesting the hippocampus supports multiple functions beyond a single neural system. The focus on distinct functions stems partly from the way research is conducted—often in controlled lab environments, where animals and humans perform repetitive tasks, enabling researchers to isolate functional correlations in neural activity. Such a setup, however, does not reflect how the hippocampus operates in real-world contexts or under conditions in which it evolved. Recent studies using more complex tasks show that the hippocampus can flexibly adapt to and encode these complexities. For instance, when neurons are analyzed for both spatial and temporal information, both can be decoded from the same population (Chen et al. 2024). This suggests that the hippocampus may not specialize in one cognitive function but rather it may integrate various types of information within behaviorally relevant contexts (Schiller et al. 2015). The contextual representations seen in rodent studies could relate to human fMRI studies which show that hippocampal and parahippocampal regions are involved in processing visual scenes (Murray et al. 2018). While spatial maps in the hippocampus may drift and change over time, contextual representations persist long term (Keinath et al. 2022), potentially forming the foundation of the memory-related functions attributed to the hippocampus. Expanding the definition of "context" to include not only the sensory environment but also emotional valence allows such contextual representations to generate predictions and guide future goal-directed behavior in both animals and humans. Understanding how the hippocampus encodes and stores information during complex, naturalistic behaviors—where experiences are highly variable and full of confounding factors—will be challenging. This is, however, a key path to uncovering what the hippocampus is truly "meant to do" in real-world functioning, beyond what it "can do" under controlled lab conditions.

4 The Hippocampus During Emotional and Motivational Conflict

An animal's survival depends on its ability to find rewards in the form of food or shelter. Obtaining rewards is frequently associated with potential dangers, such as predators or harmful environments, and the decision to seek rewards in the face of danger can induce emotional and motivational conflicts. Emotional conflicts, in this context, can be described as the internal psychological struggles that arise when individuals are confronted with competing emotions or incompatible emotional responses. Resolution of these conflicts is a fundamental and evolutionarily conserved process within the animal kingdom, as it promotes the survival of living organisms through adaptive behavior. Behaviors and neuronal circuits have therefore coevolved by exploiting information from the environment to recognize, predict, and effectively resolve emotional and motivational conflicts.

Emotional and motivational representations generated in the ventral hippocampus are thought to influence other brain areas to guide behavior, particularly through connections with the prefrontal cortex, which is involved in decision-making processes (Piquet et al. 2024). Changes in network interactions, such as the coherence of local field potential theta oscillations between the ventral hippocampus and prefrontal cortex, have been observed during decision making in approach-avoidance conflicts, suggesting functional communication between these regions (Padilla-Coreano et al. 2019). This raises the question: Are hippocampal representations simply relayed to decision-making areas, or is the hippocampus itself directly involved in decision making, thus broadening its cognitive roles? The idea that the hippocampus may play a role in behavioral inhibition was among the earliest functional discoveries about the hippocampus in the 1960s, but this concept has largely been forgotten over time. As research into its role in emotional processing has grown, however, the hippocampus has been reexamined, particularly in the context of emotions experienced during conflict situations that require decision making. Studies of approach-avoidance conflicts in rodents suggest that the ventral hippocampus may resolve conflicts by either reducing the salience of positive cues or increasing the salience of negative ones (Ito and Lee 2016). These findings align with human studies on approach-avoidance conflicts, which show activation of the anterior hippocampus during threat, and highlight its role in conflict resolution and behavioral inhibition (Bach et al. 2014).

If the hippocampus is indeed involved in decision making, its activity should not only reflect past memories or familiar locations but also predict future choices. In rodent experiments using T- or Y-mazes, researchers have identified trajectory-dependent or "splitter" cells that activate differently depending on whether an animal will turn left or right at the maze's end, even though the animal runs through the same common path (Kinsky et al. 2020; Wood et al. 2000). These findings support the notion that the hippocampus integrates nonspatial cognitive information and may even be directly involved in decision making. Similarly, during spatial navigation, place cells do more than just represent current locations: they exhibit sequences of activation in time with network oscillations (a phenomenon known as phase precession), which may provide information about both past and future trajectories (Dragoi 2013; Sanders et al. 2015). In the ventral hippocampus, neural activity can predict how far an animal will explore an anxiogenic environment, with ramping activity patterns emerging in relation to anticipated rewards (Malagon-Vina et al. 2023). Thus, the hippocampus demonstrates several mechanisms by which predictions could be generated. This raises the crucial question of how these predictive processes—particularly during goal-directed behavior involving emotional conflict—operate along the dorsoventral axis and how they influence behavioral decisions.

Once again, the hippocampus's role in decision making highlights the need for future research to view it as part of a broader neural circuit. This circuit does not perform distinct cognitive functions in isolation but instead gives rise to these functions through extensive network interactions.

5 Disruption of Hippocampal Circuits Leads to Emotional Disorders

The recognition of the hippocampus's critical role in memory as well as in emotions and motivation has heightened focus on this structure concerning mental illness, linking structural and functional abnormalities of the hippocampus to schizophrenia, anxiety-related disorders, and mood disorders (Lieberman et al. 2018). Substantial evidence from *in vivo* imaging and postmortem analysis has associated hypermetabolism that drives hyperactivity in the CA1 area and subiculum of the anterior hippocampus with the development of schizophrenia. This hyperactivity results in decreased task engagement of the hippocampus, ultimately leading to volume reduction. It has been postulated that elevated levels of glutamate support this syndrome (Coyle 2006), although the exact cause of these increased glutamate levels remains unclear. A dysfunctional balance between inhibition and excitation, potentially involving impaired local GABAergic interneurons, may contribute to this condition.

Numerous genes and factors have been linked to schizophrenia, including alterations in the neuregulin to ErbB4 receptor signaling pathway, which is disrupted in many patients with schizophrenia (Norton et al. 2006). The ErbB4 receptor is predominantly expressed in parvalbumin-expressing interneurons, and reduced populations of these interneurons in the prefrontal cortex and ventral hippocampus have also been associated with schizophrenia. Moreover, conditional deletion of ErbB4 receptors in interneurons can replicate psychosis-like conditions in mice that have also been observed in neuroimaging studies of humans (Kiemes et al. 2023).

Furthermore, hyperexcitability of the hippocampus has been linked to anxiety (Ghasemi et al. 2022), and changes in hippocampal volume have been identified in various anxiety disorders (Shin and Liberzon 2010). Alterations in anterior hippocampal connectivity have been observed in veterans with post-traumatic stress disorder (Abdallah et al. 2017) as well as in patients suffering from depression (Hu et al. 2021).

The detailed mechanisms underlying these circuit changes, however, are less well understood, necessitating future investigations in animal models exhibiting abnormal anxiety or schizophrenia-like behavior. Growing evidence indicates that interneurons—and the network oscillations they regulate—play a key role in the neurobiological mechanisms underlying emotion-related disorders (Brambilla et al. 2003; Fee et al. 2017; Vivien et al. 2023), yet much more research and new methodologies are required to investigate the specific roles of distinct interneuron types. The utilization of innovative animal models, such as Neuroligin-2 (a protein at inhibitory synapses) knockout mice that exhibit disrupted anxiety-related behavior and altered network oscillations (Cruces-Solis et al. 2021), along with new molecules targeting ventral hippocampal interneurons, such as the psychedelic 2,5-dimethoxy-4-iodoamphetamine, could pave the way for new and more effective therapeutics (Tiwari et al. 2024).

6 Conclusions and Future Perspectives

Emotions can guide navigation toward rewarding and safe outcomes by modulating memory processes. In particular, emotionally salient experiences are preferentially consolidated, which strengthens hippocampal representations of environments and events. To facilitate this process, the hippocampus is integrated into circuits that process emotions and motivation, including the amygdala and nucleus accumbens. The mechanisms underlying this memory modulation, however, are still poorly understood. One hypothesis that has gained traction is that emotion-modulated memories are better consolidated, making places and events associated with strong emotions easier to recall and more enduring in our memories (Anderson et al. 2006; Sharot et al. 2004). Memory consolidation has been shown to occur during rest and sleep and is intricately linked to the occurrence of high-frequency oscillations in the hippocampus, known as sharp-wave ripple events. During these events, sequences of place cells or previously experienced events are replayed, which is believed to strengthen the connections among the neurons that belong to those assemblies (Zhou and Norimoto 2023). In studies involving humans, the frequency and coherence of ripples in the amygdala and hippocampus following emotional experiences have been correlated with the strength of memory formation (Qasim et al. 2023; Zhang et al. 2024). Dopaminergic and noradrenergic inputs may influence these memory formations (Kleinman and Foster 2024), but the translation of these inputs into enhanced ripple initiation and coordination among different brain areas, as well as the differences across the dorsoventral axis, has only recently begun to be explored (Sosa et al. 2020).

Research on the hippocampus in a broader range of species could potentially reveal general neural mechanisms underlying hippocampal functions. To what extent are the functions of the hippocampus influenced by the specific species being investigated? Different species utilize various sensory systems and navigate distinct social structures, each presenting unique challenges. For example, studies in bats underscore the importance of social dynamics and how these factors impact spatial coding (Forli and Yartsev 2023). In addition, it has been found that bats do not exhibit prominent theta oscillations; however, phase coding and synchrony of activity still play a significant role in neural coding, highlighting some common principles of hippocampal function (Eliav et al. 2018). It is important that these comparisons be extended to more distantly related species, such as birds, which have evolved a hippocampus independent of mammals and exhibit a different anatomical organization (Madison et al. 2024). Nonetheless, similar functions have been associated with the avian hippocampus, where spatial coding as well as emotional and motivational information and behaviors are represented and regulated. Furthermore, visuospatial information and valence-related information regarding anxiety and reward also show biases toward different parts of the avian hippocampus, similar to mammals. These findings further reinforce the notion that the hippocampus functions as an integrative structure, encoding memories while actively guiding behavior through interactions with emotions and motivation.

The insight that emotions and goals are intertwined within the hippocampus, influencing spatial maps and memories, represents a crucial finding that may not only assist in targeting mental disorders in innovative ways but also reveal fundamental principles of brain function. As societies increasingly implement artificial intelligence systems capable of learning and memory, it is essential that we determine how these systems assess valence and what goals or motivations should be incorporated. For further discussion, see Milford (Chapter 3).

References

Abdallah CG, Wrocklage KM, Averill CL, et al. (2017) Anterior Hippocampal Dysconnectivity in Posttraumatic Stress Disorder: A Dimensional and Multimodal Approach. Transl Psychiat 7:e1045. https://doi.org/10.1038/tp.2017.12

Anderson A, Wais P, and Gabrieli J (2006) Emotion Enhances Remembrance of Neutral Events Past. Neurosci 103 (5):1599–1604. https://doi.org/10.1073/pnas.0506308103

Bach DR, Guitart-Masip M, Packard PA, et al. (2014) Human Hippocampus Arbitrates Approach-Avoidance Conflict. Curr Biol 24 (12):1435. https://doi.org/10.1016/j.cub.2014.05.051

Bannerman DM, Grubb M, Deacon RM, et al. (2003) Ventral Hippocampal Lesions Affect Anxiety but Not Spatial Learning. Behav Brain Res 139 (1-2):197–213. https://doi.org/10.1016/s0166-4328(02)00268-1

Bergmann E, Zur G, Bershadsky G, and Kahn I (2016) The Organization of Mouse and Human Cortico-Hippocampal Networks Estimated by Intrinsic Functional Connectivity. Cereb Cortex 26 (12):4497–4512. https://doi.org/10.1093/cercor/bhw327

Billig AJ, Lad M, Sedley W, and Griffiths TD (2022) The Hearing Hippocampus. Prog Neurobiol 218:102326. https://doi.org/10.1016/j.pneurobio.2022.102326

Brambilla P, Perez J, Barale F, Schettini G, and Soares J (2003) GABAergic Dysfunction in Mood Disorders. Mol Psychiatr 8 (8):721–737. https://doi.org/10.1038/sj.mp.4001362

Burgess N, Maguire EA, and O'Keefe J (2002) The Human Hippocampus and Spatial and Episodic Memory. Neuron 35 (4):625–641. https://doi.org/10.1016/s0896-6273(02)00830-9

Burwell RD (2000) The Parahippocampal Region: Corticocortical Connectivity. Ann N Y Acad Sci 911:25–42. https://doi.org/10.1111/j.1749-6632.2000.tb06717.x

Buzsáki G (2010) Neural Syntax: Cell Assemblies, Synapsembles, and Readers. Neuron 68 (3):362–385. https://doi.org/10.1016/j.neuron.2010.09.023

Buzsaki G, and Tingley D (2018) Space and Time: The Hippocampus as a Sequence Generator. Trends Cogn Sci 22 (10):853–869. https://doi.org/10.1016/j.tics.2018.07.006

Cenquizca LA, and Swanson LW (2007) Spatial Organization of Direct Hippocampal Field CA1 Axonal Projections to the Rest of the Cerebral Cortex. Brain Res Rev 56 (1):1–26. https://doi.org/10.1016/j.brainresrev.2007.05.002

Chen S, Cheng N, Chen X, and Wang C (2024) Integration and Competition between Space and Time in the Hippocampus. Neuron 112 (21):3651–3664. https://doi.org/10.1016/j.neuron.2024.08.007

Ciocchi S, Passecker J, Malagon-Vina H, Mikus N, and Klausberger T (2015) Brain Computation: Selective Information Routing by Ventral Hippocampal CA1 Projection Neurons. Science 348 (6234):560–563. https://doi.org/10.1126/science.aaa3245

Coyle JT (2006) Glutamate and Schizophrenia: Beyond the Dopamine Hypothesis. Cell Mol Neurobiol 26 (4-6):365-384. https://doi.org/10.1007/s10571-006-9062-8

Cruces-Solis H, Babaev O, Ali H, et al. (2021) Altered Theta and Beta Oscillatory Synchrony in a Genetic Mouse Model of Pathological Anxiety. Faseb J 35 (6):e21585. https://doi.org/10.1096/fj.202002028rr

Dalton MA, D'Souza A, Lv JL, and Calamante F (2022) New Insights into Anatomical Connectivity Along the Anterior-Posterior Axis of the Human Hippocampus Using Quantitative Fibre Tracking. eLife 11:e76143. https://doi.org/10.7554/elife.76143

Dalton MA, McCormick C, and Maguire EA (2019) Differences in Functional Connectivity Along the Anterior-Posterior Axis of Human Hippocampal Subfields. Neuroimage 192:38–51. https://doi.org/10.1016/j.neuroimage.2019.02.066

Décarie-Spain L, Liu CM, Lauer LT, et al. (2022) Ventral Hippocampus-Lateral Septum Circuitry Promotes Foraging-Related Memory. Cell Rep 40 (13):111402. https://doi.org/10.1016/j.celrep.2022.111402

Deshmukh SS, and Knierim JJ (2011) Representation of Non-Spatial and Spatial Information in the Lateral Entorhinal Cortex. Front Behav Neurosci 5:00069. https://doi.org/10.3389/fnbeh.2011.00069

Dragoi G (2013) Internal Operations in the Hippocampus: Single Cell and Ensemble Temporal Coding. Front Syst Neurosci 7:46. https://doi.org/10.3389/fnsys.2013.00046

Duarte JM, Nguyen R, Kyprou M, et al. (2024) Hippocampal Contextualization of Social Rewards in Mice. Nat Commun 15 (1):9493. https://doi.org/10.1038/s41467-024-53866-2

Eichenbaum H (2017) The Role of the Hippocampus in Navigation Is Memory. J Neurophysiol 117 (4):1785–1796. https://doi.org/10.1152/jn.00005.2017

——— (2018) What Versus Where: Non-Spatial Aspects of Memory Representation by the Hippocampus. In: Clark RE, Martin SJ (eds) Behavioral Neuroscience of Learning and Memory. Springer, Cham, pp 101–117

Ekstrom AD, Caplan JB, Fields TA, et al. (2003) Cellular Networks Underlying Human Spatial Navigation. Nature 425:184–188. https://doi.org/10.1038/nature01964

Eliav T, Geva-Sagiv M, Yartsev MM, et al. (2018) Nonoscillatory Phase Coding and Synchronization in the Bat Hippocampal Formation. Cell 175 (4):1119–1130.e1115. https://doi.org/10.1016/j.cell.2018.09.017

Fanselow MS, and Dong H-W (2010) Are the Dorsal and Ventral Hippocampus Functionally Distinct Structures? Neuron 65 (1):7–19. https://doi.org/10.1016/j.neuron.2009.11.031

Fee C, Banasr M, and Sibille E (2017) Somatostatin-Positive Gamma-Aminobutyric Acid Interneuron Deficits in Depression: Cortical Microcircuit and Therapeutic Perspectives. Biol Psychiatry 82 (8):549–559. https://doi.org/10.1016/j.biopsych.2017.05.024

Forli A, and Yartsev MM (2023) Hippocampal Representation during Collective Spatial Behaviour in Bats. Nature 621 (7980):796–803. https://doi.org/10.1038/s41586-023-06478-7

Forro T, and Klausberger T (2023) Differential Behavior-Related Activity of Distinct Hippocampal Interneuron Types during Odor-Associated Spatial Navigation. Neuron 111 (15):2399–2413.e2395. https://doi.org/10.1016/j.neuron.2023.05.007

Forro T, Volitaki E, Malagon-Vina H, et al. (2022) Anxiety-Related Activity of Ventral Hippocampal Interneurons. Prog Neurobiol 219:102368. https://doi.org/10.1016/j.pneurobio.2022.102368

Gauthier JL, and Tank DW (2018) A Dedicated Population for Reward Coding in the Hippocampus. Neuron 99 (1):179–193 e177. https://doi.org/10.1016/j.neuron.2018.06.008

Ghasemi M, Navidhamidi M, Rezaei F, Azizikia A, and Mehranfard N (2022) Anxiety and Hippocampal Neuronal Activity: Relationship and Potential Mechanisms. Cogn Affect Behav Neurosci 22 (3):431–449. https://doi.org/10.3758/s13415-021-00973-y

Gray JA, and McNaughton N (2003) The Neuropsychology of Anxiety: An Enquiry into the Function of the Septo-Hippocampal System. Oxford Univ Press, Oxford

Hu J, Liu JH, Liu Y, et al. (2021) Dysfunction of the Anterior and Intermediate Hippocampal Functional Network in Major Depressive Disorders across the Adult Lifespan. Biol Psychol 165:108192. https://doi.org/10.1016/j.biopsycho.2021.108192

Huang CC, Rolls ET, Hsu CCH, Feng JF, and Lin CP (2021) Extensive Cortical Connectivity of the Human Hippocampal Memory System: Beyond the "What" and "Where" Dual Stream Model. Cereb Cortex 31 (10):4652–4669. https://doi.org/10.1093/cercor/bhab113

Ito R, and Lee ACH (2016) The Role of the Hippocampus in Approach-Avoidance Conflict Decision-Making: Evidence from Rodent and Human Studies. Behav Brain Res 313:345–357. https://doi.org/10.1016/j.bbr.2016.07.039

Jackson AD, Cohen JL, Phensy AJ, et al. (2024) Amygdala-Hippocampus Somatostatin Interneuron Beta-Synchrony Underlies a Cross-Species Biomarker of Emotional State. Neuron 112 (7):1182–1195 e1185. https://doi.org/10.1016/j.neuron.2023.12.017

Jarzebowski P, Hay YA, Grewe BF, and Paulsen O (2022) Different Encoding of Reward Location in Dorsal and Intermediate Hippocampus. Curr Biol 32 (4):834–841.e835. https://doi.org/10.1016/j.cub.2021.12.024

Jimenez JC, Su K, Goldberg AR, et al. (2018) Anxiety Cells in a Hippocampal-Hypothalamic Circuit. Neuron 97 (3):670–683.e676. https://doi.org/10.1016/j.neuron.2018.01.016

Jung K, Kruessel S, Yoo S, et al. (2024) Dopamine-Mediated Formation of a Memory Module in the Nucleus Accumbens for Goal-Directed Navigation. Nat Neurosci 27 (11):2178–2192. https://doi.org/10.1038/s41593-024-01770-9

Kaufman AM, Geiller T, and Losonczy A (2020) A Role for the Locus Coeruleus in Hippocampal CA1 Place Cell Reorganization during Spatial Reward Learning. Neuron 105 (6):1018–1026.e1014. https://doi.org/10.1016/j.neuron.2019.12.029

Keinath AT, Mosser CA, and Brandon MP (2022) The Representation of Context in Mouse Hippocampus Is Preserved Despite Neural Drift. Nat Commun 13 (1):2415. https://doi.org/10.1038/s41467-022-30198-7

Kempermann G, Song H, and Gage FH (2015) Neurogenesis in the Adult Hippocampus. Cold Spring Harb Perspect Biol 7 (9):a018812. https://doi.org/10.1101/cshperspect.a018812

Kiemes A, Navacerrada MES, Kim E, et al. (2023) *Erbb4* Deletion from Inhibitory Interneurons Causes Psychosis-Relevant Neuroimaging Phenotypes. Schizophr Bull 49 (3):569–580. https://doi.org/10.1093/schbul/sbac192

Kier EL, Kim JH, Fulbright RK, and Bronen RA (1997) Embryology of the Human Fetal Hippocampus: MR Imaging, Anatomy, and Histology. AJNR Am J Neuroradiol 18 (3):525–532. https://pubmed.ncbi.nlm.nih.gov/9090416/

Kim EJ, Park M, Kong MS, et al. (2015) Alterations of Hippocampal Place Cells in Foraging Rats Facing a "Predatory" Threat. Curr Biol 25 (10):1362–1367. https://doi.org/10.1016/j.cub.2015.03.048

Kinsky NR, Maw W, Sullivan DW, et al. (2020) Trajectory-Modulated Hippocampal Neurons Persist Throughout Memory-Guided Navigation. Nat Commun 11 (1):2443. https://doi.org/10.1038/s41467-020-16226-4

Kjelstrup KB, Solstad T, Brun VH, et al. (2008) Finite Scale of Spatial Representation in the Hippocampus. Science 321 (5885):140–143. https://doi.org/10.1126/science.1157086

Kleinman MR, and Foster DJ (2024) Spatial Localization of Hippocampal Replay Requires Dopamine Signaling. bioRxiv Preprint:597435. https://doi.org/10.1101/2024.06.04.597435

Knierim JJ (2015) The Hippocampus. Curr Biol 25 (23):R1116–1121. https://doi.org/10.1016/j.cub.2015.10.049

Kong MS, Kim EJ, Park S, et al. (2021) "Fearful-Place" Coding in the Amygdala-Hippocampal Network. eLife 10:e72040. https://doi.org/10.7554/elife.72040

Krishnan S, Heer C, Cherian C, and Sheffield MEJ (2022) Reward Expectation Extinction Restructures and Degrades CA1 Spatial Maps through Loss of a Dopaminergic Reward Proximity Signal. Nat Commun 13 (1):6662. https://doi.org/10.1038/s41467-022-34465-5

Lavenex P, and Lavenex PB (2013) Building Hippocampal Circuits to Learn and Remember: Insights into the Development of Human Memory. Behav Brain Res 254:8–21. https://doi.org/10.1016/j.bbr.2013.02.007

Li K, Koukoutselos K, Sakaguchi M, and Ciocchi S (2024) Distinct Ventral Hippocampal Inhibitory Microcircuits Regulating Anxiety and Fear Behaviors. Nat Commun 15 (1):8228. https://doi.org/10.1038/s41467-024-52466-4

Lieberman JA, Girgis RR, Brucato G, et al. (2018) Hippocampal Dysfunction in the Pathophysiology of Schizophrenia: A Selective Review and Hypothesis for Early Detection and Intervention. Mol Psychiatr 23 (8):1764–1772. https://doi.org/10.1038/mp.2017.249

Lisman J, Buzsaki G, Eichenbaum H, et al. (2017) Viewpoints: How the Hippocampus Contributes to Memory, Navigation and Cognition. Nat Neurosci 20 (11):1434–1447. https://doi.org/10.1038/nn.4661

MacDonald C, Lepage K, Eden U, and Eichenbaum H (2011) Hippocampal "Time Cells" Bridge the Gap in Memory for Discontiguous Events. Neuron 71 (4):737–749. https://doi.org/10.1016/j.neuron.2011.07.012

Madison FN, Bingman VP, Smulders T, and Lattin CR (2024) A Bird's Eye View of the Hippocampus Beyond Space: Behavioral, Neuroanatomical, and Neuroendocrine Perspectives. Horm Behav 157:105451. https://doi.org/10.1016/j.yhbeh.2023.105451

Malagon-Vina H, Ciocchi S, and Klausberger T (2023) Firing Patterns of Ventral Hippocampal Neurons Predict the Exploration of Anxiogenic Locations. eLife 12:e83012. https://doi.org/10.7554/eLife.83012

Mao D, Neumann AR, Sun J, et al. (2018) Hippocampus-Dependent Emergence of Spatial Sequence Coding in Retrosplenial Cortex. PNAS 115 (31):8015–8018. https://doi.org/10.1073/pnas.1803224115

Moita MA, Rosis S, Zhou Y, LeDoux JE, and Blair HT (2004) Putting Fear in Its Place: Remapping of Hippocampal Place Cells during Fear Conditioning. J Neurosci 24 (31):7015–7023. https://doi.org/10.1523/jneurosci.5492-03.2004

Moser EI, Kropff E, and Moser MB (2008) Place Cells, Grid Cells, and the Brain's Spatial Representation System. Annu Rev Neurosci 31:69–89. https://doi.org/10.1146/annurev.neuro.31.061307.090723

Murray EA, Wise SP, and Graham KS (2018) Representational Specializations of the Hippocampus in Phylogenetic Perspective. Neurosci Lett 680:4–12. https://doi.org/10.1016/j.neulet.2017.04.065

Norton N, Moskvina V, Morris DW, et al. (2006) Evidence That Interaction between Neuregulin 1 and Its Receptor Erbb4 Increases Susceptibility to Schizophrenia. Am J Med Genet B 141b (1):96–101. https://doi.org/10.1002/ajmg.b.30236

Ohara S, Rannap M, Tsutsui KI, et al. (2023) Hippocampal-Medial Entorhinal Circuit Is Differently Organized Along the Dorsoventral Axis in Rodents. Cell Rep 42 (1):112001. https://doi.org/10.1016/j.celrep.2023.112001

O'Keefe J, and Krupic J (2021) Do Hippocampal Pyramidal Cells Respond to Nonspatial Stimuli? Physiol Rev 101 (3):1427–1456. https://doi.org/10.1152/physrev.00014.2020

O'Keefe J, and Nadel L (1978) The Hippocampus as a Cognitive Map. Oxford Univ Press, London

Padilla-Coreano N, Canetta S, Mikofsky RM, et al. (2019) Hippocampal-Prefrontal Theta Transmission Regulates Avoidance Behavior. Neuron 104 (3):601–610.e604. https://doi.org/10.1016/j.neuron.2019.08.006

Piquet R, Faugère A, and Parkes SL (2024) A Hippocampo-Cortical Pathway Detects Changes in the Validity of an Action as a Predictor of Reward. Curr Biol 34 (1):24–35.e24. https://doi.org/10.1016/j.cub.2023.11.036

Qasim SE, Mohan UR, Stein JM, and Jacobs J (2023) Neuronal Activity in the Human Amygdala and Hippocampus Enhances Emotional Memory Encoding. Nat Hum Behav 7 (5):754–764. https://doi.org/10.1038/s41562-022-01502-8

Rolls ET (1989) Functions of Neuronal Networks in the Hippocampus and Neocortex in Memory. In: Byrne JH, Berry WO (eds) Neural Models of Plasticity. Academic Press, Cambridge, MA, pp 240–265. https://doi.org/10.1016/B978-0-12-148955-7.50017-5

Rolls ET, and Treves A (1994) Neural Networks in the Brain Involved in Memory and Recall. Prog Brain Res 102:335–341. https://doi.org/10.1016/s0079-6123(08)60550-6

——— (2024) A Theory of Hippocampal Function: New Developments. Prog Neurobiol 238:102636. https://doi.org/10.1016/j.pneurobio.2024.102636

Royer S, Sirota A, Patel J, and Buzsaki G (2010) Distinct Representations and Theta Dynamics in Dorsal and Ventral Hippocampus. J Neurosci 30 (5):1777–1787. https://doi.org/10.1523/jneurosci.4681-09.2010

Rueckemann JW, Sosa M, Giocomo LM, and Buffalo EA (2021) The Grid Code for Ordered Experience. Nat Rev Neurosci 22 (10):637–649. https://doi.org/10.1038/s41583-021-00499-9

Sanders H, Rennó-Costa C, Idiart M, and Lisman J (2015) Grid Cells and Place Cells: An Integrated View of Their Navigational and Memory Function. Trends Neurosci 38 (12):763–775. https://doi.org/10.1016/j.tins.2015.10.004

Schiller D, Eichenbaum H, Buffalo E, et al. (2015) Memory and Space: Towards an Understanding of the Cognitive Map. J Neurosci 35 (41):13904–13911. https://doi.org/10.1523/JNEUROSCI.2618-15.2015
Schröder TN, Haak KV, Jimenez NIZ, Beckmann CF, and Doeller CF (2015) Functional Topography of the Human Entorhinal Cortex. eLife 4:e06738. https://doi.org/10.7554/elife.06738
Scoville WB, and Milner B (1957) Loss of Recent Memory after Bilateral Hippocampal Lesions. J Neurol Neurosurg Psychiatry 20 (1):11–21. https://doi.org/10.1136/jnnp.20.1.11
Sharot T, Delgado M, and Phelps E (2004) How Emotion Enhances the Feeling of Remembering. Nat Neurosci 7 (12):1376–1380. https://doi.org/10.1038/nn1353
Shin LM, and Liberzon I (2010) The Neurocircuitry of Fear, Stress, and Anxiety Disorders. Neuropsychopharmacol 35 (1):169–191. https://doi.org/10.1038/npp.2009.83
Sosa M, Joo HR, and Frank LM (2020) Dorsal and Ventral Hippocampal Sharp-Wave Ripples Activate Distinct Nucleus Accumbens Networks. Neuron 105 (4):725–741.e728. https://doi.org/10.1016/j.neuron.2019.11.022
Squire LR (1992) Memory and the Hippocampus—A Synthesis from Findings with Rats, Monkeys, and Humans. Psychol Rev 99 (2):195–231. https://doi.org/10.1037/0033-295x.99.2.195
Tao S, Wang Y, Peng J, et al. (2021) Whole-Brain Mapping the Direct Inputs of Dorsal and Ventral CA1 Projection Neurons. Front Neural Circuits 15:643230. https://doi.org/10.3389/fncir.2021.643230
Tiwari P, Davoudian PA, Kapri D, et al. (2024) Ventral Hippocampal Parvalbumin Interneurons Gate the Acute Anxiolytic Action of the Serotonergic Psychedelic DOI. Neuron 112 (22):3697–3714.e3696. https://doi.org/10.1016/j.neuron.2024.08.016
Totty MS, Tuna T, Ramanathan KR, et al. (2023) Thalamic Nucleus Reuniens Coordinates Prefrontal-Hippocampal Synchrony to Suppress Extinguished Fear. Nat Commun 14 (1):6565. https://doi.org/10.1038/s41467-023-42315-1
Tracy AL, Jarrard LE, and Davidson TL (2001) The Hippocampus and Motivation Revisited: Appetite and Activity. Behav Brain Res 127 (1-2):13–23. https://doi.org/10.1016/s0166-4328(01)00364-3
Trouche S, Koren V, Doig NM, et al. (2019) A Hippocampus-Accumbens Tripartite Neuronal Motif Guides Appetitive Memory in Space. Cell 176 (6):1393–1406.e1316. https://doi.org/10.1016/j.cell.2018.12.037
Tsetsenis T, Broussard JI, and Dani JA (2022) Dopaminergic Regulation of Hippocampal Plasticity, Learning, and Memory. Front Behav Neurosci 16:1092420. https://doi.org/10.3389/fnbeh.2022.1092420
Turi GF, Li WK, Chavlis S, et al. (2019) Vasoactive Intestinal Polypeptide-Expressing Interneurons in the Hippocampus Support Goal-Oriented Spatial Learning. Neuron 101 (6):1150–1165.e1158. https://doi.org/10.1016/j.neuron.2019.01.009
van Groen T, Miettinen P, and Kadish I (2003) The Entorhinal Cortex of the Mouse: Organization of the Projection to the Hippocampal Formation. Hippocampus 13 (1):133–149. https://doi.org/10.1002/hipo.10037
Vinogradova O (2001) Hippocampus as Comparator: Role of the Two Input and Two Output Systems of the Hippocampus in Selection and Registration of Information. Hippocampus 11 (5):578–598. https://doi.org/10.1002/hipo.1073
Vivien J, El Azraoui A, Lheraux C, et al. (2023) Axo-Axonic Cells in Neuropsychiatric Disorders: A Systematic Review. Front Cell Neurosci 17:1212202. https://doi.org/10.3389/fncel.2023.1212202
Volitaki E, Forro T, Li K, Nevian T, and Ciocchi S (2024) Activity of Ventral Hippocampal Parvalbumin Interneurons during Anxiety. Cell Rep 43 (6):114295. https://doi.org/10.1016/j.celrep.2024.114295
Wood ER, Dudchenko PA, Robitsek RJ, and Eichenbaum H (2000) Hippocampal Neurons Encode Information about Different Types of Memory Episodes Occurring in the Same Location. Neuron 27 (3):623–633. https://doi.org/10.1016/s0896-6273(00)00071-4

Zhang HX, Skelin I, Ma ST, et al. (2024) Awake Ripples Enhance Emotional Memory Encoding in the Human Brain. Nat Commun 15 (1):215. https://doi.org/10.1038/s41467-023-44295-8

Zhou ZW, and Norimoto H (2023) Sleep Sharp Wave Ripple and Its Functions in Memory and Synaptic Plasticity. Neurosci Res 189:20–28. https://doi.org/10.1016/j.neures.2023.01.011

12

The Brain's Extended Navigation Network

Insights from Single-Neuron Studies

E. Zita Patai, Thackery I. Brown, Stéphane Ciocchi, Michael Goard, Shantanu P. Jadhav, Sepiedeh Keshavarzi, Julija Krupic, Jean Laurens, Koen Vervaeke, and Kate Jeffery

Abstract Neurobiological work on mammalian navigation has been greatly informed by studies of single neurons (individually or in ensembles) in freely exploring/navigating rodents. In seeking to understand the fundamental mechanisms underlying navigation, these studies have uncovered a panoply of "canonical neurons" (e.g., place, head-direction, and grid cells) that exhibit activity patterns that are recognizable as components of a spatial code. However, it is increasingly clear that much or even most of the brain is involved in navigation, and that many navigationally involved neurons do not have canonical space-coding properties. Understanding these other areas, their functions, and the nature of the less-obvious coding schemes is challenging but important, and so this chapter asks: How do we get traction on these more complex, integrated and distributed signals? What new technology or experimental paradigms could help? What are the next steps for neurobiological navigation research? Discussion concludes with an examination of whether the mammalian navigation network has developed, over evolutionary time, to support other types of cognition.

Keywords Hippocampus, navigation, direction, neurons, emotions, cognition

---

Kate Jeffery (✉)
School of Psychology and Neuroscience, College of Medical, Veterinary and Life Sciences, University of Glasgow, Glasgow G12 8QB, UK
Email: Kate.Jeffery@glasgow.ac.uk

Coauthor affiliations are available in the List of Contributors.

N. S. Newcombe and K. Cheng (eds.), *Challenges in Navigation Research*, Strüngmann Forum Reports, https://doi.org/10.1007/978-3-032-20563-6_12

Group photos (top left to bottom right) E. Zita Patai, Thackery Brown, Koen Vervaeke, Jean Laurens, Kate Jeffery, Michael Goard, Shantanu Jadhav, Sepiedeh Keshavarzi, Julija Krupic, Thackery Brown, Jean Laurens, E. Zita Patai, Shantanu Jadhav, Koen Vervaeke, Julija Krupic, Sepiedeh Keshavarzi, Kate Jeffery, Michael Goard

1 Introduction

Navigation is the process by which mobile agents decide where to go, and how, and then execute the necessary actions. The neural basis of this competence is beginning to be understood, and the purpose of this chapter is to review what we think we now understand and then to define some of the large gaps in our knowledge to shape future work.

In this review we define navigation broadly to mean purposeful movement of an agent in its environment. This includes standard goal-directed navigation, which entails planned movement to a specific, known region of space, but it also includes foraging (searching for potentially multiple goals at unknown locations) and exploration (searching space, but with no clear goals). Mostly, we are concerned here with goal-directed navigation, primarily in mammals, as we are, of course, particularly interested in our own selves. However, since many fundamental principles of navigation have been uncovered by studying other taxa, even artificial agents, we refer to these where appropriate.

All motile animals engage in some form of navigation, and how this is achieved by neural circuits has exercised biologists for centuries. Study of the mechanistic basis of mammalian navigation began early last century with the famous studies of Tolman, who, while researching the behavior of rats in a maze, formally articulated the concept of a *cognitive map* (Tolman 1948). This term refers to an internal representation of the environment, a concept recruited to explain observations that animals actively use stored information (memories) to guide their actions in space, rather than relying on simple stimulus-response reflexive action. This notion received great impetus in the 1970s through the discovery of the place cells in the rodent hippocampus (O'Keefe and Dostrovsky 1971). These neurons are sensitive to an animal's location in its environment, and their discovery ignited a huge research program aimed at elucidating their properties and functions, and the networks within which they sit. Fueled partly by these studies, the many subcomponents of navigation have been exhaustively described and cataloged over decades of formal research (for reviews, see Jeffery et al. 2024; Wiener et al. 2009). Great strides have been made in understanding how some of the processes are implemented in nervous systems in a variety of taxa. However, navigation is endlessly complex: almost all of the brain seems involved in this competence in one way or another. It seems that the more we learn about it, the more there is left to discover. The purpose of this chapter is to examine the neuronal underpinnings of navigation computations in the context of its many related subcomponents, sometimes referred to as the *navigation toolbox* (Jeffery et al. 2024; Wiener et al. 2011), and to look to the future: to expose new areas of inquiry going forward.

We explore the issues of neuronal navigation signals within the following frameworks:

- What can canonical navigation neurons tell us about navigational processes in the brain?
- Reference frames for the navigation neurons, and the complex problem of their integration.

- Needs, motivation, and reinforcement: How do emotions and motivational states interact and influence the spatial navigational system?
- Planning and decision making: How does the brain compute a route from its current location to the goal?
- Is the navigation system the foundation for a more general cognitive system?

2 The Canonical Navigation Neurons

The discovery of the hippocampal place cells was transformative for the field of navigation and memory because it showed, for the first time, that the brain can and does make internal representations of the external world, and that such representations incorporate space-like elements (e.g., distance, direction, dimensionality, adjacency). These neurons emit their action potentials (*fire*) when the rat is at a given location, producing a spatially focal patch of activity known as a *place field.* Place cells fire when the animal is in the place field no matter what path it is taking through that region: they are thus self-localizing semi-independently of the exact sensory inputs, meaning they are making a type of abstract inference. This property of the cells led to the suggestion that they are the neural implementation of Tolman's cognitive map (O'Keefe and Nadel 1978).

In subsequent decades it was found that other types of neurons in nearby regions also have firing patterns that are spatial and somewhat independent of exact sensory inputs. The head-direction (HD) cells were discovered around twenty years later, first in cortex and subsequently in many other regions (Taube 2025): these cells fire when the animal faces in a particular direction, producing a *tuning curve.* Each cell's tuning curve is linked to a specific direction, and the population of cells preserve their relative firing directions in every environment even though the actual global direction may change (indicating, interestingly, that the system does not use geomagnetic cues, at least in rats and mice). They thus construct an internal coordinate system that is anchored to the outside world by sensory cues (although it also preserves its internal coherence during sleep) (Peyrache et al. 2015). The cells can switch between using visual cues (landmarks) from visual cortex or self-motion cues from the vestibular and other systems (e.g., optic flow sensing, motor efference copy), again showing that the representation is a spatial inference rather than a direct report of simple sensory inputs.

Grid cells and border cells were discovered in the mid-2000s and were first found in the entorhinal cortex (Hafting et al. 2005; Solstad et al. 2008), in the layers that project directly to the place-cell system in the hippocampus. Grid cells show place fields that are more circular than hippocampal place fields, but strikingly, there are many more of them. In some settings, they are arranged in a very regular array (hence "grid"). This regular spacing indicates the ability to track distances traveled and settled the question once and for all about whether this system is truly spatial, resulting in the award of the 2014 Nobel Prize in Physiology or Medicine to the discoverers of place and grid cells (Morris 2015).

These discoveries have profoundly influenced the neurobiological study of navigation because their function is easily interpretable. On one hand, we think that this interpretability has played a key role in advancing our mechanistic understanding of the neuronal circuits in which these cells are embedded (at least as long as stringent definitions are used regarding what response properties identify a given cell type). On the other hand, there are reasons to think that these "canonical" classifications may, in certain cases, be overly simplistic, reflecting human-defined categories that overlook complex neuronal processes, such as the linear and nonlinear mixed selectivity of cells (e.g., Hardcastle et al. 2017). Experimental focus has thus tended toward clear interpretability for the experimenter as much as utility for the brain. Pure selectivity in cells can also be technically problematic to discover, relying on potentially ambiguous identification methods, or constraining of experimental design to fit predefined categories. It is thus now time to ask whether we should move on to understand the broader context in which these neurons operate, and the more opaque coding schemes that they may use.

2.1 "Canonical" Cells: Is It Time to Move On?

A canonical definition encompasses the types of information that cells appear to represent (e.g., location), thus characterizing the cells by the information they encode rather than their anatomical location. For example, several studies have reported *place cells* in multiple brain regions outside the hippocampal formation (Jankowski et al. 2015; Lomi et al. 2023; Long and Zhang 2021; Mao et al. 2018; O'Mara and Aggleton 2019). This canonical definition approach has been instrumental in spatial coding research for understanding the roles particular cells may undertake in navigation. It has also been crucial for imposing specific constraints on neural connectivity and network architecture. For example, the characterization of HD cell properties, especially their coherence, led to the formulation of the ring attractor computational model (Skaggs et al. 1994), which was subsequently validated when such a network architecture was discovered in flies more than twenty years later (Kim et al. 2019; Seelig and Jayaraman 2015).

This definitional approach to cell categorization has not been unproblematic. The lesson to be learned is that multiple perturbations are needed to confirm that the properties of interest are expressed in the relevant conditions. For example, HD cells can appear to form stable *place fields* in virtual reality (VR) (Kinkhabwala et al. 2020). If *in vivo* hippocampal research had started in head-fixed VR rather than in freely moving animals, we would have mistakenly concluded that both place cells and HD cells are place cells. Many experiments confound spatial location and direction with salient nonspatial features such as landmarks or goal and reward locations. Thus, to characterize neurons as belonging to a canonical cell type, it is crucial to test cell function under different conditions to confirm whether the relevant property really attaches to the stimulus in question and not to a confounding feature.

A canonical definition may need to be expanded, if a new function is identified for cells that otherwise almost fully fit within the existing definition. For example, one of the fundamental properties of place cells, in addition to their spatial coding, is the sequence coding of nonspatial stimuli that regularly appear in a specific order and are task relevant (O'Keefe and Krupic 2021). To expand a definition, several strategies should be followed. One relies on what is known about the system, such that new experiments are informed by this knowledge. This was the case in the discovery of HD cells (Ranck 1975; Taube et al. 1990) and grid cells (Hafting et al. 2005) in which space was expected, based on place cells, to be a relevant stimulus domain. However, new and ideally unsupervised methods of comprehensive testing in more naturalistic behaviors, which become increasingly feasible with the new techniques emerging both inside (Ho et al. 2023) and outside (Beetz and Hechavarría 2022) the lab, should be applied to better understand the system's overall capacity and the individual contributions from distinct neurons.

Rather than expanding a definition, new canonical definitions may be necessary to accommodate a greater variety of discovered properties in a particular cell type, otherwise the original definition may become too broad and lose its utility. For instance, Quirk et al. (1992) reported spatially tuned neurons in the medial entorhinal cortex. However, they refrained from referring to these cells as "place cells" because the cells showed fundamental differences in how they responded to new enclosures: the hippocampal place cells completely reshuffled their preferred firing locations (place fields) in the phenomenon called *remapping*, whereas entorhinal spatially tuned cells maintained a high degree of similarity. This led the authors to conclude that the former is used to distinguish two different environments, while the latter reflects their commonalities—both functions being necessary for efficient navigation and spatial memory.

A problem can arise from the above attempts to define cell types categorically in this way, as it limits how we think about the system. Whereas the commonality between hippocampal and entorhinal place cells that Quirk et al. (1992) identified was useful, and motivated the search for more general coding of space in regions outside the hippocampus, in some cases, such direct comparison limited further progress in our understanding of what the entorhinal cortex actually does and how. After the initial discovery of spatially tuned entorhinal cells it took fully ten years to discover grid cells. This was eventually achieved by simply extending the size of the enclosure and revealing that these cells fired at multiple fields arranged in a crystal-like hexagonal structure (Hafting et al. 2005). This is a good illustration of how constraining the experimental design to fit a specific notion of a "canonical cell" may restrict the pace of discovery. To take another example: the standard method for recording spatial cells is to have the animals randomly forage over a uniform area, whereas it seems that some types of spatial coding only emerge during repeated traversals of stereotyped routes, such as the spatial patterns of some retrosplenial cortex cells (Alexander and Nitz 2017; Miller et al. 2014). Thus, basing experimental design around the known canonical cells may be counterproductive in the long run, despite its having been essential to discovery in the first place (Moser et al. 2008).

What about other cells, which often have mixed, not so well-defined encoding properties? These are potentially very important, especially for understanding the function of the navigational system as a whole. In some cases, the existence of certain cells can be explained within the framework of existing canonical cells. For example, Krupic et al. (2012) found that grid cells were only a subset of a larger continuum of spatially periodic cells. This led to the empirical and theoretical discovery that local geometrical features fundamentally affected grid-cell symmetry

(Krupic et al. 2014. 2015), with a similar framework used in other animal models, including bats and humans (Bellmund et al. 2020; Ginosar et al. 2021; Grieves et al. 2021; Zhang et al. 2021, 2022). In other cases, expanding the same theoretical framework may not be possible, and a new one must be introduced. New computational tools will likely be used, with a focus on group dynamics rather than individual cells. Existing tools for large population analysis include various dimensionality reduction and visualization methods, such as PCA and UMAP, topological data analysis used to reveal a toroidal manifold of the grid-cell network (Gardner et al. 2022), graph theory (George et al. 2021), and others. In the future, more sophisticated AI-based approaches are likely to be implemented to model and understand the network complexity, similar to the approach by Banino et al. (2018).

Is the attempt to categorically define cells based on their response properties (or "encoding") actually completely misguided? Luo et al. (2024) have rather provocatively made this case in an experiment showing that "canonical" properties can arise in a deep neural network (DNN) trained in a spatial setting. They argue that "intuitive notions of cell type, which can be misleading and arise in any complex network, should be relegated to history." It is certainly the case that modern deep learning technology is overturning some of our notions of what neurons in concert can and cannot do. However, navigation neurons in real brains have very distinct properties: HD cells exhibit high coherence within a one-dimensional stimulus domain, in contrast to the heterogeneous and flexible responses of place cells, and there is considerable diversity in cell morphologies, physiological parameters (e.g., firing rate, burstiness) that exists in this system. It seems unlikely that the navigational brain is simply a homogeneous DNN; it seems more likely that it comprises modular subsystems with specific roles, each using a specialized cell type, albeit with varying properties that perhaps do not have sharp boundaries.

In conclusion, is it helpful to maintain the notion of canonical cells, or should we go beyond this simplistic characterization? The answer seems to be "both."

3 Reference Frames for the Navigation Neurons

The firing of spatial neurons occurs relative to some type of reference frame: a set of physical structures/objects that define a coordinate system, relative to which directions and distances can be represented. An important lesson from studies of canonical spatial neurons has been the variety of spatial reference frames used by the brain. Here, we explore what is known—and remains unknown—about how the brain represents and transforms between these reference frames.

3.1 Egocentric and Allocentric Encoding

The most salient reference-frame finding has been that there is a fundamental dichotomy between the *egocentric* frame—self-referenced and viewpoint-dependent representing the relationship between the self and external environment features (e.g., borders and boundaries)—and the *allocentric* frame—world-referenced and viewpoint-independent representing the relationships among environmental features or locations. Cells in primary sensory regions are egocentric, firing in response to activation of specific cells in the animal's own sense organs. Place cells were notable when discovered because they seemed, uniquely, to be allocentric, referencing their firing to a place in the world, similar to HD, grid, and border cells. However, the firing of these cells is arguably not *purely* allocentric, since they only fire when the (egocentric) animal is at the (allocentrically defined) relevant location: for a place cell to fire, there is thus a requirement for the self and the world to intersect.

The dichotomy between allocentric and egocentric subdivides further, since there are many possible allocentric and egocentric reference frames (e.g., rooms in a house in a street in a city or fingers on the hand on the arm on the torso). A surprising allocentric reference frame to emerge from place-cell studies has been that some cells use the identity and location of a conspecific as the reference frame for location-specific activity. This observation was reported simultaneously in both bats (Omer et al. 2018) and rats (Danjo et al. 2018; for commentary, see Duvelle and Jeffery 2018). Thus, the cells can encode a relative location to an externally relevant stimulus independently of self-location, which makes this true allocentric encoding. Such observations are rare, however. There have not, for example, been reports of cells sensitive to the relative locations of two objects with respect to each other. Determining which reference frames are or are not of interest to place cells is an ongoing enterprise.

As mentioned above, spatial firing in the primary sensory regions of the brain is purely egocentric: visual cortical neurons fire, for example, in response to stimuli falling on specific regions of the retina. Since the brain receives all its spatial information via the sense organs, this needs to be transformed to an allocentric reference frame to allow navigational computations and actions. Reference-frame transformations were first explored in the context of parietal cortex (Cohen and Andersen 2002), in which encoding could switch from one frame (e.g., eyes) to another (e.g., head), perhaps via a computational process known as gain modulation (Salinas and Thier 2000). While parietal cortex seems purely egocentrically engaged, the transformation between egocentric and allocentric frames has been attributed to adjacent retrosplenial cortex (RSC) (Burgess 2006; Byrne et al. 2007; Epstein 2008; Julian et al. 2018; Vann et al. 2009). RSC serves as a bidirectional link between more egocentric-coding structures (sensory, parietal, and secondary motor regions) and more allocentric-coding structures (anterior thalamus, entorhinal cortex, and hippocampal formation), which are themselves not very interconnected (Vann et al. 2009).

Recent findings from the study of neurons in RSC support the notion of interactions between egocentric/allocentric and local/global reference frames (for a review, see Clark et al. 2018). These may be mediated by neurons with mixed egocentric/allocentric properties. Interesting neurons with such mixed properties are the so-called *egocentric vector cells*, which are sensitive to the egocentric direction and distance to an environmental object (Wang et al. 2018), landmark (Wilber et al. 2014), boundary (Alexander et al. 2020; Gofman et al. 2019; Hinman et al. 2019), corner (Park et al. 2024), or center of a bounded space (LaChance et al. 2019). These neurons have been found in a wide variety of areas associated with the hippocampus such as RSC, postrhinal cortex, medial and lateral entorhinal cortices, parietal cortex, and striatum. In a unique single-neuron study in humans, cells in the parahippocampal cortex, amygdala, hippocampus, and temporal pole coded for egocentric bearing, either toward the center or boundary of the arena, or hidden object locations (Kunz et al. 2021). These egocentric vector neurons are intriguing because their egocentricity differs from that of a visual cortical neuron: the "ego" does not correspond to a point on a sensory surface, but to the location of the whole animal in the world—hence, they are "mixed." These observations thus point to a progression from pure egocentric, through intermediate states to (perhaps) pure allocentric (for further discussion, see Ekstrom et al. 2017). Exactly what computation each performs remains to be discovered.

How these transformations are made has yet to be elucidated. Does RSC use the same gain field modulation principles as parietal cortex? Or are there other ways of achieving these types of synthesis between different reference frames?

3.2 Local and Global Reference Frames

A second, related reference frame dichotomy is local versus global reference frames, in which responsiveness to local environmental cues, such as (in the laboratory) the walls of a box, can be decoupled from the response to global cues (e.g., the room in which the box sits). The very earliest observations of place cells suggested that they are anchored by local environmental boundaries rather than tied to the larger world. This idea has been supported by subsequent studies showing indifference of place cells to manipulations of the world beyond the walls (O'Keefe and Burgess 1996; Spiers et al. 2015).

To date, no studies have shown encoding by the larger environment independently of local boundaries. However, one can know simultaneously where one is in a room and where the room is within a building, raising the question of which brain area supports the knowledge of space beyond immediate perceptual reach. Neuroimaging studies in humans suggest that RSC may be implicated in the distinction between local and global frames. Peer and Epstein (2021) found that participants in a VR courtyard showed brain activation in the hippocampus and occipital place area that reflected the segmented nature of the space, whereas RSC showed responses more consistent with a global representation (see also Marchette et al. 2014). Neuroimaging research has found that the RSC integrates information across scenes, even those outside the immediate field of view (Robertson et al. 2016), and integrates separately explored areas (subspaces) into a global representation (Peer and Epstein 2021). Moreover, the subspaces can be nested spaces which are integrated into a global representation depending on task demands (Peer and Epstein 2025). Recent evidence suggests that multiple landmarks within the same view ("panorama") are represented more similarly in the RSC, as opposed to a single landmark seen from different viewpoints (Han and Epstein 2025), suggesting a prominent role for self-localization with respect to environmental cues.

There is some single-neuron evidence to support the local versus global distinction in RSC. Studies of HD cells in a two- or four-compartment environment found that many cells in the dysgranular RSC region had firing oriented by the local environment layout, while the classic HD cells were globally stable across the whole space (Jacob et al. 2017; Zhang et al. 2022). Neurons in RSC have also been shown to encode routes with different levels of periodicity; that is, local, smaller route segments as well as global, full route segments (Alexander and Nitz 2017), suggesting the integration of spatial scales or at least the parts of a route into the whole (Nitz 2012).

However, it is not clear that these studies have localized the storage site of the large-scale map: Is it in RSC or somewhere else, such as the prefrontal cortex (PFC)? How are the distances stored and retrieved, such that one can quickly identify, say, which is the nearest supermarket given one's current location? These studies in nonhuman animals await the development of behavioral tasks that can sample knowledge of location beyond perceptual reach and, in humans, of neuroimaging methods with good enough spatial and temporal resolution to track the passage of information through the brain's representational stages.

4 Needs, Motivation, and Reinforcement

Spatial navigation serves a fundamental biological function: enabling organisms to relocate to meet internal needs that cannot be fulfilled in their current environment. These needs, which range from immediate physiological demands to more abstract social or reproductive goals, are often multiple and dynamic, requiring the brain to continuously prioritize among them. To guide adaptive behavior, the brain must not only rank these needs in terms of urgency and value, it must also identify spatial locations where they can be satisfied and generate context-sensitive action plans that account for both internal state and external constraints. In this section, we explore how the spatial navigation system integrates internal motivational states and emotional valence to guide behavior. We focus on four interrelated components: (a) needs, which provide the underlying biological drivers of behavior; (b) motivation and emotion, which modulate the urgency, valence, and directional bias of navigational choices; (c) reinforcement, which enables learning from past outcomes to shape future decisions; and (d) model-based memory systems, which allow for flexible, goal-directed navigation based on internal representations of space, expected outcomes, and current physiological state or emotional context.

4.1 Needs: The Drivers of Navigational Goals

One of the foundational assumptions in spatial navigation is that movement through space is goal-directed—driven by unmet needs (Verschure et al. 2014). At the most basic level, the navigation system exists to help an organism relocate to environments better suited to restoring homeostasis or fulfilling biologically relevant drives. However, while we understand the broad outline of how physiological needs (e.g., hunger, thirst, thermoregulation) can bias behavior (de Vries et al. 2020; Suito and Tominaga 2024; Tan et al. 2022), major gaps remain in our understanding of how the brain identifies, prioritizes, and integrates multiple simultaneous needs to shape navigational decisions, especially in complex or ambiguous environments.

We know that needs can be organized hierarchically (Kenrick et al. 2010), from immediate survival (air, water, temperature) to longer-term needs (reproduction, social belonging, or dominance). Some of these needs arise from low-level physiological detectors, many of which are well characterized (e.g., osmoreceptors, thermoreceptors, or glucose sensors), to inform subcortical structures like the hypothalamus and brainstem (Bourque and Oliet 1997; Marty et al. 2007; Tan and Knight 2018). Yet, how these basic signals are translated exactly into spatial goals remains unclear. For instance, what are the neural mechanisms by which a blood glucose drop becomes a directional bias toward a remembered food location? What transforms a physiological state into an adaptive spatial behavior?

We also know surprisingly little about how competing needs are resolved within the navigation system. For example, in real-world scenarios, organisms often face trade-offs: an animal may be hungry but detect predator scent in the area where food was previously found. What computations underlie the decision to forage versus flee? While theoretical models propose that needs are weighted according to urgency or value (Miletić et al. 2021; Thura et al. 2012), the neural correlates of such weighting—particularly how they interact with spatial memory circuits—remain unclear.

Another significant gap concerns how abstract needs influence spatial behavior. Intuitively, we understand that thirst drives the search for water, but how do more abstract needs, such as social status, manifest in spatial navigation? Do these higher-order needs engage the same neural substrates as basic drives, or do they influence behavior through distinct pathways? The hippocampus and medial PFC are likely candidates for such integration, but the precise neural circuitry and mechanisms are still poorly understood. Furthermore, little is known about developmental and individual differences in need-based navigation. How do need priorities shift over time? Do early experiences calibrate how the brain allocates navigational resources to certain needs over others? Finally, in psychiatric or neurological disorders (e.g., depression, addiction, anxiety), does distorted need prioritization underlie maladaptive navigation (e.g., withdrawal, compulsive seeking, or avoidance)? These questions remain largely underexplored but could have profound implications.

4.2 Motivation and Emotion: Modulating Action and Memory

Needs and intentions do not translate directly into actions (Wieber et al. 2015). Instead, actions are mediated by a motivational-emotional system that prioritizes behavior (Pessoa 2009). This system enables the organism to act in a focused and adaptive manner, selecting from its behavioral repertoire to satisfy competing needs. Because only one major action can typically be performed at a time, the brain must constantly evaluate and reprioritize goals, often under dynamically changing internal and external conditions. Despite extensive research, many questions remain about how motivation and emotion interact with the brain's spatial navigation circuits to guide decisions.

Emotion and motivation serve at least two critical roles in navigation. First, they provide a mechanism for assigning value to potential outcomes, thereby shaping action selection (Tye 2018). Second, they "tag" spatial memories with emotional significance, influencing future navigation toward or away from locations (Shpokayte et al. 2022). Experimental research has focused primarily on two affective poles—reward and fear—as paradigms for studying motivational valence and related disorders (Lüthi and Lüscher 2014). The dopaminergic system, particularly involving the ventral striatum, plays a central role in encoding motivational and reward signals, while the amygdala has been widely implicated in fear learning and threat detection (Cardinal et al. 2002). Both systems project to the ventral (anterior in primates) hippocampus (Gasbarri et al. 1997; Pikkarainen et al. 1999), a region of the hippocampal formation thought to specialize in integrating emotional and spatial information (Fanselow and Dong 2010; Strange et al. 2014).

Despite this knowledge, several key gaps persist. One major area concerns how emotional and motivational signals from structures like the ventral striatum and amygdala are actually used by the hippocampus to modulate spatial representation. Place coding in the ventral hippocampus is notably coarser than in its dorsal counterpart (Ciocchi et al. 2015), suggesting that it may encode emotional valence of broad environments rather than precise spatial locations (Kjelstrup et al. 2008). However, it is unclear how this coarse coding is integrated with more fine-grained spatial information for planning and decision making. How does the brain merge the "where" with "how it felt to be there" information? Similarly, although models often assume that motivation drives movement toward rewards and away from threats, real-world navigation frequently involves complex trade-offs between competing motivational and emotional values. A location associated with food might also carry the risk of predation. How does the brain weigh these conflicting signals, and which specialized circuits integrate multivalent emotional information into spatial memory?

Another open area involves the flexibility of emotional memory in navigation. Emotions not only bias current action, they also modulate how memories are stored and recalled (Tyng et al. 2017). To date, we do not fully understand how emotion-laden memories are updated when conditions change, such as when a dangerous location becomes safe or a rewarding location is depleted. Are these memories overwritten, or does the brain maintain multiple emotional maps that are selectively activated based on context?

Collectively, we conclude that while reward and fear are well studied, other emotion-related states (e.g., novelty-seeking, frustration, or social interactions) remain understudied in the context of navigation. These states may influence exploration, avoidance, or persistence in ways not captured by traditional binary models of appetitive versus aversive motivation.

4.3 Reinforcement: Learning from Experience

The consequences of actions feed back into the system through reinforcement—a process that links outcomes to future behavioral likelihood (Lee et al. 2012).

Reinforcement shapes learning by signaling whether a given action moved the organism closer to or further from its needs. When a navigational decision results in need satisfaction, it produces a positive emotional state, such as pleasure or relief, reinforcing the preceding actions and contexts. Conversely, failure to meet a goal—or encountering punishment—can result in disappointment, frustration, or pain which weakens the association and promotes avoidance in the future (Sosa and Giocomo 2021).

While reinforcement learning is one of the most thoroughly studied mechanisms in neuroscience and machine learning, significant questions remain about how it operates in real-world, need-driven navigation. Most traditional models of reinforcement learning, especially those derived from artificial intelligence, assume discrete, well-defined goals and environments (Botvinick et al. 2020). However, in biological organisms, needs are often multiple, overlapping, and dynamically shifting. The real world is continuous, uncertain, and rarely structured in discrete episodes. How the brain implements reinforcement learning under these conditions is still poorly understood. One key gap concerns the interaction between reinforcement and spatial memory (Pedamonti et al. 2025). How are reinforcement signals used to update spatial representations stored in the hippocampus? For instance, does a rewarding outcome update only the final location of a trajectory, or does it also modulate the entire route and decision points along the path? How persistent are such updates? Do they decay over time, or are they resistant to change once established?

Another open area concerns how reinforcement signals are shaped by emotional context. Most reinforcement studies have focused on primary reinforcers such as food, water, or shock but in the natural environment, outcomes are rarely either purely positive or negative. They are often ambiguous, multilayered, and emotionally complex. Does disappointment in a navigational task produce the same neural and behavioral adjustments as outright punishment? Are emotional states like frustration, uncertainty, or anticipation encoded in ways that influence spatial choice behavior? These more nuanced forms of emotional reinforcement are not well understood, but are likely crucial for flexible, adaptive navigation.

Additionally, there is growing evidence that reinforcement learning may interact with different parts of the hippocampus in distinct ways (Dombrovski et al. 2020). For example, the ventral hippocampus, which is connected to emotional and motivational brain regions (Fanselow and Dong 2010; Gergues et al. 2020), may bias navigation based on emotional value rather than spatial accuracy or efficiency. However, how this emotional biasing interacts with classical reinforcement mechanisms (e.g., dopamine-mediated prediction error signaling) remains unclear. Does a single unified reinforcement map guide navigation, or do multiple, parallel learning systems operate simultaneously, and sometimes in conflict?

Altogether, we know relatively little about how reinforcement processes handle competing needs and conflicting goals. How does the system weigh short-term rewards against long-term benefits? Can reinforcement memories be dynamically reevaluated if the need landscape changes? These are important questions, especially for understanding maladaptive navigation seen in addiction or anxiety behaviors (Alvares et al. 2014; Robbins et al. 2008), where reinforcement patterns may become uncoupled from actual need satisfaction.

5 Planning and Decision Making: How Does the Brain Compute a Journey from a Current Location to a Goal?

This brings us to the question of navigational planning: How are goals encoded and retrieved, and how are routes planned to get to the goal? Since the early days of place-cell research, this topic has been extensively studied. Although place cells provide a "you are here" signal in the brain, this alone does not explain navigation (Poucet et al. 2004). For an overview of the hippocampal literature on this topic, we refer the reader to the detailed and informative review by Nyberg et al. (2022). Here, we focus on some of the salient issues and unsolved questions that relate to this issue.

5.1 Model-Based Learning and Memory: Navigating with Knowledge

We begin with the issue of the two major systems that underlie navigational action. Initially, spatial research adopted a reinforcement learning framework called operant conditioning in which actions are chained together based on their reinforcement history. However, Tolman's work challenged this view, suggesting that not all navigation is habitual or reflexively driven by reinforcement history. In many situations, animals behave in ways that suggest the use of internal models; that is, representations of the environment, the possible outcomes of actions, and the relationships between them (McNamee and Wolpert 2019). For example, a hungry animal may be reinforced by discovering peanuts at a particular location. However, if later offered a more preferred or varied food elsewhere, it may choose to forgo the peanut location, even though this was previously rewarding. This decision reflects not only the outcome of prior reinforcement, but the integration of broader knowledge: what is currently desirable, what alternatives exist, and how to access them. This capacity, known as model-based learning, allows for flexible decision making based on prior experience, current internal state, and predictions about future outcomes. It stands in contrast to model-free reinforcement learning, which relies on cached values from past rewards or punishments (Geerts et al. 2020; Lee et al. 2014). Such planning and evaluation requires cognitive maps, memory of past episodes, and the ability to simulate future possibilities—capabilities typically associated with the hippocampus and PFC (Behrens et al. 2018; Epstein et al. 2017; Preston and Eichenbaum 2013).

Despite significant theoretical advances, many core questions about model-based learning in navigation remain unresolved. Among them is how internal models are constructed and updated. What kinds of experiences contribute to model building? How are motivational or emotional variables, such as hunger or fear, incorporated into these models? And when do animals rely on model-based strategies over simpler, model-free ones? Although laboratory tasks have begun to probe these distinctions, real-world navigation involves far more complexity, including the need to weigh uncertain or emotionally significant outcomes.

Another major gap concerns how motivational states modulate the use of internal models. While it is clear that factors like hunger or stress can bias decision making (Nuno-Perez et al. 2021; Skrynka and Vincent 2019), we know little about how these states influence the structure of the cognitive model itself. Does a hungry animal expand its cognitive map to explore more options, or does it narrow its focus to known sources of food? Does stress shrink the current model by pruning "risky" options, even if they may be rewarding? There is some evidence that emotional states can warp memory and spatial representations (Moita et al. 2004; Nguyen et al. 2023), but how this influences model-based navigation in a dynamic way is still poorly understood. Furthermore, it is unclear how different types of memory—episodic, semantic, procedural—interact during model-based decisions. Moreover, the boundary between deliberative and habitual action is likely fluid, but the mechanisms that govern this shift remain unclear.

Finally, in terms of brain circuitry, much remains to be discovered about how model-based computations are implemented and coordinated. While the hippocampus is known to support spatial memory and the PFC to support decision making and planning, the dynamic interplay between these regions (Shin and Jadhav 2016), especially under different motivational states, is still being investigated.

5.2 Route Representation and Disambiguation

Flexible navigation involves the ability to distinguish the current navigational route from alternative routes despite their possible overlap (e.g., crossing paths or a shared stretch of road). Research indicates a key role for the hippocampus in overcoming such interference, as evidenced both in humans (Brown et al. 2010, 2012; Chanales et al. 2017; Sherrill et al. 2013) and in rodents (Ferbinteanu and Shapiro 2006; Hasselmo and Eichenbaum 2005; Lee et al. 2005; Wood et al. 2000). These findings relate to the critical question of how the brain represents routes based on where the route is going (the goal). While the neural basis for this process in humans is not known, neuroimaging studies show that the human hippocampus is involved in navigational planning (Brown et al. 2010, 2012, 2016; Crivelli-Decker et al. 2023; He et al. 2022; Javadi et al. 2017; Spiers and Maguire 2006; Viard et al. 2011). In rodents, internal, temporally compressed place-cell sequences sometimes reflect potential routes toward a goal (Kay et al. 2020; Pfeiffer and Foster 2013) or information about where the route is headed (Ferbinteanu and Shapiro 2006; Wood et al. 2000). One study that has dissociated the route from the goal found that the cells are encoding the route rather than the destination (Grieves et al. 2016). Thus, it may be that a route is encoded *en bloc* as a type of context, allowing disambiguation in situations where routes overlap. How or where the brain stores and accesses routes and goals is still unresolved, as is how multiple goals are selected and the shortest (or least-effort) paths are calculated.

5.3 Navigational Planning and the Prefrontal Cortex

Route planning does not involve only the hippocampus, as we might expect. In mammals, the PFC supports flexible, adaptive behavior in a variety of domains, including navigation. Within the PFC, the orbitofrontal cortex (OFC) is a unique prefrontal hub that is functionally tied to both the hippocampus and striatum (Haber and Knutson 2010); it processes sensory information and reward (Barbas 2000; Barbas and Blatt 1995; Cavada 2000). In particular, in humans, primates, and rodents, OFC has been implicated in representing and updating the association of specific stimuli or actions to specific reward outcomes (Ma et al. 2025).

In rodents, neurons in the OFC code paths from start to goal locations (Young and Shapiro 2011). Many of these OFC neurons distinguish routes going to different locations and commonly represent different paths leading to the same goal. These data suggest that the spatial objective, more than motor-reward associations, is an essential feature represented in path-coding neurons in rodent OFC. Critically, using a similar navigational paradigm, human fMRI research demonstrates that OFC is active and functionally connected with the hippocampus when selecting between different navigational routes headed to different locations (Brown et al. 2012, 2016). Thus, hippocampal-OFC interactions in adaptive navigation may be critical for representing and distinguishing navigational events based on their association with specific goals or outcomes.

To date, studies have not clearly dissociated the selection of spatial goals from the selection of navigational motor responses, leaving the specific contributions of the OFC and neighboring ventromedial PFC unclear. This is an important future direction, as some studies suggest that the OFC represents specific stimulus-outcome associations (e.g., Rudebeck et al. 2008; Walton et al. 2010), implying that its contribution to navigation may be weighted toward selection of goal states from the cognitive map rather than determining the actions to get there. Other evidence, however, implicates OFC in behavior-outcome associations (Mainen and Kepecs 2009). PFC is a very large, functionally heterogeneous collection of regions in mammals. In addition to OFC and ventromedial PFC, other prefrontal circuits have been associated with navigational performance (Patai and Spiers 2021), although more work is needed to disentangle their unique contributions to planning. For example, one way to disentangle the selection of spatial goals from the selection of navigational responses is to consider planning processes in contexts where there are multiple possible routes to the same goal. In such cases, evaluation of alternate routes may bring in PFC subdivisions that process abstract concepts and value signals to arrive at a decision. Ventrolateral and dorsomedial PFC have recently been found in humans to track expected costs for different routes to the same goal, particularly when these costs are based on probabilistic information. This can occur, for instance, when deciding between route options to work: the perceived value of each route depends on the navigator's estimate of the likelihood of getting lost or encountering traffic delays (He et al. 2024). Taken together, a major goal for the future is to delineate the different elements of navigational planning and to understand how the subcomponents of the decision system contribute to them.

6 Is the Navigation System Part of a More General Cognitive System?

We conclude by addressing a long-standing question: Is navigational competence supported by a specialized brain system or by a general cognitive system? For researchers studying the hippocampus, a closely related question concerns whether the hippocampus is just for space or is part of a more domain-general processor. This question has been examined in several recent reviews (Behrens et al. 2018; Bellmund et al. 2018; Epstein et al. 2017; Lisman et al. 2017), which broadly conclude that although the hippocampus is clearly essential for spatial and episodic memory, there are strong arguments for thinking that evolution has built upon these functions to solve (among other problems) the "abstract problem of mapping continuous dimensions of any sort" (Lisman et al. 2017). Evidence supporting this assertion comes from studies of place cells which show that they can also respond focally in temporal tasks to become "time cells" (MacDonald et al. 2011), or to a specific auditory tone frequency in a tone perception task (Aronov et al. 2017). In entorhinal cortex, Constantinescu et al. (2016) found that human participants learning an associative task requiring integration across two orthogonal dimensions—the neck length and leg length of birds—showed a "hexadirectional" activation pattern in the entorhinal and prefrontal cortices, thought to be a signature of grid-cell processing. This activation pattern was first found in entorhinal cortex and, interestingly, in the PFC in a virtual spatial task (Doeller et al. 2010).

We note, however, that in these experiments, the tasks took a very long time for people to learn, and the relevant stimulus dimensions were bounded by very clear start and stop conditions. It may be, therefore, that cells are organizing experience along multiple dimensions that the brain has had to learn, over repeated experience. It may even be that the recruitment of spatial circuits was a desperate last-ditch attempt by the brain to make sense of the task, rather than a natural deployment of these circuits. This contrasts with spatial firing fields, which are present from the very first experience of an environment. It also suggests that space is primary and automatic, with the other dimensions secondary and acquired from experience. This is still an open question. Caveats aside, these neuronal findings, although limited in number thus far, suggest the potential for a broader role in cognition than simply understanding physical space.

6.1 Navigating Nonspatial Spaces

The discussion above pertains more to spatial self-localization, but what about navigation, which additionally entails (as we saw) goal setting and planning? Is navigation (at least in mammals) a subset of a more general cognitive function? This is a harder question to answer, in part because the full suite of navigationally involved structures has not yet been identified. Clearly, some brain structures in some species are specialized for navigation, rather than being more domain general, such as the magnetic sense that exists in many animals (Lohmann et al. 2022; Mouritsen 2018) or systems that allow animals to use the Sun or other celestial bodies (Foster et al. 2018; Freas and Spetch 2023). It seems unlikely that these specialized signals would readily have relevance for more abstract cognition. However, other aspects of navigational processing may be useful for "navigation" in abstract domains that are not spatial, but which have space-like properties.

Navigation through (as opposed to representation of) a conceptual space was shown in the stretchy-bird experiment by Constantinescu et al. (2016) discussed above, where participants had to make their way from one configuration to another to gain reward. To proceed from place A to place B in a given space requires knowledge of the direction between them, as well as one's moment-by-moment facing direction. How this is represented in abstract conceptual space remains to be determined since the HD signal is, as far as we know, anchored to real space. Whether the signal can split, or whether a direction-signal proxy can be created in another way (e.g., by grid cells or even a more abstract way) is an interesting question for future research.

6.2 Space-Like Reasoning in the Nonspatial Domain

A feature of navigating through a space is knowing something about the ordering or adjacency relationships between items in the map. A domain in which this issue has begun to be explored is relational reasoning, in which items are processed as to how they are situated with respect to each other in a space that might be physical or more abstract (e.g., quality, size, importance). Nonspatial space-like reasoning was first explored by Eichenbaum and colleagues (Cohen and Eichenbaum 1993; Eichenbaum and Cohen 2014), who argued early on that the hippocampal cognitive map is for learning these abstract space-like relational structures. Nonspatial possibilities that have been put forward include temporal—earlier than, later than, part of sequence A vs. part of sequence B (Fortin et al. 2002; Shahbaba et al. 2022)—and familial relationships (Garvert et al. 2017) or social relationships (Montagrin et al. 2018).

Encoding these relationships enables reasoning through the use of the known relations to make predictions. Consider a simple reasoning example involving transitive inference: animals trained to prefer A over B and B over C in a choice task will spontaneously choose A over C. This task is structured as A > B > C and could theoretically be represented using a spatial framework. Some evidence shows that hippocampal lesions impair transitive inference in rodents (Dusek and Eichenbaum 1997) and even fish (Sotelo-Parrilla et al. 2025). However, the task itself is heterogeneous (Gazes and Lazareva 2021) and involves multiple brain areas (Zhang et al. 2022). Thus, whether a navigational framework generally underpins these types of reasoning is not clear. Moreover, these tasks involve discrete items rather than continuous stimulus dimensions. Thus, the structure of this abstract cognitive space has been conceptualized as a graph structure, comprising nodes and edges rather than as a continuous space (Garvert et al. 2017; Muller et al. 1996). Whether the same structure(s) underpin(s) both types of stimulus domain (discrete vs. continuous) is another open question.

A recent popular model for how the hippocampus might support nonspatial reasoning is the *predictive map theory* by Stachenfeld et al. (2017)—a hybrid model for spatial reasoning that combines concepts from cognitive map theory and reinforcement learning theory. In their model, place cells do not encode current location per se, but rather a set of possible future *states*, called successor representations (see review by Gershman 2018). In the spatial domain, successor representations encompass adjacent locations, since these have the highest probability of being the next state, given the movement constraints on the animal. The predictive map theory has generated considerable discussion but there is relatively little experimental support for it yet, and some refuting evidence (Duvelle et al. 2021).

These ideas provide an important springboard for investigations about whether the navigation system has been deployed by evolution in service of other reasoning that has a space-like logic. If proponents of this idea are correct, then we ought to see limitations in the capacity for human reasoning or human imagination that reflect the limitations that we know exist in the processing of real space. Some of these limitations arise from space itself, whereas others stem from the brain circuits that evolved to process space efficiently. In terms of real space, we know that for all practical purposes it is three-dimensional, so reasoning in four dimensions is hard for humans. Furthermore, rotations are noncommutative (order-dependent), and thus we ought to find evidence for noncommutativity in human reasoning if this involves three dimensions.

With regard to the HD system, emerging evidence has shown that the compass signal instantiated by the HD cells is essentially planar; that is, it is expressed on a one-dimensional ring attractor, even during 3D movement (Laurens and Angelaki 2018; Page et al. 2018; Williams et al. 2025). This signal appears to be encoded independently from other dimensions of 3D motion (Angelaki and Laurens 2020; Finkelstein et al. 2015). Thus, this organization may also affect the human capacity for space-like nonspatial reasoning. A mix of psychological, neuroimaging, and single-neuron studies will be needed to truly answer the question of whether the ancient navigation circuits of our early vertebrate ancestors did indeed become co-opted to support human intelligence.

7 Conclusion and Open Questions for the Future

Navigation is a complex set of interacting cognitive processes supported by a brain-wide network of structures, whose individual functions we are only just beginning to understand. In this chapter we have identified important open questions to guide future neurobiological research, which should in turn inform new behavioral and ethological studies. A comprehensive understanding of the mechanistic basis of navigation, and its relation to general cognition, will require sustained collaboration across neurobiological, behavioral, and ethological work.

References

Alexander AS, Carstensen LC, Hinman JR, et al. (2020) Egocentric Boundary Vector Tuning of the Retrosplenial Cortex. Sci Adv 6 (8):eaaz2322. https://doi.org/10.1126/sciadv.aaz2322

Alexander AS, and Nitz DA (2017) Spatially Periodic Activation Patterns of Retrosplenial Cortex Encode Route Sub-Spaces and Distance Traveled. Curr Biol 27 (11):1551–1560.e1554. https://doi.org/10.1016/j.cub.2017.04.036

Alvares GA, Balleine BW, and Guastella AJ (2014) Impairments in Goal-Directed Actions Predict Treatment Response to Cognitive-Behavioral Therapy in Social Anxiety Disorder. PLoS ONE 9 (4):e94778. https://doi.org/10.1371/journal.pone.0094778

Angelaki DE, and Laurens J (2020) The Head Direction Cell Network: Attractor Dynamics, Integration within the Navigation System, and Three-Dimensional Properties. Curr Opin Neurobiol 60:136–144. https://doi.org/10.1016/j.conb.2019.12.002

Aronov D, Nevers R, and Tank DW (2017) Mapping of a Non-Spatial Dimension by the Hippocampal-Entorhinal Circuit. Nature 543 (7647):719–722. https://doi.org/10.1038/nature21692

Banino A, Barry C, Uria B, et al. (2018) Vector-Based Navigation Using Grid-Like Representations in Artificial Agents. Nature 557:429–433. https://doi.org/10.1038/s41586-018-0102-6

Barbas H (2000) Connections Underlying the Synthesis of Cognition, Memory, and Emotion in Primate Prefrontal Cortices. Brain Res Bull 52 (5):319–330. https://doi.org/10.1016/S0361-9230(99)00245-2

Barbas H, and Blatt GJ (1995) Topographically Specific Hippocampal Projections Target Functionally Distinct Prefrontal Areas in the Rhesus Monkey. Hippocampus 5 (6):511–533. https://doi.org/10.1002/hipo.450050604

Beetz MJ, and Hechavarría JC (2022) Neural Processing of Naturalistic Echolocation Signals in Bats. Front Neural Circuits 16:899370. https://doi.org/10.3389/fncir.2022.899370

Behrens TEJ, Muller TH, Whittington JCR, et al. (2018) What Is a Cognitive Map? Organizing Knowledge for Flexible Behavior. Neuron 100 (2):490–509. https://doi.org/10.1016/j.neuron.2018.10.002

Bellmund JLS, Cothi WD, Ruiter TA, et al. (2020) Deforming the Metric of Cognitive Maps Distorts Memory. Nat Hum Behav 4 (2):177–188. https://doi.org/10.1038/s41562-019-0767-3

Bellmund JLS, Gärdenfors P, Moser EI, and Doeller CF (2018) Navigating Cognition: Spatial Codes for Human Thinking. Science 362 (6415):eaat6766. https://doi.org/10.1126/science.aat6766

Botvinick M, Wang JX, Dabney W, Miller KJ, and Kurth-Nelson Z (2020) Deep Reinforcement Learning and Its Neuroscientific Implications. Neuron 107 (4):603–616. https://doi.org/10.1016/j.neuron.2020.06.014

Bourque CW, and Oliet SHR (1997) Osmoreceptors in the Central Nervous System. Annu Rev Physiol 59 (1):601–619. https://doi.org/10.1146/annurev.physiol.59.1.601

Brown TI, Carr VA, LaRocque KF, et al. (2016) Prospective Representation of Navigational Goals in the Human Hippocampus. Science 352 (6291):1323–1326. https://doi.org/10.1126/science.aaf0784

Brown TI, Ross RS, Keller JB, Hasselmo ME, and Stern CE (2010) Which Way Was I Going? Contextual Retrieval Supports the Disambiguation of Well Learned Overlapping Navigational Routes. J Neurosci 30 (21):7414–7422. https://doi.org/10.1523/JNEUROSCI.6021-09.2010

Brown TI, Ross RS, Tobyne SM, and Stern CE (2012) Cooperative Interactions between Hippocampal and Striatal Systems Support Flexible Navigation. Neuroimage 60 (2):1316–1330. https://doi.org/10.1016/j.neuroimage.2012.01.046

Burgess N (2006) Spatial Memory: How Egocentric and Allocentric Combine. Trends Cogn Sci 10 (12):551–557. https://doi.org/10.1016/j.tics.2006.10.005

Byrne P, Becker S, and Burgess N (2007) Remembering the Past and Imagining the Future: A Neural Model of Spatial Memory and Imagery. Psychol Rev 114 (2):340–375. https://doi.org/10.1037/0033-295X.114.2.340

Cardinal RN, Parkinson JA, Hall J, and Everitt BJ (2002) Emotion and Motivation: The Role of the Amygdala, Ventral Striatum, and Prefrontal Cortex. Neurosci Biobehav Rev 26 (3):321–352. https://doi.org/10.1016/S0149-7634(02)00007-6

Cavada C (2000) The Anatomical Connections of the Macaque Monkey Orbitofrontal Cortex: A Review. Cereb Cortex 10 (3):220–242. https://doi.org/10.1093/cercor/10.3.220

Chanales AJH, Oza A, Favila SE, and Kuhl BA (2017) Overlap among Spatial Memories Triggers Repulsion of Hippocampal Representations. Curr Biol 27 (15):2307–2317. e2305. https://doi.org/10.1016/j.cub.2017.06.057

Ciocchi S, Passecker J, Malagon-Vina H, Mikus N, and Klausberger T (2015) Brain Computation: Selective Information Routing by Ventral Hippocampal CA1 Projection Neurons. Science 348 (6234):560–563. https://doi.org/10.1126/science.aaa3245

Clark BJ, Simmons CM, Berkowitz LE, and Wilber AA (2018) The Retrosplenial-Parietal Network and Reference Frame Coordination for Spatial Navigation. Behav Neurosci 132 (5):416–429. https://doi.org/10.1037/bne0000260

Cohen NJ, and Eichenbaum H (1993) Memory, Amnesia, and the Hippocampal System. MIT Press, Cambridge, MA

Cohen YE, and Andersen RA (2002) A Common Reference Frame for Movement Plans in the Posterior Parietal Cortex. Nat Rev Neurosci 3 (7):553–562. https://doi.org/10.1038/nrn873

Constantinescu AO, O'Reilly JX, and Behrens TEJ (2016) Organizing Conceptual Knowledge in Humans with a Gridlike Code. Science 352 (6292):1464–1468. https://doi.org/10.1126/science.aaf0941

Crivelli-Decker J, Clarke A, Park SA, et al. (2023) Goal-Oriented Representations in the Human Hippocampus during Planning and Navigation. Nat Commun 14 (1):2946. https://doi.org/10.1038/s41467-023-35967-6

Danjo T, Toyoizumi T, and Fujisawa S (2018) Spatial Representations of Self and Other in the Hippocampus. Science 359 (6372):213–218. https://doi.org/10.1126/science.aao3898

de Vries R, Morquecho-Campos P, de Vet E, et al. (2020) Human Spatial Memory Implicitly Prioritizes High-Calorie Foods. Sci Rep 10 (1):15174. https://doi.org/10.1038/s41598-020-72570-x

Doeller CF, Barry C, and Burgess N (2010) Evidence for Grid Cells in a Human Memory Network. Nature 463 (7281):657–661. https://doi.org/10.1038/nature08704

Dombrovski AY, Luna B, and Hallquist MN (2020) Differential Reinforcement Encoding Along the Hippocampal Long Axis Helps Resolve the Explore–Exploit Dilemma. Nat Commun 11 (1):5407. https://doi.org/10.1038/s41467-020-18864-0

Dusek JA, and Eichenbaum H (1997) The Hippocampus and Memory for Orderly Stimulus Relations. PNAS 94 (13):7109–7114. https://doi.org/10.1073/pnas.94.13.7109

Duvelle É, Grieves RM, Liu A, et al. (2021) Hippocampal Place Cells Encode Global Location but Not Connectivity in a Complex Space. Curr Biol 31 (6):1221–1233.e1229. https://doi.org/10.1016/j.cub.2021.01.005

Duvelle E, and Jeffery KJ (2018) Social Spaces: Place Cells Represent the Locations of Others. Curr Biol 28 (6):R271–R273. https://doi.org/10.1016/j.cub.2018.02.017

Eichenbaum H, and Cohen NJ (2014) Can We Reconcile the Declarative Memory and Spatial Navigation Views on Hippocampal Function? Neuron 83 (4):764–770. https://doi.org/10.1016/j.neuron.2014.07.032

Ekstrom AD, Huffman DJ, and Starrett M (2017) Interacting Networks of Brain Regions Underlie Human Spatial Navigation: A Review and Novel Synthesis of the Literature. J Neurophysiol 118 (6):3328–3344. https://doi.org/10.1152/jn.00531.2017

Epstein RA (2008) Parahippocampal and Retrosplenial Contributions to Human Spatial Navigation. Trends Cogn Sci 12 (10):388–396. https://doi.org/10.1016/j.tics.2008.07.004

Epstein RA, Patai EZ, Julian JB, and Spiers HJ (2017) The Cognitive Map in Humans: Spatial Navigation and Beyond. Nat Neurosci 20 (11):1504–1513. https://doi.org/10.1038/nn.4656

Fanselow MS, and Dong H-W (2010) Are the Dorsal and Ventral Hippocampus Functionally Distinct Structures? Neuron 65 (1):7–19. https://doi.org/10.1016/j.neuron.2009.11.031

Ferbinteanu J, and Shapiro ML (2006) Prospective and Retrospective Memory Coding in the Hippocampus. Neuron 40 (6):1227–1239. https://doi.org/10.1016/s0896-6273(03)00752-9

Finkelstein A, Derdikman D, Rubin A, et al. (2015) Three-Dimensional Head-Direction Coding in the Bat Brain. Nature 517 (7533):159–164. https://doi.org/10.1038/nature14031

Fortin NJ, Agster KL, and Eichenbaum HB (2002) Critical Role of the Hippocampus in Memory for Sequences of Events. Nat Neurosci 5 (5):458–462. https://doi.org/10.1038/nn834

Foster JJ, Smolka J, Nilsson D-E, and Dacke M (2018) How Animals Follow the Stars. Proc R Soc Lond B Biol Sci 285 (1871):20172322. https://doi.org/10.1098/rspb.2017.2322

Freas CA, and Spetch ML (2023) Varieties of Visual Navigation in Insects. Anim Cogn 26 (1):319–342. https://doi.org/10.1007/s10071-022-01720-7

Gardner RJ, Hermansen E, Pachitariu M, et al. (2022) Toroidal Topology of Population Activity in Grid Cells. Nature 602 (7895):123–128. https://doi.org/10.1038/s41586-021-04268-7

Garvert MM, Dolan RJ, and Behrens TE (2017) A Map of Abstract Relational Knowledge in the Human Hippocampal–Entorhinal Cortex. eLife 6:e17086. https://doi.org/10.7554/eLife.17086

Gasbarri A, Sulli A, and Packard MG (1997) The Dopaminergic Mesencephalic Projections to the Hippocampal Formation in the Rat. Prog Neuropsychopharmacol Biol Psychiatry 21 (1):1–22. https://doi.org/10.1016/S0278-5846(96)00157-1

Gazes RP, and Lazareva OF (2021) Does Cognition Differ across Species, and How Do We Know? Lessons from Research in Transitive Inference. J Exp Psychol Anim 47 (3):223–233. https://doi.org/10.1037/xan0000301

Geerts JP, Chersi F, Stachenfeld KL, and Burgess N (2020) A General Model of Hippocampal and Dorsal Striatal Learning and Decision Making. PNAS 117 (49):31427–31437. https://doi.org/10.1073/pnas.2007981117

George D, Rikhye RV, Gothoskar N, et al. (2021) Clone-Structured Graph Representations Enable Flexible Learning and Vicarious Evaluation of Cognitive Maps. Nat Commun 12 (1):2392. https://doi.org/10.1038/s41467-021-22559-5

Gergues MM, Han KJ, Choi HS, et al. (2020) Circuit and Molecular Architecture of a Ventral Hippocampal Network. Nat Neurosci 23 (11):1444–1452. https://doi.org/10.1038/s41593-020-0705-8

Gershman SJ (2018) The Successor Representation: Its Computational Logic and Neural Substrates. J Neurosci 38 (33):7193–7200. https://doi.org/10.1523/JNEUROSCI.0151-18.2018

Ginosar G, Aljadeff J, Burak Y, et al. (2021) Locally Ordered Representation of 3D Space in the Entorhinal Cortex. Nature 596 (7872):404–409. https://doi.org/10.1038/s41586-021-03783-x

Gofman X, Tocker G, Weiss S, et al. (2019) Dissociation between Postrhinal Cortex and Downstream Parahippocampal Regions in the Representation of Egocentric Boundaries. Curr Biol 29 (16):2751–2757.e2754. https://doi.org/10.1016/j.cub.2019.07.007

Grieves RM, Jedidi-Ayoub S, Mishchanchuk K, et al. (2021) Irregular Distribution of Grid Cell Firing Fields in Rats Exploring a 3D Volumetric Space. Nat Neurosci 24 (11):1567–1573. https://doi.org/10.1038/s41593-021-00907-4

Grieves RM, Wood ER, and Dudchenko PA (2016) Place Cells on a Maze Encode Routes Rather Than Destinations. eLife 5:e15986. https://doi.org/10.7554/eLife.15986

Haber SN, and Knutson B (2010) The Reward Circuit: Linking Primate Anatomy and Human Imaging. Neuropsychopharmacol 35 (1):4–26. https://doi.org/10.1038/npp.2009.129

Hafting T, Fyhn M, Molden S, Moser M-B, and Moser EI (2005) Microstructure of a Spatial Map in the Entorhinal Cortex. Nature 436 (7052):801–806. https://doi.org/10.1038/nature03721

Han LT, and Epstein RA (2025) Distinct Mechanisms for Panoramic and Landmark-Based View Integration in Human Scene-Selective Cortex. bioRxiv Preprint. https://doi.org/10.1101/2025.01.24.634774

Hardcastle K, Maheswaranathan N, Ganguli S, and Giocomo LM (2017) A Multiplexed, Heterogeneous, and Adaptive Code for Navigation in Medial Entorhinal Cortex. Neuron 94 (2):375–387.e377. https://doi.org/10.1016/j.neuron.2017.03.025

Hasselmo ME, and Eichenbaum H (2005) Hippocampal Mechanisms for the Context-Dependent Retrieval of Episodes. Neural Netw 18 (9):1172–1190. https://doi.org/10.1016/j.neunet.2005.08.007

He Q, Liu JL, Eschapasse L, Zagora AK, and Brown TI (2024) The Neural Correlates of Memory Integration in Value-Based Decision-Making during Human Spatial Navigation. Neuropsychologia 193:108758. https://doi.org/10.1016/j.neuropsychologia.2023.108758

He Q, Starnes J, and Brown TI (2022) Environmental Overlap Influences Goal-Oriented Coding of Spatial Sequences Differently Along the Long-Axis of Hippocampus. Hippocampus 32 (6):419–435. https://doi.org/10.1002/hipo.23416

Hinman JR, Chapman GW, and Hasselmo ME (2019) Neuronal Representation of Environmental Boundaries in Egocentric Coordinates. Nat Commun 10 (1):1–8. https://doi.org/10.1038/s41467-019-10722-y

Ho H, Kejzar N, Sasaguri H, et al. (2023) A Fully Automated Home Cage for Long-Term Continuous Phenotyping of Mouse Cognition and Behavior. Cell Rep Meth 3 (7):100532. https://doi.org/10.1016/j.crmeth.2023.100532

Jacob P-Y, Casali G, Spieser L, et al. (2017) An Independent, Landmark-Dominated Head-Direction Signal in Dysgranular Retrosplenial Cortex. Nat Neurosci 20 (2):173–175. https://doi.org/10.1038/nn.4465

Jankowski MM, Passecker J, Islam MN, et al. (2015) Evidence for Spatially-Responsive Neurons in the Rostral Thalamus. Front Behav Neurosci 9:256. https://doi.org/10.3389/fnbeh.2015.00256

Javadi A-H, Emo B, Howard LR, et al. (2017) Hippocampal and Prefrontal Processing of Network Topology to Simulate the Future. Nat Commun 8 (1):1–11. https://doi.org/10.1038/ncomms14652

Jeffery KJ, Cheng K, Newcombe NS, Bingman VP, and Menzel R (2024) Unpacking the Navigation Toolbox: Insights from Comparative Cognition. Proc R Soc Lond B Biol Sci 291 (2016):20231304. https://doi.org/10.1098/rspb.2023.1304

Julian JB, Keinath AT, Marchette SA, and Epstein RA (2018) The Neurocognitive Basis of Spatial Reorientation. Curr Biol 28 (17):R1059–R1073. https://doi.org/10.1016/j.cub.2018.04.057

Kay K, Chung JE, Sosa M, et al. (2020) Constant Sub-Second Cycling between Representations of Possible Futures in the Hippocampus. Cell 180 (3):552–567 e525. https://doi.org/10.1016/j.cell.2020.01.014

Kenrick DT, Griskevicius V, Neuberg SL, and Schaller M (2010) Renovating the Pyramid of Needs. Perspect Psychol Sci 5 (3):292–314. https://doi.org/10.1177/1745691610369469

Kim SS, Hermundstad AM, Romani S, Abbott LF, and Jayaraman V (2019) Generation of Stable Heading Representations in Diverse Visual Scenes. Nature 576 (7785):126–131. https://doi.org/10.1038/s41586-019-1767-1

Kinkhabwala AA, Gu Y, Aronov D, and Tank DW (2020) Visual Cue-Related Activity of Cells in the Medial Entorhinal Cortex during Navigation in Virtual Reality. eLife 9:e43140. https://doi.org/10.7554/elife.43140

Kjelstrup KB, Solstad T, Brun VH, et al. (2008) Finite Scale of Spatial Representation in the Hippocampus. Science 321 (5885):140–143. https://doi.org/10.1126/science.1157086

Krupic J, Bauza M, Burton S, Barry C, and O'Keefe J (2015) Grid Cell Symmetry Is Shaped by Environmental Geometry. Nature 518 (232):235. https://doi.org/10.1038/nature14153

Krupic J, Bauza M, Burton S, Lever C, and O'Keefe J (2014) How Environment Geometry Affects Grid Cell Symmetry and What We Can Learn from It. Philos Trans R Soc Lond B Biol Sci 369 (1635):20130188. https://doi.org/10.1098/rstb.2013.0188

Krupic J, Burgess N, and O'Keefe J (2012) Neural Representations of Location Composed of Spatially Periodic Bands. Science 337 (6096):853–857. https://doi.org/10.1126/science.1222403

Kunz L, Brandt A, Reinacher PC, et al. (2021) A Neural Code for Egocentric Spatial Maps in the Human Medial Temporal Lobe. Neuron 109 (17):2781–2796.e2710. https://doi.org/10.1016/j.neuron.2021.06.019

LaChance PA, Todd TP, and Taube JS (2019) A Sense of Space in Postrhinal Cortex. Science 365 (6449):eaax4192. https://doi.org/10.1126/science.aax4192

Laurens J, and Angelaki DE (2018) The Brain Compass: A Perspective on How Self-Motion Updates the Head Direction Cell Attractor. Neuron 97:275–289. https://doi.org/10.1016/j.neuron.2017.12.020

Lee D, Seo H, and Jung MW (2012) Neural Basis of Reinforcement Learning and Decision Making. Annu Rev Neurosci 35 (1):287–308. https://doi.org/10.1146/annurev-neuro-062111-150512

Lee I, Jerman TS, and Kesner RP (2005) Disruption of Delayed Memory for a Sequence of Spatial Locations Following CA1- or CA3-Lesions of the Dorsal Hippocampus. Neurobiol Learn Mem 84 (2):138–147. https://doi.org/10.1016/j.nlm.2005.06.002

Lee SW, Shimojo S, and O'Doherty JP (2014) Neural Computations Underlying Arbitration between Model-Based and Model-Free Learning. Neuron 81 (3):687–699. https://doi.org/10.1016/j.neuron.2013.11.028

Lisman J, Buzsaki G, Eichenbaum H, et al. (2017) Viewpoints: How the Hippocampus Contributes to Memory, Navigation and Cognition. Nat Neurosci 20 (11):1434–1447. https://doi.org/10.1038/nn.4661

Lohmann KJ, Goforth KM, Mackiewicz AG, Lim DS, and Lohmann CMF (2022) Magnetic Maps in Animal Navigation. J Comp Physiol A Neuroethol Sens Neural Behav Physiol 208 (1):41–67. https://doi.org/10.1007/s00359-021-01529-8

Lomi E, Jeffery KJ, and Mitchell AS (2023) Convergence of Location, Direction, and Theta in the Rat Anteroventral Thalamic Nucleus. iScience 26 (7):106993. https://doi.org/10.1016/j.isci.2023.106993

Long X, and Zhang S-J (2021) A Novel Somatosensory Spatial Navigation System Outside the Hippocampal Formation. Cell Res 31 (6):649–663. https://doi.org/10.1038/s41422-020-00448-8

Luo X, Mok RM, and Love BC (2024) The Inevitability and Superfluousness of Cell Types in Spatial Cognition. eLife 13:RP99047. https://doi.org/10.1101/2024.01.10.575026

Lüthi A, and Lüscher C (2014) Pathological Circuit Function Underlying Addiction and Anxiety Disorders. Nat Neurosci 17 (12):1635–1643. https://doi.org/10.1038/nn.3849

Ma F, Lin H, and Zhou J (2025) Prediction, Inference, and Generalization in Orbitofrontal Cortex. Curr Biol 35 (7):R266–R272. https://doi.org/10.1016/j.cub.2025.02.021

MacDonald C, Lepage K, Eden U, and Eichenbaum H (2011) Hippocampal "Time Cells" Bridge the Gap in Memory for Discontiguous Events. Neuron 71 (4):737–749. https://doi.org/10.1016/j.neuron.2011.07.012

Mainen ZF, and Kepecs A (2009) Neural Representation of Behavioral Outcomes in the Orbitofrontal Cortex. Curr Opin Neurobiol 19 (1):84–91. https://doi.org/10.1016/j.conb.2009.03.010

Mao D, Neumann AR, Sun J, et al. (2018) Hippocampus-Dependent Emergence of Spatial Sequence Coding in Retrosplenial Cortex. PNAS 115 (31):8015–8018. https://doi.org/10.1073/pnas.1803224115

Marchette SA, Vass LK, Ryan J, and Epstein RA (2014) Anchoring the Neural Compass: Coding of Local Spatial Reference Frames in Human Medial Parietal Lobe. Nat Neurosci 17 (11):1598–1606. https://doi.org/10.1038/nn.3834

Marty N, Dallaporta M, and Thorens B (2007) Brain Glucose Sensing, Counterregulation, and Energy Homeostasis. Physiology 22 (4):241–251. https://doi.org/10.1152/physiol.00010.2007

McNamee D, and Wolpert DM (2019) Internal Models in Biological Control. Annul Rev Control Robo Auton Syst 2 (1):339–364. https://doi.org/10.1146/annurev-control-060117-105206

Miletić S, Boag RJ, Trutti AC, et al. (2021) A New Model of Decision Processing in Instrumental Learning Tasks. eLife 10:e63055. https://doi.org/10.7554/eLife.63055

Miller AMP, Vedder LC, Law LM, and Smith DM (2014) Cues, Context, and Long-Term Memory: The Role of the Retrosplenial Cortex in Spatial Cognition. Front Hum Neurosci 8 (586):e00586. https://doi.org/10.3389/fnhum.2014.00586

Moita MA, Rosis S, Zhou Y, LeDoux JE, and Blair HT (2004) Putting Fear in Its Place: Remapping of Hippocampal Place Cells during Fear Conditioning. J Neurosci 24 (31):7015–7023. https://doi.org/10.1523/jneurosci.5492-03.2004

Montagrin A, Saiote C, and Schiller D (2018) The Social Hippocampus. Hippocampus 28 (9):672–679. https://doi.org/10.1002/hipo.22797

Morris RGM (2015) The Mantle of the Heavens: Reflections on the 2014 Nobel Prize for Medicine or Physiology. Hippocampus 25 (6):682–689. https://doi.org/10.1002/hipo.22455

Moser EI, Kropff E, and Moser MB (2008) Place Cells, Grid Cells, and the Brain's Spatial Representation System. Annu Rev Neurosci 31:69–89. https://doi.org/10.1146/annurev.neuro.31.061307.090723

Mouritsen H (2018) Long-Distance Navigation and Magnetoreception in Migratory Animals. Nature 558 (7708):50–59. https://doi.org/10.1038/s41586-018-0176-1

Muller RU, Stead M, and Pach J (1996) The Hippocampus as a Cognitive Graph. J Gen Physiol 107 (6):663–694. https://doi.org/10.1085/jgp.107.6.663

Nguyen R, Koukoutselos K, Forro T, and Ciocchi S (2023) Fear Extinction Relies on Ventral Hippocampal Safety Codes Shaped by the Amygdala. Sci Adv 9 (22):eadg4881. https://doi.org/10.1126/sciadv.adg4881

Nitz DA (2012) Spaces within Spaces: Rat Parietal Cortex Neurons Register Position across Three Reference Frames. Nat Neurosci 15 (10):1365–1367. https://doi.org/10.1038/nn.3213

Nuno-Perez A, Trusel M, Lalive AL, et al. (2021) Stress Undermines Reward-Guided Cognitive Performance through Synaptic Depression in the Lateral Habenula. Neuron 109 (6):947–956.e945. https://doi.org/10.1016/j.neuron.2021.01.008

Nyberg N, Duvelle E, Barry C, and Spiers HJ (2022) Spatial Goal Coding in the Hippocampal Formation. Neuron 110 (3):394–422. https://doi.org/10.1016/j.neuron.2021.12.012

O'Keefe J, and Burgess N (1996) Geometric Determinants of the Place Fields of Hippocampal Neurons. Nature 381 (6581):425–428. https://doi.org/10.1038/381425a0

O'Keefe J, and Dostrovsky J (1971) The Hippocampus as a Spatial Map: Preliminary Evidence from Unit Activity in the Freely-Moving Rat. Brain Res 34 (1):171–175. https://doi.org/10.1016/0006-8993(71)90358-1

O'Keefe J, and Krupic J (2021) Do Hippocampal Pyramidal Cells Respond to Nonspatial Stimuli? Physiol Rev 101 (3):1427–1456. https://doi.org/10.1152/physrev.00014.2020

O'Keefe J, and Nadel L (1978) The Hippocampus as a Cognitive Map. Oxford Univ Press, London

O'Mara SM, and Aggleton JP (2019) Space and Memory (Far) Beyond the Hippocampus: Many Subcortical Structures Also Support Cognitive Mapping and Mnemonic Processing. Front Neural Circuits 13. https://doi.org/10.3389/fncir.2019.00052

Omer DB, Maimon SR, Las L, and Ulanovsky N (2018) Social Place-Cells in the Bat Hippocampus. Science 359 (6372):218–224. https://doi.org/10.1126/science.aao3474

Page HJI, Wilson JJ, and Jeffery KJ (2018) A Dual-Axis Rotation Rule for Updating the Head Direction Cell Reference Frame during Movement in Three Dimensions. J Neurophysiol 119:192–208. https://doi.org/10.1152/jn.00501.2017

Park K, Yeo Y, Shin K, and Kwag J (2024) Egocentric Neural Representation of Geometric Vertex in the Retrosplenial Cortex. Nat Commun 15 (1):7156. https://doi.org/10.1038/s41467-024-51391-w

Patai EZ, and Spiers HJ (2021) The Versatile Wayfinder: Prefrontal Contributions to Spatial Navigation. Trends Cogn Sci 25 (6):520–533. https://doi.org/10.1016/j.tics.2021.02.010

Pedamonti, D, Mohinta, S, Dimitrov, MV et al. (2025) Hippocampus Supports Multi-task Reinforcement Learning under Partial Observability. Nat Commun 16:9619. https://doi.org/10.1038/s41467-025-64591-9

Peer M, and Epstein RA (2025) Cognitive Maps for Hierarchical Spaces in the Human Brain. Cereb Cortex 35 (9):bhaf261. https://doi.org/10.1093/cercor/bhaf261

——— (2021) The Human Brain Uses Spatial Schemas to Represent Segmented Environments. Curr Biol 31 (21):4677–4688.e4678. https://doi.org/10.1016/j.cub.2021.08.012

Pessoa L (2009) How Do Emotion and Motivation Direct Executive Control? Trends Cogn Sci 13 (4):160–166. https://doi.org/10.1016/j.tics.2009.01.006

Peyrache A, Lacroix MM, Petersen PC, and Buzsáki G (2015) Internally Organized Mechanisms of the Head Direction Sense. Nat Neurosci 18 (4):569–575. https://doi.org/10.1038/nn.3968

Pfeiffer BE, and Foster DJ (2013) Hippocampal Place-Cell Sequences Depict Future Paths to Remembered Goals. Nature 497 (7447):74–79. https://doi.org/10.1038/nature12112

Pikkarainen M, Rnkk S, Savander V, Insausti R, and Pitknen A (1999) Projections from the Lateral, Basal, and Accessory Basal Nuclei of the Amygdala to the Hippocampal Formation in Rat. J Comp Neurol 403 (2):229–260. https://doi.org/10.1002/(SICI)1096-9861(19990111)403:2<229::AID-CNE7>3.0.CO;2-P

Poucet B, Lenck-Santini PP, Hok V, et al. (2004) Spatial Navigation and Hippocampal Place Cell Firing: The Problem of Goal Encoding. Rev Neurosci 15 (2):89–107. https://doi.org/10.1515/REVNEURO.2004.15.2.89

Preston AR, and Eichenbaum H (2013) Interplay of Hippocampus and Prefrontal Cortex in Memory. Curr Biol 23 (17):R764–773. https://doi.org/10.1016/j.cub.2013.05.041

Quirk GJ, Muller RU, Kubie JL, and Ranck J (1992) The Positional Firing Properties of Medial Entorhinal Neurons: Description and Comparison with Hippocampal Place Cells. J Neurosci 12 (5):1945–1963. https://doi.org/10.1523/jneurosci.12-05-01945.1992

Ranck JB (1975) Behavioral Correlates and Firing Repertoires of Neurons in the Dorsal Hippocampal Formation and Septum of Unrestrained Rats. In: Isaacson R, Pribram K (eds) The Hippocampus. Springer, New York, pp 207–244. https://doi.org/10.1007/978-1-4684-2979-4_7

Robbins TW, Ersche KD, and Everitt BJ (2008) Drug Addiction and the Memory Systems of the Brain. Ann N Y Acad Sci 114 (1):1–21. https://doi.org/10.1196/annals.1441.020

Robertson CE, Hermann KL, Mynick A, Kravitz DJ, and Kanwisher N (2016) Neural Representations Integrate the Current Field of View with the Remembered 360° Panorama in Scene-Selective Cortex. Curr Biol 26 (0):2463–2468. https://doi.org/DOI: http://dx.doi.org/10.1016/j.cub.2016.07.002

Rudebeck PH, Behrens TE, Kennerley SW, et al. (2008) Frontal Cortex Subregions Play Distinct Roles in Choices between Actions and Stimuli. J Neurosci 28 (51):13775–13785. https://doi.org/10.1523/JNEUROSCI.3541-08.2008

Salinas E, and Thier P (2000) Gain Modulation: A Major Computational Principle of the Central Nervous System. Neuron 27 (1):15–21. https://doi.org/10.1016/s0896-6273(00)00004-0

Seelig JD, and Jayaraman V (2015) Neural Dynamics for Landmark Orientation and Angular Path Integration. Nature 521 (7551):186–191. https://doi.org/10.1038/nature14446

Shahbaba B, Li L, Agostinelli F, et al. (2022) Hippocampal Ensembles Represent Sequential Relationships among an Extended Sequence of Nonspatial Events. Nat Commun 13 (1):787. https://doi.org/10.1038/s41467-022-28057-6

Sherrill KR, Erdem UM, Ross RS, et al. (2013) Hippocampus and Retrosplenial Cortex Combine Path Integration Signals for Successful Navigation. J Neurosci 33 (49):19304–19313. https://doi.org/10.1523/JNEUROSCI.1825-13.2013

Shin JD, and Jadhav SP (2016) Multiple Modes of Hippocampal-Prefrontal Interactions in Memory-Guided Behavior. Curr Opin Neurobiol 40:161–169. https://doi.org/10.1016/j.conb.2016.07.015

Shpokayte M, McKissick O, Guan X, et al. (2022) Hippocampal Cells Segregate Positive and Negative Engrams. Commun Biol 5 (1):1009. https://doi.org/10.1038/s42003-022-03906-8

Skaggs W, Knierim J, Kudrimoti H, and McNaughton B (1994) A Model of the Neural Basis of the Rat's Sense of Direction. Adv Neural Inf Process Syst 7:173–180. https://pubmed.ncbi.nlm.nih.gov/11539168/

Skrynka J, and Vincent BT (2019) Hunger Increases Delay Discounting of Food and Non-Food Rewards. Psychon Bull Rev 26 (5):1729–1737. https://doi.org/10.3758/s13423-019-01655-0

Solstad T, Boccara CN, Kropff E, Moser M-B, and Moser EI (2008) Representation of Geometric Borders in the Entorhinal Cortex. Science 322 (5909):1865–1868. https://doi.org/10.1126/science.1166466

Sosa M, and Giocomo LM (2021) Navigating for Reward. Nat Rev Neurosci 22 (8):472–487. https://doi.org/10.1038/s41583-021-00479-z

Sotelo-Parrilla G, Quintero B, Trujillo I, et al. (2025) Hippocampal Pallium Lesion Impairs Transitive Inference in Goldfish. Hippocampus 35 (2):e70007. https://doi.org/10.1002/hipo.70007

Spiers HJ, Hayman RMA, Jovalekic A, Marozzi E, and Jeffery KJ (2015) Place Field Repetition and Purely Local Remapping in a Multicompartment Environment. Cereb Cortex 25 (1):10–25. https://doi.org/10.1093/cercor/bht198

Spiers HJ, and Maguire EA (2006) Thoughts, Behavior, and Brain Dynamics during Navigation in the Real World. Neuroimage 31 (4):1826–1840. https://doi.org/10.1016/j.neuroimage.2006.01.037

Stachenfeld KL, Botvinick MM, and Gershman SJ (2017) The Hippocampus as a Predictive Map. Nat Neurosci 20 (11):1643–1653. https://doi.org/10.1038/nn.4650

Strange BA, Witter MP, Lein ES, and Moser EI (2014) Functional Organization of the Hippocampal Longitudinal Axis. Nat Rev Neurosci 15 (10):655–669. https://doi.org/10.1038/nrn3785

Suito T, and Tominaga M (2024) Functional Relationship between Peripheral Thermosensation and Behavioral Thermoregulation. Front Neural Circuits 18:1435757. https://doi.org/10.3389/fncir.2024.1435757

Tan B, Nöbauer T, Browne CJ, et al. (2022) Dynamic Processing of Hunger and Thirst by Common Mesolimbic Neural Ensembles. PNAS 119 (43):e2211688119. https://doi.org/10.1073/pnas.2211688119

Tan CL, and Knight ZA (2018) Regulation of Body Temperature by the Nervous System. Neuron 98 (1):31–48. https://doi.org/10.1016/j.neuron.2018.02.022

Taube JS (2025) The Discovery of Head Direction Cells: A Personal Account. Hippocampus 35 (1):e23670. https://doi.org/10.1002/hipo.23670

Taube JS, Muller R, and Ranck J (1990) Head-Direction Cells Recorded from the Postsubiculum in Freely Moving Rats. I. Description and Quantitative Analysis. J Neurosci 10 (2):420–435. https://doi.org/10.1523/jneurosci.10-02-00420.1990

Thura D, Beauregard-Racine J, Fradet C-W, and Cisek P (2012) Decision Making by Urgency Gating: Theory and Experimental Support. J Neurophysiol 108 (11):2912–2930. https://doi.org/10.1152/jn.01071.2011

Tolman EC (1948) Cognitive Maps in Rats and Men. Psychol Rev 55 (4):189–208. https://doi.org/10.1037/h0061626

Tye KM (2018) Neural Circuit Motifs in Valence Processing. Neuron 100 (2):436–452. https://doi.org/10.1016/j.neuron.2018.10.001

Tyng CM, Amin HU, Saad MNM, and Malik AS (2017) The Influences of Emotion on Learning and Memory. Front Psychol 8:1454. https://doi.org/10.3389/fpsyg.2017.01454

Vann SD, Aggleton JP, and Maguire EA (2009) What Does the Retrosplenial Cortex Do? Nat Rev Neurosci 10 (11):792–802. https://doi.org/10.1038/nrn2733

Verschure PFMJ, Pennartz CMA, and Pezzulo G (2014) The Why, What, Where, When and How of Goal-Directed Choice: Neuronal and Computational Principles. Philos Trans R Soc Lond B Biol Sci 369 (1655):20130483. https://doi.org/10.1098/rstb.2013.0483

Viard A, Doeller CF, Hartley T, Bird CM, and Burgess N (2011) Anterior Hippocampus and Goal-Directed Spatial Decision Making. J Neurosci 31 (12):4613–4621. https://doi.org/10.1523/JNEUROSCI.4640-10.2011

Walton ME, Behrens TEJ, Buckley MJ, Rudebeck PH, and Rushworth MFS (2010) Separable Learning Systems in the Macaque Brain and the Role of Orbitofrontal Cortex in Contingent Learning. Neuron 65 (6):927–939. https://doi.org/10.1016/j.neuron.2010.02.027

Wang C, Chen X, Lee H, et al. (2018) Egocentric Coding of External Items in the Lateral Entorhinal Cortex. Science 362 (6417):945–949. https://doi.org/10.1126/science.aau4940

Wieber F, Thürmer JL, and Gollwitzer PM (2015) Promoting the Translation of Intentions into Action by Implementation Intentions: Behavioral Effects and Physiological Correlates. Front Hum Neurosci 9:e000395. https://doi.org/10.3389/fnhum.2015.00395

Wiener JM, Büchner SJ, and Hölscher C (2009) Taxonomy of Human Wayfinding Tasks: A Knowledge-Based Approach. Spat Cogn Comput 9 (2):152–165. https://doi.org/10.1080/13875860902906496

Wiener JM, Shettleworth S, Bingman VP, et al. (2011) Animal Navigation: A Synthesis. Animal Thinking: Contemporary Issues in Comparative Cognition. Strüngmann Forum Reports. MIT Press, Cambridge, MA

Wilber AA, Clark BJ, Forster TC, Tatsuno M, and McNaughton BL (2014) Interaction of Egocentric and World-Centered Reference Frames in the Rat Posterior Parietal Cortex. J Neurosci 34 (16):5431–5446. https://doi.org/10.1523/JNEUROSCI.0511-14.2014

Williams M, Street J, Burgess N, and Jeffery K (2025) Head Direction Cells Use a Head-Referenced Dual-Axis Updating Rule in 3D Space. Preprint. https://doi.org/10.1101/2025.09.17.676760

Wood ER, Dudchenko PA, Robitsek RJ, and Eichenbaum H (2000) Hippocampal Neurons Encode Information about Different Types of Memory Episodes Occurring in the Same Location. Neuron 27 (3):623–633. https://doi.org/10.1016/s0896-6273(00)00071-4

Young JJ, and Shapiro ML (2011) Dynamic Coding of Goal-Directed Paths by Orbital Prefrontal Cortex. J Neurosci 31 (16):5989–6000. https://doi.org/10.1523/JNEUROSCI.5436-10.2011

Zhang N, Grieves RM, and Jeffery KJ (2021) Environment Symmetry Drives a Multidirectional Code in Rat Retrosplenial Cortex. J Neurosci 42 (49):9227–9241. https://doi.org/10.1523/jneurosci.0619-22.2022

Zhang X, Qiu Y, Li J, et al. (2022) Neural Correlates of Transitive Inference: An SDM Meta-Analysis on 32 fMRI Studies. Neuroimage 258:119354. https://doi.org/10.1016/j.neuroimage.2022.119354

Translating Knowledge into Practice

13

How Does Aging Impact the Structure and Function of Parallel Navigation Circuits across Mammals?

Carol A. Barnes, Sahana Srivathsa, Paul F. Hill, Sabina Srokova, and Arne D. Ekstrom

Abstract This review emphasizes the importance of considering the synapse as a central driver of changes of navigation behavior in normative aging in mammals. Deeper understanding of age-related changes in synaptic plasticity and connectivity will lead to the identification of vulnerabilities in specific navigation circuits with age, as well as compensatory mechanisms that can preserve navigation function. These include age comparisons involving volumetric MRI studies, anatomical cell counts, and, where data are available, fMRI and electrophysiological measures of synaptic and network function in each of the brain structures that are thought to contribute to the complex range of behaviors associated with navigation. To better appreciate the parallel circuits and thus strategies that might be used by older individuals interchangeably during navigation, discussion in this chapter shifts away from known hierarchical levels of navigation. Methods such as virtual reality, which allow the manipulation of different sensory or motor signals available for navigation, support the idea of parallel circuits, particularly ones that are differentially affected by aging. Finally, existing computational models of navigation are examined in the context of aging for their potential contribution to understanding the observed age-related differences in navigation circuits and behaviors. Throughout, existing gaps in our understanding of how normative aging impacts navigation are highlighted.

Keywords Spatial memory, hippocampus, prefrontal cortex, retrosplenial cortex, sensory and motor cortex, striatum, cerebellum

Carol A. Barnes (✉)
Dept. of Psychology, Evelyn F. McKnight Brain Institute, University of Arizona, Tucson, AZ 85749, USA
Email: carol@nsma.arizona.edu

Coauthor affiliations are available in the List of Contributors.

N. S. Newcombe and K. Cheng (eds.), *Challenges in Navigation Research*, Strüngmann Forum Reports, https://doi.org/10.1007/978-3-032-20563-6_13

1 Introduction and Overview

Navigation is a complex cognitive process that requires the integration of multiple sensorimotor cues with internally generated representations of space. A full appreciation of this complexity requires a better understanding of the multiple brain regions and circuits that contribute to the implementation of navigational behaviors. Trying to understand navigation in the context of aging increases this challenge, as each of the navigation-related circuits may be differently affected by the aging process. Many studies conducted in aging populations on navigation examine how aging impacts only one or two of the neural pathways known to affect these behaviors. Rarely do experiments examine multiple parallel pathways that can collectively contribute to the behavioral outcomes observed. In this chapter, we emphasize the importance of considering multiple parallel circuits and their interactions when studying the impact of age on navigation.

The central focus of aging in this review requires an expansion of the navigational framework proposed by a previous Ernst Strüngmann Forum on navigation, referred to as the *navigation toolbox* (Wiener et al. 2011). Wiener et al. organized the components of the toolbox into four levels of complexity of spatial navigational information: the first level of the hierarchy consists of sensorimotor processes that combine to form a second level of spatial primitives, which can then be integrated into a third level of actual "spatial representations" or constructs. We will not consider the fourth level, which focuses on human-specific navigation and includes symbols and language, because it excludes other species. Here, we focus on mammals and consider synapses as a potential cross-species unifying factor—one that can potentially explain age-related changes in navigation in many of the circuits investigated.

One critical issue to emphasize is that at any given point in time, an animal can make use of different systems or strategies (and perhaps navigation toolbox levels) to solve a wayfinding problem. Even in the same species, strategy differences have been demonstrated in older rats and humans compared to their younger counterparts (Barnes 1979; Rodgers et al. 2012). This fits with a later update of the navigation toolbox synthesis, which included newer information than was available at the time of its creation, as well as a reassessment of whether the four levels are independent or interact in both a bottom-up and top-down fashion (Jeffery et al. 2024). It remains to be discovered how the known hierarchical processes in the toolbox (Jeffery et al. 2024; Wiener et al. 2011) can be fully integrated with the multiple parallel processes that are engaged simultaneously during navigation. One strength of considering an interactive and parallel framework is that this is consistent with how sensorimotor systems, along with neocortical and subcortical circuits as well as the cerebellum, simultaneously interact and contribute to navigation in mammals. The coordination between structures is likely instrumental in the behaviors and strategies exhibited during navigation.

We begin with a review of navigation circuits in aging. This is followed by a deeper consideration of how synaptic connections, brain regions supporting navigation, and sensorimotor functions change in aging. We then critically examine experimental methods used to study components of navigation circuits in aging, including computational models, and propose suggestions for future directions to pursue.

2 Navigation Circuits in Aging: What Changes and How We Study Them

To begin to understand how aging or disease processes affect the cells and circuits involved in navigation, we must appreciate the fundamental role that age-related synaptic changes play in the mechanisms through which age impacts navigation (see Section 3). We begin our discussion with a review of how the underlying cellular properties of the aging brain are altered and in a position to sometimes impact navigation behaviors. We know that both synaptic plasticity and synaptic connectivity are altered in the aging brain. Our direct interactions with the world modify the distribution of synaptic weights in the cells and circuits throughout the brain that are involved in those experiences. These plasticity mechanisms, which presumably rely on biological processes such as long-term potentiation (LTP), long-term depression (LTD), homeostatic scaling, and cross-region oscillatory coupling, influence how spatial representations are formed and maintained. These fundamental synaptic elements can either stabilize or ensure flexibility of spatial information and contribute to coordinated representations across participating regions. This highlights why it is critical to start with an understanding of synaptic changes in aging in order to fully appreciate the range of navigation computations and outcomes that occur across the lifespan of individuals (Burke and Barnes 2006).

Any change in these fundamental plasticity mechanisms at the synaptic level can influence how brain regions communicate, biasing the navigation system either toward or away from an optimal solution for the problem at hand. Changes with age in synaptic connections as well as intra- and interregional synchronicity between neocortical and subcortical regions can also affect the navigation strategies used (see Section 4). Among the most studied brain areas in aging mammals in the context of navigation are the hippocampus and associated temporal lobe regions. While cell loss has often been considered to be a primary driver of neural dysfunction in aging, more recent anatomical methods have demonstrated that there is virtually no cell loss in normative aging in the temporal lobes (Keuker 2003; Rasmussen et al. 1996; West et al. 1994). Cell death in these regions, however, is characteristic of Alzheimer disease (e.g., West et al. 1994). Thus, a contrast can be drawn between the type of trajectories of change that might be expected in navigation in healthy aging versus that observed in Alzheimer patients or in rodent models. It is worth considering that neurodegenerative diseases in older individuals are superimposed on an "aging brain," and therefore cell loss in disease will be combined with age-related functional synaptic changes. Here, our focus is on the change that arises across normative aging as distinct from known neurodegenerative diseases.

Aging also impacts sensorimotor systems that are fundamental to all navigation behaviors. Thus, understanding the impact of these changes is critical to appreciating navigation outcomes across the lifespan (see Section 5). There are measurable changes in each sensory system during development and aging. A 2020 report by the Lancet Commission on modifiable risk factors in aging and Alzheimer disease (Livingston et al. 2020) suggested untreated hearing loss as one of the strongest longitudinal predictors of poor cognitive outcomes in aging, conferring higher risk for Alzheimer disease. While hearing aids can attenuate these losses, they are not always adequately implemented. The latest Lancet Commission report (Livingston et al. 2024) has added untreated visual loss as an additional critical modifiable factor, as new studies suggest that vision loss is associated with increased risk of all-cause dementia. Vestibular, somatosensory, and motor functions are also known to be altered across the lifespan (e.g., Ide et al. 2023; Johnson et al. 2020; Padova et al. 2024; Spooner et al. 2019; Tulimieri and Semrau 2023; Verghese et al. 2014), and age-related alterations in proprioceptive and efference copy signals are thought to contribute to gait changes that often occur with age (Cullen 2019; Tulimieri and Semrau 2023). In turn, age-related changes in gait have been hypothesized to contribute to impaired navigation in older individuals (Hill and Ekstrom 2025).

Through the use of virtual reality (VR) experiments, it is possible to dissect out how individual sensorimotor variables contribute to different spatial metrics with respect to navigation performance (see Section 6). Studies have manipulated sensorimotor systems in VR in humans and shown that age effects on navigation performance are often attenuated in enriched sensorimotor conditions (Adamo et al. 2012; Allen et al. 2021; Hill et al. 2024; Yu et al. 2021). One limitation of stationary VR, which is often used in fMRI studies, is that ambulatory signals are restricted; this can complicate comparisons with freely moving rodent tasks and prevents systematic manipulation of sensorimotor signals to test their unique contributions during navigation. In experiments that compare immersive versus stationary VR conditions, older adults benefit from the increased sensorimotor feedback present in immersive VR (Hill and Ekstrom 2024). This emphasizes the importance of experimental design to the outcomes observed in navigation performance across the lifespan. An advantage of immersive VR in human or rodent studies that incorporate ambulatory signals is that elimination or distortion of individual sensorimotor variables in VR can be examined while simultaneously monitoring spatially tuned neural activity. Thus, mobile VR studies can help identify which synaptic networks are modified in normative aging.

While it is important to consider how the techniques used to investigate navigational differences might inadvertently lead to age-associated biases, the method used can also differentially recruit navigation circuits. The task demands of the experiment can require optimizing the use of one navigational strategy, which could rely preferentially on a specific brain circuit. For example, a strategy that uses a body-based reference frame (egocentric) for navigation may rely on the prefrontal cortex (PFC) and striatum, whereas a navigational strategy that uses a reference frame external to the individual (allocentric) might depend on hippocampus and PFC circuits. Thus, it is important to identify how each experimental method may elicit a different balance in the use of navigation circuits and how this balance may shift in aging.

Existing computational models of spatial navigation may also help us better predict how age-related changes within specific brain regions affect navigation behaviors (see Section 7). Here, we consider two classes of models that aim to explain the circuit computations involved in navigation. The first class employs attractor network dynamics to explain how cells and circuits within different brain regions process spatial information from sensorimotor inputs. The second class uses reinforcement-learning concepts to explain circuit computations during goal-directed navigation behaviors. As the construction of both classes of models is influenced by synaptic modification within relevant brain structures, they may be key to facilitating the explanation of age-related differences observed in navigation outcomes.

To understand the mechanisms that underlie how navigation changes with aging requires an integrated approach. First, age-related synaptic changes within each relevant brain region must be considered, as these changes can impact network dynamics. Second, changes in the balance of circuits recruited during navigation should also be taken into account in explanations for differential strategy use in aging. Third, it is important to identify the specific biases of various experimental methods with respect to age. Careful consideration of these factors is necessary before accurate interpretations can be drawn.

3 Synaptic Changes in Aging

To understand the impact of age on navigation, it is critical to first understand how age impacts synapse structure and function. This key communication point regulates how each of the parallel navigation circuits in the brain functions and interacts with other circuits. Below we describe a number of circuits known to participate in navigation behaviors. Each of these systems depends on intact interregional anatomical synaptic connections and on the ability to modify these connections through plastic changes that allow the region-specific computations required to adaptively support navigation behaviors.

Details of how aging impacts the anatomical and functional connectivity of navigation circuits are not completely known for any region. There are answers to some basic questions on interregional connectivity, synaptic numbers, and plasticity across age, with additional evidence for cross-species correspondence. But there are many gaps in this knowledge. The data on volume, cell, and synapse numbers across aging in these navigation circuits are reported below in Section 4. Here we describe several relevant synaptic plasticity mechanisms where the data are available.

To summarize, with respect to cell loss, which implies synaptic connection loss as well, there is little to no cell loss in medial temporal lobe or prefrontal cortical navigation circuit structures (i.e., from the entorhinal cortex to the hippocampus to the PFC). There is, however, synapse loss and plasticity changes in some hippocampal and prefrontal cortical areas in aging. For example, in the dentate gyrus (DG), there is a reduction in synapse number in the axonal projection from the medial entorhinal cortex (MEC) to hippocampus granule cells (Geinisman et al. 1992). Interestingly, there are plasticity changes in aging at this synapse as well. That is, LTP is more difficult to induce (Bliss and Lømo 1973) and is not as durable at these aged synapses (e.g., Bach et al. 1999; Barnes 1979; Barnes and McNaughton 1985). Increases of synaptic strength that follow the experimental induction of NMDA receptor-dependent LTP is hypothesized to reflect the biological mechanism through which durable memories are formed in the brain. Hebb (1949) suggested that changes in synaptic weights across the brain, following a specific experience, *is* the process the brain uses to store information.

The MEC synaptic projection to CA3 in old rats also shows faster decay of this form of LTP (Dieguez and Barea-Rodriguez 2004). In CA1, the CA3-to-CA1 Schaffer collateral synapse has been examined across the lifespan in mice, and older animals showed significantly faster decay of LTP at this synapse as well (Bach et al. 1999). However, LTD (an experimental treatment that weakens synaptic strength) is easier to induce in this CA1 synapse in the aged hippocampus (Norris et al. 1996). For a more detailed review of the plasticity changes at aged hippocampus synapses, see Burke and Barnes (2024).

While widespread cell loss does not appear to occur in the PFC across aging, focal regional PFC cell loss has been reported (Smith et al. 2004; Stranahan et al. 2012). In normal older humans, preserved synaptic number was found in layers 3 or 5 of the superior-middle frontal cortex (Scheff et al. 2001). A loss in synapse number has been reported, however, in area 46 of the PFC in aged rhesus macaques, with lower numbers for older monkeys ranging from 20% in layer 5 to 30% in layers 2/3 (Peters et al. 2008). Synaptic plasticity mechanisms such as LTP or LTD in the aging PFC have not been examined; thus, changes in these processes remain to be discovered.

The striatum also makes important contributions to navigation that may change with age. This structure, however, does not appear to show cell loss with age (Haug and Eggers 1991), nor are there studies suggesting synapse loss (Walsh and Akopian 2019). Corticostriatal synapses of older animals, however, show reduced expression of NMDA receptor-dependent LTP (Akopian and Walsh 2006). Interestingly, LTD processes appear to be intact at older ages in the striatum. Clearly, there are many gaps in our knowledge of axonal projections, synapse numbers, and plasticity mechanisms within and between critical navigation circuits across the brain. Elucidating the mechanisms that contribute to navigation change observed in both nonhuman and human behavioral experiments will be important to our understanding of how age-related changes in the striatum impact navigation.

There are contradictory reports on whether the cerebellum shows generalized or more localized loss in cell number. One study reported widespread loss of neurons in the cerebellum (Viswasom and Jobby 2013), whereas another suggests that any loss of Purkinje cells is most likely to be restricted to specific subregions with age (Tang et al. 2001). Comparisons of LTD at parallel fiber synapses onto Purkinje cells were conducted across age in crossbred albino mice. A reduction in cerebellar LTD arose in middle age, suggesting an early decline across the lifespan in cerebellar plasticity. Alterations in cerebellar computations produced by transgenic manipulation of cerebellar-dependent mechanisms of synapse modification does have an important impact on hippocampus place-cell representations. Disruption of cerebellar plasticity through a protein kinase C (PKC1) mechanism that interferes with LTD at parallel fiber Purkinje cell synapses in transgenic mice leads to plasticity and behavior changes. The PKC1 transgenic mice that show impaired LTD in the cerebellum also show disrupted self-motion-based pyramidal cell firing in the hippocampus. This suggests that cerebellar LTD provides an important computation contributing to hippocampus representations that support body movements important to navigation (Rochefort et al. 2013).

Additionally, interference with the protein phosphatase 2B (PP2B) mechanism in Purkinje cells of transgenic mice, which affects synaptic and intrinsic forms of cerebellar potentiation, also impacts navigation circuits. When the calcineurin plasticity system (PP2B) is disrupted in Purkinje cells, hippocampal place cells become less stable within a familiar environment, and these transgenic mice were impaired on both allocentric and self-motion navigation tasks (Lefort et al. 2019). A similar instability of place-field maps between one session and the next has been described to occur in CA1 of aged rats (Barnes et al. 1997; Schimanski et al. 2013). It is possible that the cerebellum contributes to this map instability that occurs at older ages in the rat hippocampus. Rondi-Reig et al. (2022) suggest that altering specific plasticity mechanisms at parallel-fiber-to-cerebellar-Purkinje cell synapses can alter the balance between allothetic and idiothetic determinants of place-cell activity. This highlights the importance of better understanding the interactions between these parallel navigation systems, the cerebellum, and the hippocampus.

4 Brain Regional Contributions to Navigation and Changes in Aging

We believe that continued exploration of how and when different brain circuits operate in parallel on a given navigation problem will yield the deepest insight into which components of the brain's navigation systems are most susceptible to changes in normative aging (see Figure 13.1). These neural changes will also be present in older individuals with neurodegenerative diseases. Depending on the disease, the changes may be accentuated or occur independently of those in the normal aging brain. Below we examine several circuits known to be important to navigation and review some of the changes observed in the structure or function of these circuits during aging.

4.1 Entorhinal Cortex–Hippocampus–Prefrontal Cortex

The entorhinal–hippocampus–prefrontal cortex (EC–HC–PFC) circuit plays an important role in age-related changes in navigation across mammalian species. Each structure has been extensively studied in the context of their contributions to wayfinding and path integration. From single-cell properties to anatomical architecture to the resulting computations performed, each region makes a distinct contribution to navigation and is discussed in turn below.

Overall, this navigation circuit receives sensorimotor information, including self-motion and environmental cues, enabling the generation of a spatial representation that can be used to guide navigation behaviors (O'Keefe and Nadel 1978). Visuospatial, auditory, and olfactory information enters the entorhinal cortex (EC) (e.g., Amaral et al. 1987; Suzuki and Amaral 2003). These inputs are predominantly in the superficial layers of the EC (L2 and L3), which project to the hippocampal formation, and the deep layers (L5 and L6) receive projections back from the hippocampus and subiculum.

Functionally and anatomically, inputs to the EC are segregated into two streams. The lateral region (LEC) receives inputs from the perirhinal cortex (Agster and Burwell 2013; Burwell and Amaral 1998), which includes visuospatial, auditory, somatosensory, and olfactory information (Suzuki and Amaral 2003). The medial region (MEC) receives inputs from the postrhinal cortex (Burwell and Amaral

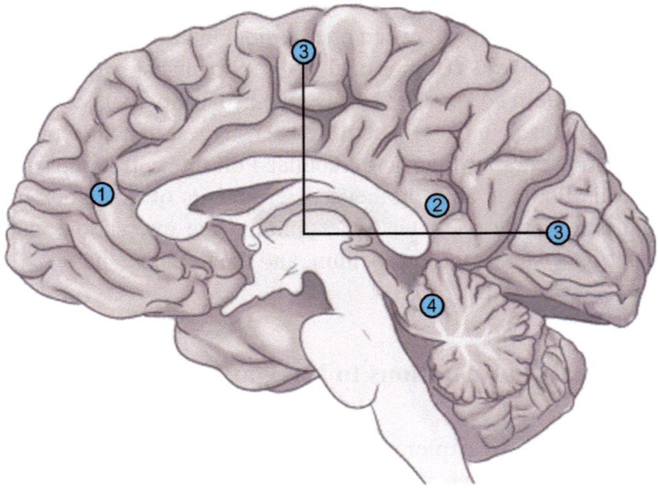

① **Prefrontal cortex**
Clear age-related decline in the volume of the prefrontal cortex, likely linked to dendritic spine loss, lower synaptic density, and possibly focal regional cell loss. Reduced functional and white matter connectivity with the hippocampus impacts route-based strategies and wayfinding. Prefrontal cortex gamma oscillations and neuronal firing are altered in aged rats during goal-directed navigation.

② **Retrosplenial cortex**
Preliminary evidence suggests that boundary vector cells and head-direction cells have less stable and broader tuning in older age. Retrosplenial cortex also shows altered connectivity with other regions important in navigation, likely impacting the region's ability to switch reference frames between egocentric and allocentric representations.

③ **Sensory and motor cortex**
Less precise (dedifferentiated) sensory representations in aging contributes to degraded visual and directional information and increases the likelihood of disorientation. Reduced vestibular and motor function generates dedifferentiated self-motion signals that reach the entorhinal cortex, which could contribute to the dysfunction of grid cells.

④ **Cerebellum**
Volumetric declines, fewer Purkinje and granule cells are primarily evident in the anterior lobe, a region implicated in posture, muscle tone, and gait. Impairments in vestibular function and inefficient estimates of body movements alter the firing of place cells in the hippocampus and head-direction cells in the thalamus, postsubiculum, entorhinal, and retrosplenial cortex.

Figure 13.1 Summary of age-related changes to specific brain regions involved in navigation behaviors.

1998; Insausti et al. 1997) or the parahippocampal region in humans (Witter et al. 2000), which primarily contains visuospatial information (Scaplen et al. 2017). Lesion studies in the EC of rodents are consistent with the idea that this structure contributes to important aspects of navigation. Specifically, inactivation of the MEC disrupts idiothetic (Van Cauter et al. 2013) and allocentric navigation (Ferbinteanu et al. 1999; Hales et al. 2014; Steffenach et al. 2005), suggesting that MEC may contribute to more global aspects of spatial representation (e.g., Kuruvilla and Ainge 2017). LEC inactivation selectively disrupts the recognition of novel objects in an environment (Kuruvilla and Ainge 2017; Rodo et al. 2017; Van Cauter et al. 2013; Wilson et al. 2013). Studies in humans utilizing fMRI methods corroborate

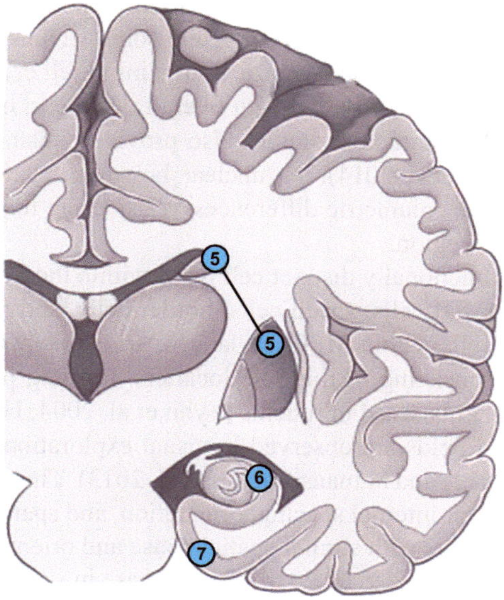

(5) **Striatum**

Robust age-related volumetric declines and accumulation of nonheme iron are evident in several striatal nuclei. No cell loss is evident, but altered connectivity and reduced NMDA receptor-dependent long-term potentiation within cortico-striatal circuits in older age may account for reduced behavioral flexibility and inefficient strategy selection during navigation.

(6) **Hippocampus**

Navigation impairments have been linked to lower hippocampal volume in humans, but in the absence of cell loss. Animal models reveal preserved principal cell counts, along with synapse loss and plasticity changes. Long-term potentiation is more difficult to induce and decays faster, while there is an age-related increase in the susceptibility to long-term depression. The computational processes of pattern separation and pattem completion appear to be disrupted, consistent with impairments in place cell remapping. There appears to be hyperexcitability in CA3 in aged rats, monkeys, and humans.

(7) **Entorhinal cortex**

Minimal to no loss of entorhinal neurons. Stability and tuning of grid and speed cells is impaired in old mice, which may disrupt inputs into the hippocampus. Rodent models show a reduced number of synapses onto the dentate gyrus granule cells, and human models point to lower integrity of the entorhinal–parahippocampal–dentate white matter tracts.

Figure 13.1 (continued)

these findings with object-associated LEC signals and spatially related MEC activity (Maass et al. 2015; Navarro Schröder et al. 2015; Reagh and Yassa 2014). The dichotomy between MEC and LEC contributions to spatial navigation performance (Knierim et al. 2014) persists in aging; in fact, age-related impairments in performance in behavioral tasks associated with object discrimination may be greater under some circumstances than the deficits in spatial discrimination (Johnson et al. 2017; Maurer et al. 2017; Reagh and Yassa 2017; Reagh et al. 2016).

With respect to aging, there is minimal to no loss of EC neurons across species in studies that utilize stereological cell-count methods. This includes rodents (Merrill et al. 2001; Rapp et al. 2002), nonhuman primates (Gazzaley et al. 1997; Merrill et al. 2000), and humans (West 1999). Consistent with this, cross-sectional volumetric

MRI studies conducted in humans found minimal differences in EC volume across age (Insausti et al. 1997; Raz et al. 2005). While some studies noted a change in both volume (Westlye et al. 2010) and cortical thinning (Fjell et al. 2009) of the EC with age, this difference usually scaled with overall decline in intracranial volume (Knoops et al. 2012). Longitudinal studies also provide evidence of considerable thinning of the EC (Fjell et al. 2014). It is unclear, however, what structural changes are responsible for these volumetric differences, considering that it is unlikely that there is cell loss in this region.

There are several functionally distinct cell types within the MEC, including grid cells, head-direction (HD) cells, speed cells, border cells, and object vector cells. The most prominent cells within MEC are layer 2 grid cells, which are characterized by firing fields consisting of multiple locations forming periodically spaced triangular grids and are observed in rodents (Fyhn et al. 2004; Hafting et al. 2005). Similar triangular grid fields are observed in visual exploration in nonhuman primates (Killian et al. 2012) and humans (Jacobs et al. 2013). The rodent grid cells are primarily defined by their internal spacing, orientation, and spatial phase relative to the environment. Grid cells with similar spatial phase and orientation are organized into distinct modules (Stensola et al. 2012) that increase in size along the dorsoventral axis of the MEC (Brun et al. 2008; Fyhn et al. 2004; Hafting et al. 2005).

In the MEC, HD cells, which reflect the direction of the head, primarily receive direct inputs from the thalamus and presubiculum carrying vestibular information (Peyrache et al. 2015; Preston-Ferrer et al. 2016; Tukker et al. 2015). It has been shown that these HD cells influence grid-cell activity, as disruption of this input to the MEC disrupts the periodicity of grid cells (Winter et al. 2015). Along with HD cells, speed cells, which modulate their firing with respect to the running speed of the animal, enable computation of the animal's velocity (Kropff et al. 2015). A potential origin of the speed signal might be the brainstem pedunculopontine tegmental nucleus (PPN), as optogenetic stimulation of the PPN modulates the activity of speed cells in the MEC (Carvalho et al. 2020).

Border cells in the MEC fire when the animal is at a certain position relative to the geometric border of the environment (Lever et al. 2009; Solstad et al. 2008). One possibility is that they help anchor the grid cells to the environment and prevent drift during path integration computations (Savelli et al. 2008). Object-vector cells have also been observed in the MEC when an animal is at a particular orientation with respect to an object (Høydal et al. 2019; Kinkhabwala et al. 2020). Object-vector cells are thought to be important for representing the relative location of local cues to the animal that can be used for navigation. Some cells in the MEC have more complex representations, reflecting the integration of inputs from many of the cell types described above (Tocker et al. 2015).

Unlike the MEC, which has multiple neuron types that clearly represent spatial variables, LEC neurons have shown stable firing responses to sensory stimuli (e.g., Igarashi et al. 2014; Suter et al. 2019), objects in the environment (Ahn and Lee 2015; Bogacz et al. 2001; Burke et al. 2012; Deshmukh and Knierim 2011; Tsao et al. 2013), and reward-centric information (Issa et al. 2024). Specifically, LEC neurons reflect the current location of objects (Deshmukh and Knierim 2013; Keene et al. 2016) and the location in the environment where objects were previously experienced (Tsao et al. 2013; Tsao et al. 2018). LEC neurons robustly respond to egocentric features of the environment and utilize more local cues than do MEC neurons, which respond more to allocentric features (Kuruvilla et al. 2020; Vandrey et al. 2020; Wang et al. 2020). These findings support the proposed dichotomy of representation between these two regions of the EC.

With respect to aging, only MEC neurons have been investigated in human or nonhuman animals. The stability and updating across contexts of grid cells is lowered in older head-fixed mice, and speed cells show an instability in tuning (Herber et al. 2024). The latter might be caused by the neuronal loss in the PPN observed in aged rats (Lolova et al. 1997) and humans (Ransmayr et al. 2000). In healthy human participants, exploration of a virtual environment produces BOLD fMRI signals with hexadirectional symmetry in the EC and associated areas (e.g., Doeller et al. 2010). An attenuation in the strength of these signals, which are considered a proxy for grid-cell firing, correlate with age-related impairments in navigation performance (Kunz et al. 2015; Stangl et al. 2018). Magnetoencephalography (MEG) and intracranial recordings in humans also report similar triangular firing fields in the MEC (Jacobs et al. 2010; Maidenbaum et al. 2018; Staudigl et al. 2018), strengthening the arguments for translation of the MEC properties derived from rodent electrophysiological studies to human fMRI studies (Bellmund et al. 2018). We will return to this topic in Section 6, where we provide a more critical examination of rodent grid-cell and human fMRI studies.

Neurons in layers 2/3 of the EC specifically project reciprocally to all subregions of the hippocampal formation (Agster and Burwell 2013). In rodent models of aging, there is a reduction in the number of synapses from the MEC to the DG (Geinisman et al. 1992; Smith et al. 2000), likely due to axonal pruning (Barnes et al. 1989). Consistent with the idea of axonal pruning in rodents, several studies report age-related changes in the human EC–parahippocampal white matter tracts to the DG (Rogalski et al. 2009; Stoub et al. 2012; Yassa et al. 2010), suggesting a change in the input to the hippocampus from the EC in aging humans as well.

Many studies in rodents have shown that lesions of the hippocampus result in learning and memory deficits on spatial navigation tasks (e.g., Jarrard 1989). Spatially tuned cells in the CA region of the hippocampus, first discovered in rodents by O'Keefe and Dostrovsky (1971), are referred to as *place cells* and preferentially fire within localized areas of an environment (place fields). The location of place fields are primarily influenced by allocentric features of an environment (Muller and Kubie 1987; O'Keefe 1976; O'Keefe and Nadel 1978), although a number of factors can alter place-cell firing, such as changes in motor and vestibular information or reward location (Foster et al. 1989; Hollup et al. 2001; Terrazas et al. 2005). Environmental changes are reflected by modifications to the allocentric framework in several ways. Large changes to the environment induce "global remapping," where a new configuration of place cells emerges in both CA1 (Muller and Kubie 1987) and CA3 (Leutgeb et al. 2005). Systematic small changes in the environments can induce "rate remapping" in CA3, where the same cells change their firing rates but not the location of their firing fields (Knierim 2002; Leutgeb et al. 2005). In nonhuman primates, it has been more difficult to observe place-specific firing of CA1 neurons in VR tasks. These cells tend to fire in response to salient task variables that are not always spatially localized (Meister and Buffalo 2016). This may be consistent with observations of single-cell recordings in the

CA fields of human patients in which the neurons were driven by task requirements (Ekstrom et al. 2003; Wicks et al. 2020).

Single-cell recordings of CA3 neurons of aged rats exposed to two very different environments sometimes do not globally remap between environments. Rather, aged rat CA3 cells tend to retrieve the same map for both, potentially suggesting a failure of a pattern separation computation (Wilson et al. 2005). On the other hand, CA1 pyramidal cells in old rats remap occasionally even between two exposures to the same familiar environment (Barnes et al. 1997; Schimanski et al. 2013), which suggests a disruption of a pattern completion computation. While there is evidence for age-related changes in network properties of CA1 principal cells, there are no major differences in the basic biophysical characteristics of these cells (e.g., average and peak firing rates, interspike intervals, spike amplitude, and width) between young and old rats (Barnes 1983; Markus et al. 1994; Mizumori and Kalyani 1997; Mizumori et al. 1996). Still, individual CA3 cells in the aged rat show hyperexcitability (Wilson et al. 2005), as do cells in the CA3 region of the aged nonhuman primate (Thome et al. 2017). Remarkably, when the CA3 region is examined in humans using fMRI methods, a similar increase in the BOLD signal is observed in older individuals (Yassa et al. 2011). This suggests that the excitability in CA3 is a consistent marker of normative aging across these species.

Granule cells in the DG are also spatially tuned, but it is notoriously difficult to record from them (Alme et al. 2010; Leutgeb et al. 2007) as they have the lowest firing rates of any hippocampus place cell. If a granule cell is active, it tends to have multiple, very small place fields. When rats are exposed to a series of gradually changing ("morphed") environments, granule cell population activity is highly sensitive to small changes in environment shape. When granule cells are exposed to very different environments, however, they do not exhibit the same kind of global remapping that occurs in both CA1 and CA3 during similar treatments but instead are active in multiple different environments (Alme et al. 2014; Leutgeb et al. 2007). While there are no electrophysiological recordings of granule cells in aged rats, fMRI activity in macaques as well as immediate early gene activity in rats suggests that DG activity, compared to CA fields, is reduced in older animals (Small et al. 2004).

In humans, hippocampus atrophy has frequently been associated with older ages, which may relate to synaptic, vascular, or glial cell changes as there is no reported neuron loss. This change tends to follow a nonlinear pattern, characterized by minimal decline until around the age of 50–60 years, after which there is a sharp decrease (Fjell et al. 2014; Fraser et al. 2015; Raz et al. 2010). When examining sex differences, a cross-sectional study with over 19,700 participants suggests that hippocampus volume decline accelerates earlier in males, occurring around 50 years, as compared to females, in whom decline tends to occur around 60 years (Nobis et al. 2019). These findings are in agreement with previous cross-sectional (Wellington et al. 2012; Woolard and Heckers 2012; Zhang et al. 2010) and longitudinal studies (Fraser et al. 2015). Nobis et al. (2019) also found that the right hippocampus was slightly larger than the left. This is consistent with the observation that the longitudinally assessed rate of atrophy has been reported to be greater in the left than in the right hippocampus (Fraser et al. 2015). Highlighting the potential importance of hippocampus lateralization, studies in rodents and humans suggest that the right hippocampus may play a greater role in spatial navigation (Howard et al. 2014; Klur et al. 2009; Maguire et al. 1997; Poe et al. 2000; Spiers et al. 2001), although meta-analyses have not always supported these differences in humans (Lee et al. 2002; Teghil et al. 2021).

Subregion-specific age-related anatomical differences have also been observed. Cross-sectional human studies have shown reduced volume of the DG/CA3 and subiculum but not of CA1 with age (Malykhin et al. 2017; Shing et al. 2011; Stark and Stark 2017). Diffusion tensor imaging, fractional anisotropy, and volumetric measurements of the DG–CA3 connection show changes across age (Pereira et al. 2014). As noted above, while hippocampal volume may change with age, neuron loss cannot account for any observed reduction, as stereological studies have shown that hippocampus principal cell numbers are preserved (Merrill et al. 2001; West 1993).

Age-related impairments in spatial navigation, such as wayfinding in the virtual Morris water maze, have been linked to lower hippocampal volume (Bohbot et al. 2007; Head and Isom 2010; Konishi and Bohbot 2013; Konishi et al. 2017). Regardless of age, lower volumes of the CA1 and subicular regions, but not CA3–DG, correspond to worse performance on the virtual Morris water maze task (Daugherty and Raz 2015). In a meta-analysis of 11 studies, a strong association was found between hippocampal volume and performance in allocentric, but not egocentric, navigation in eight of these studies (Li and King 2019). The association between atrophy and performance is first observed in middle age (40–59), which corresponds to when hippocampus volume changes are first detected as described above. In Section 6, we will return to a more in-depth consideration of these results.

Like the EC, the hippocampus is functionally organized along a longitudinal axis (Amaral and Witter 1989): the dorsal (posterior in humans) section is proposed to be responsible for detailed spatial information, with the ventral (anterior in humans) responsible for representing larger features of the environment and context. The size of the place fields increases between dorsal and ventral CA1 in rodents, suggesting higher spatial resolution in the dorsal regions (Jung et al. 1994; Kjelstrup et al. 2008). Although there are no studies that compare potential age differences along the longitudinal axis in rodents, studies have compared this axis in humans. Older adults compared to younger individuals show a change in volume in the posterior, but not anterior, hippocampus that is correlated with reduced performance on allocentric spatial tasks (Driscoll et al. 2003; Malykhin et al. 2008; Pruessner et al. 2001; Rajah et al. 2010a; Stark and Stark 2017).

Within the EC–HC–PFC circuit, the anterior HC–PFC connection appears to be involved in route-based strategies and decision making, whereas the MEC–posterior HC connection is primarily involved in wayfinding (Hartley et al. 2003; Hirshhorn et al. 2012; Woollett and Maguire 2011). The EC–HC–PFC loop occurs through the ventral anterior hippocampus, which sends direct projections to the PFC (Jay and Witter 1991). With respect to aging, studies indicate that both functional and white matter connectivity is reduced between the hippocampal and PFC regions in older adults, compared to younger individuals (Blum et al. 2014), and these changes may be associated with reduced navigation skills (Dalton et al. 2019). This change in the aHC–PFC connection may result in reduced ability to utilize contextual cues necessary for allocentric navigation (Stark and Stark 2017), causing older individuals to use less efficient strategies (Harris and Wolbers 2014).

Volumetric change or decline and cortical thinning with age have been observed in the PFC in both cross-sectional and longitudinal studies (Fjell et al. 2009; Lemaitre et al. 2012; Østby et al. 2009; Salat et al. 2005; Thambisetty et al. 2010; van Velsen et al. 2013). The longitudinal studies show a disproportionate age-related decline in the PFC volume relative to the rest of the brain (Fjell et al. 2014; Thambisetty et al. 2010). Subregional differences are observed across age, suggesting smaller surface areas in the dorsolateral prefrontal (dlPFC) and orbitofrontal (OFC) cortices but an increase in cortical thickness in the anterior cingulate cortex (Dotson et al. 2016; Fjell et al. 2009; Fjell et al. 2014; Tisserand et al. 2002). A review of fMRI highlights the differences between these PFC subregions and their role with respect to spatial navigation studies (Patai and Spiers 2021). The dorsomedial PFC (dmPFC) in conjunction with the hippocampus is important for goal retrieval whereas the anterior cingulate cortex and dmPFC appear to be important for encoding relative distance to a goal (Spiers and Maguire 2007; Viard et al. 2011; Wirth et al. 2017). The OFC is involved in assessing the relative benefit of potential routes to a goal (Kaplan et al. 2017; Schuck et al. 2016), whereas the dlPFC is generally involved in route planning and path integration. With respect to age-related fMRI changes that correlate with spatial navigation performance, studies have shown that older adults exhibit reduced activation in the lateral PFC during spatial working-memory tasks that correlate with poorer navigation performance (Moffat et al. 2007). A longitudinal study also found that reductions in lateral PFC volume were associated with utilization of less effective strategies in older individuals in a virtual Morris water maze task (Daugherty and Raz 2017). Other studies report that older adults show bilateral PFC activation on spatial working-memory tasks, where younger adults only activate a single hemisphere (Cabeza 2001, 2002). Performance on this task was comparable between age groups, which suggests that bilateral recruitment of the PFC might compensate for age-related anatomical changes in the dlPFC that affect working memory (Cabeza 2001, 2002; Dolcos et al. 2002; Piefke et al. 2012).

While a clear consensus has not been reached on the anatomical comparison of the PFC between rodents and primates (Laubach et al. 2018; Uylings et al. 2003), studies on rodents (Alexander et al. 2020b) and nonhuman primates show similar results with respect to changes in PFC volume with age (Alexander et al. 2008; Shamy et al. 2011). In rodents there are also age-related changes in neuronal architecture, including dendritic spine loss and decreased synaptic density and cell loss (Bloss et al. 2011; Dickstein et al. 2013; Stranahan et al. 2012). Cell loss is also seen in nonhuman primates (Smith et al. 2004). The modulation of firing rate of PFC neurons in response to stimuli is reduced in aged rats (Caetano et al. 2012; Kaczorowski et al. 2012) and macaques (Wang et al. 2011). These age-related PFC changes are associated with impairments in tasks examining spatial learning and memory (Mizoguchi et al. 2009; Rajah et al. 2010b; Wang et al. 2011).

Lesion studies in rodents also indicate that the infralimbic and prelimbic regions of the mPFC are necessary for complex spatial navigation, such as in tasks involving navigation across multiple environments (Churchwell et al. 2010), strategy switching (Guise and Shapiro 2017), and updating reward values (Laskowski et al. 2016). Extending beyond lesion studies, neurons in the mPFC are spatially tuned only in active goal-directed navigation (Hok et al. 2005; Jung et al. 1998) but not while freely foraging (Gemmell et al. 2002). Specifically, these neurons are modulated in spatial working memory tasks, representing task phases such as goal approach, arm selection, reward anticipation, or task rules (Guise and Shapiro 2017; Hok et al. 2005; Jadhav et al. 2016). In nonhuman primates, dlPFC neurons have been shown to exhibit sustained firing in response to the delay period in working-memory tasks (e.g., Funahashi et al. 1989), amount of reward (e.g., Wallis and Miller 2003), and location of task-relevant objects (Everling et al., 2006). In humans, intracranial recordings from PFC neurons show modulation in firing rate in response to a navigation task (Ekstrom et al. 2003). Furthermore, temporal coordination between mPFC and hippocampal neurons is necessary for route planning and spatial decision making in rodents (Floresco et al. 1997; Guise and Shapiro 2017; Jadhav et al. 2016; Wang and Cai 2006).

An important facet of the EC–HC–PFC circuit is the coordination that occurs between these regions via oscillations. Oscillations at different frequencies facilitate the flow of information between the different regions of this circuit. Theta oscillations in the 4–12 Hz range occur when a rat moves or primates visually explore the environment. These oscillations are present throughout the EC–HC–PFC circuit and are thought to help organize faster oscillations (30–100 Hz) and facilitate the organization of the spatially tuned neurons into sequences representing behavioral trajectories (Skaggs et al. 1996). In the PFC, these oscillations tend to occur during decision making, enabling coordination between hippocampal and PFC neurons (Jones and Wilson 2005). The peak frequency of theta oscillations in the hippocampus scales with movement speed, which is typically faster in young animals compared to older animals. There are no age-related changes in the properties of theta oscillations, however, when animals are moving at the same speed (Crown et al. 2022; Schimanski et al. 2013; Shen et al. 1996). Faster oscillations in the 30–100 Hz band are also present in all regions of the EC–HC–PFC loop. While there are no differences in mean gamma frequency between age groups in the hippocampus, the increase in gamma frequency that occurs with increased running speeds is attenuated in old compared to young rats (Crown et al. 2022). In the PFC there is a reduction in peak gamma frequency in aged rats compared to young (Insel et al. 2012). Although it is clear that oscillations are essential for facilitating the coordination of information between regions, it remains to be discovered exactly how this coordination is affected in aging.

Age-related changes in EC–HC–PFC circuits are complex and occur at multiple levels. While the EC to dorsal/posterior hippocampus circuit is associated with wayfinding, the ventral/anterior hippocampus to PFC circuit is involved in route-finding navigation. The strategy utilized during a task can alternate depending on task complexity and demands, drawing upon both pathways. Age-related changes occur nonuniformly at various levels of this circuit, and may be consistent with either compromised navigation or compensatory processes that might mitigate navigational deficits in task solutions.

4.2 Frontostriatal Circuits

The striatum is thought to be an important contributor to navigation, classically viewed as an area that could support egocentric or "response-based" learning. The striatum is organized into functionally discrete and parallel frontostriatal circuits that are associated with motor, cognitive, and limbic subdivisions among striatal nuclei (Alexander et al. 1986; Balleine and O'Doherty 2010; Di Martino et al. 2008). A motor circuit involving caudal-dorsal putamen and frontal motor cortices is often associated with tasks that require learning and executing complex motor sequences (Balleine and O'Doherty 2010; Lehéricy et al. 2005; Wymbs et al. 2012). These, in turn, may support more habit-based forms of navigation that require learning sequences of outcome stimulus-action associations (e.g., turn right at the lamppost) (Jog et al. 1999; Yin et al. 2004). An emerging literature, however, suggests that the distinction between the striatal habit-based system and the allocentric hippocampus-based system is more blurred than previously appreciated (Ekstrom and Hill 2023; Murray et al. 2016). A separate cognitive circuit involving the dorsal caudate, dlPFC, and lateral parietal cortex is frequently associated with tasks that demand increased top-down cognitive control and behavioral flexibility (Buckner 2004; Di Martino et al. 2008; Morris et al. 2016; Scimeca and Badre 2012). This circuit has been proposed to support the ability to switch flexibly between striatal- and hippocampus-dependent navigation systems and spatial reference frames during goal-directed navigation in dynamically changing environments (Goodroe et al. 2018; Patai and Spiers 2021).

Volumetric studies indicate that prefrontal and striatal regions are among the brain regions most profoundly affected by aging (Raz 2000). A seminal study tracked volumetric changes in 53 healthy adults aged 20–77 over the course of five years (Raz et al. 2004). Cross-sectional caudate and putamen volumes, but not globus pallidus, correlated negatively with age at baseline and at five-year follow-up. In spite of these cross-sectional differences in volume across age, the rate of volumetric decline did not differ longitudinally between age groups. The rates of decline, however, did differ between the nuclei, with the fastest rates observed in the caudate (0.83% per year), followed by the putamen (0.73%) and globus pallidus (0.51%) (Raz et al. 2004). A later study also showed both cross-sectional and longitudinal volumetric declines across all striatal nuclei (Fjell et al. 2013). Together, these two powerful longitudinal studies support the idea that multiple regions of the striatum are altered in normative aging.

Studies examining the association between striatal volume and navigation performance in older age do not provide a clear consensus on the contribution of this structure to age-related changes in navigation. Caudate volume has been shown by some studies to correlate with individual differences in response-based navigation strategies, in both young and healthy older adults (Bohbot et al. 2007; Head and Isom 2010). Reduced caudate volume in older age has also been linked to less efficient navigation performance on virtual adaptations of the Morris water maze task (Daugherty and Raz 2017; Moffat et al. 2007). Other studies have failed to identify a reliable relationship between striatal volume and navigation performance measures (Daugherty and Raz 2015; Konishi and Bohbot 2013; Konishi et al. 2017; Korthauer et al. 2016). Functional neuroimaging studies of aging and navigation often report increased brain activity in frontal and striatal regions in older, compared to younger, individuals (Antonova et al. 2009; Iaria et al. 2003; Konishi and Bohbot 2013; Meulenbroek et al. 2004; Moffat 2006; Schuck et al. 2015). Enhanced fronto-striatal activity is often observed to coincide with a commensurate decrease in brain activity in regions typically associated with allocentric-based navigation strategies, including the hippocampus and parahippocampal cortex (Antonova et al. 2009;

Konishi and Bohbot 2013; Moffat 2006; Schuck et al. 2015). This age-related shift in brain activity is generally interpreted to reflect a compensatory transition toward more egocentric or response-based navigation strategies in older age (Colombo et al. 2017; Li and King 2019).

Postmortem studies suggest disproportionate concentrations of intracellular nonheme iron levels in the striatum (Hallgren and Sourander 1958). Although essential for normal cellular metabolism, excessive concentrations of nonheme iron could contribute to oxidative stress and inflammation, which can lead to cell dysfunction. MRI contrasts that are sensitive to the high magnetic susceptibility of iron make it possible to quantify *in vivo* estimates of regional iron content in healthy aging adults. A meta-analysis of twenty cross-sectional MRI studies reported that striatal iron content increases with age, with the largest age-related differences evident in the putamen ($d = 0.88$), followed by the caudate ($d = 0.74$) and globus pallidus ($d = 0.61$) (Daugherty and Raz 2013; Li and King 2019). A two-year longitudinal study in 78 adults aged 19–77 showed that increases in iron concentrations predict commensurate volume reductions within the caudate nucleus, suggesting a candidate source of striatal-related cognitive decline. Indeed, striatal iron concentrations are associated with declines in working and spatial memory in humans (Daugherty and Raz 2017), and introducing iron to striatal structures in adult rats produces impairments in spatial navigation performance on the Morris water maze (Maaroufi et al. 2009).

Advancing age has also been shown to affect the structural and functional integrity of frontostriatal circuitry. Studies using diffusion MR techniques provide strong support for both cross-sectional (Webb et al. 2020; Ystad et al. 2011) and longitudinal (Vik et al. 2015) degradation of white matter tracts that connect the caudate and lateral prefrontal cortices. Age differences in frontostriatal white matter integrity have been associated with age-related variance in cognitive performance on measures of executive function and memory (Hernandez et al. 2015; Rodrigue et al. 2013; Webb et al. 2020; Ystad et al. 2011). Studies using resting-state fMRI provide convergent evidence of cross-sectional differences in the strength of corticostriatal functional connectivity in older age. Advancing age appears to weaken endogenous connections between the caudate and regions comprising the frontoparietal control network (including dorsolateral prefrontal, dorsomedial prefrontal, and lateral parietal areas). This loss of network coupling has been shown to predict age-related deficits in episodic memory and executive function (Buckner 2004; Choi et al. 2012; Fjell et al. 2016; Gordon et al. 2022; Guitart-Masip et al. 2016; Nyberg et al. 2016). Whether age differences in this circuit extend to navigation behaviors is yet to be explored.

Allocentric wayfinding strategies are often assessed as the ability to employ novel paths and shortcuts to reach a goal location. Compared to younger individuals, older navigators often show a preference for traveling familiar but less efficient routes in lieu of taking novel shortcuts. One prominent view is that older adults may rely more on egocentric (response-based) navigation strategies to counteract allocentric deficits stemming from gray matter loss in the medial temporal lobes (MTLs). This age-related bias for egocentric navigation strategies has been linked to enhanced volume (Head and Isom 2010) and fMRI BOLD signal change (Hartley et al. 2003; Iaria et al. 2003; Konishi and Bohbot 2013) in subregions of the dorsal striatum, including the caudate nucleus and putamen (for a review, see Colombo et al. 2017; Goodroe et al. 2018; Li and King 2019; Zhong and Moffat 2018). One potential account for increased striatal involvement in navigation is that older adults (and rodents; Barnes et al. 1980) rely to a greater extent on stimulus-response associations (e.g., turn right at the lamppost) that permit route-based navigation to compensate for impaired hippocampal-based allocentric navigation. Additionally, experimental inactivation of the striatum in older (but not younger) rats leads to improvements in putative allocentric-based navigation (Gardner et al. 2020).

In the paradigms most often used to assess navigation strategies, the *familiarity* of the routes typically preferred by older navigators presents a critical and often overlooked confound. From the human episodic memory literature, there is substantial evidence that familiarity-based memory judgments are preserved in older age (Koen and Yonelinas 2014). In almost all navigation paradigms, familiarity is not dissociated from path efficiency. In other words, allocentric navigation typically involves both novel shortcuts and greater efficiency. In contrast, stimulus-response and egocentric-based navigation are typically more familiar but less direct in terms of finding the target, and thus less efficient. A major issue here is that older adults may simply be more risk averse (Albert and Duffy 2012; Frank and Seaman 2023), and what appears to be an allocentric deficit is in fact a bias toward more familiar routes but may be independent of their path efficiency. Developing novel experimental methods that can deconvolve familiarity and efficiency may provide better insights into the functional role of the frontostriatal circuits as they relate to navigation.

4.3 The Head-Direction System

A central feature of the navigation system is the ability to maintain orientation while moving about in an environment. Stated differently, animals need something like a "compass" to maintain orientation within the space in which they navigate and learn. One key element of this global sense of direction is thought to involve HD cells in the brain. First discovered in rats (Taube et al. 1990), HD cells have also been observed in nonhuman primates (Robertson et al. 1999), and evidence for HD signals has been found in humans (Doeller et al. 2010; Jacobs et al. 2010). HD cell firing in nonhuman animals is tuned to the particular direction the head is facing, and is largely independent of whether the animal is moving. HD cells are found in a number of brainstem and thalamic areas, including the dorsal tegmental nucleus (DTN) (Sharp et al. 2001), the lateral mammillary nucleus (LMN) (Stackman and Taube 1998; Taube 2007), and the anterior dorsal thalamic nucleus (ADN) (Taube 1995). In each of these nuclei, the directional tuning curves of HD cells are largely similar, although each has differentiating features. Among these areas, the ADN shows the highest proportion of HD cells (60% exhibit directional firing), but HD cells are also found in the subiculum and cortex. The dorsal part of the presubiculum is where HD cells were first discovered, and this subregion is now referred to as the postsubiculum (Taube 2007).

HD cells have also been observed in a number of cortical regions, including the retrosplenial cortex (RSC) (Chen et al. 1994) and MEC (Sargolini et al. 2006) as well as in the medial prestriate cortex (Chen et al. 1994) and deep layers of primary visual cortex V1 (Vélez-Fort et al. 2018). Another important cell type in the HD system are the cells that reflect angular head velocity found in brainstem structures (e.g., DTN, LMN, and thalamic regions such as the lateral habenula); some cells in the postsubiculum also show this property, as do some cells in the RSC. Angular head velocity cells signal the rate at which the head is rotating and requires vestibular input, which is critical for updating the sense of direction as animals navigate. As discussed in Section 4.4 below, cerebellar circuits are updated to learn new matches when the expected correspondence between active head motion and the sensory consequence of this movement is altered, which may be one mechanism that underlies the stable representation of directional heading in the HD system.

Recent experiments also support the idea that the cerebellum exerts critical control of the HD system through its connections with the ADN and RSC. By manipulating cerebellar plasticity mechanisms using two different cerebellar-specific transgenic mouse models, Fallahnezhad et al. (2023) demonstrated that transgenic interference with PKC-mediated LTD disrupts HD cell firing in the ADN and RSC in response to self-motion cues. In addition, the transgenic interruption of PP2B-mediated potentiation of Purkinje cells disrupts HD cell anchoring in the ADN and RSC to external cues. While these manipulations did disrupt HD cell firing properties, they did not disrupt angular head velocity cells or linear speed cells in these structures (Fallahnezhad et al. 2023).

Rat lesion studies also support the role of the RSC in spatial navigation and episodic memory. These interactions are thought to occur through its powerful reciprocal connections with the anterior thalamic nuclei, the lateral dorsal thalamic nucleus, and the postsubiculum. But, only about 10% of the cells in the rat RSC are actual HD cells (Cho and Sharp 2001). Human studies also provide evidence for a role for the RSC in navigation and spatial memory (e.g., Epstein 2008; Maguire 2001); humans with damage to this structure are unable to derive directional information from landmark cues (e.g., Gainotti et al. 1998). Furthermore, models of the systems involved in navigation behavior have proposed that the RSC may act as a "translator" between the allocentric representations that reside within the temporal lobe and idiothetic representations within the parietal lobe. This would put the RSC in a position to help switch frames of reference between egocentric, viewpoint-dependent representations mediated by the posterior parietal cortex and allocentric, viewpoint-independent representations, mediated by the MTL (Burgess et al. 2001; Byrne et al. 2007). In this view, the RSC is key to integrating these components of the navigation system (Vann et al. 2009).

Not only does the RSC have HD cells, it also has cells in the dysgranular region that fire in a bidirectional pattern within a single environment as well as between environments. These cells are hypothesized to be involved in visual landmark representation (Dudchenko et al. 2019; Jacob et al. 2017). The RSC also contains cells—egocentric boundary vector cells—that have spatial receptive fields, which signal egocentric coordinates in relation to environmental boundaries that are in a specific orientation and distance with respect to the animal (Alexander et al. 2020a). These cells reside in the dysgranular region of the RSC, and preliminary data (Patel 2024) suggest that these boundary vector cells and HD cells, as measured with one photon imaging in an open arena, are less stable and have broader tuning, respectively, in older (9 mo) compared to younger (4 mo) mice. This is a potentially very interesting study in that the measurements were obtained longitudinally, but the "older" group was only middle-aged (i.e., 9 mo). Clearly, more work needs to be done to understand the details of the impact of age on this cortical area, but extant data suggest that changes which do occur during aging in RSC may occur relatively early during the lifespan.

The exact gross anatomical designation of the RSC within the posterior cingulate cortex (PCC) is still under some debate, even in humans. One recent perspective (Foster et al. 2023) suggests that the PCC contains three primary subregions: the dorsal PCC, the ventral PCC, and the RSC. The validity of Brodmann's RSC designations of areas 29 and 30 for humans has been questioned by Vogt et al. (2001), who among other things suggest that the rostral extent is underestimated and its caudoventral extent overestimated in that delineation. Currently, the boundaries are undergoing reassessment. Additionally, in nonhuman primates, the RSC does not extend to the medial surface, and the rat PCC is primarily composed of the RSC. While these anatomical differences between species need to be appreciated and studied further with respect to the role that RSC plays in behavior, strong similarities in functional connectivity appear to remain across species.

The RSC is also well known for the role it plays in the MTL "subsystem" of the default mode network (Andrews-Hanna et al. 2010, 2014; Raichle et al. 2001), but it also shows strong functional coherence with both the MTL and the dmPFC subsystems of the default mode network. Thus, the RSC is proposed to act as a functional hub in the default mode network, with important functional connections to the parietal cortex, as mentioned above. Of course, the RSC is also proposed to be important to cerebellar loops that carry vestibular information (Rochefort et al. 2013). Because of its clear participation in the navigation system, it is somewhat surprising that there have been no structural MRI studies that compare RSC volume across age, nor have there been postmortem cell counts in this region across age. This structure has, however, been examined in humans with respect to individual differences in navigation ability within a young cohort (~23 years). The volume of the RSC in this study was related to the young individual's accuracy in a complex path integration task (Chrastil et al. 2017): those with larger RSC volumes also showed better navigation behavior. Between young and older adults, a structural RSC difference might be found, since path integration ability is known to be altered at older ages (Allen et al. 2016; Harris and Wolbers 2012; Mapstone et al. 2006) and could be related to gray matter volume.

Callaert et al. (2014) suggest a number of reasons why gray matter volume estimates based on voxel-based morphometry may pose problems for accurate age comparisons. These include more motion artifacts observed in older participants, the fact that different segmentation procedures may interact with age, and the uncertainty in the field about the boundaries of the RSC (Foster et al. 2023). Fjell et al. (2014), however, have examined cortical thickness from MR experiments in large samples obtained both cross-sectionally and longitudinally in adults aged 18–94. Although the lifespan trajectory of the thickness of the PCC/RSC area was nonlinear in the cross-sectional sample, it did decline across age, and longitudinal results from this study generally support these findings.

The RSC has been examined in studies of resting-state functional connectivity of the default mode network across the lifespan. When low-frequency functional correlations across two regions of the default mode network, including the mPFC and the PCC/RSC, were measured, older participants had lower functional correlations between these regions than younger participants. When investigators placed a seed in the PCC/RSC and examined broader regional connections, they found some evidence of reduced functional correlations between activity in the RSC region and parietal and MTL structures in older participants. Furthermore, the reduced connectivity measure was related to white matter integrity of the individual brains as well as the participant's cognitive scores (Andrews-Hanna et al. 2007). Thus, the RSC region shows age-related alterations in communication with other brain regions important to navigation behaviors. It has been argued, however, that the changes observed between young and older brains may actually reflect adaptive rather than maladaptive alterations (e.g., Andrews-Hanna et al. 2019; Maillet and Schacter 2016). This should serve as a reminder that not all changes in the brain of older individuals (human or nonhuman) should necessarily be considered negative. Rather, a high degree of adaptive plasticity remains throughout the lifespan that can help maintain or optimize effective behavioral outcomes (Gray and Barnes 2015).

4.4 Cerebellar Circuits

A number of MRI studies conducted longitudinally consistently report a volumetric decline of the cerebellum with age (e.g., Raz et al. 2005; Tang et al. 2001) when measuring the anterior, posterior, and flocularmodular lobes along with the vermis. In the Tang et al. (2001) study, the overall decreases in volume between 79 and 84 years of age amounted to about 1.2% per year. More recently, Fjell et al. (2014) compared both cross-sectional (ages 18–94 years) and longitudinal (ages 60–90 years) measurements of total cerebellar volume and cerebellar white matter volume across age. Data from this study also support a decline in cerebellar volume with age, as does a recent longitudinal study by Han et al. (2020). These investigators examined cognitively normal participants from the Baltimore Longitudinal Study of Aging (n = 822, aged 50–92 years), who were followed for up to nine years. Both cross-sectional and longitudinal evaluation of the results again show age differences in cerebellar volume, with longitudinal declines in volume that varied across cerebellar subregions. Because the subregion designation in the Han et al. (2020) study differed from prior work, it is difficult to determine whether there are anatomical consistencies in the regions suggested to be impacted by age across studies.

While some experiments have reported a decline in cerebellar cell numbers across the lifespan, from 2 to 92 years of age (Viswasom and Jobby 2013), a carefully conducted stereological study of the cerebellum in postmortem tissue found that most regions of the cerebellum did not differ in numbers of cells from 19 to 84 years even though there was reduced overall volume (Andersen et al. 2003). In this study, the anterior subregion of the cerebellum did, however, show significantly fewer Purkinje and granule cells, reduced white matter, and total volume.

This anterior region is mainly thought to be concerned with posture, muscle tone, and gait. More recent imaging studies suggest that while the greatest impact of age is on the anterior lobule, the Crus I region of the posterior cerebellum is also affected (Bernard and Seidler 2013, 2014). Although the majority of the posterior cerebellum was not specifically impacted by aging, Crus I was an area known to be involved in the performance of sensorimotor tasks.

In fact, Crus I has been linked to two cerebro-cerebellar loops, thought to be involved in different navigation strategies that are not "strictly motor," which have been examined using fMRI. One is place navigation, with evidence suggesting that this strategy engages coherent activity between left cerebellar lobule VIIA Crus I, medial parietal cortex, and right hippocampus activity. The other loop is engaged when the strategy is sequence-based navigation, which is supported by coherent activity between right cerebellar lobule VIIA Crus I, mPFC, and left hippocampus activity (Iglói et al. 2015). While it is not known what the impact of age is on these loops, it would be interesting to determine whether one of these circuits is impacted more dramatically than another.

In contrast to the different "loops engaged" when specific navigation strategies are used, as suggested by the MRI studies reviewed above, the classic view of cerebro-cerebellar interconnections was that the cerebellum receives widespread information from neocortical areas and then directs this information through the ventrolateral nucleus of the thalamus back to the primary motor cortex (region M1), generating motor control signals in M1 (Evarts and Thach 1969). More recent ideas, however, suggest that the output of the cerebellum is directed to multiple different thalamic nuclei, which in turn project to multiple neocortical areas in addition to M1. It has been proposed that the anatomical organization of the cerebellum has multiple "closed-loop" connections that support both "motor" and "cognitive" domains. This idea has been directly examined anatomically, for example, by Kelly and Strick (2003), who showed closed-loop reciprocal connections between distinct motor and cognitive areas in macaques. Regions of cerebellar cortex that receive input from primary motor area M1 are those that project to M1, and those regions of cerebellar cortex that receive input from prefrontal cortical area 46 are those that project to area 46. Furthermore, these M1 and area 46 loops reside in separate regions of the cerebellar cortex. The posterior parietal cortex has been proposed to be an important area for integrating self-motion and visuospatial information (e.g., Save and Poucet 2009), which is supported by the fact that the firing of cells in this structure appear to reflect an egocentric reference frame, being tuned to self-motion and acceleration, independent of location or heading (Whitlock et al. 2012). Anatomical tracing studies in monkeys (Amino et al. 2001) suggest an important reciprocal cerebellar-parietal cortex loop via thalamic relay nuclei that could follow the architecture of other cerebro-cerebellar closed-loop interactions, contributing significantly to multisensory integration.

Additionally, the cerebellum and basal ganglia are anatomically densely interconnected (Bostan et al. 2013; Hoshi 2006). Consistent with the observation of multiple loops found between neocortex and the cerebellum, there are also synaptic loops between the cerebellum, thalamus, and the striatum. In macaques, projections to the input stage of the striatum (caudate, putamen) originate from motor and cognitive areas within the cerebellar dentate nucleus, and projections from the striatum run through the motor and cognitive domain regions of the subthalamic nucleus to the output stage of the cerebellum (Hoshi 2006). Thus, it will be important for future studies to investigate whether the interconnections between the striatum and cerebellum that link the motor and nonmotor domains of these structures are preserved or altered during the aging process.

Vestibular signals are crucial for spatial representations, and transient inactivation of the vestibular apparatus severely impacts behavior. Additionally, such inactivation alters rat hippocampal place cells and HD cells in the postsubiculum (Stackman et al. 2002), which are involved in navigational circuits. Vestibular information itself is not sufficient for accurate spatial localization, but requires a head-to-body reference-frame transformation, which appears to occur in the cerebellar fastigial nucleus (Rochefort et al. 2013). This nucleus contains cells that respond to both vestibular and neck proprioceptive information to produce an accurate estimate of body movement. These observations suggest that the cerebellar cortex and fastigial nuclei are good candidates for where the transformation of the vestibular and proprioceptive signals is performed. This raises the possibility that the hippocampus and cerebellum, while not having direct anatomical connections, may have indirect ones, such as multisynaptic contacts involving thalamic relay nuclei, connections with the posterior parietal, and retrosplenial cortices (e.g., Rochefort et al. 2013), that allow synchronization between these structures. This idea is at least partly supported by data from trace eyeblink conditioning studies in which the interpositus nucleus of the cerebellum was found to be synchronized with the hippocampal theta rhythm (Hoffmann and Berry 2009; Wikgren et al. 2010). Aging impacts vestibular function (e.g., Allen et al. 2016; Johnson et al. 2020), and because this sensory input is critical to the function of the HD system (examined in more detail in the next section), vestibular changes may also contribute to navigation difficulties in older individuals (Coughlan et al. 2023; Xie et al. 2017).

5 Sensorimotor Changes in Aging

Accurate navigation relies on sensory information processing to interpret and respond to environmental cues. Neuroimaging evidence in human and animal models demonstrates that neural firing in the higher-level sensory cortices becomes less selective and "dedifferentiated" at older ages, a phenomenon termed "age-related neural dedifferentiation" (Grady 2012; Koen and Rugg 2019; Koen et al. 2020; Park et al. 2004). At the single-cell level, electrophysiological recordings in the visual cortex of aged rhesus macaques reveal that tuning curves of visual orientation become broadened with age, whereby individual neurons respond to a wider range of visual orientation angles (Leventhal et al. 2003; Schmolesky et al. 2000). Similarly, VR tasks during human MRI have been employed to reveal that visual movement signals arising from the early visual cortex exhibit broadened tuning in older age (Koch et al. 2020). As a result, less precise visual information could lead to degraded directional information and increase the likelihood of disorientation.

Evidence from the human episodic memory literature consistently demonstrates that age-related neural dedifferentiation is especially prominent in the posterior parahippocampal cortex and retrosplenial complex during scene viewing, which is in turn associated with impaired subsequent memory for the stimuli (Koen and Rugg 2019; Srokova et al. 2024; Srokova et al. 2020). Consequently, dedifferentiation of the sensory information in early and higher-level visual cortices may hinder the ability to precisely encode the conjunction of features present in an environment, thus degrading the spatial computations in downstream navigation areas. This is supported by evidence of reduced neural specificity in the auditory cortex (Lalwani et al. 2019; Recanzone 2017). Additionally, motor and somatosensory representations appear dedifferentiated in old age as well (Heuninckx et al. 2005, 2008; Mattay et al. 2002). It is worth noting that most of the neuroimaging evidence for neural dedifferentiation in the visual domain arises from studies that employ static visual stimuli, and studies focusing on motor and somatosensory function are largely restricted to activation patterns elicited by simple motor tasks, such as uni- and bimanual hand or foot movements.

Multimodal spatial representations of the environment arise from the integration of sensory information derived from the external environment in combination with idiothetic sensory information that emerge as one's body moves through space (vestibular, proprioceptive, efference copy cues). Given the constraints imposed by MRI and electrophysiology methodologies, little is known about the effects of age on the specificity of sensorimotor representations during active locomotion and the potential impact of active locomotion on visuospatial representations. Indeed, although human neuroimaging research may provide insights into the effects of age on the processing of external sensory information (e.g., visual features, landmarks), it has failed to capture potential age differences in idiothetic cue processing during naturalistic navigation. In support of this point, Engle et al. (2016) demonstrated, in a sample of young rhesus macaques, that restrained-passive, constrained, and free-active locomotion engage distinct brain networks. In contrast, network activity in older macaques does not appear to robustly dissociate between different navigation conditions, suggestive of reduced network specificity in older macaques during navigation-based locomotion.

Path integration is a computational process that relies on idiothetic cues to continuously update and track one's position in space (e.g., Etienne and Jeffery 2004; Segen et al. 2022). In humans, path integration is commonly assessed using the triangle completion task, which involves guiding the individual along two legs of a triangle, then having them navigate back to the starting point. Older adults are impaired at the triangle completion task, exhibiting greater error between the estimated and true location of the starting point, particularly when cues can be derived from only one source (e.g., only vestibular feedback), or as is the case in the desktop-based variant of the task, only optic flow (Adamo et al. 2012; Allen et al. 2004; Mahmood et al. 2009; Skolimowska et al. 2011).

One proposed neurobiological mechanism of path integration deficits in aging posits dysfunctional firing of the grid cells and HD cells in the EC (see Section 4). An important factor that potentially contributes to age differences in path integration is loss of vestibular function (Agrawal et al. 2012; Hsieh et al. 2014; Xie et al. 2017). Reduced vestibular and motor function in aging may lead to the generation of dedifferentiated self-motion signals that input into the EC and contribute to the apparent dysfunction of EC cellular mechanisms (Campbell and

Giocomo 2018). In support of this notion, rats with vestibular lesions have been shown to exhibit less specific place-cell firing in the hippocampus (Stackman et al. 2002) and show greater reliance on distal visual landmarks for path integration (Stackman and Herbert 2002). Additionally, age differences in vestibular function also contribute to loss of functional mobility (e.g., reduced dynamic postural control and gait). Some aspects of functional mobility are believed to draw on top-down attentional resources (Lacour et al. 2008; Woollacott and Shumway-Cook 2002), which have been shown to be reduced in old age (e.g., Grady 2012; Hasher and Zacks 1988). Therefore, maintaining gait and balance may draw resources away from goal-directed navigation and impact spatial memory (Agathos et al. 2020). Nevertheless, it is important to note that age differences in path integration are weak, if present at all, when multiple sources of self-motion cues are available for navigation (i.e., when vestibular and proprioceptive information can be integrated together) (Adamo et al. 2012; Allen et al. 2004; Xie et al. 2017). This suggests that the availability of multimodal information may enable older adults to compensate for deficits in the vestibular or visual domain, thus ameliorating deficits in path integration.

6 Experimental Methods to Isolate Components of the Navigation Circuit in Aging

Studies of nonhuman mammals (particularly rodents) that require learning and recalling locations in tasks within novel environments have been fundamental to understanding how changes in neural circuitry and computations affect navigation behavior across the lifespan (e.g., Barnes 1979; Morris et al. 1982). These paradigms allow researchers to isolate the specific cognitive and sensorimotor processes required to navigate complex environments while simultaneously recording single- or multiunit neuronal activity in behaving rodents.

Many of the classic maze paradigms initially developed to track navigation performance in rodents have been adapted to study navigation in humans using VR technology. In typical VR-based studies, participants learn to navigate through computer-generated environments, often using a joystick or keyboard/mouse interface. These types of "desktop VR" paradigms enable researchers to precisely control and manipulate specific features within the environment in order to isolate many of the cognitive and computational mechanisms that support complex navigation behavior. Studies have compared performance on desktop VR navigation tests with metrics obtained from more standard spatial memory assessments in an effort to establish the internal validity of VR navigation paradigms. For example, Moffat et al. (1998) compared navigation performance on a desktop VR problem with verbal performance and psychometric measures of spatial performance in 74 young males and females. Unlike verbal tests, which were not associated with VR maze performance, the correlation of performance between three-dimensional mental rotation and the VR maze suggests that these two types of tests can measure at least some common features of spatial processing. Critically, however, there is some evidence that VR-based desktop studies may capture only a modest amount of variance in real-world navigation skills that enable unrestricted locomotion and enriched sensorimotor feedback during navigation (Hejtmanek et al. 2020; Van Der Ham et al. 2015).

Another notable advantage of desktop VR is that it can be paired with noninvasive brain-imaging techniques, such as fMRI, to assess the neuroanatomical and functional correlates of navigation behavior in humans. fMRI studies of navigation often reveal brain activity patterns that mirror those seen in rodent studies of navigation using physical mazes. Virtual maze tasks that rely on allocentric navigation strategies, for instance, often elicit an increased fMRI BOLD response in brain regions that are also observed in rodents navigating physical mazes, such as the hippocampus, parahippocampal cortex, and RSC. Critically, fMRI BOLD effects in these regions are often attenuated in older adults compared to younger individuals, potentially linked to age differences in allocentric navigation (Colombo et al. 2017; Li and King 2019). Recent analytic techniques have also revealed a sixfold spatial periodicity in the fMRI BOLD signal that is modulated by heading direction and movement speed as participants navigate a virtual environment, possibly reflecting the hexadirectional firing patterns of spatially tuned grid cells originally identified in rodents (Hafting et al. 2005). These spatially modulated signals, which are primarily observed within the EC, correlate with spatial memory and path integration performance (Doeller et al. 2010; Kunz et al. 2015; Stangl et al. 2018). Taken together, common patterns of brain activity between human and nonhuman populations reinforce the value of using virtual navigation tasks to gain insights into the neural circuitry and computations that support navigation across species.

It is worth noting, however, that extracellular recordings measured in rodents as they freely navigate their physical environment are fundamentally different from the types of fMRI measures obtained from supine humans as they navigate a virtual environment in a scanner bore. Single-neuron recordings that reveal grid cells in freely navigating rats, for example, are based on quite a different spatial resolution than voxels which may show sixfold symmetry in the fMRI BOLD signal (Ekstrom et al. 2020; Kunz et al. 2015) (i.e., electrical potentials in an area of tens of microns vs. hemodynamic activity within a 2000–3000 micron "voxel"). fMRI also involves hemodynamics (i.e., blood flow and oxygenation), whereas single-neuron recordings reflect action potentials of neurons. The timescale of these two signals are orders of magnitude different from each other. Such issues highlight the difficulty in specifying how the single-cell recordings and fMRI BOLD signals may reflect the same neurobiological process. Nevertheless, noninvasive MRI methods have played a critical role in developing brain-based models of navigation impairments in older humans.

Experimental MRI paradigms favor primarily visual input and require a person to use a joystick or button box to navigate while lying motionless in a supine position. Consequently, studies using MRI fail to capture potentially meaningful sources of age-related variance in multisensory integration of visual and body-based idiothetic cues during navigation (see Section 5). At the same time, results of these studies are potentially confounded by known age differences in prior experience with computerized gaming platforms as well as reduced visual and sensorimotor acuity (Charness and Boot 2022). Recent progress in immersive and mobile VR technology presents a unique opportunity to examine the impact of cognitive aging on spatial behavior in freely moving humans as they navigate ecologically enriched environments. Immersive VR technology also provides experimenters an opportunity to precisely control and experimentally manipulate test environments while simultaneously collecting a wealth of real-time behavioral, biometric, and psychophysiological data in behaving humans. Immersive VR is thus a powerful tool to help reconcile the biological significance of MRI findings obtained under restricted locomotion while also bridging a critical gap between the rodent and human literatures.

A recent study examined spatial performance in healthy young and older adults as they navigated a Morris water maze in each of two virtual environments. A desktop environment required using a keyboard and mouse to navigate, and an immersive and mobile environment enabled unrestricted locomotion and the full complement of sensorimotor cues during navigation (Hill et al. 2024). Participants were trained to learn and subsequently recall the location of hidden objects within the environment in relation to four distal landmarks. Robust age differences in spatial memory were evident when navigating the desktop environment, replicating numerous prior studies (e.g., Colombo et al. 2017; Daugherty and Raz 2017; Iaria et al. 2003; Moffat 2009). These age differences were significantly attenuated in the immersive environment, and this effect could not be explained by behavioral or motion-related confounds. Critically, the interaction between age and VR condition was driven by improved spatial memory performance among older adults in the immersive condition.

Studies that employ immersive VR techniques in patients with hippocampus lesions offer valuable insights into the biological significance of MRI findings, which have consistently identified the hippocampus as the primary source of age-related allocentric deficits on tasks such as the Morris water maze (e.g., Colombo et al. 2017; Daugherty and Raz 2017; Iaria et al. 2003; Moffat 2009). In a recent study, 11 patients with unilateral partial resection of right hippocampus and 22 control participants matched for age, sex, and education performed a virtual Morris water maze task on a desktop VR setup and an immersive VR setup that allowed for the full range of sensorimotor cues (Iggena et al. 2023). Relative to controls, the hippocampal patients exhibited deficits in spatial memory accuracy and putative allocentric strategy use, comparable to those typically observed in healthy older adults when navigating in the virtual desktop environment. These deficits were reduced or outright eliminated when tested in the immersive virtual environment. It is worth noting, however, that the left hippocampus was intact in all patients and may have been sufficient to support some aspects of spatial memory, especially when sensorimotor cues were available. In one case study, a severely amnestic 80-year-old patient with bilateral hippocampal lesions and nine cognitively unimpaired age-matched controls performed a Morris water maze in immersive VR (McAvan et al. 2022). Spatial memory in the patient was reliably above chance and allocentric strategy use was comparable to that observed in the control group. In a similar set of findings, egocentric deficits, presumed to be supported by the striatum, were eliminated in patients with Parkinson disease when navigating an immersive virtual environment (Tuena et al. 2022). Collectively, these results suggest that access to enriched sensorimotor cues while navigating an immersive environment may engage parallel navigation circuits such as the cerebellum, parietal cortex, RSC, PFC, and striatum (see Section 4). The engagement of these extrahippocampal navigation circuits may play a compensatory role in both brain-damaged individuals and potentially in normative aging.

While there are studies in rodents and nonhuman primates in navigational contexts (e.g., Diersch and Wolbers 2019; Dombeck et al. 2010; Killian et al. 2012; Matsumura et al. 1999), to date there have been no age comparisons in VR-based experiments. Future studies of this sort could facilitate a better understanding of how specific sensorimotor circuits change with age. The restraint conditions used in VR in nonhuman animals have similar limitations to those described above for humans. Newer approaches have been developed in animal models, however, that preserve the vestibular and motor efference copy information typically available during real-world navigation by allowing the animal to move freely while recording neural activity (Madhav et al. 2022). This kind of advance parallels those made in human immersive VR, which have revealed differences in the use of navigation circuits between age groups.

In a final note regarding methodology, it is rare to have very large numbers of participants in any study of behavior performance. A strong advancement in behavioral science is the web-based approaches to recruitment and data collection that enable large numbers of individuals to be tested, who are otherwise difficult to reach in university laboratory settings. In May 2016, the Sea Hero Quest test (Coutrot et al. 2018; Spiers et al. 2023) was launched as a platform to investigate navigation abilities and has been performed by 4.3 million individuals as of October 2021. Another web-based platform, Mind Crowd (Talboom et al. 2021), has also successfully recruited large numbers of individuals to perform cognitive tests and portends a future with larger, more geographically and racially inclusive samples across countries, which has not been possible with other recruiting methods. However, smaller-sample, face-to-face, investigator-driven experiments will always play an important role in hypothesis testing, as they allow in-depth measurements of freely moving individuals that are not yet feasible through web testing. Combining these methods will likely provide the strongest approach for hypothesis testing (Huentelman et al. 2020). Hypotheses that are generated will need clarification in both directions—large to small and small to large samples—to facilitate a deeper mechanistic understanding of both age- and disease-related changes in navigation.

7 Computational Models of Navigation and Their Potential Use for Understanding Age Differences in Behavior

Unfortunately, there are no current models of how known age-related changes in synaptic connectivity, plasticity, and network properties impact the function of circuit computations in the different brain regions involved in navigation described above (see Section 4). While we do not describe the intricacies of various computational approaches, we will suggest how existing computational models of spatial representations, often conceptualized as a cognitive map, can be utilized to explain age-related changes observed experimentally.

Explanation of flexible navigation behaviors often refer to cognitive maps, which are neural representations of the relative spatial relationship between features of the environment. Variables that contribute to building the cognitive map are thought to be generated in relation to an allocentric framework involving idiothetic (vestibular and proprioceptive-based) and allothetic (external environment and landmark-based) cues, or an egocentric framework (based on the relative location of environmental cues to the individual). The HD and EC–HC–PFC circuits are most strongly implicated in the representation of cues within both frameworks. Here, we separate influential models of navigation into two distinct approaches to facilitate our suggestions of how these models can be used to explain age-related changes in navigation circuits and behavior. The first utilizes continuous attractor network (CAN) dynamics to describe how the MEC and hippocampal system uses idiothetic and allothetic cues to estimate the position of an animal in space (e.g., McNaughton et al. 2006; Widloski and Fiete 2014). The second utilizes reinforcement-learning algorithms to describe how MEC and LEC inputs to the hippocampus and downstream PFC represent environmental states that explain the dynamics of behavior during navigation (Whittington et al. 2022).

In CANs, putative MEC grid cells are periodically organized in continuous two-dimensional manifolds through recurrent synaptic connections surrounded by feedforward inhibition. Local "bumps" of excitatory activity, representing grid-cell firing, flow through the network guided by angular velocity information provided by HD and speed cells, forming the basis of a path integration process (Redish and Touretzky 1997; Samsonovich and McNaughton 1997; Skaggs et al. 1994; Zhang 1996). The triangular periodicity of grid cells, which is independent of size or shape of the environment (Fyhn et al. 2004; Hafting et al. 2005), enables the cells to extend their firing fields across large spaces or to shift boundaries, thus facilitating flexible representations across environments.

In contrast, hippocampal place cells generate new firing fields when exposed to changing environments (e.g., O'Keefe and Conway 1978). Many models describe how hippocampal place cells generate stable allocentric representations of the environment through weighted summation or competitive dynamics of inputs from the MEC–CAN grid cells (e.g., Redish and Touretzky 1998; Rolls et al. 2006; Solstad et al. 2006). Experimental studies have shown that the stability of this representation in CA1 is affected in aged animals (Barnes et al. 1997; Schimanski et al. 2013). The known age-related changes in vestibular function (see Section 5) could arise from multiple brain regions that carry this information, thereby affecting the path integration computations necessary for the stability of allocentric representations. For example, age-related changes in the tuning of HD cells in the RSC could reduce the correlation between the movement of the animal through the environment and the activity bump across the CAN (Patel 2024). Consistent with this observation, preliminary data indicate that aged rodents show reduced stability and updating of grid cells across contexts (Herber et al. 2024). Some experiments suggest that a subset of HD cells in the RSC and subiculum are driven by external landmarks (Jacob et al. 2017; Kornienko et al. 2018; Olson et al. 2017), not by internal dynamics. These HD cells may not be involved in the traditional HD–CANs but may instead interact dynamically with them (Angelaki and Laurens 2020). The externally tuned HD cells may also be part of a parallel representation of directionality across different brain regions (Dudchenko et al. 2019), and the parallel HD circuits may be differentially recruited during navigation tasks in aging.

In another class of models, cognitive maps are often represented through graphs known as successor representations (SR) (Dayan 1993). In SRs, physical locations are represented as "states" or nodes on a graph, with the edges denoting a physical path between these locations (e.g., Foster et al. 2000). These formulations parallel theoretical cognitive-graph accounts in humans, in which navigators learn environments as sets of nodes (distinct locations/junctions) connected by edges (routes), yielding a flexible, topological representation that does not require explicit Euclidean metric knowledge (Chrastil and Warren 2014; Warren 2019). While their cognitive graphs use path length and angular information to determine the relationship between nodes, the SR models employ reinforcement-learning algorithms that use rewards and actions to link behavior to the nodes in the graph (Gustafson and Daw 2011; Stachenfeld et al. 2017). As an animal explores the environment, the relationships between states are updated using a transition matrix that represents all possible locations and the connections between them. The rows of this matrix are considered to represent hippocampal place fields, while the eigenvectors of the matrix represent grid cells (Garvert et al. 2017; Momennejad and Howard 2018; Schapiro et al. 2017). SRs can capture properties of rodent MEC and hippocampal neural firing as well as human hippocampal and PFC fMRI activity in spatial and nonspatial tasks (De Cothi et al. 2022; Garvert et al. 2017; Momennejad and Howard 2018; Schapiro et al. 2017; Stachenfeld et al. 2017). Learning within these SR graphs utilizes sequences of sensory observations that influence "latent states" in these graphs (Gershman and Niv 2010; Niv 2019). These latent states enable inferences about spatial properties of the environment which, in turn, influence neural representations. For example, in novel environments, existing states of grid-cell representations extend to novel locations, whereas place cells require unique states for each new location (Nieh et al. 2021).

When looking at anatomical contributions to the SR models and their role in generalizing spatial information across contexts, MEC grid cells that can generalize across contexts are thought to provide a framework for generating an abstract map. The LEC, on the other hand, provides sensory and context-specific information about the environment. The interaction between these two streams is thought to be represented in the hippocampus (Behrens et al. 2018) through probabilistic computational models that integrate both CAN (path integration) and SR (sensory map) models (Whittington et al. 2020). Goal and reward states, as represented by the PFC, can also be incorporated into these models (Mattar and Daw 2018; Whittington et al. 2020).

Models that enable generalization across spatial contexts are very useful in making predictions about the role of networks within the EC–HC–PFC circuit in spatial learning and navigation. In addition, the connections between the states in the models are not predetermined but based on sensory information that can be gathered by performing a task. Changing the sensory inputs in these models to reflect the reduced sensory signals that can occur during aging may better clarify the contributions made by elements of the EC–HC–PFC circuit to navigation behaviors. Taken out of the context of aging, Brunec and Momennejad (2022) demonstrated how applying SR model-based analysis to fMRI data during virtual navigation can reveal anatomical differences in how brain regions represent upcoming spatial trajectories. They found that anterior regions of the PFC and hippocampus were predictive of upcoming paths at larger spatial scales, while posterior regions were more finely tuned. Applying these methods to fMRI experiments that compare age-related differences across brain regions during spatial navigation performance (e.g., Li and King 2019) may enable us to predict how task-relevant computations are altered in aging. This type of analysis could be useful in understanding how strategy differences at various stages of learning might vary across age and help identify which parallel circuits in navigation are responsible for them.

8 Future Directions

Throughout this review, we have emphasized the need to consider how parallel navigation circuits might contribute to age-related differences in navigation behaviors. A primary obstacle that remains is that most studies conducted across mammals to examine aging in a navigation context have not considered changes in synaptic connections and plasticity in these circuits. While the synapse will always be central to understanding age-related changes within these complex systems, this is a difficult experimental goal to meet. Other challenges must also be overcome, if we are to solve the puzzle of how and under what testing conditions navigation circuits change with age.

First, multiple sensorimotor systems are known to change with age and thus need to be considered when studying the impact of age on navigation. While fMRI methods in humans play an important role in our understanding of age effects in the hemodynamic activity elicited by visual or simple motor tasks, they tell us little about how such activity may be modulated by vestibular and proprioceptive input, which are lacking in fMRI studies. Future research will benefit from integrating immersive VR with novel brain-imaging technologies that permit the recording of brain activity or gait as individuals freely navigate through space. Until recently, mobile imaging methods have often been constrained by increased noise in the signal during active locomotion. Exciting methodological developments, however, include MEG systems with optically pumped magnetometer sensors (Boto et al. 2018) that are small and light, allowing recording during movement. Accompanying these advancements are improved software technologies to localize and reduce noise in data collected through these systems (e.g., Bardouille et al. 2024; Duque-Muñoz et al. 2019). While human intracranial EEG recordings provide better spatial resolution and resistance to motion artifacts, these invasive methods are currently only recorded from young and middle-aged epilepsy patients, thus limiting the realistic application of these techniques to the study of healthy aging. Current and future advancements in mobile imaging technologies may offer more ecologically valid methods to examine the interplay between sensory inputs and brain activity during naturalistic movement that will be crucial for developing new theories on how aging affects spatial navigation abilities (Stangl et al. 2023).

Another major issue for the field is the lack of representative population diversity in most laboratory studies. Only sufficiently powered, large-scale studies have the potential to identify all critical variables that can impact navigation. For example, the previously observed difference in navigation performance between males and females in smaller samples was confirmed in the large Sea Hero Quest (SHQ) study (Coutrot et al. 2018; Spiers et al. 2023). While the underlying cause of sex differences in spatial navigation performance is still under debate, the SHQ study found that the magnitude of gender differences in navigation performance in a given country is correlated with the level of gender inequality in that country (Coutrot et al. 2018). This suggests societal factors also need to be considered in explanations of the observed performance differences between males and females.

An important consideration for these large studies, however, is that tests developed for phones and tablets need to be directly compared to performance on well-established neuropsychological tests known to be associated with real-world navigation. Garg et al. (2024) explicitly compared performance on tasks obtained from the SHQ navigation game with validated in-person tests of mental rotation ability, visuospatial processing, and spatial working memory. They found that performance on the SHQ wayfinding task did correlate with performance on the standard visuospatial processing tasks. These tasks, however, were not associated with performance on the SHQ path integration or spatial working-memory tasks. While this may suggest that some SHQ task measures are not correlated with real-world navigation performance, it is important to note that the validation study by Garg et al. (2024) included only 78 individuals and was not balanced for variables such as sex and race. Once again, this illustrates the confounds that can arise in small experiments and emphasizes the advantage of population-representative samples. It is only in these large samples that contributions from potentially confounding variables can be ruled out using methods such as propensity matching (Lewis et al. 2021).

A number of investigators have suggested that the function of different navigation circuits may be reasonably intact in aging individuals, but that switching between strategies poses a major problem for older participants (Ekstrom and Hill 2023; Harris et al. 2012; Harris and Wolbers 2014). Real-world navigation often requires optimizing multiple, appropriate strategy switches within a complex task to solve the problem efficiently. When Harris and Wolbers (2014) compared strategy switching between route-based and wayfinding navigation in young and aged adults, the results were striking. All young participants switched reliably from a learned egocentric route to a novel shortcut, but none of the older participants reliably switched from the egocentric to allocentric strategy. The authors suggest that this deficit in switching could result from age-related dysfunction of the PFC (Pfefferbaum et al. 2005; West 1996), known to play a role in strategy switching (e.g., Braver et al. 2003; Rossi et al. 2009). Another hypothesis is that weaker noradrenergic inputs from the locus coeruleus (LC) to the PFC may contribute to this change (e.g., Aston-Jones and Cohen 2005; Bouret and Sara 2005). While there is clear LC cell loss in Alzheimer disease, there may not be actual loss of LC cells in normative aging in humans (Mouton et al. 1994; Theofilas et al. 2017; Tomlinson et al. 1981) or non-human primates (McDermott and Barnes 2025). Additionally, Harris and Wolbers (2014) reliably observed an age-related change only in switching from egocentric to allocentric strategies, not the reverse. It cannot be ruled out that this effect results from a deficit in allocentric processing that could arise from a weaker connection between the hippocampus and the PFC (e.g., Hartley et al. 2003).

Throughout this review, we have emphasized the need to consider parallel circuits that could lead to differences in the strategies used by young and older adults in completing the navigation task at hand. Among the possible circuits involved in strategy switching, three examples are illustrated in Figure 13.2:

- The RSC (Figure 13.2a) encodes HD signals generated from multimodal sensorimotor cues that it receives from the parietal cortex and spatial information through reciprocal interactions with MTL regions (including the EC and the hippocampus) (e.g., Bicanski and Burgess 2018). This circuit has been hypothesized to influence switching between egocentric (parietal) and allocentric (MTL) strategies. In human literature, this distinction is under debate and still being refined (Ekstrom and Hill 2023).
- Frontostriatal circuits are also thought to play a role in strategy switching. This involves reciprocal connections between the PFC and the dorsal striatum, which are thought to be involved in stimulus-response navigation in familiar environments (Figure 13.2b), as well as the PFC–HC–ventral striatum circuit thought to be engaged in goal-direction navigation behaviors (e.g., Goodroe et al. 2018).
- With respect to the cerebellum (Figure 13.2c), a loop from the left cerebellum engages with posterior parietal cortex and premotor cortex via the thalamus and is thought to be involved in place-based navigation. In contrast, a loop from the right cerebellum engages the mPFC and hippocampus via the RSC and thalamus, thought to be involved in sequence-based navigation (Iglói et al. 2015; Zhang et al. 2023). Modifications to any of these circuits may contribute to the different navigation strategies observed in aging.

Given that the process of strategy switching may be key to understanding age-related differences in complex navigation, it is also important to consider other networks that have been proposed to contribute to this process (see Figure 13.2). For example, the RSC has been hypothesized to enable translation between viewpoint-dependent and -independent frames of reference (Vann et al. 2009). The posterior parietal cortex and RSC are also well positioned to contribute to switching between allocentric and egocentric strategies as they contain HD signals relative to both internal dynamics and external landmarks (e.g., Bicanski and Burgess 2018; Burgess et al. 2001; Byrne et al. 2007). Another critical navigation circuit involves the cerebellum, and alterations to specific plasticity mechanisms within this structure are known to alter the balance between allothetic and idiothetic representations of space in the hippocampus (Rochefort et al. 2013; Rondi-Reig et al. 2022). Although there are no known direct anatomical connections between the hippocampus and cerebellum, indirect connections via multisynaptic relays (Rochefort et al. 2013) may provide a mechanism for cerebellar contribution to strategy selection.

The PFC–striatal circuit has also been implicated in strategy switching during navigation tasks (e.g., Cools et al. 2004; Ragozzino and Rozman 2007; Robbins 2007). A separate cognitive circuit between the dorsal caudate, dorsolateral prefrontal, and lateral parietal cortex is thought to be involved in switching between striatal- and hippocampus-dependent strategies (Goodroe et al. 2018; Patai and Spiers 2021) in tasks with a higher demand for top-down control (Buckner 2004; Di Martino et al. 2008; Morris et al. 2016; Scimeca and Badre 2012).

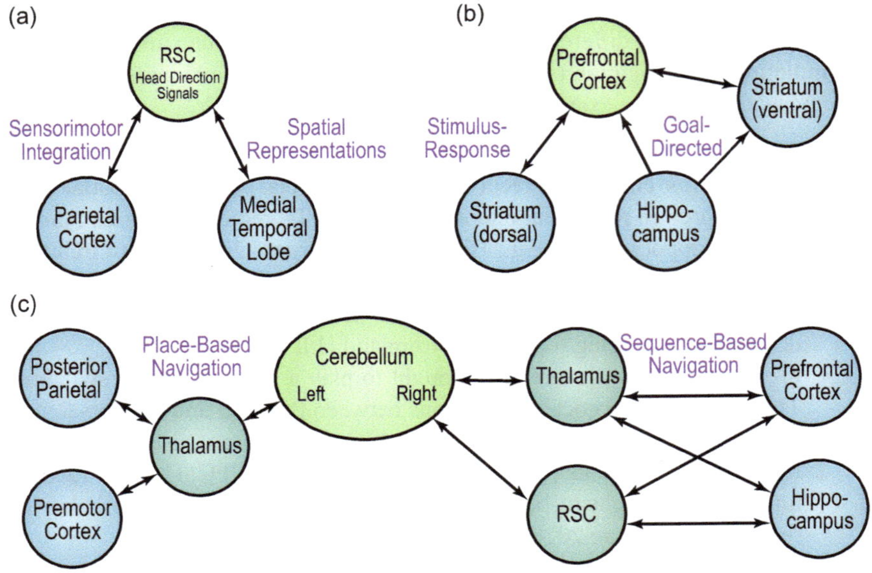

Figure 13.2 Examples of parallel circuits that can influence strategy selection in navigation.

The parallel circuits described above can support sensorimotor navigation processing at multiple levels. While the dynamic interplay between these circuits can lead to flexible behaviors, it can also result in competition between strategies that impact navigation performance. It is important for future research to better characterize the contributions made by these parallel circuits as they apply to the aging brain. This might be accomplished through experiments specifically designed to investigate how these circuits are recruited by different navigation tasks and how their function changes with age.

References

Adamo DE, Briceño EM, Sindone JA, Alexander NB, and Moffat SD (2012) Age Differences in Virtual Environment and Real World Path Integration. Front Aging Neurosci 4:e00026. https://doi.org/10.3389/fnagi.2012.00026

Agathos CP, Ramanoël S, Bécu M, et al. (2020) Postural Control While Walking Interferes with Spatial Learning in Older Adults Navigating in a Real Environment. Front Aging Neurosci 12:588653. https://doi.org/10.3389/fnagi.2020.588653

Agrawal Y, Zuniga MG, Davalos-Bichara M, et al. (2012) Decline in Semicircular Canal and Otolith Function with Age. Otol Neurotol 33 (5):832–839. https://doi.org/10.1097/MAO.0b013e3182545061

Agster KL, and Burwell RD (2013) Hippocampal and Subicular Efferents and Afferents of the Perirhinal, Postrhinal, and Entorhinal Cortices of the Rat. Behav Brain Res 254:50–64. https://doi.org/10.1016/j.bbr.2013.07.005

Ahn J-R, and Lee I (2015) Neural Correlates of Object-Associated Choice Behavior in the Perirhinal Cortex of Rats. J Neurosci 35 (4):1692–1705. https://doi.org/10.1523/jneurosci.3160-14.2015

Akopian G, and Walsh JP (2006) Pre- and Postsynaptic Contributions to Age-Related Alterations in Corticostriatal Synaptic Plasticity. Synapse 60 (3):223–238. https://doi.org/10.1002/syn.20289

Albert S, and Duffy J (2012) Differences in Risk Aversion between Young and Older Adults. Neurosci Neuroecon 1:3–9. https://doi.org/10.2147/NAN.S27184

Alexander AS, Robinson JC, Dannenberg H, et al. (2020a) Neurophysiological Coding of Space and Time in the Hippocampus, Entorhinal Cortex, and Retrosplenial Cortex. Brain Neurosci Adv 4:2398212820972871. https://doi.org/10.1177/2398212820972871

Alexander GE, Chen K, Aschenbrenner M, et al. (2008) Age-Related Regional Network of Magnetic Resonance Imaging Gray Matter in the Rhesus Macaque. J Neurosci 28 (11):2710–2718. https://doi.org/10.1523/JNEUROSCI.1852-07.2008

Alexander GE, DeLong MR, and Strick PL (1986) Parallel Organization of Functionally Segregated Circuits Linking Basal Ganglia and Cortex. Annu Rev Neurosci 9 (1):357–381. https://doi.org/10.1146/annurev.ne.09.030186.002041

Alexander GE, Lin L, Yoshimaru ES, et al. (2020b) Age-Related Regional Network Covariance of Magnetic Resonance Imaging Gray Matter in the Rat. Front Aging Neurosci 12:e00267. https://doi.org/10.3389/fnagi.2020.00267

Allen D, Ribeiro L, Arshad Q, and Seemungal BM (2016) Age-Related Vestibular Loss: Current Understanding and Future Research Directions. Front Neurol 7:231. https://doi.org/10.3389/fneur.2016.00231

Allen PA, Murphy MD, Kaufman M, Groth KE, and Begovic A (2004) Age Differences in Central (Semantic) and Peripheral Processing: The Importance of Considering Both Response Times and Errors. J Gerontol B Psychol Sci Soc Sci 59 (5):210–219. https://doi.org/10.1093/geronb/59.5.p210

Allen SL, Marshall RN, Edwards SJ, et al. (2021) The Effect of Young and Old Ex Vivo Human Serum on Cellular Protein Synthesis and Growth in an In Vitro Model of Aging. Am J Physiol Cell Physiol 321 (1):C26–C37. https://doi.org/10.1152/ajpcell.00093.2021

Alme CB, Buzzetti RA, Marrone DF, et al. (2010) Hippocampal Granule Cells Opt for Early Retirement. Hippocampus 20 (10):1109–1123. https://doi.org/10.1002/hipo.20810

Alme CB, Miao C, Jezek K, et al. (2014) Place Cells in the Hippocampus: Eleven Maps for Eleven Rooms. PNAS 111 (52):18428–18435. https://doi.org/10.1073/pnas.1421056111

Amaral DG, Insausti R, and Cowan WM (1987) The Entorhinal Cortex of the Monkey: I. Cytoarchitectonic Organization. J Comp Neurol 264 (3):326–355. https://doi.org/10.1002/cne.902640305

Amaral DG, and Witter MP (1989) The Three-Dimensional Organization of the Hippocampal Formation: A Review of Anatomical Data. Neurosci 31 (3):571–591. https://doi.org/10.1016/0306-4522(89)90424-7

Amino Y, Kyuhou S-i, Matsuzaki R, and Gemba H (2001) Cerebello–Thalamo–Cortical Projections to the Posterior Parietal Cortex in the Macaque Monkey. Neurosci Lett 309 (1):29–32. https://doi.org/10.1016/S0304-3940(01)02018-3

Andersen BB, Gundersen HJG, and Pakkenberg B (2003) Aging of the Human Cerebellum: A Stereological Study. J Comp Neurol 466 (3):356–365. https://doi.org/10.1002/cne.10884

Andrews-Hanna JR, Grilli MD, and Irish M (2019) A Review and Reappraisal of the Default Network in Normal Aging and Dementia. Oxford Research Encyclopedia of Psychology. Oxford Univ Press, Oxford. https://doi.org/10.1093/acrefore/9780190236557.013.384

Andrews-Hanna JR, Reidler JS, Sepulcre J, Poulin R, and Buckner RL (2010) Functional-Anatomic Fractionation of the Brain's Default Network. Neuron 65 (4):550–562. https://doi.org/10.1016/j.neuron.2010.02.005

Andrews-Hanna JR, Saxe R, and Yarkoni T (2014) Contributions of Episodic Retrieval and Mentalizing to Autobiographical Thought: Evidence from Functional Neuroimaging, Resting-State Connectivity, and fMRI Meta-Analyses. Neuroimage 91:324–335. https://doi.org/10.1016/j.neuroimage.2014.01.032

Andrews-Hanna JR, Snyder AZ, Vincent JL, et al. (2007) Disruption of Large-Scale Brain Systems in Advanced Aging. Neuron 56 (5):924–935. https://doi.org/10.1016/j.neuron.2007.10.038

Angelaki DE, and Laurens J (2020) The Head Direction Cell Network: Attractor Dynamics, Integration within the Navigation System, and Three-Dimensional Properties. Curr Opin Neurobiol 60:136–144. https://doi.org/10.1016/j.conb.2019.12.002

Antonova E, Parslow D, Brammer M, et al. (2009) Age-Related Neural Activity during Allocentric Spatial Memory. Memory 17 (2):125–143. https://doi.org/10.1080/09658210802077348

Aston-Jones G, and Cohen JD (2005) An Integrative Theory of Locus Coeruleus-Norepinephrine Function: Adaptive Gain and Optimal Performance. Annu Rev Neurosci 28:403–450. https://doi.org/10.1146/annurev.neuro.28.061604.135709

Bach ME, Barad M, Son H, et al. (1999) Age-Related Defects in Spatial Memory Are Correlated with Defects in the Late Phase of Hippocampal Long-Term Potentiation In Vitro and Are Attenuated by Drugs That Enhance the cAMP Signaling Pathway. PNAS 96 (9):5280–5285. https://doi.org/10.1073/pnas.96.9.5280

Balleine BW, and O'Doherty JP (2010) Human and Rodent Homologies in Action Control: Corticostriatal Determinants of Goal-Directed and Habitual Action. Neuropsychopharmacol 35 (1):48–69. https://doi.org/10.1038/npp.2009.131

Bardouille T, Smith V, Vajda E, Leslie CD, and Holmes N (2024) Noise Reduction and Localization Accuracy in a Mobile Magnetoencephalography System. Sensors 24 (11):3503. https://doi.org/10.3390/s24113503

Barnes CA (1979) Memory Deficits Associated with Senescence: A Neurophysiological and Behavioral Study in the Rat. J Comp Physiol Psychol 93 (1):74–104. https://doi.org/10.1037/h0077579

——— (1983) The Physiology of the Senescent Hippocampus. In: Seifert W (ed) Neurobiology of the Hippocampus. Academic Press, New York, pp 87–108

Barnes CA, Eppich C, and Rao G (1989) Selective Improvement of Aged Rat Short-Term Spatial Memory by 3,4-Diaminopyridine. Neurobiol Aging 10 (4):337–341. https://doi.org/10.1016/0197-4580(89)90045-6

Barnes CA, and McNaughton BL (1985) Spatial Information: How and Where Is It Stored? In: Weinberger NM, McGaugh JL, Lynch G (eds) Memory Systems of the Brain: Animal and Human Cognitive Processes. Guilford, New York, pp 49–61

Barnes CA, Nadel L, and Honig WK (1980) Spatial Memory Deficit in Senescent Rats. Can J Exp Psychol 34 (1):29–39. https://doi.org/10.1037/h0081022

Barnes CA, Suster MS, Shen J, and McNaughton BL (1997) Multistability of Cognitive Maps in the Hippocampus of Old Rats. Nature 388 (6639):272–275. https://doi.org/10.1038/40859

Behrens TEJ, Muller TH, Whittington JCR, et al. (2018) What Is a Cognitive Map? Organizing Knowledge for Flexible Behavior. Neuron 100 (2):490–509. https://doi.org/10.1016/j.neuron.2018.10.002

Bellmund JLS, Gärdenfors P, Moser EI, and Doeller CF (2018) Navigating Cognition: Spatial Codes for Human Thinking. Science 362 (6415):eaat6766. https://doi.org/10.1126/science.aat6766

Bernard JA, and Seidler RD (2013) Relationships between Regional Cerebellar Volume and Sensorimotor and Cognitive Function in Young and Older Adults. Cerebellum 12 (5):721–737. https://doi.org/10.1007/s12311-013-0481-z

——— (2014) Moving Forward: Age Effects on the Cerebellum Underlie Cognitive and Motor Declines. Neurosci Biobehav Rev 42:193–207. https://doi.org/10.1016/j.neubiorev.2014.02.011

Bicanski A, and Burgess N (2018) A Neural-Level Model of Spatial Memory and Imagery. eLife 7 (7052):1–3. https://doi.org/10.7554/elife.33752

Bliss TVP, and Lømo T (1973) Long-Lasting Potentiation of Synaptic Transmission in the Dentate Area of the Anaesthetized Rabbit Following Stimulation of the Perforant Path. J Physiol (Lond) 232 (2):331–356. https://doi.org/10.1113/jphysiol.1973.sp010273
Bloss EB, Janssen WG, Ohm DT, et al. (2011) Evidence for Reduced Experience-Dependent Dendritic Spine Plasticity in the Aging Prefrontal Cortex. J Neurosci 31 (21):7831–7839. https://doi.org/10.1523/JNEUROSCI.0839-11.2011
Blum S, Habeck C, Steffener J, Razlighi Q, and Stern Y (2014) Functional Connectivity of the Posterior Hippocampus Is More Dominant as We Age. Cogn Neurosci 5 (3–4):150–159. https://doi.org/10.1080/17588928.2014.975680
Bogacz R, Brown MW, and Giraud-Carrier C (2001) Model of Familiarity Discrimination in the Perirhinal Cortex. J Comput Neurosci 10 (1):5–23. https://doi.org/10.1023/a:1008925909305
Bohbot VD, Lerch J, Thorndycraft B, Iaria G, and Zijdenbos AP (2007) Gray Matter Differences Correlate with Spontaneous Strategies in a Human Virtual Navigation Task. J Neurosci 27 (38):10078–10083. https://doi.org/10.1523/JNEUROSCI.1763-07.2007
Bostan AC, Dum RP, and Strick PL (2013) Cerebellar Networks with the Cerebral Cortex and Basal Ganglia. Trends Cogn Sci 17 (5):241–254. https://doi.org/10.1016/j.tics.2013.03.003
Boto E, Holmes N, Leggett J, et al. (2018) Moving Magnetoencephalography Towards Real-World Applications with a Wearable System. Nature 555 (7698):657–661. https://doi.org/10.1038/nature26147
Bouret S, and Sara SJ (2005) Network Reset: A Simplified Overarching Theory of Locus Coeruleus Noradrenaline Function. Trends Neurosci 28 (11):574–582. https://doi.org/10.1016/j.tins.2005.09.002
Braver TS, Reynolds JR, and Donaldson DI (2003) Neural Mechanisms of Transient and Sustained Cognitive Control during Task Switching. Neuron 39 (4):713–726. https://doi.org/10.1016/s0896-6273(03)00466-5
Brun VH, Leutgeb S, Wu H-Q, et al. (2008) Impaired Spatial Representation in CA1 after Lesion of Direct Input from Entorhinal Cortex. Neuron 57 (2):290–302. https://doi.org/10.1016/j.neuron.2007.11.034
Brunec IK, and Momennejad I (2022) Predictive Representations in Hippocampal and Prefrontal Hierarchies. J Neurosci 42 (2):299–312. https://doi.org/10.1016/j.neuron.2007.11.03410.1523/JNEUROSCI.1327-21.2021
Buckner RL (2004) Memory and Executive Function in Aging and AD. Neuron 44 (1):195–208. https://doi.org/10.1016/j.neuron.2007.11.03410.1016/j.neuron.2004.09.006
Burgess N, Becker S, King JA, and O'Keefe J (2001) Memory for Events and Their Spatial Context: Models and Experiments. Philos Trans R Soc Lond B Biol Sci 356 (1413):1493–1503. https://doi.org/10.1098/rstb.2001.0948
Burke SN, and Barnes CA (2006) Neural Plasticity in the Ageing Brain. Nat Rev Neurosci 7 (1):30–40. https://doi.org/10.1038/nrn1809
——— (2024) Aging and the Hippocampus. In: The Hippocampus Book. Second edn. Oxford Univ Press, Oxford, pp 723–808
Burke SN, Ryan L, and Barnes CA (2012) Characterizing Cognitive Aging of Recognition Memory and Related Processes in Animal Models and in Humans. Front Aging Neurosci 4:15. https://doi.org/10.3389/fnagi.2012.00015
Burwell RD, and Amaral DG (1998) Cortical Afferents of the Perirhinal, Postrhinal, and Entorhinal Cortices of the Rat. J Comp Neurol 398 (2):179–205. https://doi.org/10.1002/(SICI)1096-9861(19980824)398:2<179::AID-CNE3>3.0.CO;2-Y
Byrne P, Becker S, and Burgess N (2007) Remembering the Past and Imagining the Future: A Neural Model of Spatial Memory and Imagery. Psychol Rev 114 (2):340–375. https://doi.org/10.1037/0033-295X.114.2.340
Cabeza R (2001) Cognitive Neuroscience of Aging: Contributions of Functional Neuroimaging. Scand J Psychol 42 (3):277–286. https://doi.org/10.1111/1467-9450.00237

——— (2002) Hemispheric Asymmetry Reduction in Older Adults: The HAROLD Model. Psychol Aging 17 (1):85–100. https://doi.org/10.1037//0882-7974.17.1.85

Caetano MS, Horst NK, Harenberg L, et al. (2012) Lost in Transition: Aging-Related Changes in Executive Control by the Medial Prefrontal Cortex. J Neurosci 32 (11):3765–3777. https://doi.org/10.1037//0882-7974.17.1.8510.1523/JNEUROSCI.6011-11.2012

Callaert DV, Ribbens A, Maes F, Swinnen SP, and Wenderoth N (2014) Assessing Age-Related Gray Matter Decline with Voxel-Based Morphometry Depends Significantly on Segmentation and Normalization Procedures. Front Aging Neurosci 6:124. https://doi.org/10.3389/fnagi.2014.00124

Campbell MG, and Giocomo LM (2018) Self-Motion Processing in Visual and Entorhinal Cortices: Inputs, Integration, and Implications for Position Coding. J Neurophysiol 120 (4):2091–2106. https://doi.org/10.1152/jn.00686.2017

Carvalho MM, Tanke N, Kropff E, et al. (2020) A Brainstem Locomotor Circuit Drives the Activity of Speed Cells in the Medial Entorhinal Cortex. Cell Rep 32 (10):108123. https://doi.org/10.1016/j.celrep.2020.108123

Charness N, and Boot WR (2022) A Grand Challenge for Psychology: Reducing the Age-Related Digital Divide. Curr Dir Psychol Sci 31 (2):187–193. https://doi.org/10.1177/09637214211068144

Chen LL, Lin L-H, Barnes CA, and McNaughton BL (1994) Head-Direction Cells in the Rat Posterior Cortex. II. Contributions of Visual and Ideothetic Information to the Directional Firing. Exp Brain Res 101 (1):24–34. https://doi.org/10.1007/bf00243213

Cho J, and Sharp PE (2001) Head Direction, Place, and Movement Correlates for Cells in the Rat Retrosplenial Cortex. Behav Neurosci 115 (1):3–25. https://doi.org/10.1037/0735-7044.115.1.3

Choi EY, Yeo BTT, and Buckner RL (2012) The Organization of the Human Striatum Estimated by Intrinsic Functional Connectivity. J Neurophysiol 108 (8):2242–2263. https://doi.org/10.1152/jn.00270.2012

Chrastil ER, Sherrill KR, Aselcioglu I, Hasselmo ME, and Stern CE (2017) Individual Differences in Human Path Integration Abilities Correlate with Gray Matter Volume in Retrosplenial Cortex, Hippocampus, and Medial Prefrontal Cortex. eNeuro 4 (2):ENEURO.0346–0316.2017. https://doi.org/10.1523/ENEURO.0346-16.2017

Chrastil ER, and Warren WH (2014) From Cognitive Maps to Cognitive Graphs. PLoS ONE 9 (11):e112544. https://doi.org/10.1371/journal.pone.0112544

Churchwell JC, Lopez-Larson M, and Yurgelun-Todd DA (2010) Altered Frontal Cortical Volume and Decision Making in Adolescent Cannabis Users. Front Psychol 1:225. https://doi.org/10.3389/fpsyg.2010.00225

Colombo D, Serino S, Tuena C, et al. (2017) Egocentric and Allocentric Spatial Reference Frames in Aging: A Systematic Review. Neurosci Biobehav Rev 80:605–621. https://doi.org/10.1016/j.neubiorev.2017.07.012

Cools R, Clark L, and Robbins TW (2004) Differential Responses in Human Striatum and Prefrontal Cortex to Changes in Object and Rule Relevance. J Neurosci 24 (5):1129–1135. https://doi.org/10.1523/JNEUROSCI.4312-03.2004

Coughlan G, Plumb W, Zhukovsky P, Aung MH, and Hornberger M (2023) Vestibular Contribution to Path Integration Deficits in "At-Genetic-Risk" for Alzheimer's Disease. PLoS ONE 18 (1):e0278239. https://doi.org/10.1371/journal.pone.0278239

Coutrot A, Silva R, Manley E, et al. (2018) Global Determinants of Navigation Ability. Curr Biol 28 (17):2861–2866.e2864. https://doi.org/10.1016/j.cub.2018.06.009

Crown LM, Gray DT, Schimanski LA, Barnes CA, and Cowen SL (2022) Aged Rats Exhibit Altered Behavior-Induced Oscillatory Activity, Place Cell Firing Rates, and Spatial Information Content in the CA1 Region of the Hippocampus. J Neurosci 42 (22):1855–1821. https://doi.org/10.1523/JNEUROSCI.1855-21.2022

Cullen KE (2019) Vestibular Processing during Natural Self-Motion: Implications for Perception and Action. Nat Rev Neurosci 20 (6):346–363. https://doi.org/10.1038/s41583-019-0153-1

Dalton MA, McCormick C, De Luca F, Clark IA, and Maguire EA (2019) Functional Connectivity Along the Anterior–Posterior Axis of Hippocampal Subfields in the Ageing Human Brain. Hippocampus 29 (11):1049–1062. https://doi.org/10.1002/hipo.23097

Daugherty A, and Raz N (2013) Age-Related Differences in Iron Content of Subcortical Nuclei Observed *In vivo*: A Meta-Analysis. Neuroimage 70:113–121. https://doi.org/10.1016/j.neuroimage.2012.12.040

——— (2015) Appraising the Role of Iron in Brain Aging and Cognition: Promises and Limitations of MRI Methods. Neuropsychol Rev 25 (3):272–287. https://doi.org/10.1007/s11065-015-9292-y

——— (2017) A Virtual Water Maze Revisited: Two-Year Changes in Navigation Performance and Their Neural Correlates in Healthy Adults. Neuroimage 146:492–506. https://doi.org/10.1016/j.neuroimage.2016.09.044

Dayan P (1993) Improving Generalization for Temporal Difference Learning: The Successor Representation. Neural Comput 5 (4):613–624. https://doi.org/10.1162/neco.1993.5.4.613

De Cothi W, Nyberg N, Griesbauer E-M, et al. (2022) Predictive Maps in Rats and Humans for Spatial Navigation. Curr Biol 32 (17):3676–3689. https://doi.org/10.1016/j.cub.2022.06.090

Deshmukh SS, and Knierim JJ (2011) Representation of Non-Spatial and Spatial Information in the Lateral Entorhinal Cortex. Front Behav Neurosci 5:00069. https://doi.org/10.3389/fnbeh.2011.00069

——— (2013) Influence of Local Objects on Hippocampal Representations: Landmark Vectors and Memory. Hippocampus 23 (4):253–267. https://doi.org/10.1002/hipo.22101

Dickstein DL, Weaver CM, Luebke JI, and Hof PR (2013) Dendritic Spine Changes Associated with Normal Aging. Neurosci 251:21–32. https://doi.org/10.1016/j.neuroscience.2012.09.077

Dieguez D, and Barea-Rodriguez EJ (2004) Aging Impairs the Late Phase of Long-Term Potentiation at the Medial Perforant Path-CA3 Synapse in Awake Rats. Synapse 52 (1):53–61. https://doi.org/10.1002/syn.20004

Diersch N, and Wolbers T (2019) The Potential of Virtual Reality for Spatial Navigation Research across the Adult Lifespan. J Exp Biol 222 (Suppl_1):jeb187252. https://doi.org/10.1242/jeb.187252

Di Martino A, Scheres A, Margulies DS, et al. (2008) Functional Connectivity of Human Striatum: A Resting State fMRI Study. Cereb Cortex 18 (12):2735–2747. https://doi.org/10.1093/cercor/bhn041

Doeller CF, Barry C, and Burgess N (2010) Evidence for Grid Cells in a Human Memory Network. Nature 463 (7281):657–661. https://doi.org/10.1038/nature08704

Dolcos F, Rice HJ, and Cabeza R (2002) Hemispheric Asymmetry and Aging: Right Hemisphere Decline or Asymmetry Reduction. Neurosci Biobehav Rev 26 (7):819–825. https://doi.org/10.1016/S0149-7634(02)00068-4

Dombeck DA, Harvey CD, Tian L, Looger LL, and Tank DW (2010) Functional Imaging of Hippocampal Place Cells at Cellular Resolution during Virtual Navigation. Nat Neurosci 13 (11):1433–1440. https://doi.org/10.1038/nn.2648

Dotson VM, Szymkowicz SM, Sozda CN, et al. (2016) Age Differences in Prefrontal Surface Area and Thickness in Middle Aged to Older Adults. Front Aging Neurosci 7:250. https://doi.org/10.3389/fnagi.2015.00250

Driscoll I, Hamilton DA, Petropoulos H, et al. (2003) The Aging Hippocampus: Cognitive, Biochemical and Structural Findings. Cereb Cortex 13 (12):1344–1351. https://doi.org/10.1093/cercor/bhg081

Dudchenko PA, Wood ER, and Smith A (2019) A New Perspective on the Head Direction Cell System and Spatial Behavior. Neurosci Biobehav Rev 105:24–33. https://doi.org/10.1016/j.neubiorev.2019.06.036

Duque-Muñoz L, Tierney TM, Meyer SS, et al. (2019) Data-Driven Model Optimization for Optically Pumped Magnetometer Sensor Arrays. Hum Brain Mapp 40 (15):4357–4369. https://doi.org/10.1002/hbm.24707

Ekstrom AD, Harootonian SK, and Huffman DJ (2020) Grid Coding, Spatial Representation, and Navigation: Should We Assume an Isomorphism? Hippocampus 30 (4):422–432. https://doi.org/10.1002/hipo.23175
Ekstrom AD, and Hill PF (2023) Spatial Navigation and Memory: A Review of the Similarities and Differences Relevant to Brain Models and Age. Neuron 111 (7):1037–1049. https://doi.org/10.1016/j.neuron.2023.03.001
Ekstrom AD, Kahana MJ, Caplan JB, et al. (2003) Cellular Networks Underlying Human Spatial Navigation. Nature 425:184–188. https://doi.org/10.1038/nature01964
Engle JR, Machado CJ, Permenter MR, et al. (2016) Network Patterns Associated with Navigation Behaviors Are Altered in Aged Nonhuman Primates. J Neurosci 36 (48):12217–12227. https://doi.org/10.1523/JNEUROSCI.4116-15.2016
Epstein RA (2008) Parahippocampal and Retrosplenial Contributions to Human Spatial Navigation. Trends Cogn Sci 12 (10):388–396. https://doi.org/10.1016/j.tics.2008.07.004
Etienne AS, and Jeffery KJ (2004) Path Integration in Mammals. Hippocampus 14 (2):180–192. https://doi.org/10.1002/hipo.10173
Evarts EV, and Thach WT (1969) Motor Mechanisms of the CNS: Cerebrocerebellar Interrelations. Annu Rev Physiol 31:451–498. https://doi.org/10.1146/annurev.ph.31.030169.002315
Fallahnezhad M, Le Mero J, Zenelaj X, et al. (2023) Cerebellar Control of a Unitary Head Direction Sense. PNAS 120 (9):e2214539120. https://doi.org/10.1073/pnas.2214539120
Ferbinteanu J, Holsinger RMD, and McDonald RJ (1999) Lesions of the Medial or Lateral Perforant Path Have Different Effects on Hippocampal Contributions to Place Learning and on Fear Conditioning to Context. Behav Brain Res 101 (1):65–84. https://doi.org/10.1016/S0166-4328(98)00144-2
Fjell AM, Sneve MH, Storsve AB, et al. (2016) Brain Events Underlying Episodic Memory Changes in Aging: A Longitudinal Investigation of Structural and Functional Connectivity. Cereb Cortex 26 (3):1272–1286. https://doi.org/10.1093/cercor/bhv102
Fjell AM, Walhovd KB, Fennema-Notestine C, et al. (2009) One-Year Brain Atrophy Evident in Healthy Aging. J Neurosci 29 (48):15223–15231. https://doi.org/10.1523/JNEUROSCI.3252-09.2009
Fjell AM, Westlye LT, Grydeland H, et al. (2013) Critical Ages in the Life Course of the Adult Brain: Nonlinear Subcortical Aging. Neurobiol Aging 34 (10):2239–2247. https://doi.org/10.1016/j.neurobiolaging.2013.04.006
——— (2014) Accelerating Cortical Thinning: Unique to Dementia or Universal in Aging? Cereb Cortex 24 (4):919–934. https://doi.org/10.1093/cercor/bhs379
Floresco SB, Seamans JK, and Phillips AG (1997) Selective Roles for Hippocampal, Prefrontal Cortical, and Ventral Striatal Circuits in Radial-Arm Maze Tasks with or without a Delay. J Neurosci 17 (5):1880–1890. https://doi.org/10.1523/jneurosci.17-05-01880.1997
Foster BL, Koslov SR, Aponik-Gremillion L, et al. (2023) A Tripartite View of the Posterior Cingulate Cortex. Nat Rev Neurosci 24 (3):173–189. https://doi.org/10.1038/s41583-022-00661-x
Foster DJ, Morris RG, and Dayan P (2000) A Model of Hippocampally Dependent Navigation, Using the Temporal Difference Learning Rule. Hippocampus 10 (1):1–16. https://doi.org/10.1002/(sici)1098-1063(2000)10:1%3C1::aid-hipo1%3E3.0.co;2-1
Foster TC, Castro CA, and McNaughton BL (1989) Spatial Selectivity of Rat Hippocampal Neurons: Dependence on Preparedness for Movement. Science 244 (4912):1580–1582. https://doi.org/10.1126/science.2740902
Frank CC, and Seaman KL (2023) Aging, Uncertainty, and Decision Making—a Review. Cogn Affect Behav Neurosci 23 (3):773–787. https://doi.org/10.3758/s13415-023-01064-w
Fraser MA, Shaw ME, and Cherbuin N (2015) A Systematic Review and Meta-Analysis of Longitudinal Hippocampal Atrophy in Healthy Human Ageing. Neuroimage 112:364–374. https://doi.org/10.1016/j.neuroimage.2015.03.035

Funahashi S, Bruce C, and Goldman-Rakic P (1989) Mnemonic Coding of Visual Space in the Monkey's Dorsolateral Prefrontal Cortex. J Neurophysiol 61 (2):331–349. https://doi.org/10.1152/jn.1989.61.2.331

Fyhn M, Molden S, Witter MP, Moser EI, and Moser MB (2004) Spatial Representation in the Entorhinal Cortex. Science 305 (5688):1258–1264. https://doi.org/10.1126/science.1099901

Gainotti G, Almonti S, Di Betta AM, and Silveri MC (1998) Retrograde Amnesia in a Patient with Retrosplenial Tumour. Neurocase 4 (6):519–526. https://doi.org/10.1080/13554799808410644

Gardner RS, Gold PE, and Korol DL (2020) Inactivation of the Striatum in Aged Rats Rescues Their Ability to Learn a Hippocampus-Sensitive Spatial Navigation Task. Neurobiol Learn Mem 172:107231. https://doi.org/10.1016/j.nlm.2020.107231

Garg T, Velasco PF, Patai EZ, et al. (2024) The Relationship between Object-Based Spatial Ability and Virtual Navigation Performance. PLoS ONE 19 (5):e0298116. https://doi.org/10.1371/journal.pone.0298116

Garvert MM, Dolan RJ, and Behrens TE (2017) A Map of Abstract Relational Knowledge in the Human Hippocampal–Entorhinal Cortex. eLife 6:e17086. https://doi.org/10.7554/eLife.17086

Gazzaley AH, Thakker MM, Hof PR, and Morrison JH (1997) Preserved Number of Entorhinal Cortex Layer II Neurons in Aged Macaque Monkeys. Neurobiol Aging 18 (5):549–553. https://doi.org/10.1016/s0197-4580(97)00112-7

Geinisman Y, de Toledo-Morrell L, Morrell F, Persina IS, and Rossi M (1992) Age-Related Loss of Axospinous Synapses Formed by Two Afferent Systems in the Rat Dentate Gyrus as Revealed by the Unbiased Stereological Dissector Technique. Hippocampus 2 (4):437–444. https://doi.org/10.1002/hipo.450020411

Gemmell C, Anderson M, and O'Mara SM (2002) Deep Layer Prefrontal Cortex Unit Discharge in a Cue-Controlled Open-Field Environment in the Freely-Moving Rat. Behav Brain Res 133 (1):1–10. https://doi.org/10.1016/s0166-4328(01)00402-8

Gershman SJ, and Niv Y (2010) Learning Latent Structure: Carving Nature at Its Joints. Curr Opin Neurobiol 20 (2):251–256. https://doi.org/10.1016/j.conb.2010.02.008

Goodroe SC, Starnes J, and Brown TI (2018) The Complex Nature of Hippocampal-Striatal Interactions in Spatial Navigation. Front Hum Neurosci 12:250. https://doi.org/10.3389/fnhum.2018.00250

Gordon EM, Laumann TO, Marek S, et al. (2022) Individualized Functional Subnetworks Connect Human Striatum and Frontal Cortex. Cereb Cortex 32 (13):2868–2884. https://doi.org/10.1093/cercor/bhab387

Grady C (2012) The Cognitive Neuroscience of Ageing. Nat Rev Neurosci 13 (7):491–505. https://doi.org/10.1038/nrn3256

Gray DT, and Barnes CA (2015) Distinguishing Adaptive Plasticity from Vulnerability in the Aging Hippocampus. Neurosci 309:17–28. https://doi.org/10.1016/j.neuroscience.2015.08.001

Guise KG, and Shapiro ML (2017) Medial Prefrontal Cortex Reduces Memory Interference by Modifying Hippocampal Encoding. Neuron 94 (1):183–192.e188. https://doi.org/10.1016/j.neuron.2017.03.011

Guitart-Masip M, Salami A, Garrett D, et al. (2016) BOLD Variability Is Related to Dopaminergic Neurotransmission and Cognitive Aging. Cereb Cortex 26 (5):2074–2083. https://doi.org/10.1093/cercor/bhv029

Gustafson NJ, and Daw ND (2011) Grid Cells, Place Cells, and Geodesic Generalization for Spatial Reinforcement Learning. PLoS Comput Biol 7 (10):e1002235. https://doi.org/10.1371/journal.pcbi.1002235

Hafting T, Fyhn M, Molden S, Moser M-B, and Moser EI (2005) Microstructure of a Spatial Map in the Entorhinal Cortex. Nature 436 (7052):801–806. https://doi.org/10.1038/nature03721

Hales JB, Schlesiger MI, Leutgeb JK, et al. (2014) Medial Entorhinal Cortex Lesions Only Partially Disrupt Hippocampal Place Cells and Hippocampus-Dependent Place Memory. Cell Rep 9 (3):893–901. https://doi.org/10.1016/j.celrep.2014.10.009

Hallgren B, and Sourander P (1958) The Effect of Age on the Non-Haemin Iron in the Human Brain. J Neurochem 3 (1):41–51. https://doi.org/10.1111/j.1471-4159.1958.tb12607.x

Han S, An Y, Carass A, Prince JL, and Resnick SM (2020) Longitudinal Analysis of Regional Cerebellum Volumes during Normal Aging. Neuroimage 220:117062. https://doi.org/10.1016/j.neuroimage.2020.117062

Harris M, Wiener J, and Wolbers T (2012) Aging Specifically Impairs Switching to an Allocentric Navigational Strategy. Front Aging Neurosci 4:29. https://doi.org/10.3389/fnagi.2012.00029

Harris MA, and Wolbers T (2012) Ageing Effects on Path Integration and Landmark Navigation. Hippocampus 22 (8):1770–1780. https://doi.org/10.1002/hipo.22011

——— (2014) How Age-Related Strategy Switching Deficits Affect Wayfinding in Complex Environments. Neurobiol Aging 35 (5):1095–1102. https://doi.org/10.1016/j.neurobiolaging.2013.10.086

Hartley T, Maguire EA, Spiers HJ, and Burgess N (2003) The Well-Worn Route and the Path Less Traveled: Distinct Neural Bases of Route Following and Wayfinding in Humans. Neuron 37 (5):877–888. https://doi.org/10.1016/s0896-6273(03)00095-3

Hasher L, and Zacks RT (1988) Working Memory, Comprehension, and Aging: A Review and a New View. In: Psychology of Learning and Motivation, vol 22. Academic Press, Cambridge, MA, pp 193–225

Haug H, and Eggers R (1991) Morphometry of the Human Cortex Cerebri and Corpus Striatum during Aging. Neurobiol Aging 12 (4):336–338; discussion 352–355. https://doi.org/10.1016/0197-4580(91)90013-a

Head D, and Isom M (2010) Age Effects on Wayfinding and Route Learning Skills. Behav Brain Res 209 (1):49–58. https://doi.org/10.1016/j.bbr.2010.01.012

Hebb DO (1949) The Organization of Behavior. Wiley, New York

Hejtmanek L, Starrett M, Ferrer E, and Ekstrom Arne D (2020) How Much of What We Learn in Virtual Reality Transfers to Real-World Navigation? Multisens Res 33 (4–5):479–503. https://doi.org/10.1163/22134808-20201445

Herber CS, Pratt KJB, Shea JM, Villeda SA, and Giocomo LM (2024) Spatial Coding Dysfunction and Network Instability in the Aging Medial Entorhinal Cortex. bioRxiv:588890. https://doi.org/10.1101/2024.04.12.588890

Hernandez AR, Maurer AP, Reasor JE, et al. (2015) Age-Related Impairments in Object-Place Associations Are Not Due to Hippocampal Dysfunction. Behav Neurosci 129 (5):599–610. https://doi.org/10.1037/bne0000093

Heuninckx S, Wenderoth N, Debaere F, Peeters R, and Swinnen SP (2005) Neural Basis of Aging: The Penetration of Cognition into Action Control. J Neurosci 25 (29):6787–6796. https://doi.org/10.1523/jneurosci.1263-05.2005

Heuninckx S, Wenderoth N, and Swinnen SP (2008) Systems Neuroplasticity in the Aging Brain: Recruiting Additional Neural Resources for Successful Motor Performance in Elderly Persons. J Neurosci 28 (1):91–99. https://doi.org/10.1523/jneurosci.3300-07.2008

Hill PF, Bermudez S, McAvan AS, et al. (2024) Age Differences in Spatial Memory Are Mitigated during Naturalistic Navigation. Aging Neuropsychol Cogn 31:2023.2001.2023.525279. https://doi.org/10.1080/13825585.2024.2326244

Hill PF, and Ekstrom AD (2025) A Cognitive-Motor Framework for Spatial Navigation in Aging and Early-Stage Alzheimer's Disease. Cortex 185:133–150. https://doi.org/10.1016/j.cortex.2025.02.003

Hirshhorn M, Grady C, Rosenbaum RS, Winocur G, and Moscovitch M (2012) Brain Regions Involved in the Retrieval of Spatial and Episodic Details Associated with a Familiar Environment: An fMRI Study. Neuropsychologia 50 (13):3094–3106. https://doi.org/10.1016/j.neuropsychologia.2012.08.008

Hoffmann LC, and Berry SD (2009) Cerebellar Theta Oscillations Are Synchronized during Hippocampal Theta-Contingent Trace Conditioning. PNAS 106 (50):21371–21376. https://doi.org/10.1073/pnas.0908403106

Hok V, Save E, Lenck-Santini PP, and Poucet B (2005) Coding for Spatial Goals in the Prelimbic/Infralimbic Area of the Rat Frontal Cortex. PNAS 102 (12):4602–4607. https://doi.org/10.1073/pnas.0407332102

Hollup SA, Molden S, Donnett JG, Moser M-B, and Moser EI (2001) Accumulation of Hippocampal Place Fields at the Goal Location in an Annular Watermaze Task. J Neurosci 21 (5):1635–1644. https://doi.org/10.1523/jneurosci.21-05-01635.2001

Hoshi E (2006) Functional Specialization within the Dorsolateral Prefrontal Cortex: A Review of Anatomical and Physiological Studies of Non-Human Primates. Neurosci Res 54 (2):73–84. https://doi.org/10.1016/j.neures.2005.10.013

Howard LR, Javadi AH, Yu Y, et al. (2014) The Hippocampus and Entorhinal Cortex Encode the Path and Euclidean Distances to Goals during Navigation. Curr Biol 24 (12):1331–1340. https://doi.org/10.1016/j.cub.2014.05.001

Høydal ØA, Skytøen ER, Andersson SO, Moser MB, and Moser EI (2019) Object-Vector Coding in the Medial Entorhinal Cortex. Nature 568:400–404. https://doi.org/10.1038/s41586-019-1077-7

Hsieh L-T, Gruber MJ, Jenkins LJ, and Ranganath C (2014) Hippocampal Activity Patterns Carry Information about Objects in Temporal Context. Neuron 81 (5):1165–1178. https://doi.org/10.1016/j.neuron.2014.01.015

Huentelman MJ, Talboom JS, Lewis CR, Chen Z, and Barnes CA (2020) Reinventing Neuroaging Research in the Digital Age. Trends Neurosci 43 (1):17–23. https://doi.org/10.1016/j.tins.2019.11.004

Iaria G, Petrides M, Dagher A, Pike B, and Bohbot VD (2003) Cognitive Strategies Dependent on the Hippocampus and Caudate Nucleus in Human Navigation: Variability and Change with Practice. J Neurosci 23 (13):5945–5952. https://doi.org/10.1523/jneurosci.23-13-05945.2003

Ide R, Ota M, Hada Y, et al. (2023) Relationship between Hippocampal Subfields Volume and Balance Function in Healthy Older Adults. Gait Posture 101:90–94. https://doi.org/10.1016/j.gaitpost.2023.02.003

Igarashi KM, Lu L, Colgin LL, Moser M-B, and Moser EI (2014) Coordination of Entorhinal-Hippocampal Ensemble Activity during Associative Learning. Nature 510 (7503):143–147. https://doi.org/10.1038/nature13162

Iggena D, Jeung S, Maier PM, et al. (2023) Multisensory Input Modulates Memory-Guided Spatial Navigation in Humans. Commun Biol 6 (1):1167. https://doi.org/10.1038/s42003-023-05522-6

Iglói K, Doeller CF, Paradis A-L, et al. (2015) Interaction between Hippocampus and Cerebellum Crus I in Sequence-Based but Not Place-Based Navigation. Cereb Cortex 25 (11):4146–4154. https://doi.org/10.1093/cercor/bhu132

Insausti R, Herrero MT, and Witter MP (1997) Entorhinal Cortex of the Rat: Cytoarchitectonic Subdivisions and the Origin and Distribution of Cortical Efferents. Hippocampus 7 (2):146–183. https://doi.org/10.1002/(sici)1098-1063(1997)7:2%3C146::aid-hipo4%3E3.0.co;2-l

Insel N, Patron LA, Hoang LT, et al. (2012) Reduced Gamma Frequency in the Medial Frontal Cortex of Aged Rats during Behavior and Rest: Implications for Age-Related Behavioral Slowing. J Neurosci 32 (46):16331–16344. https://doi.org/10.1523/jneurosci.1577-12.2012

Issa JB, Radvansky BA, Xuan F, and Dombeck DA (2024) Lateral Entorhinal Cortex Subpopulations Represent Experiential Epochs Surrounding Reward. Nat Neurosci 27 (3):536–546. https://doi.org/10.1101/2023.10.09.561557

Jacob P-Y, Casali G, Spieser L, et al. (2017) An Independent, Landmark-Dominated Head-Direction Signal in Dysgranular Retrosplenial Cortex. Nat Neurosci 20 (2):173–175. https://doi.org/10.1038/nn.4465

Jacobs J, Kahana MJ, Ekstrom AD, Mollison MV, and Fried I (2010) A Sense of Direction in Human Entorhinal Cortex. PNAS 107 (14):6487–6492. https://doi.org/10.1073/pnas.0911213107

Jacobs NS, Allen TA, Nguyen N, and Fortin NJ (2013) Critical Role of the Hippocampus in Memory for Elapsed Time. J Neurosci 33 (34):13888–13893. https://doi.org/10.1523/jneurosci.1733-13.2013

Jadhav SP, Rothschild G, Roumis DK, and Frank LM (2016) Coordinated Excitation and Inhibition of Prefrontal Ensembles during Awake Hippocampal Sharp-Wave Ripple Events. Neuron 90 (1):113–127. https://doi.org/10.1016/j.neuron.2016.02.010

Jarrard LE (1989) On the Use of Ibotenic Acid to Lesion Selectively Different Components of the Hippocampal Formation. J Neurosci Methods 29 (3):251–259. https://doi.org/10.1016/0165-0270(89)90149-0

Jay TM, and Witter MP (1991) Distribution of Hippocampal CA1 and Subicular Efferents in the Prefrontal Cortex of the Rat Studied by Means of Anterograde Transport of Phaseolus Vulgaris-Leucoagglutinin. J Comp Neurol 313 (4):574–586. https://doi.org/10.1002/cne.903130404

Jeffery KJ, Cheng K, Newcombe NS, Bingman VP, and Menzel R (2024) Unpacking the Navigation Toolbox: Insights from Comparative Cognition. Proc R Soc Lond B Biol Sci 291 (2016):20231304. https://doi.org/10.1098/rspb.2023.1304

Jog MS, Kubota Y, Connolly CI, Hillegaart V, and Graybiel AM (1999) Building Neural Representations of Habits. Science 286 (5445):1745–1749. https://doi.org/10.1126/science.286.5445.1745

Johnson C, Hallemans A, Verbecque E, et al. (2020) Aging and the Relationship between Balance Performance, Vestibular Function and Somatosensory Thresholds. Int Adv Otol 16 (3):328–337. https://doi.org/10.5152/iao.2020.8287

Johnson SA, Turner SM, Santacroce LA, et al. (2017) Rodent Age-Related Impairments in Discriminating Perceptually Similar Objects Parallel Those Observed in Humans. Hippocampus 27 (7):759–776. https://doi.org/10.1002/hipo.22729

Jones MW, and Wilson MA (2005) Theta Rhythms Coordinate Hippocampal-Prefrontal Interactions in a Spatial Memory Task. PLoS Biol 3 (12):e402. https://doi.org/10.1371/journal.pbio.0030402

Jung MW, Qin Y, McNaughton BL, and Barnes CA (1998) Firing Characteristics of Deep Layer Neurons in Prefrontal Cortex in Rats Performing Spatial Working Memory Tasks. Cereb Cortex 8 (5):437–450. https://doi.org/10.1093/cercor/8.5.437

Jung MW, Wiener SI, and McNaughton BL (1994) Comparison of Spatial Firing Characteristics of Units in Dorsal and Ventral Hippocampus of the Rat. J Neurosci 14 (12):7347–7356. https://doi.org/10.1523/jneurosci.14-12-07347.1994

Kaczorowski CC, Davis SJ, and Moyer JR (2012) Aging Redistributes Medial Prefrontal Neuronal Excitability and Impedes Extinction of Trace Fear Conditioning. Neurobiol Aging 33 (8):1744–1757. https://doi.org/10.1016/j.neurobiolaging.2011.03.020

Kaplan R, King J, Koster R, et al. (2017) The Neural Representation of Prospective Choice during Spatial Planning and Decisions. PLoS Biol 15 (1):e1002588. https://doi.org/10.1371/journal.pbio.1002588

Keene CS, Bladon J, McKenzie S, et al. (2016) Complementary Functional Organization of Neuronal Activity Patterns in the Perirhinal, Lateral Entorhinal, and Medial Entorhinal Cortices. J Neurosci 36 (13):3660–3675. https://doi.org/10.1523/jneurosci.4368-15.2016

Kelly RM, and Strick PL (2003) Cerebellar Loops with Motor Cortex and Prefrontal Cortex of a Nonhuman Primate. J Neurosci 23 (23):8432–8444. https://doi.org/10.1523/jneurosci.23-23-08432.2003

Keuker J (2003) Preservation of Hippocampal Neuron Numbers in Aged Rhesus Monkeys. Neurobiol Aging 24 (1):157–165. https://doi.org/10.1016/s0197-4580(02)00062-3

Killian NJ, Jutras MJ, and Buffalo EA (2012) A Map of Visual Space in the Primate Entorhinal Cortex. Nature 491 (7426):761–764. https://doi.org/10.1038/nature11587

Kinkhabwala AA, Gu Y, Aronov D, and Tank DW (2020) Visual Cue-Related Activity of Cells in the Medial Entorhinal Cortex during Navigation in Virtual Reality. eLife 9:e43140. https://doi.org/10.7554/elife.43140

Kjelstrup KB, Solstad T, Brun VH, et al. (2008) Finite Scale of Spatial Representation in the Hippocampus. Science 321 (5885):140–143. https://doi.org/10.1126/science.1157086

Klur S, Muller C, Pereira de Vasconcelos A, et al. (2009) Hippocampal-Dependent Spatial Memory Functions Might Be Lateralized in Rats: An Approach Combining Gene Expression Profiling and Reversible Inactivation. Hippocampus 19 (9):800–816. https://doi.org/10.1002/hipo.20562

Knierim JJ (2002) Dynamic Interactions between Local Surface Cues, Distal Landmarks, and Intrinsic Circuitry in Hippocampal Place Cells. J Neurosci 22 (14):6254–6264. https://doi.org/10.1523/jneurosci.22-14-06254.2002

Knierim JJ, Neunuebel JP, and Deshmukh SS (2014) Functional Correlates of the Lateral and Medial Entorhinal Cortex: Objects, Path Integration and Local–Global Reference Frames. Philos Trans R Soc Lond B Biol Sci 369 (1635):20130369. https://doi.org/10.1098/rstb.2013.0369

Knoops AJG, Gerritsen L, Graaf Y, Mali WPTM, and Geerlings MI (2012) Loss of Entorhinal Cortex and Hippocampal Volumes Compared to Whole Brain Volume in Normal Aging: The SMART-Medea Study. Psychiatry Res Neuroimaging 203 (1):31–37. https://doi.org/10.1016/j.pscychresns.2011.12.002

Koch C, Li S-C, Polk TA, and Schuck NW (2020) Effects of Aging on Encoding of Walking Direction in the Human Brain. Neuropsychologia 141:107379. https://doi.org/10.1016/j.neuropsychologia.2020.107379

Koen JD, and Rugg MD (2019) Neural Dedifferentiation in the Aging Brain. Trends Cogn Sci 23 (7):547–559. https://doi.org/10.1016/j.tics.2019.04.012

Koen JD, Srokova S, and Rugg MD (2020) Age-Related Neural Dedifferentiation and Cognition. Curr Opin Behav Sci 32:7–14. https://doi.org/10.1016/j.cobeha.2020.01.006

Koen JD, and Yonelinas AP (2014) The Effects of Healthy Aging, Amnestic Mild Cognitive Impairment, and Alzheimer's Disease on Recollection and Familiarity: A Meta-Analytic Review. Neuropsychol Rev 24 (3):332–354. https://doi.org/10.1007/s11065-014-9266-5

Konishi K, and Bohbot VD (2013) Spatial Navigational Strategies Correlate with Gray Matter in the Hippocampus of Healthy Older Adults Tested in a Virtual Maze. Front Aging Neurosci 5:1. https://doi.org/10.3389/fnagi.2013.00001

Konishi K, McKenzie S, Etchamendy N, Roy S, and Bohbot VD (2017) Hippocampus-Dependent Spatial Learning Is Associated with Higher Global Cognition among Healthy Older Adults. Neuropsychologia 106:310–321. https://doi.org/10.1016/j.neuropsychologia.2017.09.025

Kornienko O, Latuske P, Bassler M, Kohler L, and Allen K (2018) Non-Rhythmic Head-Direction Cells in the Parahippocampal Region Are Not Constrained by Attractor Network Dynamics. eLife 7:e35949. https://doi.org/10.7554/elife.35949

Korthauer LE, Nowak NT, Moffat SD, et al. (2016) Correlates of Virtual Navigation Performance in Older Adults. Neurobiol Aging 39:118–127. https://doi.org/10.1016/j.neurobiolaging.2015.12.003

Kropff E, Carmichael JE, Moser M-B, and Moser EI (2015) Speed Cells in the Medial Entorhinal Cortex. Nature 523 (7561):419–424. https://doi.org/10.1038/nature14622

Kunz L, Schröder TN, Lee H, et al. (2015) Reduced Grid-Cell-Like Representations in Adults at Genetic Risk for Alzheimer's Disease. Science 350 (6259):430–433. https://doi.org/10.1126/science.aac8128

Kuruvilla MV, and Ainge JA (2017) Lateral Entorhinal Cortex Lesions Impair Local Spatial Frameworks. Front Syst Neurosci 11:30. https://doi.org/10.3389/fnsys.2017.00030

Kuruvilla MV, Wilson DIG, and Ainge JA (2020) Lateral Entorhinal Cortex Lesions Impair Both Egocentric and Allocentric Object–Place Associations. Brain Neurosci Adv 4:239821282093946. https://doi.org/10.1177/2398212820939463

Lacour M, Bernard-Demanze L, and Dumitrescu M (2008) Posture Control, Aging, and Attention Resources: Models and Posture-Analysis Methods. Neurophysiol Clin 38 (6):411–421. https://doi.org/10.1016/j.neucli.2008.09.005

Lalwani P, Gagnon H, Cassady K, et al. (2019) Neural Distinctiveness Declines with Age in Auditory Cortex and Is Associated with Auditory GABA Levels. Neuroimage 201:116033. https://doi.org/10.1016/j.neuroimage.2019.116033

Laskowski CS, Williams RJ, Martens KM, et al. (2016) The Role of the Medial Prefrontal Cortex in Updating Reward Value and Avoiding Perseveration. Behav Brain Res 306:52–63. https://doi.org/10.1016/j.bbr.2016.03.007

Laubach M, Amarante LM, Swanson K, and White SR (2018) What, If Anything, Is Rodent Prefrontal Cortex? eNeuro 5 (5):ENEURO.0315–0318.2018. https://doi.org/10.1523/eneuro.0315-18.2018

Lee TMC, Yip JTH, and Jones-Gotman M (2002) Memory Deficits after Resection from Left or Right Anterior Temporal Lobe in Humans: A Meta-Analytic Review. Epilepsia 43 (3):283–291. https://doi.org/10.1046/j.1528-1157.2002.09901.x

Lefort JM, Vincent J, Tallot L, et al. (2019) Impaired Cerebellar Purkinje Cell Potentiation Generates Unstable Spatial Map Orientation and Inaccurate Navigation. Nat Commun 10 (1):2251. https://doi.org/10.1038/s41467-019-09958-5

Lehéricy S, Benali H, Van De Moortele P-F, et al. (2005) Distinct Basal Ganglia Territories Are Engaged in Early and Advanced Motor Sequence Learning. PNAS 102 (35):12566–12571. https://doi.org/10.1073/pnas.050276210210.1073/pnas.0502762102

Lemaitre H, Goldman AL, Sambataro F, et al. (2012) Normal Age-Related Brain Morphometric Changes: Nonuniformity across Cortical Thickness, Surface Area and Gray Matter Volume? Neurobiol Aging 33 (3):617.e611–619. https://doi.org/10.1016/j.neurobiolaging.2010.07.013

Leutgeb JK, Leutgeb S, Moser M-B, and Moser EI (2007) Pattern Separation in the Dentate Gyrus and CA3 of the Hippocampus. Science 315 (5814):961–966. https://doi.org/10.1126/science.1135801

Leutgeb S, Leutgeb JK, Barnes CA, et al. (2005) Independent Codes for Spatial and Episodic Memory in Hippocampal Neuronal Ensembles. Science 309 (5734):619–623. https://doi.org/10.1126/science.1114037

Leventhal AG, Wang Y, Pu M, Zhou Y, and Ma Y (2003) GABA and Its Agonists Improved Visual Cortical Function in Senescent Monkeys. Science 300 (5620):812–815. https://doi.org/10.1126/science.1082874

Lever C, Burton S, Jeewajee A, O'Keefe J, and Burgess N (2009) Boundary Vector Cells in the Subiculum of the Hippocampal Formation. J Neurosci 29 (31):9771–9777. https://doi.org/10.1523/jneurosci.1319-09.2009

Lewis C, Talboom JS, De Both MD, et al. (2021) Smoking Impairs Women's Verbal Learning and Memory Performance More Than Men's: An International Web-Cohort Study of 70,000 Participants. Sci Rep 11:10248. https://doi.org/10.1038/s41598-021-88923-z

Li AW, and King J (2019) Spatial Memory and Navigation in Ageing: A Systematic Review of MRI and fMRI Studies in Healthy Participants. Neurosci Biobehav Rev 103:33–49. https://doi.org/10.1016/j.neubiorev.2019.05.005

Livingston G, Huntley J, Liu K, et al. (2024) Dementia Prevention, Intervention, and Care: 2024 Report of the Lancet Standing Commission. Lancet 404 (10452):572–628. https://doi.org/10.1016/s0140-6736(24)01296-0

Livingston G, Huntley J, Sommerlad A, et al. (2020) Dementia Prevention, Intervention, and Care: 2020 Report of the Lancet Commission. Lancet 396 (10248):413–446. https://doi.org/10.1016/s0140-6736(20)30367-6

Lolova IS, Lolov SR, and Itzev DE (1997) Aging and the Dendritic Morphology of the Rat Laterodorsal and Pedunculopontine Tegmental Nuclei. Mech Ageing Dev 97 (3):193–205. https://doi.org/10.1016/s0047-6374(97)00062-6

Maaroufi K, Ammari M, Jeljeli M, et al. (2009) Impairment of Emotional Behavior and Spatial Learning in Adult Wistar Rats by Ferrous Sulfate. Physiol Behav 96 (2):343–349. https://doi.org/10.1016/j.physbeh.2008.10.019

Maass A, Berron D, Libby LA, Ranganath C, and Düzel E (2015) Functional Subregions of the Human Entorhinal Cortex. eLife 4:e06426. https://doi.org/10.7554/elife.06426

Madhav MS, Jayakumar RP, Lashkari SG, et al. (2022) The Dome: A Virtual Reality Apparatus for Freely Locomoting Rodents. J Neurosci Methods 368:109336. https://doi.org/10.1016/j.jneumeth.2021.109336

Maguire EA (2001) The Retrosplenial Contribution to Human Navigation: A Review of Lesion and Neuroimaging Findings. Scand J Psychol 42 (3):225–238. https://doi.org/10.1111/1467-9450.00233

Maguire EA, Frackowiak RSJ, and Frith CD (1997) Recalling Routes around London: Activation of the Right Hippocampus in Taxi Drivers. J Neurosci 17 (18):7103–7110. https://doi.org/10.1523/jneurosci.17-18-07103.1997

Mahmood O, Adamo D, Briceno E, and Moffat SD (2009) Age Differences in Visual Path Integration. Behav Brain Res 205 (1):88–95. https://doi.org/10.1016/j.bbr.2009.08.001

Maidenbaum S, Miller J, Stein JM, and Jacobs J (2018) Grid-Like Hexadirectional Modulation of Human Entorhinal Theta Oscillations. PNAS 115 (42):10798–10803. https://doi.org/10.1073/pnas.1805007115

Maillet D, and Schacter DL (2016) Default Network and Aging: Beyond the Task-Negative Perspective. Trends Cogn Sci 20 (9):646–648. https://doi.org/10.1016/j.tics.2016.05.009

Malykhin NV, Bouchard TP, Camicioli R, and Coupland NJ (2008) Aging Hippocampus and Amygdala. Neuroreport 19 (5):543. https://doi.org/10.1097/WNR.0b013e3282f8b18c

Malykhin NV, Huang Y, Hrybouski S, and Olsen F (2017) Differential Vulnerability of Hippocampal Subfields and Anteroposterior Hippocampal Subregions in Healthy Cognitive Aging. Neurobiol Aging 59:121–134. https://doi.org/10.1016/j.neurobiolaging.2017.08.001

Mapstone M, Logan D, and Duffy CJ (2006) Cue Integration for the Perception and Control of Self-Movement in Ageing and Alzheimer's Disease. Brain 129 (11):2931–2944. https://doi.org/10.1093/brain/awl201

Markus EJ, Barnes CA, McNaughton BL, Gladden VL, and Skaggs WE (1994) Spatial Information Content and Reliability of Hippocampal CA1 Neurons: Effects of Visual Input. Hippocampus 4 (4):410–421. https://doi.org/10.1002/hipo.450040404

Matsumura N, Nishijo H, Tamura R, et al. (1999) Spatial- and Task-Dependent Neuronal Responses during Real and Virtual Translocation in the Monkey Hippocampal Formation. J Neurosci 19 (6):2381–2393. https://doi.org/10.1523/jneurosci.19-06-02381.1999

Mattar MG, and Daw ND (2018) Prioritized Memory Access Explains Planning and Hippocampal Replay. Nat Neurosci 21 (11):1609–1617. https://doi.org/10.1038/s41593-018-0232-z

Mattay VS, Fera F, Tessitore A, et al. (2002) Neurophysiological Correlates of Age-Related Changes in Human Motor Function. Neurology 58 (4):630–635. https://doi.org/10.1212/wnl.58.4.630

Maurer AP, Johnson SA, Hernandez AR, et al. (2017) Age-Related Changes in Lateral Entorhinal and CA3 Neuron Allocation Predict Poor Performance on Object Discrimination. Front Syst Neurosci 11:00049. https://doi.org/10.3389/fnsys.2017.00049

McAvan AS, Wank AA, Rapcsak SZ, Grilli MD, and Ekstrom AD (2022) Largely Intact Memory for Spatial Locations during Navigation in an Individual with Dense Amnesia. Neuropsychologia 170:108225. https://doi.org/10.1016/j.neuropsychologia.2022.108225

McDermott KE, and Barnes CA (2025) Stability of Locus Coeruleus Cell Counts Despite Volume Loss in Cognitively Impaired Aged Rhesus Macaques. Neurobiol Aging 148:41–49. https://doi.org/10.1016/j.neurobiolaging.2025.01.007

McNaughton BL, Battaglia FP, Jensen O, Moser EI, and Moser M-B (2006) Path Integration and the Neural Basis of the "Cognitive Map." Nat Rev Neurosci 7 (8):663–678. https://doi.org/10.1038/nrn1932

Meister MLR, and Buffalo EA (2016) Getting Directions from the Hippocampus: The Neural Connection between Looking and Memory. Neurobiol Learn Mem 134 Pt A (Pt A):135–144. https://doi.org/10.1016/j.nlm.2015.12.004

Merrill DA, Chiba AA, and Tuszynski MH (2001) Conservation of Neuronal Number and Size in the Entorhinal Cortex of Behaviorally Characterized Aged Rats. J Comp Neurol 438 (4):445–456. https://doi.org/10.1002/cne.1327

Merrill DA, Roberts JA, and Tuszynski MH (2000) Conservation of Neuron Number and Size in Entorhinal Cortex Layers II, III, and V/VI of Aged Primates. J Comp Neurol 422 (3):396–401. https://doi.org/10.1002/1096-9861(20000703)422:3%3C396::aid-cne6%3E3.0.co;2-r

Meulenbroek O, Petersson KM, Voermans N, Weber B, and Fernández G (2004) Age Differences in Neural Correlates of Route Encoding and Route Recognition. Neuroimage 22 (4):1503–1514. https://doi.org/10.1016/j.neuroimage.2004.04.007

Mizoguchi K, Shoji H, Tanaka Y, Maruyama W, and Tabira T (2009) Age-Related Spatial Working Memory Impairment Is Caused by Prefrontal Cortical Dopaminergic Dysfunction in Rats. Neurosci 162 (4):1192–1201. https://doi.org/10.1016/j.neuroscience.2009.05.023

Mizumori SJ, and Kalyani A (1997) Age and Experience-Dependent Representational Reorganization during Spatial Learning. Neurobiol Aging 18 (6):651–659. https://doi.org/10.1016/s0197-4580(97)00150-4

Mizumori SJY, Lavoie AM, and Kalyani A (1996) Redistribution of Spatial Representation in the Hippocampus of Aged Rats Performing a Spatial Memory Task. Behav Neurosci 110 (5):1006–1016. https://doi.org/10.1037//0735-7044.110.5.1006

Moffat SD (2006) Does Testosterone Mediate Cognitive Decline in Elderly Men? J Gerontol A Biol Sci Med Sci 61 (5):521. https://doi.org/10.1093/gerona/61.5.521

——— (2009) Aging and Spatial Navigation: What Do We Know and Where Do We Go? Neuropsychol Rev 19 (4):478–489. https://doi.org/10.1007/s11065-009-9120-3

Moffat SD, Hampson E, and Hatzipantelis M (1998) Navigation in a "Virtual" Maze: Sex Differences and Correlation with Psychometric Measures of Spatial Ability in Humans. Evol Hum Behav 19 (2):73–87. https://doi.org/10.1016/S1090-5138(97)00104-9

Moffat SD, Kennedy KM, Rodrigue KM, and Raz N (2007) Extrahippocampal Contributions to Age Differences in Human Spatial Navigation. Cereb Cortex 17 (6):1274–1282. https://doi.org/10.1093/cercor/bhl036

Momennejad I, and Howard MW (2018) Predicting the Future with Multi-Scale Successor Representations. https://www.biorxiv.org/content/10.1101/449470v1

Morris LS, Kundu P, Dowell N, et al. (2016) Fronto-Striatal Organization: Defining Functional and Microstructural Substrates of Behavioural Flexibility. Cortex 74:118–133. https://doi.org/10.1016/j.cortex.2015.11.004

Morris RG, Garrud P, Rawlins JN, and O'Keefe J (1982) Place Navigation Impaired in Rats with Hippocampal Lesions. Nature 297 (5868):681–683. https://doi.org/10.1038/297681a0

Mouton PR, Pakkenberg B, Gundersen HJG, and Price DL (1994) Absolute Number and Size of Pigmented Locus Coeruleus Neurons in Young and Aged Individuals. J Chem Neuroanat 7 (3):185–190. https://doi.org/10.1016/0891-0618(94)90028-0

Muller R, and Kubie J (1987) The Effects of Changes in the Environment on the Spatial Firing of Hippocampal Complex-Spike Cells. J Neurosci 7 (7):1951–1968. https://doi.org/10.1523/JNEUROSCI.07-07-01951.1987

Murray EA, Wise SP, and Graham KS (2016) The Evolution of Memory Systems. Oxford Univ Press, Oxford

Navarro Schröder T, Haak K, Zaragoza Jimenez N, Beckmann CF, and Doeller CF (2015) Functional Topography of the Human Entorhinal Cortex. eLife 4:e06738. https://doi.org/10.7554/eLife.06738

Nieh EH, Schottdorf M, Freeman NW, et al. (2021) Geometry of Abstract Learned Knowledge in the Hippocampus. Nature 595 (7865):80–84. https://doi.org/10.1038/s41586-021-03652-7

Niv Y (2019) Learning Task-State Representations. Nat Neurosci 22 (10):1544–1553. https://doi.org/10.1038/s41593-019-0470-8
Nobis L, Manohar SG, Smith SM, et al. (2019) Hippocampal Volume across Age: Nomograms Derived from over 19,700 People in UK Biobank. Neuroimage Clin 23:101904. https://doi.org/10.1016/j.nicl.2019.101904
Norris CM, Korol DL, and Foster TC (1996) Increased Susceptibility to Induction of Long-Term Depression and Long-Term Potentiation Reversal during Aging. J Neurosci 16 (17):5382–5392. https://doi.org/10.1523/jneurosci.16-17-05382.1996
Nyberg L, Karalija N, Salami A, et al. (2016) Dopamine D2 Receptor Availability Is Linked to Hippocampal–Caudate Functional Connectivity and Episodic Memory. PNAS 113 (28):7918–7923. https://doi.org/10.1073/pnas.1606309113
O'Keefe J (1976) Place Units in the Hippocampus of the Freely Moving Rat. Exp Neurol 51 (1):78–109. https://doi.org/10.1016/0014-4886(76)90055-8
O'Keefe J, and Conway DH (1978) Hippocampal Place Units in the Freely Moving Rat: Why They Fire Where They Fire. Exp Brain Res 31 (4):573–590. https://doi.org/10.1007/bf00239813
O'Keefe J, and Dostrovsky J (1971) The Hippocampus as a Spatial Map: Preliminary Evidence from Unit Activity in the Freely-Moving Rat. Brain Res 34 (1):171–175. https://doi.org/10.1016/0006-8993(71)90358-1
O'Keefe J, and Nadel L (1978) The Hippocampus as a Cognitive Map. Oxford Univ Press, London
Olson JM, Tongprasearth K, and Nitz DA (2017) Subiculum Neurons Map the Current Axis of Travel. Nat Neurosci 20 (2):170–172. https://doi.org/10.1038/nn.4464
Østby Y, Tamnes CK, Fjell AM, et al. (2009) Heterogeneity in Subcortical Brain Development: A Structural Magnetic Resonance Imaging Study of Brain Maturation from 8 to 30 Years. J Neurosci 29 (38):11772–11782. https://doi.org/10.1523/jneurosci.1242-09.2009
Padova D, Faria A, Ratnanather JT, et al. (2024) Vestibular Function Is Associated with Prefrontal and Sensorimotor Cortical Gray Matter Volumes in a Cross-Sectional Study of Healthy, Older Adults. Apert Neuro 4:116785. https://doi.org/10.52294/001c.116785
Park DC, Polk TA, Park R, et al. (2004) Aging Reduces Neural Specialization in Ventral Visual Cortex. PNAS 101 (35):13091–13095. https://doi.org/10.52294/001c.116785
Patai EZ, and Spiers HJ (2021) The Versatile Wayfinder: Prefrontal Contributions to Spatial Navigation. Trends Cogn Sci 25 (6):520–533. https://doi.org/10.1016/j.tics.2021.02.010
Patel KD (2024) Aging Destabilizes Egocentric Boundary Cells in the Retrosplenial Cortex of Mice. Master's Thesis, Boston Univ, Boston
Pereira JB, Valls-Pedret C, Ros E, et al. (2014) Regional Vulnerability of Hippocampal Subfields to Aging Measured by Structural and Diffusion MRI. Hippocampus 24 (4):403–414. https://doi.org/10.1002/hipo.22234
Peters A, Sethares C, and Luebke JI (2008) Synapses Are Lost during Aging in the Primate Prefrontal Cortex. Neurosci 152 (4):970–981. https://doi.org/10.1016/j.neuroscience.2007.07.014
Peyrache A, Lacroix MM, Petersen PC, and Buzsáki G (2015) Internally Organized Mechanisms of the Head Direction Sense. Nat Neurosci 18 (4):569–575. https://doi.org/10.1038/nn.3968
Pfefferbaum A, Adalsteinsson E, and Sullivan EV (2005) Frontal Circuitry Degradation Marks Healthy Adult Aging: Evidence from Diffusion Tensor Imaging. Neuroimage 26 (3):891–899. https://doi.org/10.1016/j.neuroimage.2005.02.034
Piefke M, Onur ÖA, and Fink GR (2012) Aging-Related Changes of Neural Mechanisms Underlying Visual-Spatial Working Memory. Neurobiol Aging 33 (7):1284–1297. https://doi.org/10.1016/j.neurobiolaging.2010.10.014
Poe GR, Teed RGW, Insel N, et al. (2000) Partial Hippocampal Inactivation: Effects on Spatial Memory Performance in Aged and Young Rats. Behav Neurosci 114 (5):940–949. https://doi.org/10.1037//0735-7044.114.5.940

Preston-Ferrer P, Coletta S, Frey M, and Burgalossi A (2016) Anatomical Organization of Presubicular Head-Direction Circuits. eLife 5:e14592. https://doi.org/10.7554/elife.14592
Pruessner JC, Collins DL, Pruessner M, and Evans AC (2001) Age and Gender Predict Volume Decline in the Anterior and Posterior Hippocampus in Early Adulthood. J Neurosci 21 (1):194–200. https://doi.org/10.1523/jneurosci.21-01-00194.2001
Ragozzino ME, and Rozman S (2007) The Effect of Rat Anterior Cingulate Inactivation on Cognitive Flexibility. Behav Neurosci 121 (4):698–706. https://doi.org/10.1037/0735-7044.121.4.698
Raichle ME, MacLeod AM, Snyder AZ, et al. (2001) A Default Mode of Brain Function. PNAS 98 (2):676–682. https://doi.org/10.1073/pnas.98.2.676
Rajah MN, Kromas M, Han JE, and Pruessner JC (2010a) Group Differences in Anterior Hippocampal Volume and in the Retrieval of Spatial and Temporal Context Memory in Healthy Young versus Older Adults. Neuropsychologia 48 (14):4020–4030. https://doi.org/10.1016/j.neuropsychologia.2010.10.010
Rajah MN, Languay R, and Valiquette L (2010b) Age-Related Changes in Prefrontal Cortex Activity Are Associated with Behavioural Deficits in Both Temporal And spatial Context Memory Retrieval in Older Adults. Cortex 46 (4):535–549. https://doi.org/10.1016/j.cortex.2009.07.006
Ransmayr G, Faucheux B, Nowakowski C, et al. (2000) Age-Related Changes of Neuronal Counts in the Human Pedunculopontine Nucleus. Neurosci Lett 288 (3):195–198. https://doi.org/10.1016/s0304-3940(00)01244-1
Rapp PR, Deroche PS, Mao Y, and Burwell RD (2002) Neuron Number in the Parahippocampal Region Is Preserved in Aged Rats with Spatial Learning Deficits. Cereb Cortex 12 (11):1171–1179. https://doi.org/10.1093/cercor/12.11.1171
Rasmussen T, Schliemann T, Sørensen JC, Zimmer J, and West MJ (1996) Memory Impaired Aged Rats: No Loss of Principal Hippocampal and Subicular Neurons. Neurobiol Aging 17 (1):143–147. https://doi.org/10.1016/0197-4580(95)02032-2
Raz N (2000) Aging of the Brain and Its Impact on Cognitive Performance: Integration of Structural and Functional Findings. In: The Handbook of Aging and Cognition, 2nd ed. Lawrence Erlbaum, Mahwah, pp 1–90
Raz N, Ghisletta P, Rodrigue KM, Kennedy KM, and Lindenberger U (2010) Trajectories of Brain Aging in Middle-Aged and Older Adults: Regional and Individual Differences. Neuroimage 51 (2):501–511. https://doi.org/10.1016/j.neuroimage.2010.03.020
Raz N, Lindenberger U, Rodrigue KM, et al. (2005) Regional Brain Changes in Aging Healthy Adults: General Trends, Individual Differences and Modifiers. Cereb Cortex 15 (11):1676–1689. https://doi.org/10.1093/cercor/bhi044
Raz N, Rodrigue KM, Head D, Kennedy KM, and Acker JD (2004) Differential Aging of the Medial Temporal Lobe: A Study of a Five-Year Change. Neurology 62 (3):433–438. https://doi.org/10.1212/01.wnl.0000106466.09835.46
Reagh ZM, Ho HD, Leal SL, et al. (2016) Greater Loss of Object Than Spatial Mnemonic Discrimination in Aged Adults. Hippocampus 26 (4):417–422. https://doi.org/10.1002/hipo.22562
Reagh ZM, and Yassa MA (2014) Object and Spatial Mnemonic Interference Differentially Engage Lateral and Medial Entorhinal Cortex in Humans. PNAS 111 (40):E4264–E4273. https://doi.org/10.1073/pnas.1411250111
——— (2017) Selective Vulnerabilities and Biomarkers in Neurocognitive Aging. F1000Res 6:491. https://pubmed.ncbi.nlm.nih.gov/28491288/
Recanzone G (2017) Auditory Processing in the Aging Brain. Oxford Research Encyclopedia of Neuroscience. Oxford Univ Press, Oxford. https://doi.org/10.1093/acrefore/9780190264086.013.89
Redish AD, and Touretzky DS (1997) Cognitive Maps Beyond the Hippocampus. Hippocampus 7 (1):15–35. https://doi.org/10.1002/(sici)1098-1063(1997)7:1%3C15::aid-hipo3%3E3.0.co;2-6

——— (1998) The Role of the Hippocampus in Solving the Morris Water Maze. Neural Comput 10 (1):73–111. https://doi.org/10.1162/089976698300017908

Robbins TW (2007) Shifting and Stopping: Fronto-Striatal Substrates, Neurochemical Modulation and Clinical Implications. Philos Trans R Soc Lond B Biol Sci 362 (1481):917–932. https://doi.org/10.1098/rstb.2007.2097

Robertson RG, Rolls ET, Georges-François P, and Panzeri S (1999) Head Direction Cells in the Primate Pre-Subiculum. Hippocampus 9 (3):206–219. https://doi.org/10.1002/(sici)1098-1063(1999)9:3%3C206::aid-hipo2%3E3.0.co;2-h

Rochefort C, Lefort J, and Rondi-Reig L (2013) The Cerebellum: A New Key Structure in the Navigation System. Front Neural Circuits 7:35. https://doi.org/10.3389/fncir.2013.00035

Rodgers MK, Sindone JA, and Moffat SD (2012) Effects of Age on Navigation Strategy. Neurobiol Aging 33 (1):202.e215–202.e222. https://doi.org/10.1016/j.neurobiolaging.2010.07.021

Rodo C, Sargolini F, and Save E (2017) Processing of Spatial and Non-Spatial Information in Rats with Lesions of the Medial and Lateral Entorhinal Cortex: Environmental Complexity Matters. Behav Brain Res 320:200–209. https://doi.org/10.1016/j.bbr.2016.12.009

Rodrigue KM, Daugherty AM, Haacke EM, and Raz N (2013) The Role of Hippocampal Iron Concentration and Hippocampal Volume in Age-Related Differences in Memory. Cereb Cortex 23 (7):1533–1541. https://doi.org/10.1093/cercor/bhs139

Rogalski EJ, Murphy CM, Morrell L, et al. (2009) Changes in Parahippocampal White Matter Integrity in Amnestic Mild Cognitive Impairment: A Diffusion Tensor Imaging Study. Behav Neurol 21 (1):51–61. https://doi.org/10.3233/ben-2009-0235

Rolls ET, Stringer SM, and Elliot T (2006) Entorhinal Cortex Grid Cells Can Map to Hippocampal Place Cells by Competitive Learning. Network 17 (4):447–465. https://doi.org/10.1080/09548980601064846

Rondi-Reig L, Paradis A-L, and Fallahnezhad M (2022) A Liaison Brought to Light: Cerebellum-Hippocampus, Partners for Spatial Cognition. Cerebellum 21 (5):826–837. https://doi.org/10.1007/s12311-022-01422-3

Rossi AF, Pessoa L, Desimone R, and Ungerleider LG (2009) The Prefrontal Cortex and the Executive Control of Attention. Exp Brain Res 192 (3):489–497. https://doi.org/10.1007/s00221-008-1642-z

Salat DH, Tuch DS, Greve DN, et al. (2005) Age-Related Alterations in White Matter Microstructure Measured by Diffusion Tensor Imaging. Neurobiol Aging 26 (8):1215–1227. https://doi.org/10.1016/j.neurobiolaging.2004.09.017

Samsonovich A, and McNaughton BL (1997) Path Integration and Cognitive Mapping in a Continuous Attractor Neural Network Model. J Neurosci 17 (15):5900–5920. https://doi.org/10.1523/jneurosci.17-15-05900.1997

Sargolini F, Fyhn M, Hafting T, et al. (2006) Conjunctive Representation of Position, Direction, and Velocity in Entorhinal Cortex. Science 312 (5774):758–762. https://doi.org/10.1126/science.1125572

Save E, and Poucet B (2009) Role of the Parietal Cortex in Long-Term Representation of Spatial Information in the Rat. Neurobiol Learn Mem 91 (2):172–178. https://doi.org/10.1016/j.nlm.2008.08.005

Savelli F, Yoganarasimha D, and Knierim JJ (2008) Influence of Boundary Removal on the Spatial Representations of the Medial Entorhinal Cortex. Hippocampus 18 (12):1270–1282. https://doi.org/10.1002/hipo.20511

Scaplen KM, Ramesh RN, Nadvar N, Ahmed OJ, and Burwell RD (2017) Inactivation of the Lateral Entorhinal Area Increases the Influence of Visual Cues on Hippocampal Place Cell Activity. Front Syst Neurosci 11:40. https://doi.org/10.3389/fnsys.2017.00040

Schapiro AC, Turk-Browne NB, Botvinick MM, and Norman KA (2017) Complementary Learning Systems within the Hippocampus: A Neural Network Modelling Approach to Reconciling Episodic Memory with Statistical Learning. Philos Trans R Soc Lond B Biol Sci 372 (1711):20160049. https://doi.org/10.1098/rstb.2016.0049

Scheff SW, Price DA, and Sparks DL (2001) Quantitative Assessment of Possible Age-Related Change in Synaptic Numbers in the Human Frontal Cortex. Neurobiol Aging 22 (3):355–365. https://doi.org/10.1016/s0197-4580(01)00222-6

Schimanski LA, Lipa P, and Barnes CA (2013) Tracking the Course of Hippocampal Representations during Learning: When Is the Map Required? J Neurosci 33 (7):3094–3106. https://doi.org/10.1523/jneurosci.1348-12.2013

Schmolesky MT, Wang Y, Pu M, and Leventhal AG (2000) Degradation of Stimulus Selectivity of Visual Cortical Cells in Senescent Rhesus Monkeys. Nat Neurosci 3 (4):384–390. https://doi.org/10.1038/73957

Schuck NW, Cai MB, Wilson RC, and Niv Y (2016) Human Orbitofrontal Cortex Represents a Cognitive Map of State Space. Neuron 91 (6):1402–1412. https://doi.org/10.1016/j.neuron.2016.08.019

Schuck NW, Doeller CF, Polk TA, Lindenberger U, and Li S-C (2015) Human Aging Alters the Neural Computation and Representation of Space. Neuroimage 117:141–150. https://doi.org/10.1016/j.neuroimage.2015.05.031

Scimeca Jason M, and Badre D (2012) Striatal Contributions to Declarative Memory Retrieval. Neuron 75 (3):380–392. https://doi.org/10.1016/j.neuron.2012.07.014

Segen V, Ying J, Morgan E, Brandon M, and Wolbers T (2022) Path Integration in Normal Aging and Alzheimer's Disease. Trends Cogn Sci 26 (2):142–158. https://doi.org/10.1016/j.tics.2021.11.001

Shamy JL, Habeck C, Hof PR, et al. (2011) Volumetric Correlates of Spatiotemporal Working and Recognition Memory Impairment in Aged Rhesus Monkeys. Cereb Cortex 21 (7):1559–1573. https://doi.org/10.1093/cercor/bhq210

Sharp PE, Tinkelman A, and Cho J (2001) Angular Velocity and Head Direction Signals Recorded from the Dorsal Tegmental Nucleus of Gudden in the Rat: Implications for Path Integration in the Head Direction Cell Circuit. Behav Neurosci 115 (3):571–588. https://doi.org/10.1037/0735-7044.115.3.571

Shen J, Barnes CA, Wenk GL, and McNaughton BL (1996) Differential Effects of Selective Immunotoxic Lesions of Medial Septal Cholinergic Cells on Spatial Working and Reference Memory. Behav Neurosci 110 (5):1181–1186. https://doi.org/10.1037//0735-7044.110.5.1181

Shing YL, Rodrigue KM, Kennedy KM, et al. (2011) Hippocampal Subfield Volumes: Age, Vascular Risk, and Correlation with Associative Memory. Front Aging Neurosci 3:2. https://doi.org/10.3389/fnagi.2011.00002

Skaggs W, Knierim J, Kudrimoti H, and McNaughton B (1994) A Model of the Neural Basis of the Rat's Sense of Direction. Adv Neural Inf Process Syst 7:173–180. https://pubmed.ncbi.nlm.nih.gov/11539168/

Skaggs WE, McNaughton BL, Wilson MA, and Barnes CA (1996) Theta Phase Precession in Hippocampal Neuronal Populations and the Compression of Temporal Sequences. Hippocampus 6 (2):149–172. https://doi.org/10.1002/(sici)1098-1063(1996)6:2%3C149::aid-hipo6%3E3.0.co;2-k

Skolimowska J, Wesierska M, Lewandowska M, Szymaszek A, and Szelag E (2011) Divergent Effects of Age on Performance in Spatial Associative Learning and Real Idiothetic Memory in Humans. Behav Brain Res 218 (1):87–93. https://doi.org/10.1016/j.bbr.2010.11.035

Small SA, Chawla MK, Buonocore M, Rapp PR, and Barnes CA (2004) Imaging Correlates of Brain Function in Monkeys and Rats Isolates a Hippocampal Subregion Differentially Vulnerable to Aging. PNAS 101 (18):7181–7186. https://doi.org/10.1073/pnas.0400285101

Smith AC, Gerrard JL, Barnes CA, and McNaughton BL (2000) Effect of Age on Burst Firing Characteristics of Rat Hippocampal Pyramidal Cells. Neuroreport 11 (17):3865–3871. https://doi.org/10.1097/00001756-200011270-00052

Smith DE, Rapp PR, McKay HM, Roberts JA, and Tuszynski MH (2004) Memory Impairment in Aged Primates Is Associated with Focal Death of Cortical Neurons and Atrophy of Subcortical Neurons. J Neurosci 24 (18):4373–4381. https://doi.org/10.1523/jneurosci.4289-03.2004

Solstad T, Boccara CN, Kropff E, Moser M-B, and Moser EI (2008) Representation of Geometric Borders in the Entorhinal Cortex. Science 322 (5909):1865–1868. https://doi.org/10.1126/science.1166466

Solstad T, Moser EI, and Einevoll GT (2006) From Grid Cells to Place Cells: A Mathematical Model. Hippocampus 16 (12):1026–1031. https://doi.org/10.1002/hipo.20244

Spiers HJ, Burgess N, Hartley T, Vargha-Khadem F, and O'Keefe J (2001) Bilateral Hippocampal Pathology Impairs Topographical and Episodic Memory but Not Visual Pattern Matching. Hippocampus 11 (6):715–725. https://doi.org/10.1002/hipo.1087

Spiers HJ, Coutrot A, and Hornberger M (2023) Explaining World-Wide Variation in Navigation Ability from Millions of People: Citizen Science Project Sea Hero Quest. Top Cogn Sci 15 (1):120–138. https://doi.org/10.1111/tops.12590

Spiers HJ, and Maguire EA (2007) A Navigational Guidance System in the Human Brain. Hippocampus 17 (8):618–626. https://doi.org/10.1002/hipo.20298

Spooner RK, Wiesman AI, Proskovec AL, Heinrichs-Graham E, and Wilson TW (2019) Rhythmic Spontaneous Activity Mediates the Age-Related Decline in Somatosensory Function. Cereb Cortex 29 (2):680–688. https://doi.org/10.1093/cercor/bhx349

Srokova S, Aktas ANZ, Koen JD, and Rugg MD (2024) Dissociative Effects of Age on Neural Differentiation at the Category and Item Levels. J Neurosci 44 (4):e0959232023. https://doi.org/10.1523/jneurosci.0959-23.2023

Srokova S, Hill PF, Koen JD, King DR, and Rugg MD (2020) Neural Differentiation Is Moderated by Age in Scene-Selective, but Not Face-Selective, Cortical Regions. eNeuro 7 (3):ENEURO.0142–0120.2020. https://doi.org/10.1523/eneuro.0142-20.2020

Stachenfeld KL, Botvinick MM, and Gershman SJ (2017) The Hippocampus as a Predictive Map. Nat Neurosci 20 (11):1643–1653. https://doi.org/10.1038/nn.4650

Stackman RW, Clark AS, and Taube JS (2002) Hippocampal Spatial Representations Require Vestibular Input. Hippocampus 12 (3):291–303. https://doi.org/10.1002/hipo.1112

Stackman RW, and Herbert AM (2002) Rats with Lesions of the Vestibular System Require a Visual Landmark for Spatial Navigation. Behav Brain Res 128 (1):27–40. https://doi.org/10.1016/s0166-4328(01)00270-4

Stackman RW, and Taube JS (1998) Firing Properties of Rat Lateral Mammillary Single Units: Head Direction, Head Pitch, and Angular Head Velocity. J Neurosci 18 (21):9020–9037. https://doi.org/10.1523/jneurosci.18-21-09020.1998

Stangl M, Achtzehn J, Huber K, et al. (2018) Compromised Grid-Cell-Like Representations in Old Age as a Key Mechanism to Explain Age-Related Navigational Deficits. Curr Biol 28 (7):1108–1115. e1106. https://doi.org/10.1016/j.cub.2018.02.038

Stangl M, Maoz SL, and Suthana N (2023) Mobile Cognition: Imaging the Human Brain in the "Real World." Nat Rev Neurosci 24 (6):347–362. https://doi.org/10.1038/s41583-023-00692-y

Stark SM, and Stark CEL (2017) Age-Related Deficits in the Mnemonic Similarity Task for Objects and Scenes. Behav Brain Res 333:109–117. https://doi.org/10.1016/j.bbr.2017.06.049

Staudigl T, Leszczynski M, Jacobs J, et al. (2018) Hexadirectional Modulation of High-Frequency Electrophysiological Activity in the Human Anterior Medial Temporal Lobe Maps Visual Space. Curr Biol 28 (20):3325–3329.e3324. https://doi.org/10.1016/j.cub.2018.09.035

Steffenach H-A, Witter M, Moser M-B, and Moser EI (2005) Spatial Memory in the Rat Requires the Dorsolateral Band of the Entorhinal Cortex. Neuron 45 (2):301–313. https://doi.org/10.1016/j.neuron.2004.12.044

Stensola H, Stensola T, Solstad T, et al. (2012) The Entorhinal Grid Map Is Discretized. Nature 492 (7427):72–78. https://doi.org/10.1038/nature11649

Stoub TR, Barnes CA, Shah RC, et al. (2012) Age-Related Changes in the Mesial Temporal Lobe: The Parahippocampal White Matter Region. Neurobiol Aging 33 (7):1168–1176. https://doi.org/10.1016/j.neurobiolaging.2011.02.010

Stranahan AM, Jiam NT, Spiegel AM, and Gallagher M (2012) Aging Reduces Total Neuron Number in the Dorsal Component of the Rodent Prefrontal Cortex. J Comp Neurol 520 (6):1318–1326. https://doi.org/10.1002/cne.22790

Suter EE, Weiss C, and Disterhoft JF (2019) Differential Responsivity of Neurons in Perirhinal Cortex, Lateral Entorhinal Cortex, and Dentate Gyrus during Time-Bridging Learning. Hippocampus 29 (6):511–526. https://doi.org/10.1002/hipo.23041

Suzuki WA, and Amaral DG (2003) Where Are the Perirhinal and Parahippocampal Cortices? A Historical Overview of the Nomenclature and Boundaries Applied to the Primate Medial Temporal Lobe. Neurosci 120 (4):893–906. https://doi.org/10.1016/s0306-4522(03)00281-1

Talboom JS, De Both MD, Naymic MA, et al. (2021) Two Separate, Large Cohorts Reveal Modifiers of Age-Related Change in Visual Reaction Time Performance. NPJ Aging Mech Dis 7 (1):14. https://doi.org/10.1038/s41514-021-00067-6

Tang Y, Whitman GT, Lopez I, and Baloh RW (2001) Brain Volume Changes on Longitudinal Magnetic Resonance Imaging in Normal Older People. J Neuroimaging 11 (4):393–400. https://doi.org/10.1111/j.1552-6569.2001.tb00068.x

Taube JS (1995) Head Direction Cells Recorded in the Anterior Thalamic Nuclei of Freely Moving Rats. J Neurosci 15 (1):70–86. https://doi.org/10.1523/jneurosci.15-01-00070.1995

Taube JS (2007) The Head Direction Signal: Origins and Sensory-Motor Integration. Annu Rev Neurosci 30:181–207. https://doi.org/10.1146/annurev.neuro.29.051605.112854

Taube JS, Muller R, and Ranck J (1990) Head-Direction Cells Recorded from the Postsubiculum in Freely Moving Rats. I. Description and Quantitative Analysis. J Neurosci 10 (2):420–435. https://doi.org/10.1523/jneurosci.10-02-00420.1990

Teghil A, Bonavita A, Guariglia C, and Boccia M (2021) Commonalities and Specificities between Environmental Navigation and Autobiographical Memory: A Synthesis and a Theoretical Perspective. Neurosci Biobehav Rev 127:928–945. https://doi.org/10.1016/j.neubiorev.2021.06.012

Terrazas A, Krause M, Lipa P, et al. (2005) Self-Motion and the Hippocampal Spatial Metric. J Neurosci 25 (35):8085–8096. https://doi.org/10.1523/jneurosci.0693-05.2005

Thambisetty M, Wan J, Carass A, et al. (2010) Longitudinal Changes in Cortical Thickness Associated with Normal Aging. Neuroimage 52 (4):1215–1223. https://doi.org/10.1016/j.neuroimage.2010.04.258

Theofilas P, Ehrenberg AJ, Dunlop S, et al. (2017) Locus Coeruleus Volume and Cell Population Changes during Alzheimer's Disease Progression: A Stereological Study in Human Postmortem Brains with Potential Implication for Early-Stage Biomarker Discovery. Alzheimers Dement (N Y) 13 (3):236–246. https://doi.org/10.1016/j.jalz.2016.06.2362

Thome A, Marrone DF, Ellmore TM, et al. (2017) Evidence for an Evolutionarily Conserved Memory Coding Scheme in the Mammalian Hippocampus. J Neurosci 37 (10):2795–2801. https://doi.org/10.1523/JNEUROSCI.3057-16.2017

Tisserand DJ, Pruessner JC, Sanz Arigita EJ, et al. (2002) Regional Frontal Cortical Volumes Decrease Differentially in Aging: An MRI Study to Compare Volumetric Approaches and Voxel-Based Morphometry. Neuroimage 17 (2):657–669. https://doi.org/10.1006/nimg.2002.1173

Tocker G, Barak O, and Derdikman D (2015) Grid Cells Correlation Structure Suggests Organized Feedforward Projections into Superficial Layers of the Medial Entorhinal Cortex: Grid Cells Correlation Structure. Hippocampus 25 (12):1599–1613. https://doi.org/10.1002/hipo.22481

Tomlinson BE, Irving D, and Blessed G (1981) Cell Loss in the Locus Coeruleus in Senile Dementia of Alzheimer Type. J Neurol Sci 49 (3):419–428. https://doi.org/10.1016/0022-510x(81)90031-9

Tsao A, Moser M-B, and Moser EI (2013) Traces of Experience in the Lateral Entorhinal Cortex. Curr Biol 23 (5):399–405. https://doi.org/10.1016/j.cub.2013.01.036

Tsao A, Sugar J, Lu L, et al. (2018) Integrating Time from Experience in the Lateral Entorhinal Cortex. Nature 561 (7721):57–62. https://doi.org/10.1038/s41586-018-0459-6

Tuena C, Riva G, Murru I, et al. (2022) Contribution of Cognitive and Bodily Navigation Cues to Egocentric and Allocentric Spatial Memory in Hallucinations Due to Parkinson's Disease: A Case Report. Front Behav Neurosci 16:992498. https://doi.org/10.3389/fnbeh.2022.992498

Tukker JJ, Tang Q, Burgalossi A, and Brecht M (2015) Head-Directional Tuning and Theta Modulation of Anatomically Identified Neurons in the Presubiculum. J Neurosci 35 (46):15391–15395. https://doi.org/10.1523/jneurosci.0685-15.2015

Tulimieri DT, and Semrau JA (2023) Aging Increases Proprioceptive Error for a Broad Range of Movement Speed and Distance Estimates in the Upper Limb. Front Hum Neurosci 17:1217105. https://doi.org/10.3389/fnhum.2023.1217105

Uylings HBM, Groenewegen HJ, and Kolb B (2003) Do Rats Have a Prefrontal Cortex? Behav Brain Res 146 (1-2):3–17. https://doi.org/10.1016/j.bbr.2003.09.028

Van Cauter T, Camon J, Alvernhe A, et al. (2013) Distinct Roles of Medial and Lateral Entorhinal Cortex in Spatial Cognition. Cereb Cortex 23 (2):451–459. https://doi.org/10.1093/cercor/bhs033

Van Der Ham IJM, Faber AME, Venselaar M, Van Kreveld MJ, and Löffler M (2015) Ecological Validity of Virtual Environments to Assess Human Navigation Ability. Front Psychol 6:637. https://doi.org/10.3389/fpsyg.2015.00637

Vandrey B, Garden DLF, Ambrozova V, et al. (2020) Fan Cells in Layer 2 of the Lateral Entorhinal Cortex Are Critical for Episodic-Like Memory. Curr Biol 30 (1):169–175. https://doi.org/10.1016/j.cub.2019.11.027

Vann SD, Aggleton JP, and Maguire EA (2009) What Does the Retrosplenial Cortex Do? Nat Rev Neurosci 10 (11):792–802. https://doi.org/10.1038/nrn2733

van Velsen EFS, Vernooij MW, Vrooman HA, et al. (2013) Brain Cortical Thickness in the General Elderly Population: The Rotterdam Scan Study. Neurosci Lett 550:189–194. https://doi.org/10.1016/j.neulet.2013.06.063

Vélez-Fort M, Bracey EF, Keshavarzi S, et al. (2018) A Circuit for Integration of Head- and Visual-Motion Signals in Layer 6 of Mouse Primary Visual Cortex. Neuron 98 (1):179–191.e176. https://doi.org/10.1016/j.neuron.2018.02.023

Verghese J, Ayers E, Barzilai N, et al. (2014) Motoric Cognitive Risk Syndrome: Multicenter Incidence Study. Neurology 83 (24):2278–2284. https://doi.org/10.1212/wnl.0000000000001084

Viard A, Doeller CF, Hartley T, Bird CM, and Burgess N (2011) Anterior Hippocampus and Goal-Directed Spatial Decision Making. J Neurosci 31 (12):4613–4621. https://doi.org/10.1523/jneurosci.4640-10.2011

Vik A, Hodneland E, Haasz J, et al. (2015) Fractional Anisotropy Shows Differential Reduction in Frontal-Subcortical Fiber Bundles—A Longitudinal MRI Study of 76 Middle-Aged and Older Adults. Front Aging Neurosci 7:81. https://doi.org/10.3389/fnagi.2015.00081

Viswasom AA, and Jobby A (2013) Age Related Changes in the Purkinje Cells in Human Cerebellar Cortex. J Evol Med Dent Sci 2 (31):2698–2705. https://jemds.com/abstract.php?at_id=1740

Vogt BA, Vogt LJ, Perl DP, and Hof PR (2001) Cytology of Human Caudomedial Cingulate, Retrosplenial, and Caudal Parahippocampal Cortices. J Comp Neurol 438 (3):353–376. https://doi.org/10.1002/cne.1320

Wallis JD, and Miller EK (2003) Neuronal Activity in Primate Dorsolateral and Orbital Prefrontal Cortex during Performance of a Reward Preference Task. Eur J Neurosci 18 (7):2069–2081. https://doi.org/10.1046/j.1460-9568.2003.02922.x

Walsh JP, and Akopian G (2019) Physiological Aging at Striatal Synapses. J Neurosci Res 97 (12):1720–1727. https://doi.org/10.1002/jnr.24484

Wang C, Chen X, and Knierim JJ (2020) Egocentric and Allocentric Representations of Space in the Rodent Brain. Curr Opin Neurobiol 60:12–20. https://doi.org/10.1016/j.conb.2019.11.005
Wang G-W, and Cai J-X (2006) Disconnection of the Hippocampal-Prefrontal Cortical Circuits Impairs Spatial Working Memory Performance in Rats. Behav Brain Res 175 (2):329–336. https://doi.org/10.1016/j.bbr.2006.09.002
Wang M, Gamo NJ, Yang Y, et al. (2011) Neuronal Basis of Age-Related Working Memory Decline. Nature 476 (7359):210–213. https://doi.org/10.1038/nature10243
Warren WH (2019) Non-Euclidean Navigation. J Exp Biol 222:jeb187971. https://doi.org/10.1242/jeb.187971
Webb CE, Hoagey DA, Rodrigue KM, and Kennedy KM (2020) Frontostriatal White Matter Connectivity: Age Differences and Associations with Cognition and BOLD Modulation. Neurobiol Aging 94:154–163. https://doi.org/10.1016/j.neurobiolaging.2020.05.014
Wellington RL, Bilder RM, Napolitano B, and Szeszko PR (2012) Effects of Age on Prefrontal Subregions and Hippocampal Volumes in Young and Middle-Aged Healthy Humans. Hum Brain Mapp 34 (9):2129–2140. https://doi.org/10.1002/hbm.22054
West MJ (1993) Regionally Specific Loss of Neurons in the Aging Human Hippocampus. Neurobiol Aging 14 (4):287–293. https://doi.org/10.1016/0197-4580(93)90113-p
West MJ, Coleman PD, Flood DG, and Troncoso JC (1994) Differences in the Pattern of Hippocampal Neuronal Loss in Normal Ageing and Alzheimer's Disease. Lancet 344 (8925):769–772. https://doi.org/10.1016/s0140-6736(94)92338-8
West RL (1996) An Application of Prefrontal Cortex Function Theory to Cognitive Aging. Psychol Bull 120 (2):272–292. https://doi.org/10.1037/0033-2909.120.2.272
——— (1999) Age Differences in Lapses of Intention in the Stroop Task. J Gerontol B Psychol Sci Soc Sci 54 (1):P34–43. https://doi.org/10.1093/geronb/54b.1.p34
Westlye LT, Walhovd KB, Dale AM, et al. (2010) Differentiating Maturational and Aging-Related Changes of the Cerebral Cortex by Use of Thickness and Signal Intensity. Neuroimage 52 (1):172–185. https://doi.org/10.1016/j.neuroimage.2010.03.056
Whitlock Jonathan R, Pfuhl G, Dagslott N, Moser M-B, and Moser Edvard I (2012) Functional Split between Parietal and Entorhinal Cortices in the Rat. Neuron 73 (4):789–802. https://doi.org/10.1016/j.neuron.2011.12.028
Whittington JCR, McCaffary D, Bakermans JJW, and Behrens TEJ (2022) How to Build a Cognitive Map. Nat Neurosci 25 (10):1257–1272. https://doi.org/10.1038/s41593-022-01153-y
Whittington JCR, Muller TH, Mark S, et al. (2020) The Tolman-Eichenbaum Machine: Unifying Space and Relational Memory through Generalization in the Hippocampal Formation. Cell 183 (5):1249–1263.e1223. https://doi.org/10.1016/j.cell.2020.10.024
Wicks RT, Witcher MR, Couture DE, et al. (2020) Hippocampal CA1 and CA3 Neural Recording in the Human Brain: Validation of Depth Electrode Placement through High-Resolution Imaging and Electrophysiology. Neurosurg Focus 49 (1):E5. https://doi.org/10.3171/2020.4.focus20164
Widloski J, and Fiete I (2014) How Does the Brain Solve the Computational Problems of Spatial Navigation? In: Derdikman D, Knierim JJ (eds) Space, Time and Memory in the Hippocampal Formation. Springer, Vienna, pp 373–407
Wiener J, Shettleworth S, Bingman VP, et al. (2011) Animal Navigation. In: Menzel R, Fischer J (eds) Animal Thinking. MIT Press, Boston, pp 51–76
Wikgren J, Nokia MS, and Penttonen M (2010) Hippocampo–Cerebellar Theta Band Phase Synchrony in Rabbits. Neurosci 165 (4):1538–1545. https://doi.org/10.1016/j.neuroscience.2009.11.044
Wilson IA, Ikonen S, Gallagher M, Eichenbaum H, and Tanila H (2005) Age-Associated Alterations of Hippocampal Place Cells Are Subregion Specific. J Neurosci 25 (29):6877–6886. https://doi.org/10.1523/jneurosci.1744-05.2005

Wilson RS, Boyle PA, Yu L, et al. (2013) Life-Span Cognitive Activity, Neuropathologic Burden, and Cognitive Aging. Neurology 81 (4):314–321. https://doi.org/10.1212/WNL.0b013e31829c5e8a

Winter SS, Clark BJ, and Taube JS (2015) Disruption of the Head Direction Cell Network Impairs the Parahippocampal Grid Cell Signal. Science 347 (6224):870–874. https://doi.org/10.1126/science.1259591

Wirth S, Baraduc P, Planté A, Pinède S, and Duhamel J-R (2017) Gaze-Informed, Task-Situated Representation of Space in Primate Hippocampus during Virtual Navigation. PLoS Biol 15 (2):e2001045. https://doi.org/10.1371/journal.pbio.2001045

Witter MP, Wouterlood FG, Naber PA, and Van Haeften T (2000) Anatomical Organization of the Parahippocampal-Hippocampal Network. Ann N Y Acad Sci 911 (1):1–24. https://doi.org/10.1111/j.1749-6632.2000.tb06716.x

Woolard AA, and Heckers S (2012) Anatomical and Functional Correlates of Human Hippocampal Volume Asymmetry. Psychiatry Res Neuroimaging 201 (1):48–53. https://doi.org/10.1016/j.pscychresns.2011.07.016

Woollacott M, and Shumway-Cook A (2002) Attention and the Control of Posture and Gait: A Review of an Emerging Area of Research. Gait Posture 16 (1):1–14. https://doi.org/10.1016/s0966-6362(01)00156-4

Woollett K, and Maguire EA (2011) Acquiring "the Knowledge" of London's Layout Drives Structural Brain Changes. Curr Biol 21 (24–2):2109–2114. https://doi.org/10.1016/j.cub.2011.11.018

Wymbs Nicholas F, Bassett Danielle S, Mucha Peter J, Porter Mason A, and Grafton Scott T (2012) Differential Recruitment of the Sensorimotor Putamen and Frontoparietal Cortex during Motor Chunking in Humans. Neuron 74 (5):936–946. https://doi.org/10.1016/j.neuron.2012.03.038

Xie Y, Bigelow RT, Frankenthaler SF, et al. (2017) Vestibular Loss in Older Adults Is Associated with Impaired Spatial Navigation: Data from the Triangle Completion Task. Front Neurol 8:173. https://doi.org/10.3389/fneur.2017.00173

Yassa MA, Lacy JW, Stark SM, et al. (2011) Pattern Separation Deficits Associated with Increased Hippocampal CA3 and Dentate Gyrus Activity in Nondemented Older Adults. Hippocampus 21 (9):968–979. https://doi.org/10.1002/hipo.20808

Yassa MA, Muftuler LT, and Stark CEL (2010) Ultrahigh-Resolution Microstructural Diffusion Tensor Imaging Reveals Perforant Path Degradation in Aged Humans *In vivo*. PNAS 107 (28):12687–12691. https://doi.org/10.1073/pnas.1002113107

Yin HH, Knowlton BJ, and Balleine BW (2004) Lesions of Dorsolateral Striatum Preserve Outcome Expectancy but Disrupt Habit Formation in Instrumental Learning. Eur J Neurosci 19 (1):181–189. https://doi.org/10.1111/j.1460-9568.2004.03095.x

Ystad M, Hodneland E, Adolfsdottir S, et al. (2011) Cortico-Striatal Connectivity and Cognition in Normal Aging: A Combined DTI and Resting State fMRI Study. Neuroimage 55 (1):24–31. https://doi.org/10.1016/j.neuroimage.2010.11.016

Yu M, Zhang H, Wang B, et al. (2021) Key Signaling Pathways in Aging and Potential Interventions for Healthy Aging. Cells 10 (3):660. https://doi.org/10.3390/cells10030660

Zhang K (1996) Representation of Spatial Orientation by the Intrinsic Dynamics of the Head-Direction Cell Ensemble: A Theory. J Neurosci 16 (6):2112–2126. https://doi.org/10.1523/jneurosci.16-06-02112.1996

Zhang L, Fournier J, Fallahnezhad M, et al. (2023) The Cerebellum Promotes Sequential Foraging Strategies and Contributes to the Directional Modulation of Hippocampal Place Cells. iScience 26 (3):106200. https://doi.org/10.1016/j.isci.2023.106200

Zhang Y, Qiu C, Lindberg O, et al. (2010) Acceleration of Hippocampal Atrophy in a Non-Demented Elderly Population: The SNAC-K Study. Int Psychogeriatr 22 (1):14–25. https://doi.org/10.1017/s1041610209991396

Zhong JY, and Moffat SD (2018) Extrahippocampal Contributions to Age-Related Changes in Spatial Navigation Ability. Front Hum Neurosci 12:272. https://doi.org/10.3389/fnhum.2018.00272

14

How Do Aging and Neurodegenerative Diseases Affect Spatial Navigation at the Behavioral Level?

Jan Wiener and Michael Hornberger

Abstract In a globally aging society, developing an in-depth understanding of cognitive changes, particularly in spatial navigation, is crucial both for healthy aging and in managing neurodegenerative diseases like Alzheimer disease. This chapter reviews current understanding of how cognitive aging and Alzheimer disease affect spatial navigation at the behavioral level and discusses neuronal changes in the brain that contribute to these changes. While an extensive body of behavioral and neuroscientific research offers a detailed understanding of the aging navigational system, our knowledge of how age-related navigational deficits impact everyday navigation and people's experience of living with declining navigation abilities is surprisingly limited. Similarly, despite the prevalence of spatial disorientation in Alzheimer disease, which leads to an increased risk of getting lost, our understanding of how these deficits progress and their real-world implications is limited. To support independence in aging and improve safety for individuals with Alzheimer disease, future research should focus on real-world everyday navigation, develop targeted interventions, and explore the role of spatial navigation in diagnostics and everyday functioning.

Keywords Aging, Alzheimer disease, lab-based spatial navigation, real-world spatial navigation, path integration

---

Jan Wiener (✉)
School of Psychology, University of Bournemouth, Bournemouth BH12 5BB, UK
Email: jwiener@bournemouth.ac.uk

Michael Hornberger (✉)
Clinical Neurosciences, Department for Clinical and Experimental Sciences,
Faculty of Medicine, University of Southampton, SO14 3DT, UK
Email: M.Hornberger@soton.ac.uk

N. S. Newcombe and K. Cheng (eds.), *Challenges in Navigation Research*,
Strüngmann Forum Reports, https://doi.org/10.1007/978-3-032-20563-6_14

1 Introduction

We live in an aging society. Over the last five decades, global life expectancy has risen significantly in low-, middle-, and high-income countries, albeit at different rates, as a result of medical advances, improved sanitation, nutrition, and reduction in child mortality. Global life expectancy reached 73 years in 2024, representing substantial progress from previous decades. By 2030, one in six people in the world will be aged 60 years or over, increasing from one billion in 2020 to 1.4 billion. By 2050, the world's population of people aged 60 years and older will double to 2.1 billion, with the number of persons aged 80 years or older expected to triple between 2020 and 2050 to reach 426 million. The demographic transformation is particularly pronounced in low- and middle-income countries. While population aging started in high-income countries (e.g., in Japan, 30% of the population is already over 60 years old), low- and middle-income countries are now experiencing the greatest change. By 2050, two-thirds of the world's population over 60 years of age will live in low- and middle-income countries (Livingston et al. 2024). We are facing, therefore, a global aging population, for which it is important to understand how their cognition changes throughout their aging life. It is not only important for healthy aging, but also for neurodegenerative conditions, such as Alzheimer disease, which have a higher incidence in older age.

The global burden of dementia is substantial and increasing rapidly. In 2020, there were over 55 million people worldwide who were living with dementia, and this number is expected to almost double every 20 years, reaching 78 million in 2030 and 139 million in 2050. The regional distribution of this burden reflects global demographic patterns; currently, 60% of people with dementia live in low- and middle-income countries, and this is expected to rise to 71% by 2050. Age remains the strongest risk factor for dementia. The percentage of people with Alzheimer disease increases dramatically with age: 5.0% of people age 65–74 yr, 13.2% of people age 75 to 84, and 33.4% of people age 85 and older suffer from Alzheimer disease (Livingston et al. 2024). The age-related increase in prevalence highlights the urgent need to understand cognitive changes across the lifespan.

Over the last several years, spatial navigation/orientation has emerged as a key cognitive domain that is understood to be affected by aging and neurodegeneration. Historically, spatial navigation deficits in aging and neurodegeneration were regarded more as a collateral effect of episodic memory deficits. This is particularly true for earlier spatial navigation/orientation tests, which were more reliant on episodic memory, creating in essence spatial memory tasks. More recently, it has emerged that episodic memory is not the sole contributor to spatial navigation/orientation changes in aging and neurodegeneration. Instead, there are separate, though overlapping, spatial processes which are unique to aging and neurodegeneration.

This development has led to a plethora of new aging and neurodegeneration research, using spatial navigation measures as diagnostic tools, functional outcome measures, and disease progression trackers. This development is not trivial, since spatial navigation/orientation allows nonverbal cognitive assessments to be taken in diverse populations where there is variance in socioeconomic levels, educational backgrounds, and environmental contexts. Given the projected emergence of aging populations in low- and middle-income countries, spatial navigation/orientation has the potential to be an important tool for future cognitive assessments in aging and neurodegeneration.

In this chapter, we present an overview of current understanding of spatial navigation/orientation in aging and neurodegeneration. For in-depth reviews of the field, the reader is directed to Coughlan et al. (2018) and Lester et al. (2017). Here, our main focus is to highlight the gaps and opportunities for future research to pursue.

2 Aging

2.1 Brain Changes during Aging

As we age, the brain undergoes a series of structural, biochemical, metabolic, cellular, and molecular changes (for an overview, see Lee and Kim 2022). Brain volume and weight decrease around 5% per decade as of the age of 40 years, and this decline increases substantially after 70 years of age. In the context of navigation and wayfinding, it is important to note that the frontal and temporal lobe volumes decrease from about 12% to 9% between middle age and old age, while the occipital and parietal lobes do not show significant volume loss.

The temporal lobe, in particular, includes several areas that contain neurons that code spatial information. Age-related functional and structural changes in these areas have been linked to declining navigational abilities (for an overview, see Lester et al. 2017). Specifically, the hippocampal–entorhinal circuit contains a number of specialized neurons (e.g., place, grid, and border cells) that code spatial information and are thought to be crucial for successful spatial learning, orientation, and navigation. As animals and humans navigate through space, grid cells in the medial entorhinal cortex fire at a regular interval in a grid-like activity pattern (Moser et al. 2015). Grid cells provide metric information and integrate various spatial cues, such as the distance traveled and the direction of movement. They also work in conjunction with other types of spatially responsive neurons, such as place cells and head-direction cells in the medial entorhinal cortex. Hippocampal place cells fire when the animal is in a specific location of the environment, creating a so-called place field. Border cells respond to the presence of environmental boundaries such as walls, and head-direction cells encode the orientation of the head in the environment.

While much aging research focuses on the role of the hippocampal–entorhinal circuit for navigation, there are also extrahippocampal contributions to age-related changes (Zhong and Moffat 2018). For example, age-related differences in the ability to effectively switch between different navigation strategies (Harris and Wolbers 2014) has been related to degradation of the axonal circuitry in the prefrontal cortex and/or age-related changes in the anatomical pathways that link the locus coeruleus to the prefrontal cortex. A detailed review of all age-related brain changes and how they affect navigation is beyond the scope of this chapter; for an in-depth review of the aging navigational system, see Lester et al. (2017). It is important to note, however, the strong overlap of brain areas recruited during navigation and those that undergo functional and structural changes during aging, contributing to age-related declines in navigational abilities. Spatial navigation decline in aging represents a complex interaction between specific spatial processing deficits and general cognitive changes. The nonuniform pattern of impairment—with preserved static perception but impaired dynamic processing, intact recognition but impaired integration, and biased strategy selection—distinguishes navigational aging from general cognitive decline. As our understanding of the aging navigational system continues to evolve, it promises to yield both theoretical insights into brain aging mechanisms and practical tools for maintaining cognitive health in our increasingly aging society.

2.2 Laboratory Findings

Much of the contemporary research that focuses on the effects of cognitive aging on navigation at the behavioral level is laboratory based and often makes use of virtual environment technology (Lester et al. 2017). Such experiments show that cognitive aging affects both fundamental navigation mechanisms, path integration, and landmark-based navigation. Overall, navigation abilities in humans peak at around 20 years of age and then start to decline (Coutrot et al. 2018), with the rate of decline increasing in older age.

Path integration, which is thought to depend critically on the grid-cell system, refers to the integration of perceived self-motion to update position and orientation in the environment during travel (Mittelstaedt and Mittelstaedt 1980). In experiments, participants are typically guided along a path in a featureless environment and then asked to point or navigate back to the start of the path. In such tasks, path-integration performance in older adults is comparable to that of young adults—a stage when vestibular and proprioceptive information or when visual and proprioceptive information is available. However, when older adults relied solely on vestibular or solely on visual information, performance deteriorated. These results suggest age-related declines if path integration relies on a single source of self-motion information, but not when multiple sources are integrated (for an overview, see Segen et al. 2022).

Landmark-based navigation is an umbrella term for navigation tasks and abilities in which navigators use recognizable features in the environment (i.e., landmarks) as reference points to guide movement or to determine location. When learning novel routes, landmarks at decision points are often associated with the required movement direction to follow the route. Aging negatively impacts the processing, memorization, and retrieval of landmarks during route learning. Compared to younger adults, older adults typically perform worse on tasks involving free recall of landmarks (Head and Isom 2010), landmark recognition, and the association of landmarks with directions (Zhong and Moffat 2018). However, when given time to first learn novel routes, knowledge about the landmarks encountered along the route and the landmark-direction associations to follow the route are similar to that of younger participants, whereas knowledge about the sequence in which landmarks were encountered along the route was impaired (Hilton et al. 2021). When retracing a recently traveled route (e.g., from destination back to the start), younger participants outperform older adults (Wiener et al. 2020). Cognitive aging impacts people's ability to learn unfamiliar virtual environments and then use the resulting environmental knowledge (often referred to as cognitive maps) to plan novel routes (Head and Isom 2010).

Cognitive aging also affects preferences for navigation strategies. In experiments in which younger and older participants can choose between different strategies when solving navigation tasks, older adults reliably showed preferences for egocentric or response strategies over allocentric or place strategies (Zhong and Moffat 2018). This shift away from allocentric or place strategies is thought to be due to the structural and functional changes to the hippocampal–entorhinal pathway (described above), as these structures are crucial for these strategies and could constitute a compensation for declining navigational abilities.

It should be noted that the virtual reality (VR) setups used in laboratory studies vary greatly, from desktop-based setups that only provide visual stimulation to immersive and ambulatory setups that permit unrestricted locomotion through virtual environments. Recent findings suggest that age-related differences in spatial memory and navigation are less pronounced when using VR setups that allow for more naturalistic and ecologically valid navigation measures (Hill et al. 2024). Such findings highlight the importance of considering how results from simple laboratory and VR studies relate to navigation in complex real-world navigation. To give another example, representations similar to grid cells have been shown to be less stable in older age and have been associated with declining path-integration/navigation performance in laboratory settings (Stangl et al. 2018). Yet it is unclear how such findings apply to real-world navigation, thus raising important questions: Does the fact that the real world is much noisier and more complex than the VR environments typically used for laboratory-based research present additional challenges to the stability of spatial codes? If so, when they navigate in the real world, do older adults employ compensatory cognitive strategies and behaviors instead of relying on navigation mechanisms that require grid cell-like representations?

2.3 Real-World Navigation

While age-related declines in navigational abilities have been firmly established in laboratory and virtual environment experiments, it is not always clear how these findings translate to everyday real-world navigation for a number of reasons:

- Experiments are often carried out in minimalistic laboratory or VR environments designed to meet experimental requirements, rather than simulating the real world.
- Experiments typically aim to isolate single factors such as specific environmental features or navigation tasks, whereas everyday real-world navigation involves an interplay of environmental variables and strategies.
- Experimental approaches often focus on the individual, typically ignoring the importance of other people and external artifacts, such as signage and maps, as active constituents of the navigation process.
- To preserve higher levels of functioning in daily life, older adults adopt compensation strategies which are difficult, if not impossible, to capture in experiments.

Navigation experiments conducted in real-world environments do mitigate some of the limitations of laboratory and VR environments experiments, as they allow insights into how the complexity and often dynamic nature of the real world affect navigation behavior. Early research into the effects of cognitive aging on navigation and wayfinding was carried out in real-world experimental settings. Some studies reported age-related differences similar to the ones obtained in laboratory-based VR environments experiments, such as declines in route learning performance and in learning of unfamiliar environments (Kirasic 1991). When asked to freely recall landmarks of a real-world route, older adults' recall was based on landmark distinctiveness or saliency rather than temporal order in which landmarks were encountered along the route, and included information not relevant to the spatial task (Lipman 1991). Such differences in landmark-selection strategies, when remembering real-world routes, are likely to contribute to age-related differences in route learning.

Despite these findings, which highlight age-related deficits in unfamiliar environments, older adults seem to be performing similarly to younger adults on spatial tasks in familiar environments (Kirasic 1991). In other words, cognitive aging seems to affect performance in more abstract laboratory tasks, which often use VR environments, as well as when navigation is required in unfamiliar environments. Older adults perform adequately, however, in realistic/ecological tasks (De Beni et al. 2006), when navigating in familiar environments. These findings suggest that older adults can make use of experience-based coping strategies in real-world everyday tasks.

To explore strategies that people employ during navigation in more detail, experimental approaches can be enriched with qualitative approaches such as reflective interviews or walking interviews (Spiers and Maguire 2006). Such mixed-method approaches provide additional insights into how the complexity and often dynamic nature of real-world environments affect navigation. For example, in a recent real-world mixed-method navigation experiment, participants reported to have frequently probed, switched between, and amended their strategies during wayfinding, often based on environmental characteristics (Alcock et al. 2024). Importantly, navigation behavior was affected by aesthetic considerations and the presence of other people in the environment, factors difficult to study under laboratory or VR environment settings.

Bracci et al. (under review) recently used a mixed methods approach to study the strategies that older adults employ when repeating and retracing routes in unfamiliar real-world environments. While behavioral results closely resembled those of VR environment studies conducted in the laboratory (Wiener et al. 2020), older participants reported using navigation strategies that differed from those reported in the laboratory experiments. Specifically, older participants reported using simpler strategies, such as moving to decision points and then looking for the next landmark along the route rather than associating landmarks with movement directions or using viewpoint-independent mechanisms during retracing. This study also highlights that landmark selection is a highly idiosyncratic process driven by people's experiences, knowledge, and interests rather than being solely based on object saliency. In interviews with older adults after they learned a novel route through an unfamiliar residential development, O'Malley et al. (2022) explored reasons for disorientation.

Older adults reported that a lack of landmarks, the repetitive design of the environment, long corridors, and lack of windows to the outside world affected their ability to remain oriented. These results are in line with interview studies showing that older adults who live in communal retirement developments frequently report experiencing spatial disorientation due to the uniformity of their living environment (O'Malley et al. 2018).

While real-world mixed-method studies help to address the impact of complex environments on navigation behavior and abilities, they are not optimally situated to capture everyday navigation behavior and the experiences of living with declining navigation abilities.

2.4 Everyday Navigation

Age-related changes in everyday navigation behaviors are evident in driving surveys, which show that older adults undertake fewer journeys and tend to avoid driving in unfamiliar environments (Burns 1999). In a recent interview study, Heward et al. (2023) explored the experiences of older adults living with declining navigation abilities. Participants reported being aware that aging impacted their ability to navigate and resulted in greater feelings of vulnerability. Moreover, they reported taking steps to avoid disorientation by preparing meticulously for trips, or by avoiding visits to unfamiliar places unless accompanied by someone who took responsibility for navigation. Specifically, the fear of getting lost was a strong negative incentive, and interviewees were prepared to give up independence and accept that their "world was getting smaller" to avoid the psychological harm associated with the risk of disorientation. These perspectives are in line with research showing that spatial anxiety increases in older age (van der Ham et al. 2021). Critically, age-related loss of independence and autonomy have the potential to affect activity levels and social inclusion in older age, thus negatively impacting health and well-being.

Everyday outdoor wayfinding in the community has been investigated in surprisingly few studies. In a study by Marquez et al. (2017), older adults were asked to plan a route in their local community and then the authors walked the route with each participant. Interviews conducted before and during the walk explored participants' route planning and wayfinding strategies. In the study, 33 of the 35 participants sought information from other people as the primary method of route planning. Eleven participants reported exploring on their own, eight reported using maps, and only two participants reported using satellite navigation (SatNav) systems. Not surprisingly, street signs and landmarks were the primary environmental wayfinding features used.

It is important to note that both studies reviewed above mentioned (a) the importance of social interactions for wayfinding in older age and (b) that older adults rarely seem to make use of SatNav systems during navigation (Heward et al. 2023). While social aspects of navigation are understudied, the topic has recently received more traction in navigation research (for an overview, see Dalton et al. 2019). Still, it remains unclear why older adults do not make use of SatNav systems that, in principle, could compensate for navigational difficulties, particularly when navigating unfamiliar outdoor environments. One possible explanation is that these systems currently require participants to divide limited attentional resources when driving or walking.

Overall, there is a dearth of studies investigating everyday spatial navigation and the influence that cognitive aging has on people's everyday navigation experiences and their navigation behaviors. Future research should focus on these issues with the aim of developing interventions and age-friendly (design) guidelines to support independence in older age.

3 Neurodegeneration

Neurodegeneration is a very broad topic, encompassing any condition that progressively affects the peripheral and central nervous system. Since it is clearly beyond the scope of this chapter to cover all neurodegenerative conditions, we will focus on the most common neurodegenerative condition in aging, dementia. Dementia itself encompasses multiple conditions, with around 15 recognized clinical diagnoses: Alzheimer disease is the most common (~60–70%), followed by vascular dementia (~10–20%) and mixed dementia (combined Alzheimer and vascular pathophysiology). The remaining percentage mainly consists of dementia with Lewy bodies, which overlaps in symptomology with Parkinson disease, the frontotemporal dementia syndrome (which itself consists of three clinical subtypes), and finally some even rarer sporadic or genetic dementia subtypes (Prince et al. 2013).

From a spatial navigation/orientation, perspective, the vast majority of research has been conducted in the clinical, prodromal, and at-genetic risk forms of Alzheimer disease (Coughlan et al. 2018). By contrast, there is very little research on vascular dementia, mixed dementia, or frontotemporal dementia, and spatial navigation/orientation research in the remaining rarer dementia syndromes is virtually absent. This highlights the first gap for neurodegeneration in that spatial navigation/orientation performance in many dementia syndromes is unknown to date. Even for Alzheimer disease, for which the majority of spatial navigation/orientation research has been conducted, we are only now starting to understand how the underlying pathophysiological processes affect spatial navigation/orientation performance in laboratory settings. Less is known about how those spatial navigation/orientation performances change over time as the disease progresses, and how this can affect everyday life in people with dementia.

Thus, our discussion here is mainly on Alzheimer disease, and vascular dementia or frontotemporal dementia where relevant. It is important to understand, however, that the disease pathophysiologies of vascular and frontotemporal dementia differ not only from each other but also from Alzheimer disease. Thus, spatial navigation/orientation processes can be differentially affected. To understand the spatial navigation/orientation changes in Alzheimer disease, we first need to understand the underlying pathophysiology changes, which are often poorly understood by behavioral research investigating spatial navigation/orientation in this disease.

3.1 Alzheimer Disease Pathophysiology

Alzheimer disease is thought to be caused by the accumulation of two proteins: amyloid and tau. Both proteins are important for normal cellular function: amyloid as part of cell membrane transport functions, and tau as part of the cytoskeleton and intracellular transport mechanism. Amyloid is well known to accumulate as beta-amyloid during aging outside of nerve cells, forming beta-amyloid plaques. Historically, amyloid was seen as the "cause" for Alzheimer disease; however, currently it is regarded as "only" one component of disease emergence. The reasons for this shift in thinking are that older, healthy people with high levels of beta-amyloid in their brains do not necessarily develop Alzheimer disease; beta-amyloid accumulates mostly in cortical regions and not the medial temporal lobe, at least at the beginning of the disease; and beta-amyloid disease treatments do not stop the disease, but only slow it down. Beta-amyloid plaques can be seen, therefore, as a milieu which significantly increases the risk for Alzheimer disease, but not its cause. For Alzheimer disease to emerge, tau also needs to accumulate. Currently, the confluence of beta-amyloid and tau is considered to trigger and drive the progression of Alzheimer disease. Tau has three critical characteristics which make it central to Alzheimer disease:

1. It is highly concentrated in the axons of neurons; hence when it malfunctions, tau affects neuronal communication pathways and thus cognitive function directly (whereas beta-amyloid is extracellular).
2. It begins, for yet unknown reasons, to accumulate in medial temporal lobe regions, initiating deficits in episodic memory and spatial navigation/orientation.
3. It has "prion-like" characteristics; that is, tau pathophysiology spreads along synaptic pathways, "infecting" neighboring neurons and promoting disease progression.

As mentioned, for Alzheimer disease to start, there seems to be a confluence of amyloid and tau in the medial temporal lobe regions, specifically the medial entorhinal cortex. From there, tau spreads to connected regions in the hippocampus, retrosplenial cortex, lateral entorhinal cortex, and perirhinal cortex, before spreading even further to connected regions. Thus, it is important to regard Alzheimer disease as a neuronal network disease that is mostly driven by tau propagation across synapses. The exact reason why the medial entorhinal cortex seems to be the first region affected by the disease is unknown. Note, competing theories suggest that the disease begins in the locus coeruleus, spreading via the noradrenergic system until the entorhinal cortex is affected (Galgani and Giorgi 2023). Regardless, the majority of Alzheimer disease pathology is centered around Papez circuit regions, at least at the beginning of the disease before gradually all cortical and subcortical areas become affected. Mapping the disease pathophysiology is, therefore, important to establish which spatial navigation/orientation processes are affected over the disease progression, ranging from subtle path-integration deficits to complete spatial disorientation even in highly familiar places.

3.2 Spatial Memory in Alzheimer Disease

Historically, most investigations into spatial navigation used some form of spatial memory task that required participants to learn either the location of objects in a landscape or to find their way back after having been led through a virtual landscape or maze. Due to significant episodic memory deficits, it is perhaps not surprising that people with Alzheimer disease showed significant deficits on these tasks, which did not disentangle whether the deficits were due to spatial navigation or more general episodic memory deficits. Thus, despite being pioneering, these tasks were of limited utility in advancing our understanding of spatial navigation/orientation deficits in Alzheimer disease, as well as in the cognitive evaluation or diagnostics of patients. Indeed, these spatial memory tasks showed similar sensitivity and specificity in detecting Alzheimer disease when compared against commonly used neuropsychological tasks, such as the Rey Complex Figure Test. Because such tasks were recruiting a number of brain regions, they were unspecific to specialized navigation processes in the brain. It is beyond the scope of this chapter to explore all brain regions involved in those changes, in particular related to amyloid pathology which is widespread. Several excellent reviews, however, cover these topics (e.g., Coughlan et al. 2018). In brief, the evidence strongly supports spatial navigation deficits as sensitive and specific early markers for Alzheimer disease pathophysiology. Given the substantial overlap between navigation networks and regions affected at early stages by Alzheimer disease, spatial assessment offers a more effective approach to evaluate at-risk individuals, to differentiate between Alzheimer disease and other dementias, to enable outcome measurements in clinical trials, and to reduce bias in cross-cultural assessments. Neuroimaging reveals that spatial navigation deficits in Alzheimer disease result from systematic degradation of medial temporal, medial parietal, and subcortical navigation brain networks, with changes following the characteristic pattern of Alzheimer disease pathology spread. The navigation network's early vulnerability, combined with its functional specificity, makes it an ideal target for neuroimaging-based biomarker development. These findings support the clinical utility of navigation assessment and provide a neurobiological foundation for its implementation in Alzheimer disease diagnosis and monitoring (Coughlan et al. 2018).

The usage of virtual or real-world Morris water maze (MWM) tasks made the investigation into spatial memory deficits in Alzheimer disease more specific. A widely used task in animal models, both with and without Alzheimer pathophysiology, the MWM task allows for a better delineation of allo- and egocentric spatial memory changes in prodromal and clinical Alzheimer disease. Studies using virtual or even real-world MWM showed consistently that people with Alzheimer disease had significant allocentric memory deficits while their egocentric spatial memory deficits were less pronounced (Hort et al. 2014). These findings were extended to prodromal forms of Alzheimer disease; people with mild cognitive impairment showed significant deficits in allocentric MWM conditions. Those deficits also correlated with underlying biomarker changes in amyloid and tau and associated medial temporal lobe, particularly hippocampal atrophy (Laczó et al. 2012). However, the MWM has clear methodological limitations. For example, Hamilton et al. (2007) argue that MWM procedures, which are typically seen as probing allocentric knowledge (or place learning), can often be solved with simpler directional response strategies. For a detailed discussion of the challenges of inferring spatial reference usage from navigation tasks and associating neuronal mechanisms with specific reference frames, see Wolbers and Wiener (2014). Importantly, detecting more subtle spatial deficits is challenging, thus raising the question whether more traditional or paper-and-pencil tasks offer comparable sensitivity to identify prodromal or preclinical Alzheimer disease (Bird et al. 2010).

3.3 Path Integration in Alzheimer Disease

Over the past decade, path integration has emerged as the key spatial navigation/orientation process that is affected in Alzheimer disease, in genetically at-risk groups as well as in sporadic preclinical and prodromal forms of the disease. Initially, animal studies showed that medial entorhinal cortex changes, due to Alzheimer pathophysiology, disrupted grid-cell functioning in the medial entorhinal cortex, which in turn led to changes in behavior. Specifically, animals with a particular tau pathology showed disrupted grid-cell functioning in the medial entorhinal cortex, which resulted in the animal avoiding open areas and moving along the borders of enclosures (Fu et al. 2017). Similar findings have been observed in humans. However, the types of tasks employed varied in those experiments; some used more real-life scenarios whereas others employed abstract environments using VR technology.

Most human path-integration tasks use a form of a triangulation task (Loomis et al. 1993), which requires the participant to move from a starting point to a turning point, followed by another turning point. Some tasks require participants to actively move to those turning points, either on screen or through immersive VR. At the final turning point, all tasks ask for the participant to point or walk back toward the starting point, and some tasks require participants to point to the starting point via a top-down map of the environment. Although triangulation tasks in human studies are not homologous to animal tasks, they tap into functions akin to path integration. Discussion is ongoing as to whether triangulation tasks should be considered "process pure" path integration, in particular since most tasks do not have idiothetic inputs, which in the animal literature are considered essential for path integration. Still, triangulation tasks allow a readout of entorhinal and/or retrosplenial–medial parietal path-integration processes. Specifically, the angular deviation on the return path for the triangulation task taps into egocentric navigation processes that are mediated mostly via the medial parietal regions. The over- or undershooting of the starting point provides a readout of grid-cell integrity. Finally, some tasks measure whether participants thought their final destination was closer to the border or not, thus providing a "border effect" readout.

Evidently, people with clinical Alzheimer disease perform very poorly on triangulation tasks, indicating that the spatial navigation/orientation system is largely compromised and leads to stark deviations in angular deviations, over/undershooting of targets, and reliance on border cells. Therefore, these tasks are mainly used in preclinical or prodromal forms of Alzheimer disease. For example, people with mild cognitive impairment, a prodromal form of Alzheimer disease, show significant differences to healthy people of similar age and led Howett et al. (2019) to suggest that this can be used to detect incipient Alzheimer disease. Even more interesting is that healthy people with elevated Alzheimer biomarkers (Coughlan et al. 2023; Newton et al. 2024) or who are genetically at risk (*APOE* genotype) can show more subtle differences on those tasks compared to healthy people of similar age (Bierbrauer et al. 2020). Interestingly, *APOE* carriers can also show subtle vestibular changes, which might contribute to the largely unexplored vestibular contributions to path-integration changes observed in Alzheimer disease. The latter findings are of particular relevance for upcoming research, as blood biomarkers of Alzheimer disease will likely allow the identification of people with the disease long before episodic memory problems become apparent. Hence, cognitive readouts are urgently needed to track disease progression and treatment outcomes in preclinical Alzheimer disease. Path integration might emerge as a functional readout for future preclinical cohorts with blood biomarker confirmation.

Although preclinical cohorts exhibit clear but subtle effects during path-integration tasks, normative data remain outstanding and are critically needed before such deficits can be detected with adequate sensitivity and specificity (Coughlan et al. 2019). Efforts have been made but further validation is required (Coughlan et al. 2020). Understanding whether path integration or broader spatial navigation/orientation changes lead to real-world changes in navigation/orientation will help develop functional outcome measures, which are largely missing in preclinical Alzheimer disease but crucial for tracking disease progression, assessing intervention trials, and informing clinical management.

3.4 Real-World Changes

The process by which spatial navigation/orientation changes and affects real-word navigation in Alzheimer disease is unknown. This is surprising, as such information is critical if we are to safeguard people who suffer from Alzheimer disease and dementia more broadly. "Getting lost" or "going missing" are common occurrences in people with Alzheimer disease. In the United Kingdom, for example, approximately 40,000 people with dementia experience this every year, and while most are found unharmed within a short span of time, it is estimated that 10–20% are injured or die (e.g., from dehydration or exposure) from such episodes. Despite our limited understanding of spatial disorientation, data indicate that men are at greater risk of getting lost than women, for reasons unknown (Puthusseryppady et al. 2019). In addition, people with dementia restrict their own outdoor movement, most likely due to spatial navigation/orientation challenges. Compared to movement patterns of healthy individuals, people with dementia stay closer to home and prefer walking along familiar routes, unless accompanied by caregivers. This self-regulation is clearly a sign that something is amiss and a potential harbinger of spatial orientation getting worse.

Performance on VR-based spatial navigation tasks predicts how people with Alzheimer disease move outdoors in the real world. It is challenging, however, to directly predict spatial disorientation during outings. Sensor data (GPS, accelerometer, gyroscopic) show that both the complexity of the environment and the frequency of turns greater than 90 degrees contribute to increased spatial disorientation (Ghosh et al. 2022). These findings reveal that in the real world, the environment plays an important role in how spatial disorientation is manifest in Alzheimer disease. For instance, landmark density and street network entropy are critical environmental factors in healthy navigation that can influence disorientation or missing episodes (Puthusseryppady et al. 2019). Indeed, higher street network entropy with higher crossroad complexity has been linked to missing episodes, which makes sense, since crossroads are spatial decision points; more complex spatial decision points allow for greater mistakes in taking the wrong turn.

Spatial navigation/orientation has clear relevancy for car driving, both in aging individuals and those with dementia. Driving is a highly emotive subject: most people are desperate to hold onto their driving licenses to meet mobility needs, yet the number of collisions and fatalities in this group are one of the highest (Brown and Ott 2004). The impact of spatial navigation/orientation on a person's ability to drive has not been adequately explored and is not used, in any country, to assess a person's fitness to drive. Existing studies have shown that spatial navigation/orientation is related to lane maintenance and that allocentric navigation changes are particularly predictive of people reducing their driving and avoiding challenging driving situations (Morrissey et al. 2024). Interestingly, performance on allocentric navigation tasks is associated with performance for turning across oncoming traffic, one of the high-risk driving situations in aging and dementia (Morrissey et al. 2024). In addition, challenging driving situations, which older people and those with dementia actively avoid, often result from reduced visibility of landmarks, such as driving at night, in the rain, or in fog. Finally, emerging data indicate that the use of SatNav technology is related to greater driving confidence in older drivers and allows them to maintain their driving independence (Morrissey et al. 2025). Further, ameliorating spatial navigation challenges has a positive effect on driving performance and safety. Thus, further research into navigation technologies is advisable.

4 Future Directions

4.1 Aging

We believe that future research into the effects of cognitive aging on navigation behavior should focus on a number of issues, with the ultimate goal of supporting independent navigation in older age. First, while age-related decline in navigation behavior has now been firmly established in the literature as well as in laboratory- and VR-based experiments, there is a need to develop a better understanding of how this process manifests in everyday, real-world navigation. While experimental approaches are particularly suited to investigate the intricacies of the aging navigation system as well as factors underlying age-related decline in navigation behavior, the study of everyday navigation requires a different approach. Therefore, future aging and navigation research should widen the prevalent methodological approach and scope to include more real-world studies, adopt qualitative and mixed methods approaches, and include cognitive ethnography. These methods will allow for in-depth insights into the effects that cognitive aging has on everyday navigation and the experiences of living with declining navigation abilities. The resultant information is needed to develop improved age-friendly design guidelines that will improve independent navigation in older age.

Second, understanding how aging affects everyday navigation is essential for developing interventions that support independent wayfinding in older age. To this end, we need to increase our understanding of how changes that are experienced by older people influence, more causally, navigation, and how this can be ameliorated. Existing interventions, which often employ navigational training in virtual environments, have had mixed results in terms of their effectiveness and transferability of the trained components to everyday wayfinding. Thus, future research is needed to explore whether other types of interventions (e.g., instructor-led interventions informed by older adults' reports of challenging everyday navigation scenarios and their strategies and tactics to overcome these) might be better suited to improve navigational confidence and reduce spatial anxiety, thereby increasing navigational activity and independence.

Third, navigation in unfamiliar environments is often supported by SatNav systems. However, recent research (Heward et al. 2023) and anecdotal reports suggest that older adults do not often use SatNav systems to navigate or find the technology difficult. This may be due to a number of reasons; for example, limited attentional resources may make it difficult to use SatNav systems during driving or walking. Using mobile phones while walking, for example, reduces gait stability, even in young adults, and is therefore likely to increase the risk of falling in older adults. Future research into the design of navigation assistance systems should consider older adults as a key demographic and tailor navigation solutions to specific age-related navigation challenges; this would support and enhance spatial navigation in older age (Weisberg et al. 2024). Further to this point, Newton et al. (Chapter 16) provide a detailed description of a navigation framework that would bolster navigation and independence in older age, and suggest different types of interventions (e.g., support, maintenance, educational interventions) that could be pursued.

Finally, the influence of sociodemographic factors on health, well-being, and cognitive performance in later life is well established (Steptoe and Zaninotto 2020). Recent research has demonstrated that sociodemographic factors affect navigation performance. Specifically, in one of the largest citizen science projects ever conducted, Sea Hero Quest, it was shown that people who grew up in rural areas navigated better than people who grew up in cities (Coutrot et al. 2018) and that higher levels of education were associated with greater navigational skills (Fernandez-Velasco et al. 2023). However, the way in which associations between navigation abilities and sociodemographic factors translate to people's everyday wayfinding experiences remains unclear and should be considered in future research.

4.2 Neurodegeneration

Specific to spatial navigation/orientation in Alzheimer disease, three clear directions have emerged. The first involves the detection of cognitive symptoms in preclinical

Alzheimer disease before the onset of episodic memory symptoms. Clearly, path integration offers a spatial navigation measure by which entorhinal grid-cell changes can be detected in preclinical populations. To date, however, the reliability of this detection method is unknown. There is, therefore, an urgent need to establish a reliable, quick task for clinical use—one backed by large normative data to ensure high sensitivity and specificity in those populations. We do not foresee that path-integration tasks will be used as a diagnostic tool in the future, since blood biomarkers are more efficient. Instead, path-integration tasks should be used to measure disease progression and/or treatment outcomes as well as to monitor cognitive deficit development in preclinical, sporadic, and genetically at-risk individuals for Alzheimer disease. In that regard, it will be important to understand how such measures differ from other neurodegenerative diseases, including vascular dementia, frontotemporal dementia, and dementia with Lewy bodies. At present, we only have very limited data on how these forms of dementia affect path integration.

Second, there is a clear gap in understanding how spatial navigation/orientation deficits progress throughout the disease course. To our knowledge, no longitudinal studies are currently tracking how these spatial navigation symptoms change over time. This aspect is not trivial, since the pathophysiology clearly seems to affect more allocentric-reliant navigation regions in the brain, before it impacts egocentric-reliant navigation regions. Further, the contribution of head-direction cells is virtually unknown in this population, despite emerging evidence that vestibular disturbances occur at early stages of dementia. Tracking spatial navigation symptomology will allow the creation of progression measures, which in turn would help determine how cognitive reserve and resilience can potentially mitigate spatial navigation changes, which is currently unknown. Further, it will also inform future intervention studies aimed at mitigating changes that occur in dementia. Similar to aging studies, these interventions would enable us to investigate the relationship between spatial navigation deficits and associated brain network changes in Alzheimer disease. Current disease progression measures are strongly biased toward clinical assessments in high-income and high-education populations. The nonverbal, everyday nature of spatial navigation would allow disease progression measures to be established that could be used globally, transforming low- and middle-income country diagnostics and clinical care.

Finally, the impact of spatial navigation changes on everyday functioning in Alzheimer disease are only beginning to be understood, yet they are of paramount concern to people suffering from Alzheimer disease as well as their families. Specifically, spatial disorientation is highly common in people with Alzheimer disease and affects everyday functioning (e.g., outings, driving). Getting lost or missing events are still very poorly understood, and it is unclear whether there are predictive factors that could aid in risk assessment. Understanding potential risk factors would allow better safeguarding measures to be put in place, while keeping people as independent as possible within the community. Similarly, to address the challenges inherent in driving—a significant and emotive topic in aging—requires a clearer understanding of how the changes in spatial navigation affect the fitness to drive. This is important not only for people with dementia and their families, but more broadly for all citizens, driving license agencies, governments, and insurance companies.

References

Alcock M, Wiener JM, and Hardman D (2024) Searching in an Unfamiliar Environment: A Phenomenologically Informed Experiment. Phenom Cogn Sci. https://doi.org/10.1007/s11097-024-09960-3

Bierbrauer A, Kunz L, Gomes CA, et al. (2020) Unmasking Selective Path Integration Deficits in Alzheimer's Disease Risk Carriers. Sci Adv 6:eaba1394. https://doi.org/10.1126/SCIADV.ABA1394

Bird CM, Chan D, Hartley T, et al. (2010) Topographical Short-Term Memory Differentiates Alzheimer's Disease from Frontotemporal Lobar Degeneration. Hippocampus 20:1154–1169. https://doi.org/10.1002/hipo.20715

Bracci A, Hardman D, and Wiener J (under review) Route Repetition and Route Retracing: A Real-World Investigating into Effects of Cognitive Aging.

Brown L, and Ott B (2004) Driving and Dementia: A Review of the Literature. J Geriatr Psychiatry Neurol 17 (4):232–240. https://doi.org/10.1177/0891988704269825

Burns PC (1999) Navigation and the Mobility of Older Drivers. J Gerontol B Psychol Sci Soc Sci 54 (1):S49–55. https://doi.org/10.1093/geronb/54b.1.s49

Coughlan G, Coutrot A, Khondoker M, et al. (2019) Toward Personalized Cognitive Diagnostics of At-Genetic-Risk Alzheimer's Disease. PNAS 116 (19):9285–9292. https://doi.org/10.1073/pnas.1901600116

Coughlan G, DeSouza B, Zhukovsky P, et al. (2023) Spatial Cognition Is Associated with Levels of Phosphorylated-Tau and B-Amyloid in Clinically Normal Older Adults. Neurobiol Aging 130:124–134. https://doi.org/10.1016/j.neurobiolaging.2023.06.016

Coughlan G, Laczó J, Hort J, Minihane A-M, and Hornberger M (2018) Spatial Navigation Deficits—Overlooked Cognitive Marker for Preclinical Alzheimer Disease? Nat Rev Neurol 14 (8):496–506. https://doi.org/10.1038/s41582-018-0031-x

Coughlan G, Puthusseryppady V, Lowry E, et al. (2020) Test-Retest Reliability of Spatial Navigation in Adults At-Risk of Alzheimer's Disease. PLoS ONE 15:0239077. https://doi.org/10.1371/journal.pone.0239077

Coutrot A, Silva R, Manley E, et al. (2018) Global Determinants of Navigation Ability. Curr Biol 28 (17):2861–2866.e2864. https://doi.org/10.1016/j.cub.2018.06.009

Dalton RC, Hölscher C, and Montello DR (2019) Wayfinding as a Social Activity. Front Psychol 10:00142. https://doi.org/10.3389/fpsyg.2019.00142

De Beni R, Pazzaglia F, and Gardini S (2006) The Role of Mental Rotation and Age in Spatial Perspective-Taking Tasks: When Age Does Not Impair Perspective-Taking Performance. Appl Cogn Psychol 20 (6):807–821. https://doi.org/10.1002/acp.1229

Fernandez-Velasco P, Coutrot A, Oloye H, et al. (2023) No Link between Handedness and Spatial Navigation: Evidence from over 400,000 Participants in 41 Countries. Proc R Soc Lond B Biol Sci 290 (2008):20231514. https://doi.org/10.1098/rspb.2023.1514

Fu H, Rodriguez GA, Herman M, et al. (2017) Tau Pathology Induces Excitatory Neuron Loss, Grid Cell Dysfunction, and Spatial Memory Deficits Reminiscent of Early Alzheimer's Disease. Neuron 93 (3):533–541. https://doi.org/10.1016/j.neuron.2016.12.023

Galgani A, and Giorgi F (2023) Exploring the Role of Locus Coeruleus in Alzheimer's Disease: A Comprehensive Update on MRI Studies and Implications. Curr Neurol Neurosci Rep 23 (12):925–936. https://doi.org/10.1007/s11910-023-01324-9

Ghosh A, Puthusseryppady V, Chan D, Mascolo C, and Hornberger M (2022) Machine Learning Detects Altered Spatial Navigation Features in Outdoor Behaviour of Alzheimer's Disease Patients. Sci Rep 12:3160. https://doi.org/10.1038/s41598-022-06899-w

Hamilton DA, Akers KG, Weisend MP, and Sutherland RJ (2007) How Do Room and Apparatus Cues Control Navigation in the Morris Water Task? Evidence for Distinct Contributions to a Movement Vector. J Exp Psychol Anim Learn Cogn 33 (2):100–114. https://doi.org/10.1037/0097-7403.33.2.100

Harris MA, and Wolbers T (2014) How Age-Related Strategy Switching Deficits Affect Wayfinding in Complex Environments. Neurobiol Aging 35 (5):1095–1102. https://doi.org/10.1016/j.neurobiolaging.2013.10.086

Head D, and Isom M (2010) Age Effects on Wayfinding and Route Learning Skills. Behav Brain Res 209 (1):49–58. https://doi.org/10.1016/j.bbr.2010.01.012

Heward M, Hicks B, Hedges B, Gaden R, and Wiener JM (2023) Experiences of Age-Related Declining Navigation Abilities and Impact on Use of Outdoor Environments: A Qualitative Study of Young-Old Adults with Self-Reported Memory Difficulties. Ageing Soc 45 (3):573–593. https://doi.org/10.1017/S0144686X23000612

Hill PF, Bermudez S, McAvan AS, et al. (2024) Age Differences in Spatial Memory Are Mitigated during Naturalistic Navigation. Neuropsychol Dev Cogn B Aging Neuropsychol Cogn 31:2023.2001.2023.525279. https://doi.org/10.1080/13825585.2024.2326244

Hilton C, Johnson A, Slattery TJ, Miellet S, and Wiener JM (2021) The Impact of Cognitive Aging on Route Learning Rate and the Acquisition of Landmark Knowledge. Cognition 207:104524. https://doi.org/10.1016/j.cognition.2020.104524

Hort J, Andel R, Mokrisova I, et al. (2014) Effect of Donepezil in Alzheimer Disease Can Be Measured by a Computerized Human Analog of the Morris Water Maze. Neurodegener Dis 13 (2-3):192–196. https://doi.org/10.1159/000355517

Howett D, Castegnaro A, Krzywicka K, et al. (2019) Differentiation of Mild Cognitive Impairment Using an Entorhinal Cortex-Based Test of Virtual Reality Navigation. Brain 142 (6):1751–1766. https://doi.org/10.1093/brain/awz116

Kirasic KC (1991) Spatial Cognition and Behavior in Young and Elderly Adults: Implications for Learning New Environments. Psychol Aging 6 (1):10–18. https://doi.org/10.1037//0882-7974.6.1.10

Laczó J, Andel R, Vyhnalek M, et al. (2012) From Morris Water Maze to Computer Tests in the Prediction of Alzheimer's Disease. Neurodegener Dis 10 (1-4):153–157. https://doi.org/10.1159/000333121

Lee J, and Kim H-J (2022) Normal Aging Induces Changes in the Brain and Neurodegeneration Progress: Review of the Structural, Biochemical, Metabolic, Cellular, and Molecular Changes. Front Aging Neurosci 14:931536. https://doi.org/10.3389/fnagi.2022.931536

Lester AW, Moffat SD, Wiener JM, Barnes CA, and Wolbers T (2017) The Aging Navigational System. Neuron 95 (5):1019–1035. https://doi.org/10.1016/j.neuron.2017.06.037

Lipman PD (1991) Age and Exposure Differences in Acquisition of Route Information. Psychol Aging 6 (1):128–133. https://doi.org/10.1037//0882-7974.6.1.128

Livingston G, Huntley J, Liu K, et al. (2024) Dementia Prevention, Intervention, and Care: 2024 Report of the Lancet Standing Commission. Lancet 404 (10452):572–628. https://doi.org/10.1016/s0140-6736(24)01296-0

Loomis JM, Klatzky RL, Golledge RG, et al. (1993) Nonvisual Navigation by Blind and Sighted: Assessment of Path Integration Ability. J Exp Psychol 122 (1):73–91. https://doi.org/10.1037//0096-3445.122.1.73

Marquez DX, Hunter RH, Griffith MH, et al. (2017) Older Adult Strategies for Community Wayfinding. J Appl Gerontol 36 (2):213–233. https://doi.org/10.1177/0733464815581481

Mittelstaedt ML, and Mittelstaedt H (1980) Homing by Path Integration in a Mammal. Naturwiss 67 (11):566–567. https://doi.org/10.1007/BF00450672

Morrissey S, Jeffs S, Gillings R, et al. (2024) The Impact of Spatial Orientation Changes on Driving Behavior in Healthy Aging. J Gerontol B Psychol Sci Soc Sci 79:188. https://doi.org/10.1093/geronb/gbad188

——— (2025) GPS Navigation Assistance Is Associated with Driving Mobility in Older Drivers. PLoS Digit Health 4 (4):e0000768. https://doi.org/10.1371/journal.pdig.0000768

Moser M-B, Rowland DC, and Moser EI (2015) Place Cells, Grid Cells, and Memory. Cold Spring Harb Perspect Biol 7 (2):a021808. https://doi.org/10.1101/cshperspect.a021808

Newton C, Pope M, Rua C, et al. (2024) Entorhinal-Based Path Integration Selectively Predicts Midlife Risk of Alzheimer's Disease. Alzheimers Dement (N Y) 20 (4):2779–2793. https://doi.org/10.1002/alz.13733

O'Malley M, Innes A, Muir S, and Wiener JM (2018) "All the Corridors Are the Same": A Qualitative Study of the Orientation Experiences and Design Preferences of UK Older Adults Living in a Communal Retirement Development. Ageing Soc 38 (9):1791–1816. https://doi.org/10.1017/S0144686X17000277

O'Malley M, Innes A, and Wiener JM (2022) (Dis)Orientation and Design Preferences Within an Unfamiliar Care Environment: A Content Analysis of Older Adults' Qualitative Reports after Route Learning. Environ Behav 54 (1):116–142. https://doi.org/10.1177/0013916520953148

Prince M, Bryce R, Albanese E, et al. (2013) The Global Prevalence of Dementia: A Systematic Review and Metaanalysis. Alzheimers Dement (N Y) 9:63. https://doi.org/10.1016/j.jalz.2012.11.007

Puthusseryppady V, Coughlan G, Patel M, and Hornberger M (2019) Geospatial Analysis of Environmental Risk Factors for Missing Dementia Patients. J Alzheimers Dis 71 (3):1005–1013. https://doi.org/10.3233/JAD-190244

Segen V, Ying J, Morgan E, Brandon M, and Wolbers T (2022) Path Integration in Normal Aging and Alzheimer's Disease. Trends Cogn Sci 26 (2):142–158. https://doi.org/10.1016/j.tics.2021.11.001

Spiers HJ, and Maguire EA (2006) Thoughts, Behavior, and Brain Dynamics during Navigation in the Real World. Neuroimage 31 (4):1826–1840. https://doi.org/10.1016/j.neuroimage.2006.01.037

Stangl M, Achtzehn J, Huber K, et al. (2018) Compromised Grid-Cell-Like Representations in Old Age as a Key Mechanism to Explain Age-Related Navigational Deficits. Curr Biol 28 (7):1108–1115. e1106. https://doi.org/10.1016/j.cub.2018.02.038

Steptoe A, and Zaninotto P (2020) Lower Socioeconomic Status and the Acceleration of Aging: An Outcome-Wide Analysis. PNAS 117 (26):14911–14917. https://doi.org/10.1073/pnas.1915741117

van der Ham IJM, van der Kuil MNA, and Claessen MHG (2021) Quality of Self-Reported Cognition: Effects of Age and Gender on Spatial Navigation Self-Reports. Aging Ment Health 25 (5):873–878. https://doi.org/10.1080/13607863.2020.1742658

Weisberg SM, Ebner NC, and Seidler RD (2024) Getting LOST: A Conceptual Framework for Supporting and Enhancing Spatial Navigation in Aging. WIRES Cogn Sci 15 (2):e1669. https://doi.org/10.1002/wcs.1669

Wiener JM, Carroll D, Moeller S, et al. (2020) A Novel Virtual-Reality-Based Route-Learning Test Suite: Assessing the Effects of Cognitive Aging on Navigation. Behav Res Methods 52 (2):630–640. https://doi.org/10.3758/s13428-019-01264-8

Wolbers T, and Wiener JM (2014) Challenges for Identifying the Neural Mechanisms That Support Spatial Navigation: The Impact of Spatial Scale. Front Hum Neurosci 8:00571. https://doi.org/10.3389/fnhum.2014.00571

Zhong JY, and Moffat SD (2018) Extrahippocampal Contributions to Age-Related Changes in Spatial Navigation Ability. Front Hum Neurosci 12:272. https://doi.org/10.3389/fnhum.2018.00272

15

Designing Effective Clinical Tools for Measuring Navigation Deficits Associated with Cognitive Impairment

Insights from Human-Computer Interaction Research

Victor R. Schinazi, Giorgio Colombo, and Karolina Minta

Abstract Navigation deficits are increasingly being recognized as important early indicators of cognitive impairment in Alzheimer disease and other forms of dementia. Accurately measuring these deficits is critical for early diagnosis, intervention, and management of the disease. The widespread adoption of mobile technologies presents a unique opportunity to integrate more efficient and accurate methods for detecting navigation deficits. However, for a technology to be accepted by patients and incorporated into clinical workflows, merely developing a battery of tests to assess navigation ability is insufficient. To truly succeed, usability testing should be a core component of development and needs to include both patients and clinicians from the very beginning. Focusing on usability can translate innovation into a user-friendly technology that is reliable, efficient, and secure. Here, the development of a tablet-based game, Spatial Performance Assessment for Cognitive Evaluation, is used to illustrate how insights from research into human-computer interaction can guide and support the design of effective clinical tools. The authors argue that for

Victor R. Schinazi (✉)
Department of Psychology, Bond University, Robina, 4226, Queensland, Australia, and Future Health Technologies, Singapore-ETH Centre, Campus for Research Excellence and Technological Enterprise (CREATE), Singapore 138602, Singapore
Email: vschinaz@bond.edu.au

Giorgio Colombo (✉)
Future Health Technologies, Singapore-ETH Centre, Campus for Research Excellence and Technological Enterprise (CREATE), Singapore 138602, Singapore; and Chair of Cognitive Science, ETH Zürich, 8092 Zürich, Switzerland
Email: gicolombo@ethz.ch

Karolina Minta
Future Health Technologies, Singapore-ETH Centre, Campus for Research Excellence and Technological Enterprise (CREATE), Singapore 138602, Singapore

N. S. Newcombe and K. Cheng (eds.), *Challenges in Navigation Research*, Strüngmann Forum Reports, https://doi.org/10.1007/978-3-032-20563-6_15

these tools to be clinically viable, they must match the sensitivity of gold-standard neuropsychological tests and fluid or imaging biomarkers. When it comes to usability, there is no such thing as universal design since the needs and abilities of individuals differ greatly across age groups. Further, when transitioning from the lab to the clinic, modularity is essential because it enables rapid assessment and collection of feedback on different features that can shape the final product. Ultimately, it is important to evaluate the cost-effectiveness of the new technology by comparing it against existing clinical standards since a user-friendly and reliable tool alone is insufficient if it is unable to alleviate some of the burden in the health-care system.

Keywords Human-computer interaction, cognitive impairment, Alzheimer disease, spatial cognition, navigation, digital health

1 Introduction

Approximately 55 million people are affected by dementia worldwide (WHO 2023), and this number is expected to increase threefold by 2050 (Alzheimer's Association 2024; Patterson 2018). Dementia poses a significant threat to the health system (National Center for Health Statistics 2021), with an economic burden projected to reach 16.9 trillion USD over the next 25 years (Nandi et al. 2022). The costs of dementia are multifaceted and include specialized long-term medical care, caregiver support, informal caregiving, and costs associated with detection and diagnosis (Wong 2020). The lack of a definitive cure for dementia highlights the increasing need for interventions aimed at slowing the disease's progression and assisting individuals in maintaining their quality of life. For early interventions to be most effective, it is critical to have a diagnosis that is both timely and accurate.

Alzheimer disease is known to adversely affect the hippocampus and entorhinal cortex of humans through extracellular plaque deposits of the β-amyloid (Aβ) peptide (Fletcher et al. 2016; Murphy and LeVine 2010) and neurofibrillary tangles of the microtubule-binding protein tau (Bloom 2014; Bubu et al. 2015; Josephs et al. 2017). These two regions also play an instrumental role in the coding of locations in the environment and in assisting individuals to keep track of changes in their position and orientation during navigation (Epstein et al. 2017; Hafting et al. 2005; O'Keefe and Nadel 1978; Schinazi and Thrash 2018). Indeed, patients living with mild cognitive impairment and Alzheimer disease often report episodes of disorientation in addition to deficits in attention and memory (Lester et al. 2017; Li and King 2019). Targeted assessments designed to evaluate deficits in navigation may help uncover novel markers of cognitive and neural damage (Castegnaro et al. 2022; delpolyi et al. 2007; Laczó et al. 2017, 2022) and address some limitations (Dautzenberg et al. 2019) in current paper-and-pencil screening tools, such as the Montreal Cognitive Assessment (MoCA, Nasreddine et al. 2005) and the Mini-Mental State Examination (MMSE, Folstein 1983), as well as the high costs and invasive nature of imaging and fluid biomarkers (Chong et al. 2013; Coughlan et al. 2018). Specifically, the MoCA and the MMSE focus on attention, memory, and executive function but overlook other important domains, such as spatial and navigation abilities. Similarly, imaging and biomarker methods require participants to undergo procedures such as MRI scans or plasma collection, both of which can be uncomfortable and cause psychological stress. In some cases, the collection of cerebrospinal fluid through a lumbar puncture is necessary and carries considerable physical discomfort and potential risks. Virtual reality (VR) technology offers new possibilities for designing effective tools to measure navigational deficits by allowing individuals to interact with immersive and engaging environments (Allen et al. 2024; Coughlan et al. 2018). However, before these tools can be translated into clinical settings, they must undergo rigorous usability testing and validation. This is particularly the case when designing tools for older adults with diverse needs, who often have limited experience with novel technologies (Colombo et al. 2024a; Czaja et al. 2019).

In this chapter, we trace the development of a novel iPad-based serious game, the Spatial Performance Assessment for Cognitive Evaluation (SPACE), and highlight how research in human-computer interaction can inform the design of clinical tools for measuring navigational deficits. By applying principles tailored to the needs of aging populations (Nielsen 1994), we show how user-friendly, accurate, and effective tools can be developed to measure navigation deficits indicative of cognitive impairment. We argue that researchers need to balance innovation with clinical validity and practicality through user-centered, iterative, and participatory design. In addition, serious games should be modular to allow researchers and clinicians to flexibly customize the game to meet specific needs. This includes the ability to configure and deploy game setups that progressively increase in difficulty, making them adaptable for individuals of different ages and levels of cognitive impairment. We conclude with a mock cost-effectiveness analysis comparing the monetary benefits of SPACE against clinical standards, and advocate that this type of analysis is integral for translating a new technology from the lab to the clinic.

2 Measuring Navigation Deficits and Cognitive Impairment

Navigation deficits refer to impairments in the ability to orient oneself and move through the environment efficiently and effectively. These deficits can be broadly categorized as egocentric and allocentric, and they manifest in various ways, including difficulties in remembering and recognizing landmarks, route planning, keeping track of changes in position and orientation while moving, and using environmental cues to establish an orientation (Ekstrom and Isham 2017; Klatzky 1998; Tuena et al. 2021). Researchers in spatial cognition and navigation have a long history of designing spatial tests to measure navigation ability (Ekstrom et al. 2018; Kitchin and Blades 2001; Segen et al. 2022; Vlček 2011). Many of these tests have been successfully adapted to VR and used to measure navigation deficits associated with pathological aging (Bierbrauer et al. 2020; Chan et al. 2016; Coutrot et al. 2018; Howett et al. 2019; Kunz et al. 2015; Levine et al. 2020; Parizkova et al. 2018; Rekers and Finke 2024; Stangl et al. 2020; Tu et al. 2015; van der Ham et al. 2020; Wiener et al. 2020). Most of these tests, however, have been developed primarily for controlled laboratory experiments and have yet to be optimized for deployment in clinical settings (Rekers and Finke 2024; van der Ham and Claessen 2022).

Recently, researchers have combined multiple tasks to create a more encompassing assessment capable of capturing deficits in egocentric and allocentric ability and be widely distributed. For example, Sea Hero Quest (SHQ, Coutrot et al. 2018) is an unsupervised serious game deployed on mobile devices in which participants are asked to memorize a map of the environment before navigating to a series of checkpoints. On some levels of the game, participants are probed on their knowledge of their starting location (i.e., path integration) or asked to complete a radial arm maze task. Similarly, the Leiden Navigation Test (LNT, van der Ham and Claessen 2022; van der Ham et al. 2020) is a comprehensive (but supervised) desktop assessment of navigation ability in which participants watch a video of a route and are tested on their ability to recall the traveled path and remember the identity and position of landmarks along it. Although SHQ has been immensely successful in implementing various gamification elements to increase engagement and subsequently test millions of participants, the developers may have oversimplified some navigation tests and overlooked the needs of older adults when interacting with the control interface. In contrast, the LNT has smartly simplified the mechanics of the assessment (e.g., videos of routes, 1-click responses), thus reducing technological barriers and ensuring simplicity and ease of use for all ages. Nonetheless, LNT lacks modularity (e.g., no game levels that increase in difficulty) and gamification features to sustain user engagement, resulting in a less stimulating experience. Given these limitations, both SHQ and LNT may be less sensitive in detecting subtle navigation deficits associated with cognitive impairment. While neither SHQ nor LNT were originally designed for clinical use, incorporating formal usability testing during various stages of development could have addressed some of these challenges and improved their potential for clinical implementation.

3 Research in Human-Computer Interaction

3.1 Fundamental Principles

The interdisciplinary field of human-computer interactions focuses on designing, implementing, and evaluating computing systems for human use (Carroll 1997). Integrating principles from computer science, cognitive psychology, and design, the aim is to enhance the user experience (UX) and optimize the usability of digital interfaces and systems by making them user-friendly, efficient, and accessible (Carroll 1997, 2014; Olson and Olson 2003). Incorporating these principles into the design of clinical tools is essential for improving the validity and reliability of assessments (Rundo et al. 2020). Well-designed tools can reduce user frustration, enhance engagement, and provide more accurate data for clinicians to interpret. This process requires researchers to consider the user, the goal-oriented task, the interface, and the context in which the interaction occurs at every design stage (Norman 2013). Here, the context can refer to, but is not limited to, the population that will engage with the tool and the location (e.g., at home or the clinic) where it will be deployed. Indeed, previous research has found cultural differences in perceived usability (Noiwan and Norcio 2006; Santoso and Schrepp 2019; Wallace et al. 2013), although interindividual differences in usability among the same culture are significantly more pronounced than cultural influences (Santoso and Schrepp 2019; Schrepp et al. 2023).

Similarly, the setting in which the tool is used will determine the level of supervision required. In clinical settings, digital tools may benefit from supervision, which comes at the expense of clinicians' time. Unsupervised assessments may save time but must be designed to be simpler and more user-friendly (Polk et al. 2024). To further advance the usability of clinical tools, we suggest that researchers adopt agile or iterative usability techniques (Adikari et al. 2013; Alexander and Staggers 2009; Matz 2013; Méndez-Gurrola et al. 2023) that focus on users while considering accessibility, inclusivity, trust, and ethics. A user-centered approach will also ensure the correct balance between innovation and practicality to maximize UX, boost motivation, and increase receptivity (De Vito Dabbs et al. 2009; Nielsen 1994). Here, it is critical to consider the limitations of the patients and the requirements of the clinicians when interacting with the tool.

One promising approach that is gaining popularity is the codesign of mobile health (mHealth) tools. Codesigning consists of involving individuals from the target population throughout the design process (Fox et al. 2022; Noorbergen et al. 2021). Codesigning can help developers address the entirety of the context in which the product will be used and the way different factors (e.g., age of users, time constraints of clinicians) might affect the overall experience. Sanders and Stappers (2014) outline a codesign framework with four successive phases:

1. The predesign phase emphasizes understanding the context, UX, and goals through methods like interviews, personas, focus groups, and questionnaires.
2. The generative phase focuses on exploring design possibilities using methods such as sketching, storyboards, mock-ups, and paper prototyping.
3. In the evaluative phase, users assess the effectiveness of concepts through high-fidelity prototyping and interviews.
4. The post-design phase investigates real-world UXs using interviews and focus groups.

Over the last decades, UX researchers have developed a variety of instruments to measure the usability of systems. These instruments include self-report questionnaires to measure the overall quality of interactive products, such as the System Usability Scale (SUS, Brooke 1996), the Questionnaire for User Interface Satisfaction (Chin et al. 1988), and the User Experience Questionnaire (UEQ, Laugwitz et al. 2008). Researchers can also measure cognitive load (Skulmowski 2023) using single-item instruments (Paas 1992) or more comprehensive instruments that measure different types of task loads (e.g., mental, physical, temporal), such as the NASA-Task Load Index (NASA-TLX, Hart and Staveland 1988). When working with VR, researchers also typically measure the immersion of a system (Witmer and Singer 1998) and simulator sickness (Kennedy et al. 1993). One challenge posed by these questionnaires is that they typically produce a usability score rather than identify the usability issues that require remediation. In addition, these questionnaires were originally designed for software evaluation and may not apply to digital health technologies (Moreno-Ger et al. 2012). As such, qualitative methods and open-ended questions that are tailored to the specific product and context should be used to complement these questionnaires. An alternative approach is to probe the user's actual performance on the task, which can vary depending on the quality of the interaction (van der Kuil et al. 2018). Unlike self-report methods, assessing performance changes offers objective insights into the system's functional effectiveness, which may not align with subjective measurements. Combining self-report and objective usability measures becomes increasingly important when the target population is older and the technological barriers are more prominent.

3.2 Opportunities and Challenges for Use in Aging Populations

A recent report found a significant increase in smartphone and tablet ownership among individuals aged 55–64 and 65 and older, from 2012 to 2021. For smartphones, ownership among those aged 55–64 rose from 34% to 83%, while for individuals aged 65 and older, it increased from 13% to 61%. Similarly, tablet ownership among those aged 55–64 grew from 10% to 53%, and for the 65 and older age group, it increased from 6% to 44% (Faverio 2022). Notably, 45% of older adults also report playing games at least once a month (Clement 2023; Kakulla 2023). This trend has been accompanied by a proliferation of mHealth apps, with approximately 350,000 apps currently in the market (Mathews et al. 2019). By providing greater accessibility with remote assessment and minimizing the need for in-person clinical consultations, mHealth apps are revolutionizing clinical care. These apps simplify repeated testing as well as the tracking and sharing of data over time (Amagai et al. 2022; Czaja et al. 2013), while enabling users to take an active role in managing their health (Czaja 2019). Many mHealth apps rely on gamification and incorporate game elements (e.g., storytelling, scaffolding, rewards) to provide more immersive and engaging interactions that can lead to behavioral changes and treatment adherence (Alahäivälä and Oinas-Kukkonen 2016; Amagai et al. 2022). In gamified assessments, users typically receive reminders to complete a task or "compete" against others to achieve a goal that promotes a healthier lifestyle.

Despite these opportunities, designing digital tools for the clinical assessment of older adults remains challenging because of cognitive (e.g., reduced attention and memory) and physical limitations (e.g., reduced motor skills). These factors can hinder the ability of users to interact and comprehend the goals of the app and may lead to technology anxiety and disengagement (Czaja 2019; Gamberini et al. 2006; Martinho et al. 2020). Effective UX design for this demographic should leverage the flexibility inherent to mHealth apps. Here, developers need to maximize the visual appeal while providing adaptable interfaces (e.g., controls, text size) and meaningful challenges alongside features like in-app training and clear instructions (Barnard et al. 2013; Czaja et al. 2013; Flores et al. 2008; Ijsselsteijn et al. 2007). Indeed, most UX researchers agree that the golden rule of age-friendly design is to prioritize clarity and simplicity. Here, UX researchers suggest that mHealth apps need to provide training accompanied by step-by-step intuitive instructions, use familiar icons and large fonts, favor control tapping over complex gestures, and extend screen and response times when providing or requesting feedback (Gomez-Hernandez et al. 2023).

4 SPACE

SPACE is a novel serious game for the iPad designed to detect navigation deficits indicative of early signs of cognitive impairment. SPACE was developed for deployment as a supervised assessment in clinical settings for middle-aged users. In SPACE, players take on the role of an astronaut sent on a mission to uncover the potential of a foreign planet to harbor life. The game begins with an extensive training phase that allows players to familiarize themselves with the control interface while assessing basic visuospatial skills. SPACE includes five spatial tasks designed to detect deficits in egocentric and allocentric learning (Table 15.1). Players complete *visuospatial training*, *path integration*, and *pointing* tasks from a first-person perspective that simulates the astronaut's viewpoint from the helmet through a heads-up display (HUD). The HUD also provides players with essential information to complete the tasks, including pop-up instructions, screen tips, destination icons, an inventory of items collected in the virtual environment (VE), and a progress bar that indicates the percentage of completed trials. Players complete the *mapping*, *associative memory*, and *perspective-taking* tasks via a top-down two-dimensional representation of a panel within the rocket's cockpit. Throughout the game, players are guided by a companion robot known as L15A, which provides just-in-time instructions and feedback through facial expressions and reward badges. In SPACE, the pointing, mapping, and associative memory tasks depend on knowledge acquired during the path-integration task. Perspective taking is the only stand-alone task since participants are provided with a top-down map of the environment.

Table 15.1 Overview of the tasks included in SPACE.

Visuospatial training	Players learn to rotate, translate, and integrate these movements by following the robot in a circuit around the planet. During training, players are also introduced to the logic of the first task (i.e., path integration) to minimize learning effects during testing.
Path integration	Players start their journey at the rocket and are asked to follow the robot to two distinct landmarks in the environment before returning unguided back to the rocket. During the path-integration task, players are explicitly asked to remember the position of the landmarks in the environment. At each landmark, the robot scans an item that players will be asked to recall in the associative memory task.
Pointing	Players test their memory for the locations encountered during the path-integration task by pointing to and from the different landmarks.
Mapping	Players test their configurational knowledge of the environment by dragging and dropping icons representing the locations they learned during the path-integration task.
Associative memory	Players are asked to pair the items scanned by the robot with the corresponding landmarks they encountered during the path-integration task.
Perspective taking	Players are asked to imagine standing at a landmark facing another landmark by looking at the corrected map of the environment. Next, they are asked to indicate the correct bearing toward a third landmark.

4.1 Preliminary Findings

Over the last two years, we evaluated the effectiveness of SPACE in detecting cognitive impairment across three cohorts of younger and older participants in Singapore and established normative benchmarks (Figure 15.1a) for healthy individuals for all tasks in SPACE (Colombo et al. 2024b). Preliminary results reveal that performance in the various tasks in SPACE declines with age and correlates with better MoCA scores (Figure 15.1b). In addition, performance in the path-integration and mapping tasks significantly predicted atrophy in the left and right hippocampus in a group of older participants after controlling for age and education (Figure 15.1c, d). In a short version of SPACE, which included only the visuospatial training and perspective-taking tasks, we also found that SPACE can differentiate between "no cognitive impairment," "cognitive impairment no dementia," and dementia patients diagnosed by a full neuropsychological assessment (Figure 15.1e). In addition, we found that a subsample of participants who were carriers of the ε4 allele of the *APOE* gene and had a genetic predisposition to dementia committed larger errors in the path-integration task than homozygous carriers of the ε3 allele (Figure 15.1f). Finally, the results of a training intervention revealed that SPACE may be effective in improving navigation skills and sustaining cognitive function. Here, participants who played SPACE twice a week for six months showed superior performance in all SPACE tasks compared to a control group (Figure 15.1g).

4.2 Applying Principles from Human-Computer Interactions to the Design of SPACE

Designing effective tools to assess navigational deficits requires considering the needs and abilities of the target population alongside the requirements and constraints of deploying these tests in clinical settings. With respect to the first point, researchers must differentiate between ability and performance (Schinazi et al. 2016). Ability refers to an individual's inherent capacity to navigate, which may be influenced by cognitive factors such as spatial memory, strategy use, attention, and executive functioning. In contrast, performance encompasses the actual execution of navigation tasks in the assessment, which can be affected by external factors such as environmental conditions (e.g., lighting, noise), familiarity with the technology, and the tool's usability. A successful clinical assessment will be time-efficient, easy to administer, reliable, valid, culturally appropriate, and compliant with patient confidentiality and regulatory standards. Successful tools are the ones capable of closing the gap between ability and performance. As noted by the Nielsen Norman UX Group, the best designs are built for people as they really are and not who we wish them to be (Kohler 2024).

The creation of the *design document* is the most important step when developing a new tool and should include an in-depth description of every aspect of the product (e.g., story, goals, development phases, budget, graphics, music, sound effects) as well as the usability needs of the target population. Using illustrations and videos will assist in removing potential ambiguities and help protect against misunderstandings and misinterpretations from collaborators and contractors. A meticulously

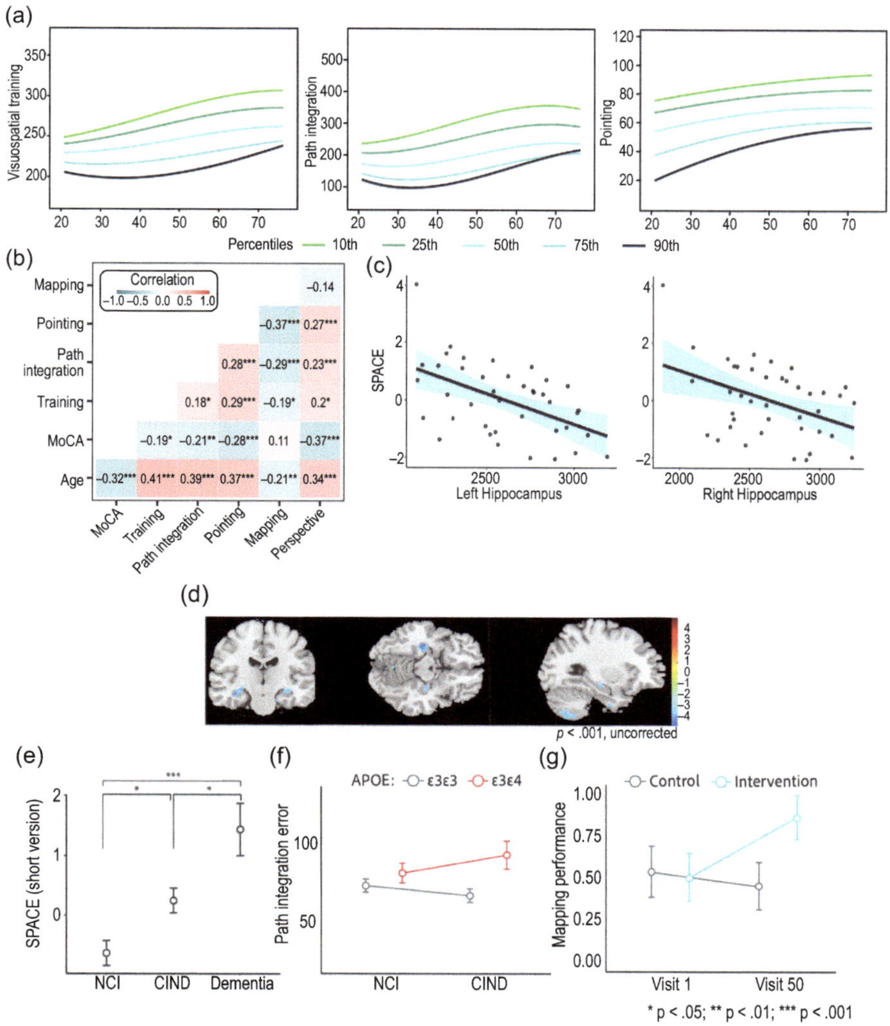

Figure 15.1 Preliminary results from SPACE. (a) Normative data for the visuospatial training, path-integration, and pointing tasks. (b) SPACE task correlations with age and MoCA. (c) Lower performance in SPACE correlates with hippocampal atrophy after correcting for age and education. (d) Voxel-Based Morphometry (VBM) results show a negative correlation between gray matter volume in the left and right hippocampus and path-integration error. (e) Performance in SPACE can discriminate between "no cognitive impairment" (NCI), "cognitive impairment no dementia" (CIND), and dementia patients. (f) Carriers of the *APOE* ε3ε4 genotype committed larger errors in the path-integration task compared to ε3ε3 carriers. (g) Participants who completed a six-month training intervention outperformed a control group in the mapping task in SPACE.

crafted design document will also ensure that usability remains a priority throughout the development process and provide a structured framework for translating high-level design concepts into actionable, user-centered features (Figure 15.2).

In preparation for our SPACE usability and cohort studies, we conducted a focus group session with four older adults and individual meetings with different stakeholders (i.e., target user groups as well as clinical and academic partners) to collect

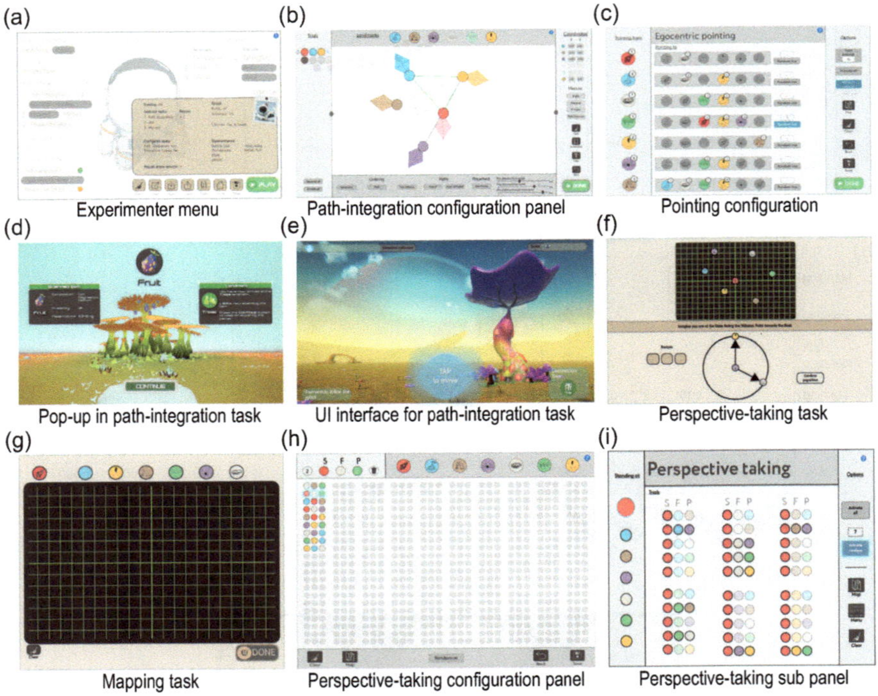

Figure 15.2 Images from the SPACE design document depicting mock-ups of different game features as references for developers.

feedback about additional features to implement during the development. Below, we trace the design and development of SPACE from the initial focus group session and individual meetings with stakeholders to the large-scale usability study before its deployment in research labs and clinics in Singapore and Australia. We also share key findings and potential pitfalls of this process. Table 15.2 summarizes how we implemented the key aspects of the *Design Guidelines of Mobile Applications for Older Adults* proposed by Gomez-Hernandez et al. (2023) during the development of SPACE.

Our focus group and individual stakeholder meetings yielded a series of critical feedback regarding the need to simplify the control interface and familiarize players with the tablet gestures before the assessment. Several players complained that they struggled with the *Tap & Swipe* and *Joystick* options for locomoting around the VE. To address this difficulty, we created a new semi-automated control interface known as the *Anchor* control, which allows players to trigger their movement in the VE by simply tapping a button on the screen. Critically, the *Anchor* control splits the players' response into smaller, manageable steps (rotations or translations) and reduces the challenges associated with precise finger movements when interacting with the screen. To familiarize players with the control interface, we created a detailed training phase to help players learn to rotate, translate, and integrate these two gestures when locomoting in the VE. At the end of the training, players also learn the steps to complete the subsequent path-integration task. Given the gestures and interactions required to complete the training, we also used this phase as a visuospatial assessment to help discriminate between levels of cognitive impairment.

Table 15.2 Usability guidelines and implementation in SPACE.

Usability Guidelines	Implementation in SPACE
Flexibility:	
System: The system should be sufficiently flexible to accommodate a wide range of users and skill levels.	*Panels*: The *Experimenter Menu* and the three configuration panels in SPACE allow users to flexibly design the tasks by adjusting the number of trials and difficulty levels to match different abilities and contexts.
Interface: The tool should provide different ways to interact to suit different user types and preferences.	*Controls*: SPACE can be played with *Tap & Swipe*, *Joystick*, and *Anchor* control interfaces. The semi-automated *Anchor* control is tailored for older adults (who may find touch screen tapping and swiping challenging) by breaking actions into more manageable steps.
Graphics: Graphical elements should capture attention with clear, actionable information.	*User Interface (UI)*: In-game instructions and critical prompts appear with zoom effects to ensure visibility and engagement.
Appeal:	
Design: The design should be visually appealing, intuitive, and accessible to a broad audience; it should also enhance the UX and consider cultural and contextual relevance.	*3D models*: Despite the fantastical setting of an alien planet with unique landmarks, the graphics in SPACE were refined through multiple iterations based on players' feedback to remain relatable and familiar, even for older players. *UI design*: The HUD mimics the helmet of an astronaut. Instructions scroll from the top, and destination icons pop into view for greater emphasis and to capture the attention of players. *Music and sound effects*: Each task has its own background music, specifically composed to fit the cultural context of the original location of deployment (Singapore). Sound effects provide real-time feedback to enrich the overall gaming experience.
Motivation: Users should be motivated with rewards and meaningful challenges to stimulate curiosity and encourage retention.	*Badges*: Performance badges reward players to sustain motivation. *Companion robot*: The robot provides real-time reactions and motivational feedback to provide constant encouragement to players. *Challenges*: The tasks in SPACE are intentionally designed to be challenging and ensure that players remain engaged and motivated throughout the experience.
Assistance:	
Training: Users should be adequately trained on controls, game mechanics, and objectives.	*Visuospatial training*: SPACE provides an extensive training phase to gradually familiarize players with game controls and required gestures.
Instructions: Users should be provided with clear and consistent instructions to ensure they understand the tasks.	*Video instructions*: Before each task, instructional videos or animations demonstrate the goals of the task and actions to improve clarity. *Step-by-step instructions*: In-game instructions are zoomed in and guide players through each step of the task to increase saliency and avoid confusion.

Stakeholders also highlighted the need to provide feedback on performance and progress, refine the way instructions are delivered, reduce the information displayed in the pop-ups during gameplay, enhance the general clarity and appeal of the game, and decrease the overall complexity and time of the assessment. We translated SPACE into 11 languages so that players could play it in their mother tongue.

Table 15.2 (continued)

Usability Guidelines	Implementation in SPACE
Accessibility:	
Screen time: Response times, feedback durations, and time-outs should be extended to accommodate slower reactions.	*UI feedback*: Pop-ups, in-game instructions, and feedback remain visible until players acknowledge them by pressing a button to accommodate slower response times.
UI elements: Key elements should be visible or highlighted as needed to avoid screen clutter.	*Icon design*: Large, clear icons are used to accommodate players with reduced visual acuity. Critical icons and instructions zoom or jiggle to capture attention when necessary. *Simplified panels*: Pop-ups are simplified to highlight key information.

To provide feedback on performance, we rewarded players with badges after each trial and provided a loading screen detailing the performance of every task. To keep players informed about their progress, we developed in-game progress bars to show the percentage of trials completed during the task. Additionally, we introduced a transition screen to mark the end of the training phase and the beginning of the spatial tasks. To improve the delivery of instructions and ensure that all players understood the requirements of the tasks, we created video instructions and user interface (UI) animations to complement the text instructions and demonstrate the actions required to complete each task. To provide players with directional information, we created a UI rotation widget on the right side of the HUD that fills up in proportion to the player's rotation angle and introduced a callout balloon that appears on either side of the screen when the robot is out of sight (Figure 15.3).

We also made a series of enhancements to existing features. For example, to reduce information clutter and improve the visibility of critical elements, we simplified in-game pop-ups to display only critical content in a larger font, increased the size of game icons, and added a zoom effect for pop-up instructions and certain UI game elements. We also changed the robot's color to create a more significant contrast with the background of the planet and implemented a shader that highlights the contour of the robot when the player is at a distance (Figure 15.4). To further enhance the overall appeal, we hired a professional graphic designer to refine the 3D graphics, added various sound effects, and engaged a composer to create a custom soundtrack for the game. Finally, to simplify the assessment and shorten testing times, we created a configuration with fewer landmarks in the VE and reduced the number of trials participants were required to complete for each task.

After adapting SPACE to the feedback from the focus group and individual stakeholder meetings, we conducted a rolling usability study with 261 young, middle-aged, and older adults to evaluate what they deemed to be the three most important changes. Specifically, we examined whether performance, usability ratings, cognitive load, and presence improved after implementing the *Anchor* control interface, the rotation widget, and a simplified configuration of landmarks in the VE. The results of the usability study are reported in detail by Colombo et al. (2024a). Here, we discuss the most important lessons.

Despite the additional features and refinements, age remained the most important predictor of performance, usability, cognitive load, and presence. However, our

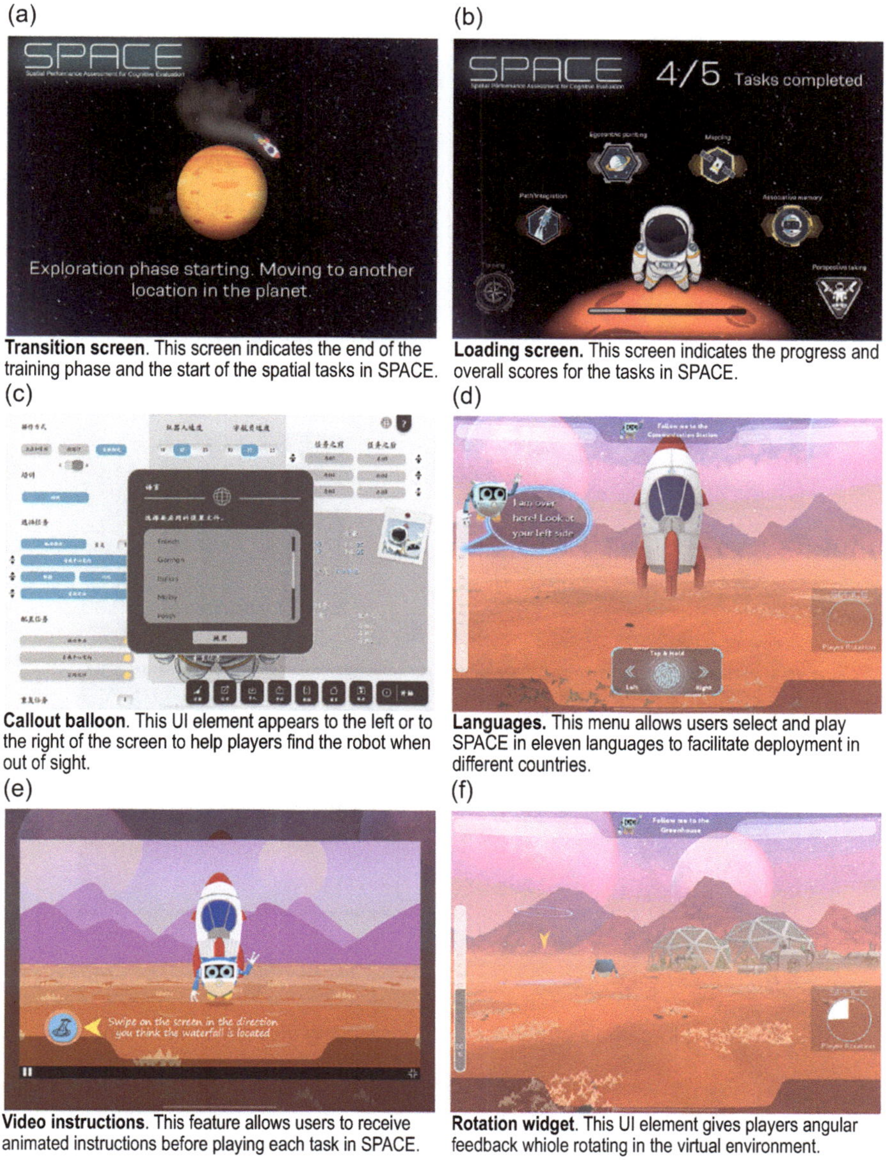

(a) **Transition screen**. This screen indicates the end of the training phase and the start of the spatial tasks in SPACE.

(b) **Loading screen.** This screen indicates the progress and overall scores for the tasks in SPACE.

(c) **Callout balloon**. This UI element appears to the left or to the right of the screen to help players find the robot when out of sight.

(d) **Languages.** This menu allows users select and play SPACE in eleven languages to facilitate deployment in different countries.

(e) **Video instructions**. This feature allows users to receive animated instructions before playing each task in SPACE.

(f) **Rotation widget**. This UI element gives players angular feedback whiole rotating in the virtual environment.

Figure 15.3 Screenshots of new features added to SPACE.

findings also revealed that the benefits of some usability changes are age specific and do not uniformly benefit all age groups (Figure 15.5). For example, middle-aged and older adults responded positively to the introduction of the *Anchor* control interface and the simplification of the landmark configuration, whereas younger participants benefited from the freedom to maneuver in the VE afforded by the *Tap & Swipe* control interface. Similarly, we found that middle-aged and older adults performed better and had lower cognitive load with the simplified configuration of landmarks, but this change did not particularly affect the performance of younger participants for all the tasks. Notably, our results revealed substantial individual

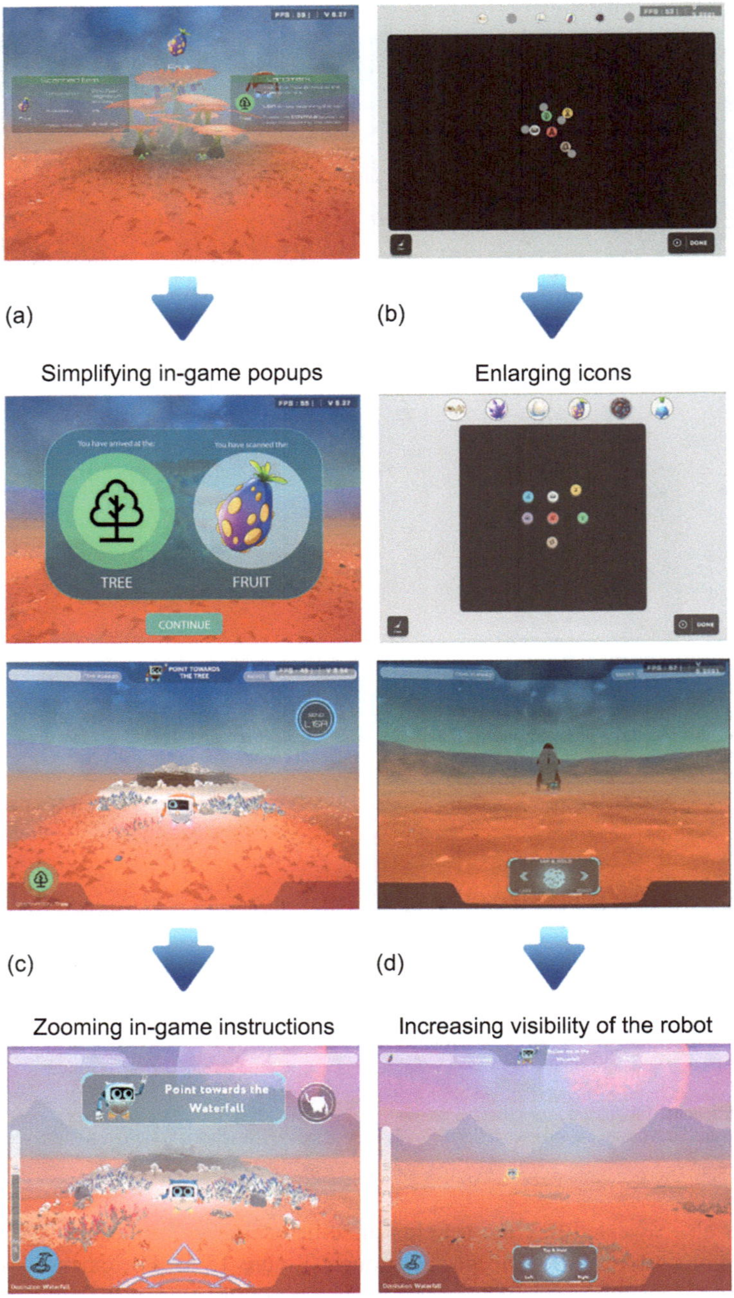

Figure 15.4 Enhancement of various features in SPACE.

differences in performance across participants of all ages, suggesting that the simplified configuration maintained an appropriate level of difficulty for picking up differences in cognitive status. In contrast, the performance of younger participants benefited more from the feedback provided by the rotation widget than older adults.

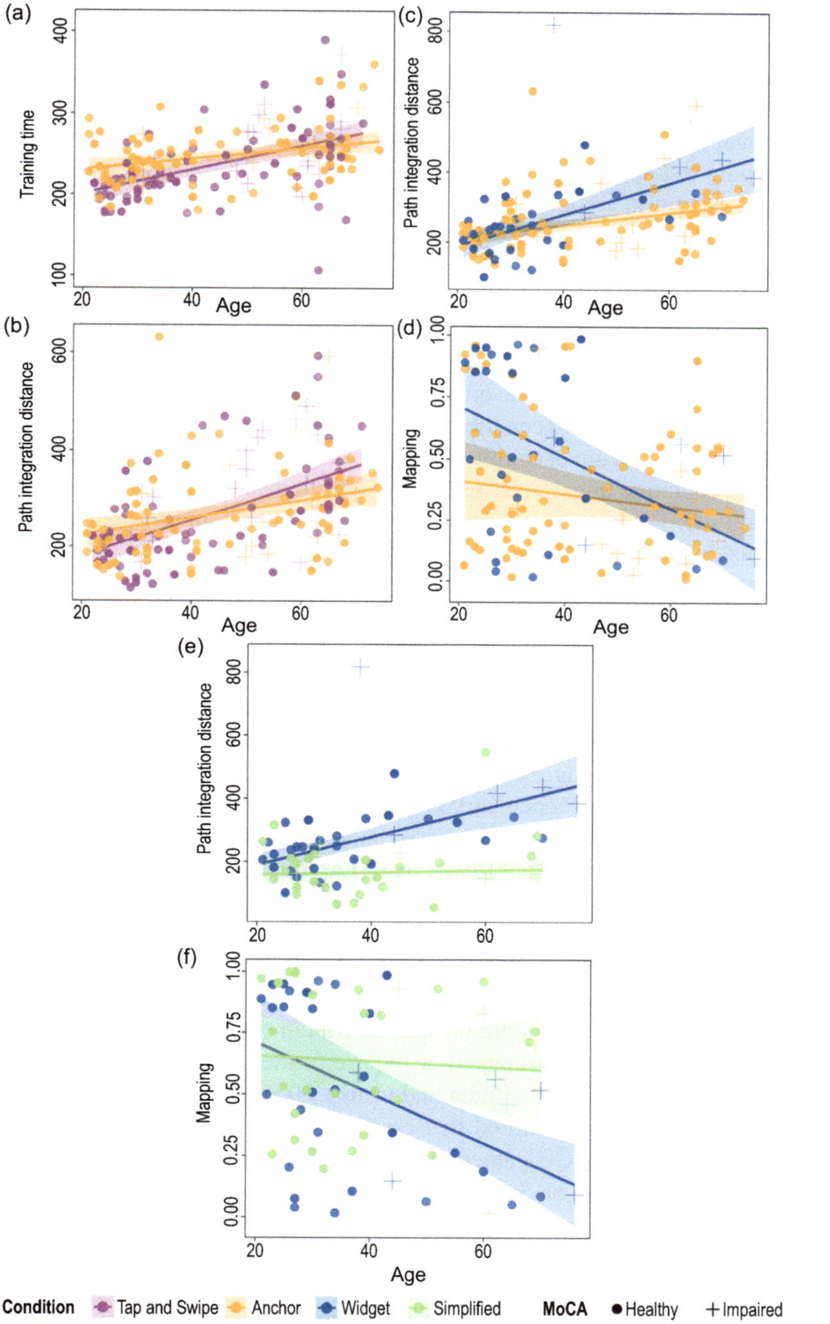

Figure 15.5 Usability changes are age specific. (a) and (b): Compared to the *Tap & Swipe*, middle-aged and older adults benefited more from *Anchor* control than young adults in the training and path-integration tasks. (c) and (d): The introduction of the rotation widget benefited younger adults more than older adults in the path-integration and mapping tasks. (e) and (f): A simplified configuration of landmarks benefited more middle-aged and older adults than young adults. Circles represent healthy participants (MoCA > 25), whereas crosses represent cognitively impaired participants (MoCA ≤ 25).

Here, the decline in performance accompanied by increased cognitive load among older adults suggests that the rotation widget may have hindered their ability to complete the navigation task. Interestingly, the feedback provided by the rotation widget was particularly beneficial for middle-aged participants in developing configurational knowledge of the VE, as evidenced by their improved performance in the mapping task.

It is important to recognize that performance improvements do not always correspond to higher usability ratings. In our usability study, participants did not report higher scores in the SUS, UEQ, and Presence questionnaires, even though their performance in the spatial tasks generally improved with each newly implemented feature. Thus, to fully capture UX, researchers should aim to blend objective and subjective measures. However, the type and extent of this triangulation will depend on the research question and the quality of the data collected. By analyzing discrepancies and alignments between these measures, researchers can better identify specific areas for improving usability. Moreover, creating accessible serious games for measuring cognitive impairment requires a delicate balance between simplification and sensitivity. The best way to achieve this balance is through systematic usability testing, as this allows researchers to iteratively refine task complexity while ensuring the measures remain engaging and are diagnostically meaningful. Some tasks in SPACE are inherently difficult because they allow subtle deficits in navigation abilities to be measured. The demanding nature of such tasks, however, often resulted in lower usability ratings since participants became frustrated when they repeatedly underperformed. When benchmarked to the UEQ, SPACE was perceived to be above the level of mature products for commercial purposes, although older adults did not find it especially engaging. Scores in the Presence and NASA-TLX questionnaires were also slightly below benchmarked values and did not improve with the various feature iterations. Notwithstanding these challenges, the Presence and NASA-TLX scores were within expectations, given that SPACE is a tablet-based game designed primarily as a diagnostic tool rather than a fully immersive experience.

In our usability study, we also collected feedback on what participants liked and disliked about each task in SPACE as well as their overall experience in playing the game. Qualitative data is essential to usability studies because it can provide detailed insights into UX and highlight specific design elements and game mechanics that were successfully implemented or require further adjustments. We conducted a reflexive thematic analysis (Braun and Clarke 2006) of qualitative data in which six major themes were identified that captured distinct attributes experienced by participants during SPACE gameplay. Three researchers independently analyzed the data, and a fourth researcher oversaw the process to ensure rigor. Themes were derived from overlapping codes identified by the individual researchers, with the final refinement and naming of the themes completed collaboratively. The results of the thematic analysis reinforced the idea that simplicity is key. Middle-aged and older participants found the SPACE setup featuring the *Anchor* control, rotation widget, and simplified configuration of landmarks to be the most enjoyable, intuitive, and stimulating while also being the less demanding, repetitive, and frustrating version of the game. Our qualitative analysis also highlighted some important discordances in feedback from younger and older adults. For example, younger adults found the *Tap & Swipe* control interface more enjoyable and the rotation widget less demanding than middle-aged and older adults. This discrepancy in results suggests that there is no *one-size-fits-all* solution and that developers should aim to design tools that are modular and sufficiently flexible for testing various age groups.

4.3 Modularity

SPACE was designed from the outset to be modular to meet the diverse requirements of researchers and clinicians along with the needs and intrinsic abilities of the end users. This flexibility in customization is achieved through the *Experimenter Menu*, which serves as a user-friendly central hub for configuring and deploying the assessment. In the *Experimenter Menu*, users can select the language and type of control interface, adjust the speed that the astronaut and the robot move, set the volume for the music and sound effects, and select which pre- and post-game questionnaires to include (e.g., demographics, SUS, UEQ) in the assessment. Users can also choose the spatial tasks (including visuospatial training) that will be part of the assessment and set the task order and number of session repetitions. The *Experimenter Menu* required clinicians to use just two button presses to deploy the game. As such, it was only informally assessed by our four clinical partners, who found it to be highly user-friendly and encountered no difficulties in its use.

In SPACE, participants always complete the path-integration task first because the information acquired about the location of the landmarks in the VE is fundamental for the majority of the subsequent tasks. However, we incorporated flexibility for usability testing and future assessments by developing separate configuration panels that allow users to adjust various characteristics of each task. For example, the configuration panel for the path-integration task allows users to select and position the landmarks that will make up the VE and includes a logic system that verifies whether the selected trials can be executed within the VE. Here, users receive real-time feedback on the angular and distance relationships between landmarks, enabling them to change the difficulty level of each trial and configuration. Additional features in all configuration panels provide innovative and efficient methods for rapidly designing and executing experiments via automated trial generation, trial randomization, and repetition. The *Experimenter Menu* was designed for research and clinical efficiency and includes a series of shortcut buttons for loading, importing, and saving configuration files in JavaScript Object Notation (JSON) format. To administer SPACE, users only need to load a predefined configuration file and hand the tablet to the participant or patient. Alternatively, users can deploy the default configuration preloaded in the splash screen. The *Data* button allows users to export the collected data to other devices or applications, further streamlining the analysis and creating a trusted platform to store and share data. Currently, all data is stored on the iPad and secured via the user's password. We are working on a pipeline in which de-identified JSON files are securely sent to our server in Singapore after participants complete the assessment. This ease of deployment ensures that assessments are conducted consistently and reliably across diverse clinical and research environments while promoting standardization and enhancing the overall quality of the collected data (Figure 15.6).

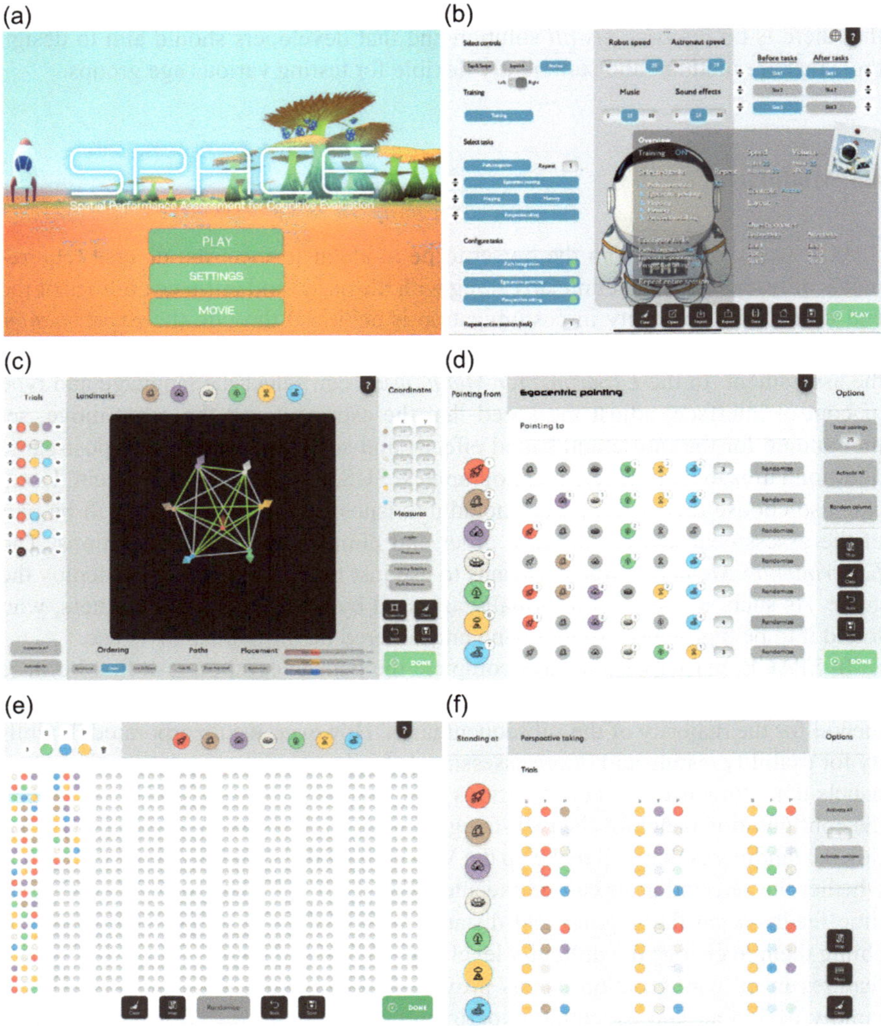

Figure 15.6 Screenshots of the different panels in SPACE: (a) splash screen, (b) experimenter menu, (c) path integration configuration panel, (d) pointing configuration panel, (e) perspective taking configuration panel, (f) perpective taking subpanel

4.4 Challenges with Construct Validity

When navigation deficits are used as indicators of cognitive status, one challenge that researchers face is that despite usability enhancements, some individuals may be poor navigators, even without cognitive impairment. While extensive usability testing and incremental UX improvements may guarantee higher reliability, the effort is futile if the tasks fail to reflect the deficits they are designed to assess accurately. Indeed, construct validity is a frequently neglected but fundamental element in usability evaluations for clinical tools.

Currently, we are implementing four different strategies to ensure construct validity in SPACE:

1. Performance is validated by comparing it against the clinical diagnoses reached by health-care professionals and common, standardized neuropsychological assessments (e.g., MoCA, Trail Making Test, Digit Cancellation Test) to identify healthy and impaired individuals. This approach does not entirely eliminate the need to conduct real-world studies (Coutrot et al. 2019; Goodroe et al. 2025) to ensure ecological validity but rather prioritizes the tool's ability to distinguish between individuals based on their cognitive profiles.
2. To establish normative benchmarks for healthy individuals with cognitive impairments, we are compiling a large amount of data from participants in Singapore and Australia with a wide age range (21–92 yr). Normative data provides a baseline to differentiate between normal navigation performance, age-related decline, and pathological impairment.
3. Longitudinal assessments are being conducted to track navigation deficits over time to further validate the sensitivity of the tasks in SPACE for identifying those at risk of developing cognitive impairment. This will allow us to determine whether SPACE can outperform standard clinical assessments in detecting early signs of cognitive impairment and predicting long-term outcomes.
4. We have established a clinical partnership to triangulate the behavioral results from SPACE and neuropsychological assessments with imaging (i.e., MRI) and fluid (e.g., saliva and plasma) biomarkers of neurodegeneration (see Figure 15.1).

Altogether, establishing construct validity for novel clinical tools can minimize reliance on extensive and expensive diagnostic procedures and lower health-care costs. However, to fully understand the economic impact of the tool, researchers should include a cost-effectiveness analysis as part of their usability assessment.

5 Assessing the Economic Impact of Digital Health Tools

Developing and implementing new digital clinical tools requires evaluating their costs and benefits against existing practices to ensure they are sustainable and provide value for money. While all digital tools will involve initial irrecoverable costs, developers should take care not to underestimate these expenses, as doing so can hinder the tool's ability to achieve the necessary standards for effectiveness, scalability, and usability required for widespread deployment. As such, project leads need to consider development (e.g., research and design, software development, quality assurance, licensing), implementation (e.g., marketing, training, distribution), and operational (e.g., maintenance, updates and customer support) costs.

Calculating the cost-effectiveness of a new digital tool relative to a standard assessment begins with assigning the probability of a person undergoing any assessment and estimating the prevalence of cognitive impairment, the probability of undergoing testing using the new digital tool, the probability of undergoing the standard cognitive testing, the sensitivity of the new digital tool and the standard cognitive testing, and the specificity of the new digital tool and the standard cognitive testing. Researchers also need to include their estimation of caregiving costs, costs related to procuring and using the new digital tool, treatment costs once the pathology is detected, costs of undergoing standard cognitive testing in the absence of pathology, and the effectiveness of treatment after cognitive impairment is detected. Finally, researchers need to consider the number of years a person lives with dementia once diagnosed (Figure 15.7).

Figure 15.8 presents a sensitivity analysis showing the net monetary benefit (NMB) of the new digital tool relative to the standard method for detecting cognitive impairment. Although our goal was to maintain plausible estimates in the context of Singapore for the various parameters in the decision tree (Azman et al. 2024; Chay et al. 2024; Chong et al. 2013; Liu et al. 2022), we intentionally assumed that the new digital tool would have a broader reach and a higher sensitivity to discriminate cognitive impairment relative to standard cognitive testing. In general, as the prevalence of cognitive impairment increases, the NMB decreases for both strategies. However, the new digital tool consistently offers a higher NMB for populations where the prevalence of cognitive impairment is modestly high. Indeed, the steeper slope of the standard testing indicates that the cost-effectiveness of this strategy diminishes more rapidly as the prevalence of cognitive impairment increases in the population. Overall, this analysis demonstrates that the new digital tool offers a more sustainable solution for detecting cognitive impairment in populations with rising prevalence.

6 Discussion

In the treatment of Alzheimer disease, the efficacy of drugs like Aducanumab (Aduhelm) and Lecanemab (Leqembi) remains a topic of considerable debate among experts (Shi et al. 2022; Tampi et al. 2021). Although the US Food and Drug Administration has fast-tracked approval of these drugs, and studies have shown their potential to reduce amyloid plaques in the brain (Niidome et al. 2024; van Dyck et al. 2023), there are growing doubts regarding their effectiveness. Indeed, recent reviews suggest that these drugs carry substantial side effects and, in the case of Leqembi, may not hold any benefits in slowing cognitive decline in women and *APOE* ε4 carriers (Kepp et al. 2023; Kurkinen 2023). Regardless whether pharmacological treatments prove to be successful, early detection of cognitive impairment is essential, as research indicates that pharmacological, cognitive, and lifestyle treatments tend to be more effective when initiated during the early stages of cognitive decline (Huang et al. 2020; van Dyck et al. 2023). Digital tools for detecting preclinical stages of cognitive impairment may help bridge this gap by facilitating early assessment. Indeed, as blood biomarkers are poised to revolutionize the diagnosis of Alzheimer disease (Palmqvist et al. 2024), digital tools may play a critical role as a noninvasive step prior to plasma collection to support clinical decision making. However, their potential may be only fully realized if they are effectively integrated into clinical workflows.

Researchers who have made their careers investigating navigation ability now face the formidable task of using their expertise to explore the link between navigation deficits and cognitive impairment. This challenge introduces a new type of responsibility and requires them to ensure that their research transcends theoretical boundaries and translates into clinical practice. While the surge in funding opportunities in this domain is promising, it also demands that researchers carefully balance the procurement of financial support with their ethical duty to produce clinically relevant and transformative research. Multidisciplinarity in academia is often championed as the ideal approach to tackle complex problems. Integrating digital tools into clinical practice will necessitate a concerted effort that draws expertise from psychologists, neuroscientists, clinicians, computer scientists, and designers (Alexander and Staggers 2009). UX researchers and designers will play a pivotal role in these collaborations, as even the most advanced knowledge and technologies will remain ineffectual if not properly tailored into user-friendly, accessible, and scalable solutions for clinicians and patients.

Despite its recognized importance, usability testing often comes as an afterthought due to constraints on time and budget (Olsen et al. 2011). In academia, this challenge is further compounded by the limited size and availability of the design teams (e.g., development and UX designers). This means that such work frequently needs to be outsourced. In addition, design teams may have excellent computer scientists but lack expertise in spatial cognition and navigation and thus will likely need support in understanding the requirements of discipline-specific tasks. Introducing new technologies must also contend with resistance from healthcare professionals (Safi et al. 2018) and patients, who may perceive the new tools as disruptive to existing clinical workflows or daily life. To counteract this and build trust, stakeholders should be involved during the development process to discuss the practical and health benefits of the new technology. Here, it will be critical to demonstrate that the new technology is efficient (Alexander and Staggers 2009) and reliable, requires minimal training, prioritizes data security and privacy, and provides clear feedback to professionals and patients. In the end, a successful tool will be one that can balance cost-effectiveness, usability, and clinical significance. Failure to address all these elements may result in the new technology becoming another addition to an already saturated digital health market.

In conclusion, we used SPACE as a case study to demonstrate how principles from human-computer interactions can be effectively applied to the design of new digital health technologies. As future technologies are developed, it is critical to involve stakeholders early in the design process and to emphasize the essential role of iterative usability testing for refining UX and clinical workflows. Relying solely on longitudinal data or biomarkers and imaging may limit the validation scope of novel digital technologies. Critically, robust validation for tools designed to support clinical decision-making should show a clear association with longitudinal decline and neurobiological indicators. Here, longitudinal associations with cognitive decline can demonstrate the ability of a tool to track changes over time and identify early signs of impairment, while associations with clinical and imaging biomarkers can help establish the tool's validity in capturing underlying pathological processes. Finally, we discuss the importance of considering the overall cost-effectiveness of the project for the large-scale integration of the product. SPACE is far from being a perfect tool. Over the last few years, we have identified ways in which some spatial tasks can be improved, combined, or removed to create a better experience and a

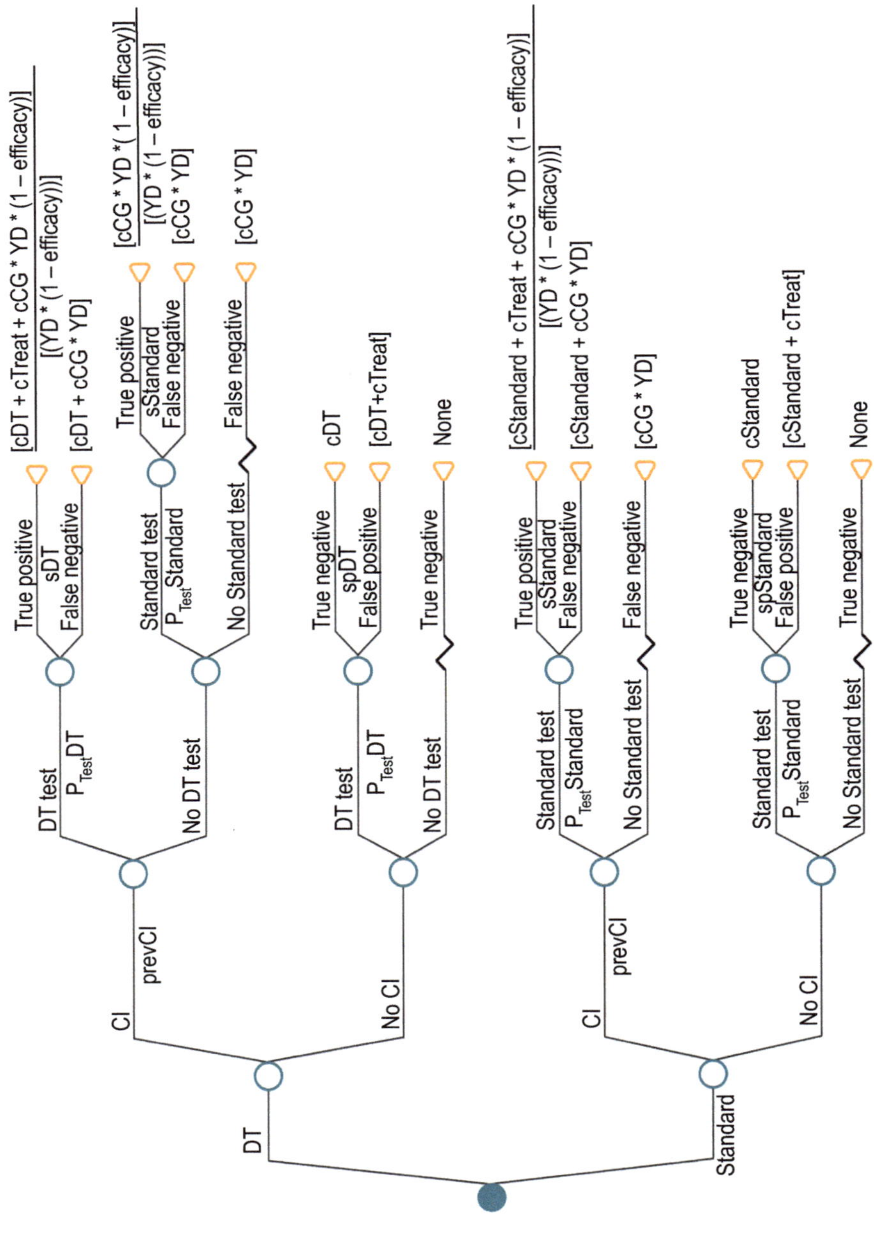
DT
Standard
CI
prevCI
No CI
DT test
$P_{Test}DT$
No DT test
Standard test
$P_{Test}Standard$
No Standard test
True positive
sDT
False negative
True negative
spDT
False positive
sStandard
spStandard
None
cDT
[cDT+cTreat]
[cDT + cTreat + cCG * YD * (1 – efficacy)]
[(YD * (1 – efficacy))]
[cDT + cCG * YD]
[cCG * YD *(1 – efficacy)]
[cCG * YD]
[cStandard + cTreat + cCG * YD * (1 – efficacy)]
[cStandard + cCG * YD]
cStandard
[cStandard + cTreat]

Key

cCG = 10,000; cDT=10; cTreat = 3,000; cStandard = .5; efficacy = .8; prevCI = .1; P_{Test} = .5; P_{Test}DT = .9; P_{Test}Standard = .2; sDT = .9; spDT = .7; sStandard = .6; spStandard = .7; YD = 5

Abbreviations: CCG = costs caregiving; cDT = costs digital tool; cTreat = costs treatment; cStandard = costs standard testing; efficacy = successful rate of treatment; prevCI = prevalence cognitive impairment; p_{Test} = probability undergoing the digital tool or the standard testing; p_{Test}DT = probability testing with the digital tool; p_{test}Standard = probability of testing with the standard testing; sDT = sensitivity digital tool; spDT = specificity digital tool; sStandard = sensitivity standard testing spStandard = specificity standard testing; YD = years of dementia

Figure 15.7 Cost-effective analysis of comparing a new digital tool (DT) against the standard cognitive testing (Standard) to detect cognitive impairment. For the new DT, the tree branches out to consider the prevalence of cognitive impairment (prevCI). For individuals with cognitive impairment, the tree branches should consider whether testing with the new DT (p_{test}DT) is performed or not. If testing with the new DT is conducted, the outcomes are a true positive or a false negative. If testing with the new DT is not conducted, individuals can undergo standard cognitive testing, resulting in a true positive or false negative. Alternatively, individuals may not undergo standard testing, leading to a false negative. Similarly, for individuals without cognitive impairment, the tree branches to consider whether testing with the new DT (p_{test}DT) is performed or not. If testing with the new DT is conducted, the outcomes are a true negative and a false positive. If testing with the new DT is not conducted, the outcome is a true negative. An equivalent branching structure applies if individuals undergo the standard cognitive testing. Each of these outcomes is associated with a different cost calculation.

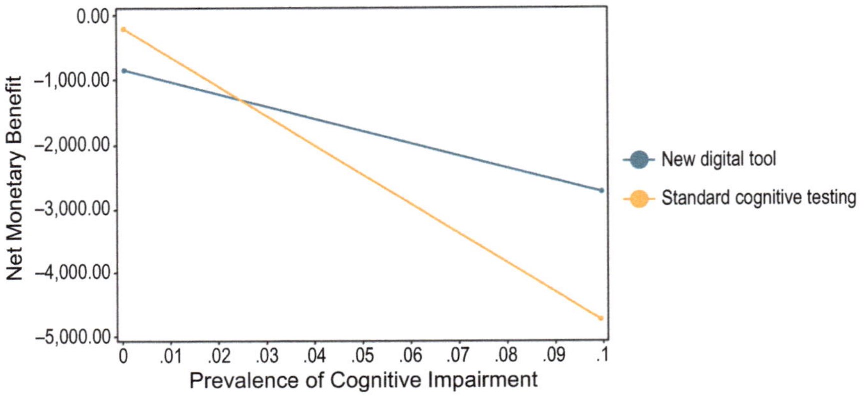

Figure 15.8 A mock sensitivity analysis comparing a new digital tool to the standard cognitive testing method. The x-axis presents the prevalence of cognitive impairment in the population. The y-axis shows the Net Monetary Benefit (NMB), with higher values indicating greater cost-effectiveness. The analysis suggests that as the prevalence of cognitive impairment increases, the new digital tool remains more cost-effective than the standard cognitive testing method.

more reliable assessment. However, the extensive usability process we engaged in allowed us to understand where SPACE stands in terms of accessibility, scalability, and meeting the demands of clinical practice.

Acknowledgments We extend our heartfelt thanks to Eunice Tai Wei Lin, Martina Neudecker, Simona Margraf, and Sophia Niklaus for their invaluable assistance with data collection at the Singapore-ETH Centre. Our deepest gratitude goes to Jascha Grübel for his technical expertise in developing SPACE. We are also sincerely grateful to Christopher Chen, Saima Hillal, Mervin Tee, and Marcus Low Yu Ming for their collaboration in the studies at the National University of Singapore (NUS) and the National University Health System (NUHS). Finally, we express our appreciation to Professor David Matchar for his insightful consultation and guidance in developing the mock cost-effectiveness analysis. The research was conducted in the Future Health Technologies programme, established collaboratively between ETH Zürich and the National Research Foundation Singapore. This research is supported by the National Research Foundation, Prime Minister's Office, Singapore, under its Campus for Research Excellence and Technological Enterprise (CREATE) programme and ETH Zürich.

References

Adikari S, McDonald C, and Campbell J (2013) Reframed Contexts: Design Thinking for Agile User Experience Design. In: Design, User Experience, and Usability. Design Philosophy, Methods, and Tools. Lecture Notes in Computer Science. Springer, Berlin, Heidelberg, pp 3–12

Alahäivälä T, and Oinas-Kukkonen H (2016) Understanding Persuasion Contexts in Health Gamification: A Systematic Analysis of Gamified Health Behavior Change Support Systems Literature. Int J Med Inform 96:62–70. https://doi.org/10.1016/j.ijmedinf.2016.02.006

Alexander G, and Staggers N (2009) A Systematic Review of the Designs of Clinical Technology: Findings and Recommendations for Future Research. ANS Adv Nurs Sci 32 (3):252–279. https://doi.org/10.1097/ans.0b013e3181b0d737

Allen K, Brändle F, Botvinick M, et al. (2024) Using Games to Understand the Mind. Nat Hum Behav 8 (6):1035–1043. https://doi.org/10.1038/s41562-024-01878-9

Alzheimer's Association (2024) Alzheimer's Disease Facts and Figures. https://doi.org/10.1002/alz.13016

Amagai S, Pila S, Kaat AJ, Nowinski CJ, and Gershon RC (2022) Challenges in Participant Engagement and Retention Using Mobile Health Apps: Literature Review. J Med Internet Res 24 (4):e35120. https://doi.org/10.2196/35120

Azman NDB, Visaria A, Goh VS, et al. (2024) Informal Caregiving Time and Its Monetary Value in the Context of Older Adults in Singapore. Aging Health Res 4 (2):100193. https://doi.org/10.1016/j.ahr.2024.100193

Barnard Y, Bradley MD, Hodgson F, and Lloyd AD (2013) Learning to Use New Technologies by Older Adults: Perceived Difficulties, Experimentation Behaviour and Usability. Comput Human Behav 29 (4):1715–1724. https://doi.org/10.1016/j.chb.2013.02.006

Bierbrauer A, Kunz L, Gomes CA, et al. (2020) Unmasking Selective Path Integration Deficits in Alzheimer's Disease Risk Carriers. Sci Adv 6:eaba1394. https://doi.org/10.1126/SCIADV.ABA1394

Bloom GS (2014) Amyloid-B and Tau: The Trigger and Bullet in Alzheimer Disease Pathogenesis. JAMA Neurol 71 (4):505–508. https://doi.org/10.1001/jamaneurol.2013.5847

Braun V, and Clarke V (2006) Using Thematic Analysis in Psychology. Qual Res Psychol 3 (2):77–101. https://doi.org/10.1191/1478088706qp063oa

Brooke J (1996) SUS: A 'Quick and Dirty' Usability Scale. In: Usability Evaluation in Industry. CRC Press, Boca Raton, pp 207–212

Bubu OM, Schwartz S, and Umasabor-Bubu OQ (2015) IC-P-083: CSF p-tau, Brain Beta-Amyloid, and Hippocampal Atrophy Predict Development of Alzheimer's Disease from MCI Interactively in ADNI Subjects. Alzheimers Dement (N Y) 11 (7):P59–P60. https://doi.org/10.1016/j.jalz.2015.06.104

Carroll JM (1997) Human–Computer Interaction: Psychology as a Science of Design. Int J Hum Comput Stud 46 (4):501–522. https://doi.org/10.1006/ijhc.1996.0101

——— (2014) Human Computer Interaction-Brief Intro. https://www.interaction-design.org/literature/book/the-encyclopedia-of-human-computer-interaction-2nd-ed/human-computer-interaction-brief-intro

Castegnaro A, Howett D, Li A, et al. (2022) Assessing Mild Cognitive Impairment Using Object-Location Memory in Immersive Virtual Environments. Hippocampus 32 (9):660–678. https://doi.org/10.1002/hipo.23458

Chan D, Gallaher LM, Moodley K, et al. (2016) The 4 Mountains Test: A Short Test of Spatial Memory with High Sensitivity for the Diagnosis of Pre-Dementia Alzheimer's Disease. J Vis Exp 116:54454. https://doi.org/10.3791/54454

Chay J, Koh WP, Tan KB, and Finkelstein EA (2024) Healthcare Burden of Cognitive Impairment: Evidence from a Singapore Chinese Health Study. Ann Acad Med Singap 53 (4):233–240. https://doi.org/10.47102/annals-acadmedsg.2023253

Chin JP, Diehl VA, and Norman LK (1988) Development of an Instrument Measuring User Satisfaction of the Human-Computer Interface. In: Proceedings of the SIGCHI Conference on Human Factors in Computing Systems CHI '88. ACM, New York, pp 213–218

Chong MS, Tan WS, Chan M, et al. (2013) Cost of Informal Care for Community-Dwelling Mild–Moderate Dementia Patients in a Developed Southeast Asian Country. Int Psychogeriatr 25 (9):1475–1483. https://doi.org/10.1017/S1041610213000707

Clement J (2023) Distribution of Gamers in Selected Gaming Markets in Europe in 2023, by Age Group. https://www.statista.com/forecasts/448467/gaming-in-europe-by-age-and-country

Colombo G, Minta K, Grübel J, et al. (2024a) Detecting Cognitive Impairment through an Age-Friendly Serious Game: The Development and Usability of the Spatial Performance Assessment for Cognitive Evaluation (SPACE). Comput Human Behav 160 (108349):108349. https://doi.org/10.1016/j.chb.2024.108349

Colombo G, Minta K, Thrash T, et al. (2024b) Beyond Traditional Assessments of Cognitive Impairment: Exploring the Potential of Spatial Navigation Tasks. https://doi.org/10.1101/2024.10.12.24315402

Coughlan G, Laczó J, Hort J, Minihane A-M, and Hornberger M (2018) Spatial Navigation Deficits—Overlooked Cognitive Marker for Preclinical Alzheimer Disease? Nat Rev Neurol 14 (8):496–506. https://doi.org/10.1038/s41582-018-0031-x

Coutrot A, Schmidt S, Coutrot L, et al. (2019) Virtual Navigation Tested on a Mobile App Is Predictive of Real-World Wayfinding Navigation Performance. PLoS ONE 14 (3):e0213272. https://doi.org/10.1371/journal.pone.0213272

Coutrot A, Silva R, Manley E, et al. (2018) Global Determinants of Navigation Ability. Curr Biol 28 (17):2861–2866.e2864. https://doi.org/10.1016/j.cub.2018.06.009

Czaja SJ (2019) Usability of Technology for Older Adults: Where Are We and Where Do We Need to Be? https://uxpajournal.org/de/usability-technology-older-adults/

Czaja SJ, Beach S, Charness N, and Schulz R (2013) Older Adults and the Adoption of Healthcare Technology: Opportunities and Challenges. In: Technologies for Active Aging. Springer, Boston, pp 27–46

Czaja SJ, Boot WR, Charness N, and Rogers WA (2019) Designing for Older Adults: Principles and Creative Human Factors Approaches, Third Edition. CRC Press, Boca Raton

Dautzenberg G, Lijmer J, and Beekman A (2019) Diagnostic Accuracy of the Montreal Cognitive Assessment (MoCA) for Cognitive Screening in Old Age Psychiatry: Determining Cutoff Scores in Clinical Practice. Avoiding Spectrum Bias Caused by Healthy Controls. Int J Geriatr Psychiatry 35 (3):261–269. https://doi.org/10.1002/gps.5227

deIpolyi AR, Rankin KP, Mucke L, Miller BL, and Gorno-Tempini ML (2007) Spatial Cognition and the Human Navigation Network in AD and MCI. Neurology 69 (10):986–997. https://doi.org/10.1212/01.wnl.0000271376.19515.c6

De Vito Dabbs A, Myers BA, Mc Curry KR, et al. (2009) User-Centered Design and Interactive Health Technologies for Patients. Comput Inform Nurs 27 (3):175–183. https://doi.org/10.1097/NCN.0b013e31819f7c7c

Ekstrom AD, and Isham EA (2017) Human Spatial Navigation: Representations across Dimensions and Scales. Curr Opin Behav Sci 17:84–89. https://doi.org/10.1016/j.cobeha.2017.06.005

Ekstrom AD, Spiers HJ, Bohbot VD, and Rosenbaum RS (2018) Human Spatial Navigation. Princeton Univ Press, Princeton

Epstein RA, Patai EZ, Julian JB, and Spiers HJ (2017) The Cognitive Map in Humans: Spatial Navigation and Beyond. Nat Neurosci 20 (11):1504–1513. https://doi.org/10.1038/nn.4656

Faverio M (2022) Share of Those 65 and Older Who Are Tech Users Has Grown in the Past Decade. https://www.pewresearch.org/short-reads/2022/01/13/share-of-those-65-and-older-who-are-tech-users-has-grown-in-the-past-decade/

Fletcher E, Villeneuve S, Maillard P, et al. (2016) B-Amyloid, Hippocampal Atrophy and Their Relation to Longitudinal Brain Change in Cognitively Normal Individuals. Neurobiol Aging 40:173–180. https://doi.org/10.1016/j.neurobiolaging.2016.01.133

Flores E, Tobon G, Cavallaro E, et al. (2008) Improving Patient Motivation in Game Development for Motor Deficit Rehabilitation. In: Proceedings of the 2008 International Conference on Advances in Computer Entertainment Technology. ACM, New York, pp 381–384

Folstein MF (1983) The Mini-Mental State Examination. Arch Gen Psychiatry 40 (7):812. https://doi.org/10.1001/archpsyc.1983.01790060110016

Fox S, Brown LJE, Antrobus S, et al. (2022) Co-Design of a Smartphone App for People Living with Dementia by Applying Agile, Iterative Co-Design Principles: Development and Usability Study. JMIR MHealth UHealth 10 (1):e24483. https://doi.org/10.2196/24483

Gamberini L, Raya MA, Barresi G, et al. (2006) Cognition, Technology and Games for the Elderly: An Introduction to ELDERGAMES Project. PsychNology J 4 (3):285–308. http://www.savie.qc.ca/BaseConnaissances/upload/pdf/3000_gamberinietal_2006_cognition_technology_and_games_for_the_elderly_Index3000.pdf

Gomez-Hernandez M, Ferre X, Moral C, and Villalba-Mora E (2023) Design Guidelines of Mobile Apps for Older Adults: Systematic Review and Thematic Analysis. JMIR MHealth UHealth 11 (1):e43186. https://doi.org/10.2196/43186

Goodroe S, Fernandez Velasco P, Gahnstrom CJ, et al. (2025) Predicting Real-World Navigation Performance from a Virtual Navigation Task in Older Adults. PLoS ONE 20 (1):e0317026. https://doi.org/10.1371/journal.pone.0317026

Hafting T, Fyhn M, Molden S, Moser M-B, and Moser EI (2005) Microstructure of a Spatial Map in the Entorhinal Cortex. Nature 436 (7052):801–806. https://doi.org/10.1038/nature03721

Hart SG, and Staveland LE (1988) Development of NASA-TLX (Task Load Index): Results of Empirical and Theoretical Research. Adv Psychol 52:139–183. https://doi.org/10.1016/S0166-4115(08)62386-9

Howett D, Castegnaro A, Krzywicka K, et al. (2019) Differentiation of Mild Cognitive Impairment Using an Entorhinal Cortex-Based Test of Virtual Reality Navigation. Brain 142 (6):1751–1766. https://doi.org/10.1093/brain/awz116

Huang L-K, Chao S-P, and Hu C-J (2020) Clinical Trials of New Drugs for Alzheimer Disease. J Biomed Sci 27 (1):18. https://doi.org/10.1186/s12929-023-00976-6

Ijsselsteijn W, Nap HH, de Kort Y, and Poels K (2007) Digital Game Design for Elderly Users. In: Proceedings of the 2007 Conference on Future Play. ACM, New York, pp 17–22

Josephs KA, Murray ME, Tosakulwong N, et al. (2017) Tau Aggregation Influences Cognition and Hippocampal Atrophy in the Absence of Beta-Amyloid: A Clinico-Imaging-Pathological Study of Primary Age-Related Tauopathy (PART). Acta Neuropathol 133 (5):705–715. https://doi.org/10.1007/s00401-017-1681-2

Kakulla B (2023) The 50-Plus Gamer of Today and Tomorrow. https://www.aarp.org/content/dam/aarp/research/surveys_statistics/technology/2023/2023-gamers-50-plus.doi10.26419-2Fres.00585.001.pdf

Kennedy RS, Lane NE, Berbaum KS, and Lilienthal MG (1993) Simulator Sickness Questionnaire: An Enhanced Method for Quantifying Simulator Sickness. Int J Aviat Psychol 3 (3):203–220. https://doi.org/10.1207/s15327108ijap0303_3

Kepp KP, Sensi SL, Johnsen KB, et al. (2023) The Anti-Amyloid Monoclonal Antibody Lecanemab: 16 Cautionary Notes. J Alzheimers Dis 94 (2):497–507. https://doi.org/10.3233/jad-230099

Kitchin R, and Blades M (2001) The Cognition of Geographic Space. International Library of Human Geography. I. B. Tauris, London

Klatzky RL (1998) Allocentric and Egocentric Spatial Representations: Definitions, Distinctions, and Interconnections. In: Freksa C, Habel C, Wender KF (eds) Spatial Cognition: An Interdisciplinary Approach to Representing and Processing Spatial Knowledge. Springer, Berlin, Heidelberg, pp 1–17

Kohler T (2024) Psychology for UX: Study Guide. https://www.nngroup.com/articles/psychology-study-guide/

Kunz L, Schröder TN, Lee H, et al. (2015) Reduced Grid-Cell-Like Representations in Adults at Genetic Risk for Alzheimer's Disease. Science 350 (6259):430–433. https://doi.org/10.1126/science.aac8128

Kurkinen M (2023) Lecanemab (Leqembi) Is Not the Right Drug for Patients with Alzheimer's Disease. Adv Clin Exp Med 32 (9):943–947. https://doi.org/10.17219/acem/171379

Laczó J, Andel R, Nedelska Z, et al. (2017) Exploring the Contribution of Spatial Navigation to Cognitive Functioning in Older Adults. Neurobiol Aging 51:67–70. https://doi.org/10.1016/j.neurobiolaging.2016.12.003
Laczó M, Martinkovic L, Lerch O, et al. (2022) Different Profiles of Spatial Navigation Deficits in Alzheimer's Disease Biomarker-Positive versus Biomarker-Negative Older Adults with Amnestic Mild Cognitive Impairment. Front Aging Neurosci 14:886778. https://doi.org/10.3389/fnagi.2022.886778
Laugwitz B, Held T, and Schrepp M (2008) Construction and Evaluation of a User Experience Questionnaire. In: Lecture Notes in Computer Science. Springer, Berlin, Heidelberg, pp 63–76
Lester AW, Moffat SD, Wiener JM, Barnes CA, and Wolbers T (2017) The Aging Navigational System. Neuron 95 (5):1019–1035. https://doi.org/10.1016/j.neuron.2017.06.037
Levine TF, Allison SL, Stojanovic M, et al. (2020) Spatial Navigation Ability Predicts Progression of Dementia Symptomatology. Alzheimers Dement (N Y) 16 (3):491–500. https://doi.org/10.1002/alz.12031
Li AW, and King J (2019) Spatial Memory and Navigation in Ageing: A Systematic Review of MRI and fMRI Studies in Healthy Participants. Neurosci Biobehav Rev 103:33–49. https://doi.org/10.1016/j.neubiorev.2019.05.005
Liu L-Y, Lu Y, Shen L, et al. (2022) Prevalence, Risk and Protective Factors for Mild Cognitive Impairment in a Population-Based Study of Singaporean Elderly. J Psychiatr Res 145:111–117. https://doi.org/10.1016/j.jpsychires.2021.11.041
Martinho D, Carneiro J, Corchado JM, and Marreiros G (2020) A Systematic Review of Gamification Techniques Applied to Elderly Care. Artif Intell Rev 53 (7):4863–4901. https://doi.org/10.1007/s10462-020-09809-6
Mathews SC, McShea MJ, Hanley CL, et al. (2019) Digital Health: A Path to Validation. NPJ Digit Med 2 (1):38. https://doi.org/10.1038/s41746-019-0111-3
Matz K (2013) Designing Usable Apps: An Agile Approach to User Experience Design. Winchelsea Press, Nanaimo, BC
Méndez-Gurrola II, Barraza-Castillo RI, Ramírez Reyes A, and Barranco-Gutiérrez AI (2023) An Approach to Mobile App Design and Development Combining Design Thinking, User Experience, and Iterative-Incremental Development. In: Lecture Notes in Networks and Systems. Springer, Cham, pp 623–638
Moreno-Ger P, Torrente J, Hsieh YG, and Lester WT (2012) Usability Testing for Serious Games: Making Informed Design Decisions with User Data. ACHI Int Conf Adv Comput Hum Interact 2012 (1):1–13. https://doi.org/10.1155/2012/369637
Murphy MP, and LeVine H, 3rd (2010) Alzheimer's Disease and the Amyloid-Beta Peptide. J Alzheimers Dis 19 (1):311–323. https://doi.org/10.3233/JAD-2010-1221
Nandi A, Counts N, Chen S, et al. (2022) Global and Regional Projections of the Economic Burden of Alzheimer's Disease and Related Dementias from 2019 to 2050: A Value of Statistical Life Approach. EClinicalMedicine 51:101580. https://doi.org/10.1016/j.eclinm.2022.101580
Nasreddine ZS, Phillips NA, Bédirian V, et al. (2005) The Montreal Cognitive Assessment, MoCA: A Brief Screening Tool for Mild Cognitive Impairment. J Am Geriatr Soc 53 (4):695–699. https://doi.org/10.1111/j.1532-5415.2005.53221.x
National Center for Health Statistics (2021) About Underlying Cause of Death, 2018–2021. http://wonder.cdc.gov/ucd-icd10-expanded.html
Nielsen J (1994) Enhancing the Explanatory Power of Usability Heuristics. In: Proceedings of the SIGCHI Conference on Human Factors in Computing Systems Celebrating Interdependence CHI '94. ACM, New York, pp 152–158
Niidome T, Ishikawa Y, Ogawa T, Nakagawa M, and Nakamura Y (2024) Mechanism of Action and Clinical Trial Results of Lecanemab (Leqembi® 200 Mg, 500 Mg for Intravenous Infusion), a Novel Treatment for Alzheimer's Disease. Nihon Shinkei Seishin Yakurigaku Zasshi 159 (3):173–181. DOI: 10.1254/fpj.24005

Noiwan J, and Norcio AF (2006) Cultural Differences on Attention and Perceived Usability: Investigating Color Combinations of Animated Graphics. Int J Hum Comput Stud 64 (2):103–122. https://doi.org/10.1016/j.ijhcs.2005.06.004

Noorbergen TJ, Adam MTP, Roxburgh M, and Teubner T (2021) Co-Design in mHealth Systems Development: Insights from a Systematic Literature Review. AIS Trans Hum Comput Interact 13 (2):175–205. https://doi.org/10.17705/1thci.00147

Norman D (2013) The Design of Everyday Things: Revised and Expanded Edition. Basic Books, London

O'Keefe J, and Nadel L (1978) The Hippocampus as a Cognitive Map. Oxford Univ Press, London

Olsen T, Procci K, and Bowers C (2011) Serious Games Usability Testing: How to Ensure Proper Usability, Playability, and Effectiveness. In: Lecture Notes in Computer Science. Springer, Berlin, Heidelberg, pp 625–634

Olson GM, and Olson JS (2003) Human-Computer Interaction: Psychological Aspects of the Human Use of Computing. Annu Rev Psychol 54 (1):491–516. https://doi.org/10.1146/annurev.psych.54.101601.145044

Paas FGWC (1992) Training Strategies for Attaining Transfer of Problem-Solving Skill in Statistics: A Cognitive-Load Approach. J Educ Psychol 84 (4):429–434. https://doi.org/10.1037/0022-0663.84.4.429

Palmqvist S, Tideman P, Mattsson-Carlgren N, et al. (2024) Blood Biomarkers to Detect Alzheimer Disease in Primary Care and Secondary Care. JAMA 332 (15):1245–1257. https://doi.org/10.1001/jama.2024.13855

Parizkova M, Lerch O, Moffat SD, et al. (2018) The Effect of Alzheimer's Disease on Spatial Navigation Strategies. Neurobiol Aging 64:107–115. https://doi.org/10.1016/j.neurobiolaging.2017.12.019

Patterson C (2018) World Alzheimer Report 2018. https://www.alzint.org/resource/world-alzheimer-report-2018/

Polk SE, Öhman F, Hassenstab J, et al. (2024) A Scoping Review of Remote and Unsupervised Digital Cognitive Assessments in Preclinical Alzheimer's Disease. https://www.medrxiv.org/content/10.1101/2024.09.25.24314349v1.full

Rekers S, and Finke C (2024) Translating Spatial Navigation Evaluation from Experimental to Clinical Settings: The Virtual Environments Navigation Assessment (Vienna). Behav Res Methods 56 (3):2033–2048. https://doi.org/10.3758/s13428-023-02134-0

Rundo L, Pirrone R, Vitabile S, Sala E, and Gambino O (2020) Recent Advances of HCI in Decision-Making Tasks for Optimized Clinical Workflows and Precision Medicine. J Biomed Inform 108:103479. https://doi.org/10.1016/j.jbi.2020.103479

Safi S, Thiessen T, and Schmailzl KJ (2018) Acceptance and Resistance of New Digital Technologies in Medicine: Qualitative Study. JMIR Res Protoc 7 (12):e11072. https://doi.org/10.2196/11072

Sanders EBN, and Stappers PJ (2014) Probes, Toolkits and Prototypes: Three Approaches to Making in Codesigning. CoDesign 10 (1):5–14. https://doi.org/10.1080/15710882.2014.888183

Santoso HB, and Schrepp M (2019) The Impact of Culture and Product on the Subjective Importance of User Experience Aspects. Heliyon 5 (9):e02434. https://doi.org/10.1016/j.heliyon.2019.e02434

Schinazi VR, and Thrash T (2018) Cognitive Neuroscience of Spatial and Geographic Thinking. In: Montello DR (ed) Handbook of Behavioral and Cognitive Geography, vol 419. Edward Elgar Publ, Northampton, MA, pp 154–174

Schinazi VR, Thrash T, and Chebat D-R (2016) Spatial Navigation by Congenitally Blind Individuals. Wiley Interdiscip Rev Cogn Sci 7 (1):37–58. https://doi.org/10.1002/wcs.1375

Schrepp M, Kollmorgen J, Meiners AL, et al. (2023) On the Importance of UX Quality Aspects for Different Product Categories. Int J Interact Multimed Artif Intell 8 (2):232–246. https://doi.org/10.9781/ijimai.2023.03.001

Segen V, Ying J, Morgan E, Brandon M, and Wolbers T (2022) Path Integration in Normal Aging and Alzheimer's Disease. Trends Cogn Sci 26 (2):142–158. https://doi.org/10.1016/j.tics.2021.11.001
Shi M, Chu F, Zhu F, and Zhu J (2022) Impact of Anti-Amyloid-B Monoclonal Antibodies on the Pathology and Clinical Profile of Alzheimer's Disease: A Focus on Aducanumab and Lecanemab. Front Aging Neurosci 14:870517. https://doi.org/10.3389/fnagi.2022.870517
Skulmowski A (2023) Guidelines for Choosing Cognitive Load Measures in Perceptually Rich Environments. Mind Brain Educ 17 (1):20–28. https://doi.org/10.1111/mbe.12342
Stangl M, Kanitscheider I, Riemer M, Fiete I, and Wolbers T (2020) Sources of Path Integration Error in Young and Aging Humans. Nat Commun 11 (1):2626. https://doi.org/10.1038/s41467-020-15805-9
Tampi RR, Forester BP, and Agronin M (2021) Aducanumab: Evidence from Clinical Trial Data and Controversies. Drugs Context 10:1–9. https://doi.org/10.7573/dic.2021-7-3
Tu S, Wong S, Hodges JR, et al. (2015) Lost in Spatial Translation: A Novel Tool to Objectively Assess Spatial Disorientation in Alzheimer's Disease and Frontotemporal Dementia. Cortex 67:83–94. https://doi.org/10.1016/j.cortex.2015.03.016
Tuena C, Mancuso V, Stramba-Badiale C, et al. (2021) Egocentric and Allocentric Spatial Memory in Mild Cognitive Impairment with Real-World and Virtual Navigation Tasks: A Systematic Review. J Alzheimers Dis 79 (1):95–116. https://doi.org/10.3233/jad-201017
van der Ham IJM, and Claessen MHG (2022) A Clinical Guide to Assessment of Navigation Impairment: Standardized Subjective and Objective Instruments and Normative Data. J Clin Exp Neuropsychol 44 (7):487–498. https://doi.org/10.1080/13803395.2022.2123895
van der Ham IJM, Claessen MHG, Evers AWM, and van der Kuil MNA (2020) Large-Scale Assessment of Human Navigation Ability across the Lifespan. Sci Rep 10 (1):3299. https://doi.org/10.1038/s41598-020-60302-0
van der Kuil MNA, Visser-Meily JMA, Evers AWM, and van der Ham IJM (2018) A Usability Study of a Serious Game in Cognitive Rehabilitation: A Compensatory Navigation Training in Acquired Brain Injury Patients. Front Psychol 9:846. https://doi.org/10.3389/fpsyg.2018.00846
van Dyck CH, Swanson CJ, Aisen P, et al. (2023) Lecanemab in Early Alzheimer's Disease. N Engl J Med 388 (1):9–21. https://doi.org/10.1056/nejmoa2212948
Vlček K (2011) Spatial Navigation Impairment in Healthy Aging and Alzheimer's Disease. In: Monte SDL (ed) The Clinical Spectrum of Alzheimer's Disease. IntechOpen, Rijeka, DOI: 10.5772/20278
Wallace S, Reid A, Clinciu D, and Kang JS (2013) Culture and the Importance of Usability Attributes. Inform Technol Peopl 26 (1):77–93. http://doi.org/10.1108/09593841311307150
WHO (2023) Dementia: Key Facts. https://www.who.int/news-room/fact-sheets/detail/dementia
Wiener JM, Carroll D, Moeller S, et al. (2020) A Novel Virtual-Reality-Based Route-Learning Test Suite: Assessing the Effects of Cognitive Aging on Navigation. Behav Res Methods 52 (2):630–640. https://doi.org/10.3758/s13428-019-01264-8
Witmer BG, and Singer MJ (1998) Measuring Presence in Virtual Environments: A Presence Questionnaire. Presence 7 (3):225–240. https://doi.org/10.1162/105474698565686
Wong W (2020) Economic Burden of Alzheimer Disease and Managed Care Considerations. Am J Manag Care 26 (8 Suppl):S177–S183. https://doi.org/10.37765/ajmc.2020.88482

16

From Fundamental Science to the Clinic (and Back Again)

A Translational Framework for Navigation and Aging

Coco Newton, Carol A. Barnes, Virginia L. Flanagin, Anna K. Gillespie, Michael Hornberger, Scott Moffat, Victor R. Schinazi, Vladislava Segen, Jan Wiener, Steven M. Weisberg, and Thomas Wolbers

Abstract Global societies face unprecedented challenges as populations age. The ability to navigate is central to nearly all meaningful activities in which older adults participate, promoting both independence and health. Loss of navigation ability is associated with early-stage age-related brain and body disorders. To address the diverse challenges associated with aging populations, tools must be developed to maintain, support, and assess navigation in real-world applications—tools grounded in reliable and valid research. This chapter presents the "Triadic Navigation Framework," a platform that can be used to guide and stimulate translational navigation research in aging. Its intent is to foster deeper consideration of research design, enhance collaboration among researchers working across species, and better identify the real-world needs of aging populations that research must address. Opportunities for future work are highlighted, as are recommendations to enhance navigation research quality.

Keywords Translation, aging, dementia, innovation, society

Coco Newton (✉)
Institute of Cognitive Neuroscience, University College London, London WC1E 6BT, UK; Engineering Design Centre, University of Cambridge, Cambridge CB2 1PZ, UK
Email: coco.newton@ucl.ac.uk

Coauthor affiliations are available in the List of Contributors.

N. S. Newcombe and K. Cheng (eds.), *Challenges in Navigation Research*, Strüngmann Forum Reports, https://doi.org/10.1007/978-3-032-20563-6_16

Group photos (top left to bottom right) Coco Newton, Carol Barnes, Jan Wiener, Michael Hornberger, Virginia Flanagin, Steven Weisberg, Anna Gillespie, Vladislava Segen, Thomas Wolbers, Victor Schinazi, Coco Newton, Carol Barnes and Anna Gillespie, Vladislava Segen, Michael Hornberger, Carol Barnes, Jan Wiener, Thomas Wolbers, Virginia Flanagin

1 Need for a Comprehensive Approach: Understanding Healthy and Unhealthy Aging Effects on Navigation

Globally, the aging population is increasing at an unprecedented rate. By 2050 the number of people aged 60 years and older is projected to exceed two billion (WHO 2025). This demographic shift is primarily driven by increased life expectancy due to advances in health care, sanitation, and reduced infant mortality (Chen et al. 2024). While these gains reflect positive societal progress, they also introduce unique challenges related to health, independence, and quality of life in older populations. Navigation is fundamental to independent living and is particularly vulnerable to cognitive decline, making it a significant area for research and intervention in aging populations, both unhealthy and healthy. In this chapter, we provide a framework to leverage navigation to help address these challenges as well as recommendations for the future of navigation research.

1.1 Defining Healthy and Unhealthy Aging

Aging is a lifelong developmental progression of changes in cells and tissues of the body and brain that occurs ubiquitously across species. It changes an organism's cognitive as well as overall ability to function effectively in the world. Both the World Health Organization and the United Nations commonly classify older adults as those above the age of 60 years. However, individuals experience aging differently, and genetic, environmental, and lifestyle factors lead to clear individual differences in aging trajectories (Figure 16.1).

Healthy aging trajectories support an individual's ability to engage in meaningful activities (Beard et al. 2016). Although functional capacities such as navigation decline naturally over time (Lester et al. 2017), adaptive processes can accommodate physiological and cognitive changes. Key changes underlying healthy aging include alterations in synaptic plasticity and connectivity, particularly within the brain regions involved in navigation, such as the hippocampus, prefrontal cortex, and the cerebellum. While there is minimal neuronal loss in these areas, age-related declines in synaptic density and efficiency affect cognitive function and sensory processing. For a comprehensive review, see Barnes et al. (Chapter 13).

In humans, aging is a primary risk factor for neurodegenerative diseases such as Alzheimer disease, which accounts for 60–70% of dementia cases (Lane et al. 2018). Dementia diseases are defined biologically by the abnormal accumulation of misfolded proteins, triggering pathophysiological changes beyond those associated with healthy aging. These unhealthy aging trajectories accelerate the loss of cognitive and physical functioning, especially in navigation abilities, and increases the risk of premature death (Coughlan et al. 2018).

While the total number of people living with dementia worldwide is expected to reach 150 million by 2050 due to the growth of the aging population, the risk of dementia has actually declined in high-income countries since the 1970s, with an average 20% decrease in incidence per decade (Satizabal et al. 2016; Wolters et al. 2020). Several factors have been proposed to contribute to this trend, including

Figure 16.1 Although there is variance among individuals and across cultural contexts, the ability to navigate is a critical skill for all individuals.

higher levels of education in the population and improved treatments that maintain cardiovascular health. If modifiable risk factors under personal or societal control can impact aging trajectories, it may be possible to reduce the impact of deleterious factors and promote brain and cognitive health resilience. Identifying ways to nudge societies toward maximizing healthy aging trajectories, while continuing to reduce rates of incident dementia, is a global health priority.

1.2 Leveraging Navigation

Understanding the effects of healthy and unhealthy aging on navigation can inform the development of tools to help maximize healthy aging trajectories. Independent living, remaining mobile, and maintaining the ability to engage socially are highly meaningful faculties among older people, all of which typically involve spatial navigation. To support independence and mobility, navigation tools might include resources for the design and planning of environments with improved wayfinding cues as well as assistive technologies to help prevent vulnerable individuals from getting lost and consequently harmed, or training and education interventions aimed at improving independent navigation behavior. Clinical tools include cognitive assessments that could help clinicians diagnose Alzheimer disease earlier for more timely interventions in care, support, and planning.

Creating most of these tools requires knowledge from across species of the link between navigation behavioral patterns and the fundamental brain computations underpinning them, as well as how these change with age and disease. When considered holistically, several key factors influence the brain–behavior link and may themselves be impacted by aging pathology (Lester et al. 2017). Core navigation brain systems can undergo age-related changes (e.g., such as synapse loss in the dentate gyrus), which may impair spatial memory and navigation accuracy. Supporting (primarily sensory) brain systems (e.g., the visual and vestibular systems essential for processing external cues and self-motion information) may also decline with age and further affect navigational abilities. Age-related impairments in switching between spatial strategies, often linked to reduced prefrontal-hippocampal connectivity, also play a role in navigational difficulties observed in older adults. In addition, the context of the navigational task an individual faces—including environmental variables such as the availability and clarity of cue types as well as the presence or absence of route knowledge—may vary in salience or relative importance with age, thereby influencing which brain systems are engaged. Finally, individual differences, such as varying levels of risk aversion, motivation, and sensory capacity, affect how people approach and perform navigational tasks.

The wealth of knowledge on the navigation brain–behavior link is sometimes conflicting and contradictory. Without considering all the different factors together, we risk misattributing behavioral changes in aging and pathology to the wrong source of altered brain computation, which would slow progress in developing accessible and effective tools to prolong healthy navigation across the lifespan.

1.3 Objective

To effectively address the challenges of an aging society, a comprehensive understanding of brain–behavior relationships in navigation is essential. We contend that the full spectrum of aging- and pathology-related effects on both brain function and behavior should be considered holistically across all relevant systems in parallel. While progress has been made in identifying how aging and pathology impact navigation-related neural systems, behaviors, and individual differences, these factors have yet to be integrated into a unified framework. At first glance, this may seem like an intractable problem with countless variables, but we believe that current insights into cellular mechanisms of aging and the computational principles of navigation allow us now to meaningfully relate the two. Specifically, we can begin to build more holistic computational models which in turn can serve as the starting point to probe navigational issues in healthy and pathological aging. This integration has the potential to significantly enhance our understanding of how aging shapes the neural mechanisms underlying navigation and the corresponding behavioral outcomes.

Below, we propose a framework—the Triadic Navigation Framework—to incorporate models of spatial navigation computations together with a taxonomy of navigation scenarios within which the model can be contextualized. It also enables established sources of variability to tailor both the model and scenario to different healthy or unhealthy aging individuals. The Triadic Navigation Framework aims

- to strengthen efforts in the navigation research community and achieve sufficient understanding to permit effective translation,
- to inform the design of improved experiments, controlling for confounds in validity and reliability,

- to facilitate an understanding of the relevance of work across species, and
- to identify key gaps in knowledge that must be addressed to facilitate effective implementation of navigation tools, from research into practice.

After presenting the framework, we highlight the broad range of different navigation tools that could be created to help address unmet societal needs across all trajectories of aging. Finally, we offer recommendations to generate higher-quality data that will accelerate our understanding of how aging, both healthy and pathological, affects the interplay between navigation and brain function.

2 The Triadic Navigation Framework: Translating Navigation Research to Address Aging and Clinical Societal Needs

Our proposed framework aims to provide a comprehensive, holistic platform for researchers to consider the full spectrum of factors relevant to understanding the brain–behavior relationship in navigation in healthy and unhealthy aging individuals. Although the framework is oriented toward human navigation, the physiological and neuronal computations used in navigation, along with a deeper understanding of the changes incurred by age, will inevitably be rooted in insights across multiple species.

The Triadic Navigation Framework (Figure 16.2) is an integrative model that explains navigation behavior through the interaction of three core components: (1) the navigation system, comprising the neural and computational mechanisms that support spatial processing; (2) the navigation context, defined by the task demands, environmental features, movement modalities, and optimization principles that constrain performance; and (3) individual differences, encompassing inter- and intraindividual variations in sensory, cognitive, motivational, and pathological factors that shape how navigation is executed.

For instance, to execute the same navigational goal, individuals may rely on distinct systems due to factors such as personal choice or declines in sensory systems. This is especially relevant in aging and disease, which are nonhomogenous processes with systems-wide effects that influence not only the brain's navigation circuits but also the non-navigation sensory and broader cognitive circuits on which successful navigation depends. It is also important to consider the nature of the navigational goal and how performance is evaluated: Is there a time or accuracy pressure, or no constraint whatsoever? Does it matter to the individual? Ultimately it is important to distinguish between the variation that matters (i.e., variation that influences the mapping between neural computations and behaviors with age) and the variation that can acceptably occur with little consequence on the mapping. This framework intends to prompt researchers to consider and discuss this question during the design and interpretation of experiments aimed at addressing challenges posed by an aging society.

Below, we describe each component in greater detail, identifying where there is scope for future work and outlining how the framework can be used to map out the simultaneous factors related to aging effects on navigation in the real-world. We conclude the section with two worked examples of experiments considered under the Triadic Navigation Framework.

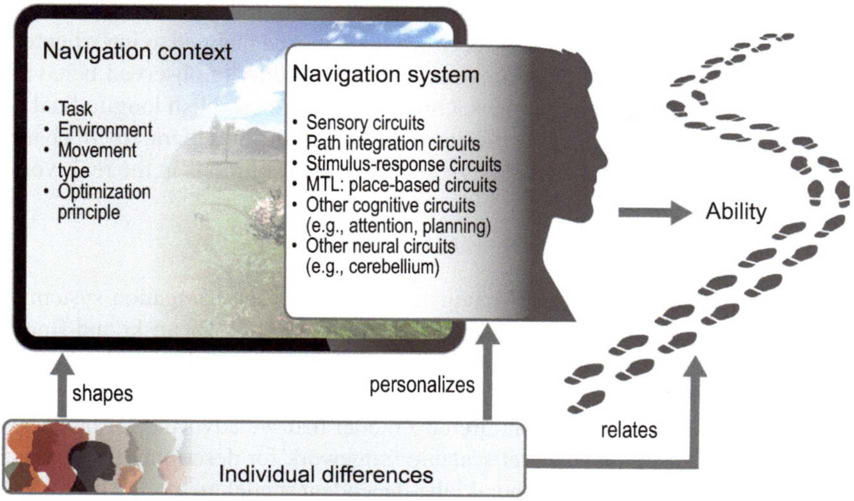

Figure 16.2 Overview of the Triadic Navigation Framework to explain navigational ability in individuals, considering the three components of navigation context, navigation system, and individual differences.

2.1 Navigation System

The *navigation system component* of the framework refers to the brain circuitry required for goal-directed navigation behavior in the form of a computational model. The framework provides a flexible structure that allows researchers to incorporate and model various navigation and non-navigation circuitries in ways that best suit their research objectives. However, four key principles should be adopted by these models:

1. The model should primarily focus on representing the computations of information required to complete a given navigation task. This is necessary to understand (at an abstract level) the interactions between systems operating in parallel.
2. The model computations should be flexibly representable across levels of cells, circuits, systems, and behaviors to enable mapping between behavioral and regional neural patterns, and to incorporate insights generated across species using the different methods of interrogation that they facilitate.
3. The model should allow for the simulation of specific perturbations that reflect aging or disease processes, such as decreased tuning curves and synaptic integrity with age or cell loss with pathology. This is necessary to understand how perturbations might affect systems computations or interactions and help assess their relative influences on behavioral output.
4. The model, while representing an "average" navigation system, should be tunable according to inter- and intraindividual differences (e.g., demographics, level of motivation, weighting of different strategies) or other characteristics of an individual that could influence the mapping between neural and behavioral patterns.

These principles enable broader thinking around the relationship between navigation behaviors and the full range of healthy and unhealthy aging brain computations and circuitries that underpin them. By considering multiple circuitries simultaneously and across levels, interactions between them can be mapped to interrogate the potential ways in which altered computations could underlie observed behavioral changes, both in animals and humans. Only then can we establish longitudinal patterns and timing of different aging effects, which will indicate a hierarchy of changes relevant for the timing of different navigation-oriented solutions in the real-world.

2.1.1 The BB Model: Medial Temporal Lobe Circuitry

For a central model of the "core" navigation circuits in the navigation system, we took inspiration from the "BB" model of spatial memory by Bicanski and Burgess (2018) (see Figure 16.3). It is a neural, circuit- and systems-level model with behavioral outcomes, based on an earlier model by Byrne et al. (2007). The BB model is not the comprehensive navigation circuitry model that we advocate for here; however, it has value as a tractable yet scalable framework for describing the computations underlying core medial temporal lobe-dependent spatial memory. This makes it a useful example architecture, as it provides a testable substrate to examine healthy and unhealthy aging in line with the principles we laid out above. For other examples of models, see Banino et al. (2018), Edvardsen et al. (2020), and Kessler et al. (2024). In brief, the BB model uses known firing properties from spatially relevant neuronal populations in distinct regions of the brain—including the parietal cortex and posterior cingulum, medial temporal lobe, as well as putative inputs from (visual) sensory and idiothetic systems—to show how these cell types can interact to explain behavior (e.g., spatial learning memory, scene construction, and imagery). If we face a corner of the environment that has clear boundaries, the sensory representation in the visual and parietal cortices will be in an egocentric reference frame. Information is then translated using a head-direction signal into an allocentric or viewpoint-independent reference frame in the retrosplenial cortex and transmitted to the medial temporal lobe, leading to boundary vector cells. This environmental information is combined with information from object vector cells, place cells, grid cells, and identity cells to embed object location in the context of the larger environment. During memory retrieval or mental imagery, information flow is reversed because reactivation of the allocentric medial temporal lobe system is followed by a back-transformation to an egocentric representation in the parietal window, allowing for actively remembering or imagining the object location.

2.1.2 Expanding the BB Model in Future Work: Other Circuitries and the Effects of Aging

To maximize the usefulness, the navigation system component of the Triadic Navigation Framework must both extend beyond a model of medial temporal lobe circuitry and additionally incorporate simulations of the physiological effects of aging and pathology grounded in established research. While the BB model hints at other circuitries, such as the sensory inputs required, we suggest that for the Triadic Navigation Framework, these hints should be explicitly expanded to include more detailed models in their own right (Figure 16.4). For instance, although the model was originally designed to capture the neural mechanisms underlying spatial

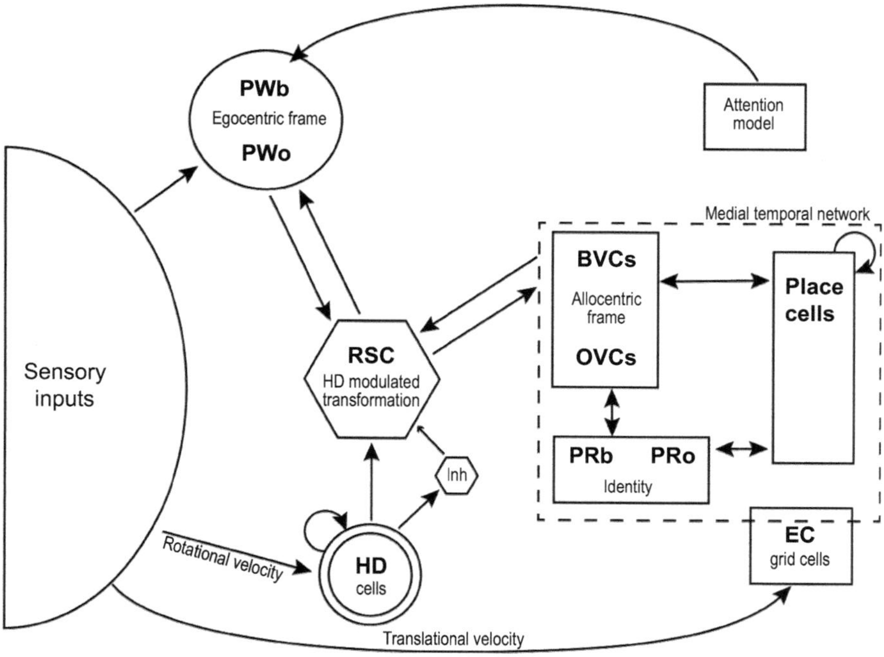

Figure 16.3 The BB model of spatial memory constitutes a suggested core of the "navigation systems" model for our framework. Adapted from Bicanski and Burgess (2018; https://creativecommons.org/licenses/by/4.0/).

memory, the incorporation of self-motion inputs from sensory systems enables it to support path integration (i.e., the continuous updating of one's position based on self-motion cues). However, for effective path integration, the model must explicitly define the nature of the self-motion inputs received by grid cells, including those originating from head-direction cells and speed cells.

Physiological changes associated with healthy aging can be simulated within various aspects of the BB model to test their influence on navigational ability; for instance, broader tuning curves or decreased synaptic activity in certain hippocampal regions. Similarly, changes associated with pathological aging can be reflected by factors such as degraded grid-cell integrity due to the spread of early tau pathology from transentorhinal to medial entorhinal cortex.

Creating this holistic and multilevel navigation system component of the Triadic Navigation Framework will require collaborative effort between cross-species researchers, clinicians, as well as behavioral and computational scientists. Below, we outline several opportunities for future work to achieve this, which involve expanding on or simulating perturbations to the BB model.

2.1.2.1 Opportunity 1: Expand the Sensory Inputs and Simulate Aging Effects

Sensory systems deliver visual and auditory external cues as well as internal self-motion cues from proprioception and vestibular systems, which can be integrated and relied upon differentially. These cues might influence the medial temporal or other navigation circuits by providing noisier, less reliable signals to the head-direction

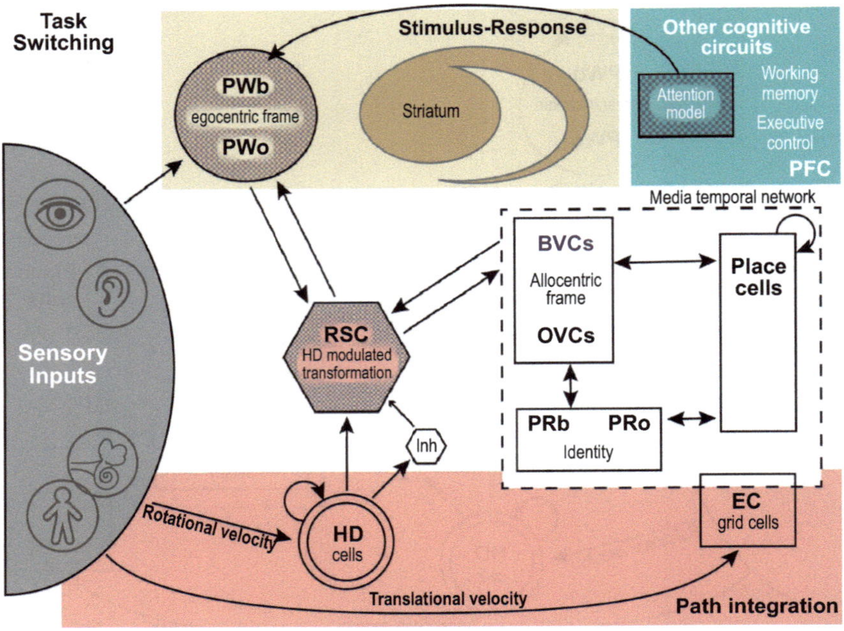

Figure 16.4 The current formulation of the navigation system component in the Triadic Navigation Framework. This "expanded" version of the BB model has suggestions for future work, including other navigation, cognitive, and sensory circuits affected by aging and which influence brain–behavior mapping. Adapted from Bicanski and Burgess (2018; https://creativecommons.org/licenses/by/4.0/).

network or grid cells, as well as noisier visual input to the parietal window. Below we summarize key sensory system changes with healthy and unhealthy aging that could be simulated within the navigation system.

Hearing loss is the most common sensory impairment among older adults. By their seventh decade, over half experience hearing loss severe enough to affect communication (Agrawal 2008), particularly impacting the ability to detect high-pitched sounds (Hunter et al. 2020). Beyond communication challenges, hearing loss can affect spatial navigation by limiting access to both extrinsic and intrinsic auditory cues essential for movement coordination. Intrinsic cues, such as the bone-conducted sound of foot strikes, contribute to the calibration of spatial-temporal aspects of gait. Research indicates that hearing-impaired individuals exhibit greater variability in their gait patterns compared to those with normal hearing (Sakurai et al. 2021). The use of hearing aids and cochlear implants, however, can improve gait stability and reduce variability by restoring access to these crucial auditory cues (Shayman et al. 2017). Additionally, auditory input influences self-motion perception, with studies showing that bilaterally deaf individuals experience a diminished sensorimotor response to forward drift when their implants are turned off (Camponogara et al. 2016). These findings suggest that hearing impairment may disrupt the real-time calibration of gait and spatial awareness, potentially leading to difficulties in estimating distances and navigating unfamiliar environments. This aligns with recent work which shows that older adults with hearing impairments underestimated distance during path integration (Grove et al. 2024).

Age-related visual changes, such as reduced visual acuity, decreased contrast sensitivity, increased glare sensitivity, and alterations in optic flow processing (Atchley and Andersen 1998; Berard et al. 2009), can significantly impact spatial navigation in several ways. These changes can make it more difficult for older adults to accurately perceive and interpret their surroundings, leading to challenges in recognizing landmarks, judging distances, and maintaining orientation within complex environments (Bécu et al. 2020; Bian and Andersen 2013). Reduced visual acuity can impair the ability to detect and focus on navigation cues, whereas decreased contrast sensitivity (Owsley et al. 1983) can hinder the differentiation of important spatial features such as edges, pathways, and obstacles, especially under low-light conditions. Changes in optic flow processing, which provides critical self-motion cues based on visual scene movement, may reduce an individual's ability to gauge speed and direction (Etienne and Jeffery 2004), potentially leading to misjudgments of movement and impaired path integration (Kavcic et al. 2011; Warren et al. 1989).

In terms of idiothetic information, structural changes in the peripheral nervous system and loss of motor neurons and neuromuscular junction efficiency with age cause a decrease in perceptual sensitivity and slowing of movement (Hunter et al. 2016). The vestibular system, which is closely linked to motor, proprioceptive, and self-inferential systems, is essential for balance and spatial orientation, contributing translation and angular velocity inputs to grid-cell stability (Jacob et al. 2014) and head-direction updating (Yoder and Taube 2014). Our vestibular perceptual ability deteriorates with age at a rate similar to hearing loss (Allen et al. 2016). Many older adults experience vestibular dysfunction due to the loss of vestibular sensory cells and microvascular changes in the inner ear, contributing to dizziness, unsteadiness, an increased likelihood of falls, and associations with cognitive decline and an increased risk of dementia. Importantly, degraded vestibular inputs comprise the stability of translation and angular velocity information (more on this in Section 2.4.1), potentially leading to increased drift in spatial location estimates in the BB model. This degradation could explain why older adults exhibit more errors in path-integration tasks and greater susceptibility to disorientation, particularly in environments with limited external spatial cues.

2.1.2.2 Opportunity 2: Expand the Stimulus-Response System

Route knowledge is often conceptualized as stimulus-response associations in which places or landmarks are associated with movement decisions (e.g., "turn left at landmark X"). This ability has been linked to the striatum, in particular the caudate nucleus, in both rodents (Packard and McGaugh 1996) and humans (Hartley et al. 2003). Route navigation also encompasses sequential information, such as knowledge about the sequences of turns or sequences of landmarks/places encountered along the route (Hilton et al. 2021). This more integrated route knowledge is supported by the retrosplenial cortex (Schinazi and Epstein 2010) and involves the interaction of hippocampus and caudate nucleus during route navigation (Buckley et al. 2024). In familiar environments, navigators can choose to follow a known route or decide to plan novel routes or shortcuts. In functional brain imaging studies in humans, these strategy choices are reflected in prefrontal, caudate nucleus, and hippocampal activation (Dahmani and Bohbot 2015). Importantly, age has been shown to systematically affect strategy choices with older adults shifting away from hippocampal-dependent place and toward striatal-dependent response or route strategies (Rodgers et al. 2012), possibly in response to age-related functional and structural changes to the hippocampal system.

2.1.2.3 Opportunity 3: Expand with Other Cognitive Functions and Brain Circuits

Except for semantic memory, most cognitive domains show age-related declines that can impact the function of navigation circuitry. Working memory, essential for remembering recent turns or landmarks and the planning of multistep routes, declines with age (Bopp and Verhaeghen 2005). Deterioration in executive functions, including problem solving, planning, and cognitive flexibility, impacts adaptability to unexpected changes and increases susceptibility to distractions (Diamond 2013). Attention, long-term memory, visuospatial processing, and multisensory integration all exhibit age-related declines, which further complicate navigational tasks (Craik and Salthouse 2011; Moffat 2009; Mozolic et al. 2012; Nyberg et al. 2012). Decision making becomes slower and more risk averse. In addition, emotional and motivational changes, such as heightened anxiety and reduced interest in exploration, can also hinder navigation performance (Mather and Carstensen 2005; Nolte and Hanoch 2024).

Circuitry within the cerebellum, often overlooked in navigation studies, plays a significant role in spatial navigation. The right cerebellum (Crus I), left hippocampus, and frontal regions are coactivated during sequence-based navigation, whereas the left cerebellum (Crus I) is part of a subnetwork involving the right hippocampus, parietal, retrosplenial, and entorhinal cortices (Iglói et al. 2015). This circuit may facilitate allocentric-to-egocentric transformations or mental navigation (Mirino et al. 2022). In healthy aging, the left cerebellum shows reduced gray matter volume, but the right cerebellum Crus I exhibits increased activation during spatial navigation, suggesting compensatory mechanisms (Ramanoël et al. 2022). This shift could explain the age-related preference for response over place-based navigation strategies. Additionally, cerebellar gray matter volume differs in individuals with Alzheimer disease compared to normal age-matched controls, and cerebellar transcranial magnetic stimulation has been shown to improve global cognition in older individuals with Alzheimer disease (Gellersen et al. 2021; Yao et al. 2022).

2.2 Navigation Context

The second component of the Triadic Navigation Framework, the *navigation context*, refers to the specific conditions surrounding the navigator at a given moment. It is treated as a static snapshot in time: if new information emerges that alters the context, the framework must be updated accordingly. The purpose of defining the context is to establish clear constraints on how the navigation system operates, ensuring that its execution is based on the information available to the navigator at that moment.

We have defined the navigational context as the "task" (based on Wiener et al. 2009, see below), the "optimization principle" (e.g., time to arrival, length of route, chance of becoming lost, visual pleasantness), the "environment" (e.g., cue-rich, urban, dark, sensory, movement constraints), and the "movement" type (e.g., a two-dimensional screen with a joystick versus a walkable physical world).

The Wiener et al. (2009) "task" taxonomy considers and classifies only unaided wayfinding; that is, wayfinding without the use of navigation aids such as maps, signage, or GPS, depending on the spatial knowledge available to navigators. Specifically, it considers knowledge about a point in space (e.g., landmark or destination knowledge), knowledge about a sequence of places (route knowledge), and knowledge about the environment (survey knowledge). For example, directed wayfinding tasks in which navigators aim to reach a specific destination or place are classified as uninformed search (no destination, route, or survey knowledge), informed search (no destination or route knowledge but survey knowledge), route following (destination and route knowledge), and route planning (destination and survey but no route knowledge). For details, see Figure 16.5.

The optimization principle is the focal parameter for the navigator. Common examples include the amount of time to get to the destination or the length of the route. If, for instance, the navigator needs to arrive at a specific destination by 12:00 p.m., they might adopt strategies such as choosing a highly familiar route or following GPS instructions. Here, the optimization principle is efficiency, with minimal uncertainty and travel time. The same "task" could, however, be pursued differently if, for example, the navigator wants to take the scenic route, in which case the optimization principle shifts to maximizing visual pleasantness. Other examples might include taking the shortest possible path (either by distance or by number of turns) or visiting interesting sites along the way. Of course, multiple optimization principles are usually at play in a single navigation scenario. For instance, a navigator might wish to maximize visual pleasantness, but only at the expense of up to 10% of additional travel time.

A final orthogonal component of navigation context is movement type. In theory, any combination of how movement through the space is generated is possible without changing the task or the optimization principle. Relevant dimensions along which movement can vary include

- whether the movement is active or passive (for a discussion of the dimensions of active/passive movement, see Chrastil and Warren 2014),
- the method of movement control (e.g., joystick, walking, wheelchair, driving, etc.), and
- the means by which movement is conveyed or sensed (e.g., on a monitor or head-mounted display, idiothetically, or auditorily).

By operationalizing the multilevel *navigation system* within a given navigational task and definition of success, the interaction between these various factors can be considered simultaneously and interdependently in the context of aging and disease. This will allow for a more precise dissociation of variation in brain–behavior relationships, which will ultimately enable greater utility of navigation assessments or other targeted interventions, although the applicability of this *navigation context* taxonomy will vary across species. It will also lead to better identification of

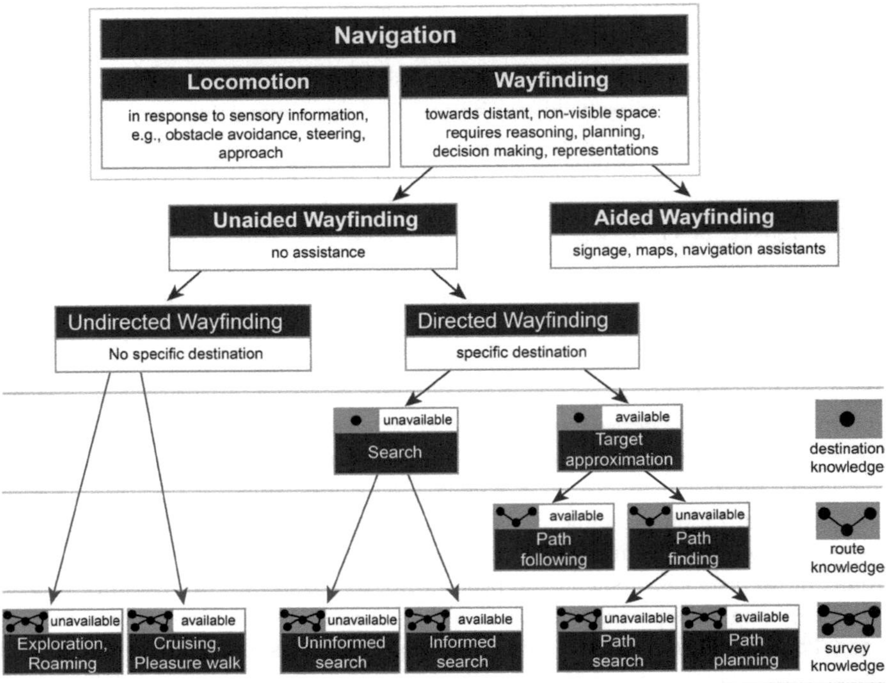

Figure 16.5 Taxonomy of wayfinding tasks, classified by the existence of an external aid, a specific destination, and the availability of different levels of knowledge. Reprinted from Wiener et al. (2009) with permission of Taylor & Francis Ltd, http://www.tandfonline.com.

research gaps to address, which may previously have been missed due to research questions being asked in isolation of other contributing factors.

2.3 Sources of Inter- and Intraindividual Differences

The circuitries, models, and tasks described above reflect a "normative" navigation system and navigation context. Yet, individual variation in ability and strategy across different navigation contexts is well documented in both humans and animals. Elucidating the mechanisms of variation and their impact on neural computations is critical to understanding the precise mapping between brain and behavior, which is needed to inform the development of better-designed navigational solutions for aging populations. Here, we describe the third component of the Triadic Navigation Framework: how individual differences can be considered and integrated into translational navigation research, and where key research gaps remain.

2.3.1 How Do Individual Differences Emerge?

To complete a given task, individuals may attend to different types of information or select alternative routes, depending on varying levels of motivation or risk aversion. Some individuals may have degraded sensory information due to age or atrophy of select navigation regions due to disease, whereas others may be using confounding medications or have comorbid physical or mental health conditions. Despite relying on similar cues and possessing an equivalent sensory apparatus, individuals may differ in their ability to construct and flexibly use different spatial representations (Weisberg and Newcombe 2016). Individuals may perform identically on behavioral navigation measures despite underlying variability in neural processing. Even if navigation tasks were designed to force a particular type of navigation behavior, there is no guarantee that a participant would actually adopt that behavior or rely on specific neural mechanisms (see Section 3).

Several factors may account for interindividual variability in both navigation outcomes and the underlying navigation processes used to generate those outcomes (Perez and Weisberg 2025; Wolbers and Hegarty 2010). These range from genetically linked variations in neural systems or combined genetic-environment interactions on personality traits to more "proxy variables" (i.e., categorizations of people based on broad characteristics that relate to constellations of more specific changes), such as sex, gender, ethnicity, and education as well as age, socioeconomic factors, and geographical experience. Although they often capture a large proportion of variance and are easy to study, proxy variables can disguise underlying causes for individual variability because correlated clusters of processes might differ or lead to cascading differences. For instance, individuals who are highly motivated and open to exploring their environment may become lost more frequently but also acquire greater spatial knowledge. By contrast, individuals with a high working-memory capacity may rely more on navigational processes related to online access to information (e.g., referring to an image of a mental map, or a list of verbal directions) than on external aids, such as handwritten directions.

Healthy aging trajectories can produce substantial variation in sensory and motor systems. Older navigators may lose a navigationally relevant process or sensory system but be able to adapt and still achieve acceptable navigation outcomes for their particular needs. In contrast, unhealthy aging, particularly in the context of disease, introduces additional sources of variability, including differences in pathology and severity. For older adults living with dementia, this may lead to profound changes beyond those associated with typical aging. Additional clinical features may include comorbidities affecting sensory and cognitive function, polypharmacy (i.e., concurrent use of more than seven medications), and alteration in sleep patterns.

2.3.2 How Do Individual Differences Influence the Triadic Navigation Framework?

There are two broad ways individual differences could impact the Triadic Navigation Framework: they could yield practical variability (i.e., contribute to functional changes in real-world behavioral outputs) or result in process variability (i.e., change the brain–behavior relation, such that two people who appear to behave the same way could use distinct cues or processes). As detailed by Perez and Weisberg (2025), individual differences can be measured in terms of navigation *outcomes* and navigation *processes*. For example, if the paths of two individuals intersect, one may predominantly follow visual optic cues whereas the other may rely more on vestibular cues. In the Triadic Navigation Framework, at the navigation systems level, the neural representations guiding path integration would be very different at a sensory level but perhaps similar at the grid-cell level. Now imagine that the individual who relies more on vestibular cues also happens to have risk factors for Alzheimer disease and performs worse than the other. One interpretation might hold that this individual has an impaired grid-cell system. However, without understanding why they prefer vestibular cues or measuring their vestibular ability, we cannot know for certain. To date, very few studies investigating path integration as a marker of early cognitive decline in Alzheimer disease have concomitantly assessed vestibular function in the same participant group, despite there being a known performance link (Allen et al. 2004).

To help discern what does and does not matter, we delineate three ways in which researchers should consider the impact of individual differences within the Triadic Navigation Framework. First, individual differences *shape* the properties of an individual's *navigation context*. What cues are available for them to use, and which optimization principle do they employ? For example, city-dwellers tend to be poorer navigators than people from rural areas who contend with more complex environments compared to simpler "grid-style" landscapes (Barhorst-Cates et al. 2021; Coutrot et al. 2022). Within younger adults, people who are more likely to explore on a gambling task are more likely to explore a novel virtual environment (Perez et al. 2025).

Second, individual differences *personalize* an individual's *navigation system*. Men tend to use place-based navigation strategies more frequently than women, suggesting a general increased weighting on one type of representation over another (Nazareth et al. 2019). *APOE* ε4 allele carriers tend to prioritize proximity to environment walls than noncarriers, which may be a compensatory mechanism for altered hippocampal circuitry (Kunz et al. 2015). People that spontaneously choose place- versus response-based strategies have more hippocampal gray matter but reduced caudate gray matter (Bohbot et al. 2007).

Finally, *individual differences* have relevance for an individual's navigation ability. When designing interventions to support older people's everyday navigation ability (see Section 3), knowing that someone prefers route-based navigation would make a place-based assistive tool irrelevant to them.

2.3.3 Research Gaps to Address

Moving forward, assessing individual variability in older adult populations continues to be a high priority. Although base rates of poor navigation function in old age and in neurodegeneration are well documented (Coughlan et al. 2018; Lester et al. 2017), other groups have shown "super-agers" who perform at or above levels of the average younger adult (Zhou et al. 2023). Yet, a large literature similarly demonstrates individual variability in healthy younger adults (Hegarty et al. 2002; Ishikawa and Montello 2006; Weisberg et al. 2014). The dearth of longitudinal research makes it difficult to know how such variability at younger ages may be protective for older adults, both in terms of developing dementia and preserving navigation function (e.g., through cognitive or neural reserve, see Section 3).

Inter- and intraindividual differences are virtually not taken into account in animal models for spatial navigation, since many animal models are constrained to specific sexes and strains. This is not only an issue for spatial navigation research in animals per se. It also has implications for the animal-human translation of spatial navigation changes during aging and in pathology.

Leveraging our framework, we can begin to design studies to ascertain at what levels in a particular navigation task individuals vary (e.g., sensory, motor, cognitive, motivational) and which aspects of the task are uniquely affected. Many of the above factors become relevant during aging when they emerge. Currently, it is often not possible to delineate spatial navigation deficits from variation due to individual factors, as the latter are too numerous. This becomes especially relevant for the detection of subtle spatial navigation changes in preclinical populations.

2.4 Example Use Cases of the Triadic Navigation Framework

2.4.1 Simulating Aging

To illustrate how the navigation system BB model (or alternative models) can help characterize the mechanisms that drive navigational decline in aging, consider an approach taken by Bernard et al. (in prep.), who investigated whether aging affects computations in the head-direction circuit and sought to identify which mechanisms might account for potential deficits. They conducted a behavioral experiment in immersive virtual reality in which participants had to monitor and report their facing direction after performing a turn, either immediately or after a brief delay. As expected, older adults made larger angular errors when reporting their orientation, but both groups were similarly affected by the delay condition. Bernard et al. then used parts of the BB model (i.e., the ring attractor network underlying the head-direction component and angular velocity inputs reflecting self-motion cues) to model the key neural computations thought to drive task performance. Critically, the authors also perturbed the model to simulate potential age-related changes (e.g., reduced synaptic efficacy within the ring attractor, neuronal loss within the ring attractor, and noise added to the sensory inputs). These simulations revealed that the activity in the head-direction circuit is differentially affected by the different perturbations. For example, a reduced representational space was primarily a consequence of neuronal loss, whereas noise in the self-motion inputs affected the stability of the attractor. Linking these insights back to the observed behavioral performance ultimately allowed the most likely causes of impaired head-direction coding in older adults to be identified. At the same time, the model could not accommodate the entire range of the observed phenomena, yielding clear implications for future studies.

2.4.2 Contextualizing with Individual Differences

The Dual Solution Paradigm—a navigational task in a virtual, maze-like environment—enables navigators to learn the location of multiple objects by viewing (or virtually traveling along) the same route multiple times. After learning, the navigators are positioned along the route and must navigate to one of the objects. Older adults and women tend to use the path they learned initially whereas younger adults and men tend to use a shortcut that is available and shorter than the learned path, but which was not taken during learning (Yu et al. 2021). Despite this finding, which was replicated several times, it is not clear how or why individuals differ in preference for familiar routes or novel shortcuts. The Triadic Navigation Framework provides insight into this question by encouraging researchers to evaluate when and where individual differences might play a role in navigational decision making. First, consider the results from a *process-focused lens*. For instance, if sex and age play a role in shifting an individual's navigation context, we might expect younger adults and men to minimize time as their optimization principle. Alternatively, women and older adults may rely more on memorizing a sequence of turns given the availability of certain cues, or women and older adults may be less familiar with desktop virtual reality or moving through an environment with a video game controller, resulting in stricter adherence to something overlearned rather than something that requires online computation. In contrast, individual differences may shape the representations used during retrieval: men and younger adults may more actively engage their medial temporal lobes during learning, resulting in better knowledge of the spatial layout.

Next, consider the results from a *functional lens*. Women and older adults use a less flexible strategy, but nevertheless find the goal approximately as often as men and younger adults. Functionally, this results in no difference in ability, measured by whether the navigator gets to their goal, but it would result in differences in path efficiency, being able to route around detours, and perhaps cognitive load (e.g., their ability to perform concurrent tasks).

Each of these hypotheses would benefit from a thorough study of individual factors and more careful specification of which factors can be and are used by people solving the Dual Solution Paradigm. They would also benefit from the application of specific computational models, which could specify which neural computations are required for task performance and which may be impacted by aging or especially in neurodegeneration.

3 Potential Ways for Navigation to Contribute to Societal and Clinical Needs

Having presented the Triadic Navigation Framework, we will now proceed to describe potential applications in navigation research and development to various unmet needs in society. Specifically, we will use the framework to identify areas where research is needed and how it can translate to practice and everyday use. We divide our discussion into *assessment*, which focuses on measuring an individual's navigation ability, and *intervention*, which explores ways to support and maintain independent navigation.

3.1 Measuring Ability

Within the Triadic Navigation Framework, there are two main goals for assessing someone's navigation ability (see Figure 16.6). The first is to index an individual's navigation system using specially designed tasks ("systems assessment"). The

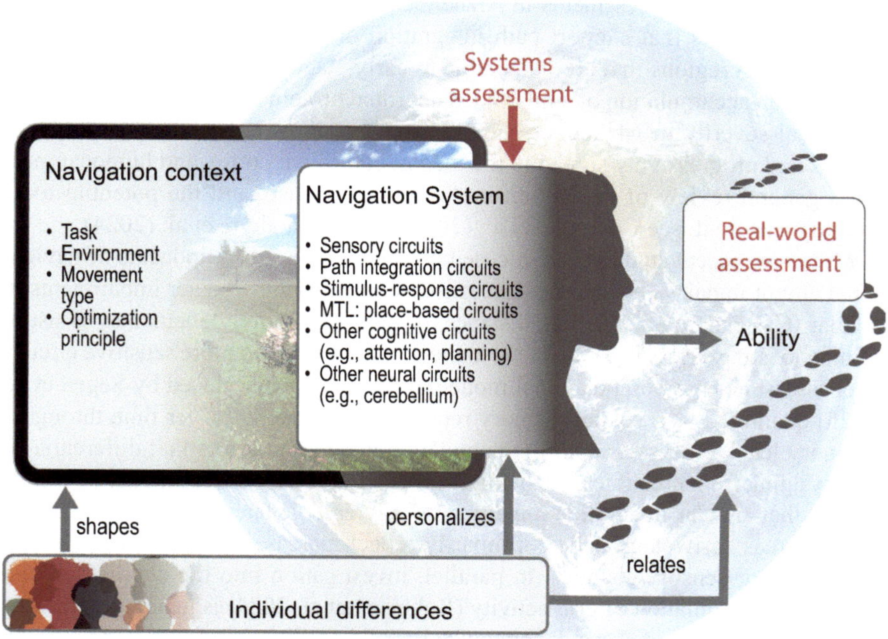

Figure 16.6 The two main types of navigation assessment tools: circuit/systems and real-world functioning. Both are assessed through behavioral tasks.

second is to measure an individual's ability in an everyday context, regardless of the specific instantiation of the navigation system used ("real-world assessment").

3.1.1 Navigation Systems Assessments

Systems assessments can be most useful in clinical contexts to detect or monitor a disease's progression. The goal of systems assessments is to infer the (mal)functioning of a specific navigation circuit or system in humans by analyzing their behavior. Highly reduced and carefully controlled behavioral paradigms that drive people to use a certain type of navigation system to perform a task can inform us how well a specific circuit is functioning in clinical and preclinical individuals.

3.1.1.1 Case Study: Alzheimer Disease

In Alzheimer disease, for example, several studies have demonstrated that tests of path integration and place-based spatial memory, learning, or wayfinding can be used to detect early-stage Alzheimer disease and accurately classify

- mildly cognitively impaired people at high versus low risk for future disease (Coughlan et al. 2020b; Howett et al. 2019; Wood et al. 2016),
- people who progress or do not progress clinically over time (Levine et al. 2020), and
- higher risk unimpaired individuals or those with subjective cognitive decline (Newton et al. 2024; Segen et al. 2025).

Navigation systems assessments in Alzheimer disease are highly relevant because the neural circuits that support path integration or place-based computations are found in brain regions that are vulnerable to early-stage pathology. Indeed, tau pathology—an accumulation of a misfolded neuronal protein that is the main correlate of clinical severity in Alzheimer patients—is thought to begin in transentorhinal regions and progressively spread to the medial entorhinal cortex and hippocampus. For a general review of navigation in Alzheimer disease, and the potential tools used to measure it, see Coughlan et al. (2018) and Čepukaitytė et al. (2024).

Moreover, recent studies have focused on the specific subcomponents of navigational circuit impairments in Alzheimer disease. Identifying whether impairments in angular (Castegnaro et al. 2023), distance, or path-integration deficits may assist in refining existing assessments and lead to the design of simpler, more sensitive circuit-level assessments. A computational modeling framework introduced by Segen et al. (2025) quantified how spatial memory representations degrade over time through a "memory leak" process—capturing cumulative integration errors that differentiated healthy aging from preclinical Alzheimer disease. Importantly, these modeling results suggest that angular integration impairments may emerge later in the disease trajectory, whereas early-stage deficits primarily reflect increased memory decay rather than motor or sensory changes. In parallel, investigation into the cholinergic and GABAergic modulation of grid activity (Robinson et al. 2024) is generating mechanistic understanding of these deficits, which could facilitate drug development and improve our understanding of key causes of Alzheimer disease pathology.

Nonetheless, it remains to be determined whether the current assessments of path integration in these studies (namely, using virtual reality triangle completion-like tasks) actually assess the integrity of the grid-cell circuitry, or whether performance variation is due to variance from sensory systems affected by aging and disease (e.g., visual and vestibular systems). By integrating behavioral modeling approaches, such as those applied in Segen et al. (2025) and Castegnaro et al. (2023), systems-level assessments can move beyond descriptive performance metrics to mechanistically isolate spatial memory integration failures from sensory or motor confounds.

3.1.1.2 Toward Clinical Application

Assuming these "systems" navigation assessments are valid and reliable for clinical decision making (for further discussion, see Section 4), the transition to clinical use can happen in several ways. In the most realistic and near-future scenario, a navigation assessment tool could complement cognitive tests that clinicians already use to evaluate cognition in their patients undergoing dementia diagnostic workups. To persuade clinicians to change their current systems, the tool would be accompanied by an explanation of the critical constraints regarding the practicality of use, time of administration, ease of setup, and demonstrations of cost-effectiveness (for more on this, see Chapter 15). The transition to the clinic requires the collaboration and codesign between navigation neuroscientists, clinicians, design engineers, human-computer interaction researchers, medical regulators, health-care decision makers, and users to balance the scientific need of the assessment with the clinical and economic constraints of the testing environment.

Further into the future, it might be possible for a battery of separate entorhinal- and hippocampal-dependent navigation tests to behaviorally track the progression of disease pathology, either under supervised clinical settings or in remote at-home unsupervised settings. For instance, supervised clinical monitoring of decline in patients could help identify the optimum window for treatment intervention. If a drug for Alzheimer disease were available, there might be a need to administer it before tau spreads from entorhinal cortex to hippocampus (Busche et al. 2019); halting tau spread can rescue hippocampal (but not entorhinal) deficits in mice, depending on when the intervention is administered. These batteries of entorhinal-hippocampal tests could also be relevant in clinical trials as an outcome measure for drugs that specifically target tau in the entorhinal cortex (Ritchie et al. 2017). For example, if a drug improves an individual's path-integration ability, that could mean that their grid-cell system is regaining function in response to the drug. It might also be relevant in preclinical drug development contexts: the grid cell–path integration brain–behavior bridge could be used as a platform to evaluate whether a drug shows both target engagement and functional improvement; in animal drug trials, the Morris water maze remains the gold standard, but since the hippocampus is affected relatively late in the progression of Alzheimer disease, it may not serve as a sensitive behavioral marker for early-stage drug development. Outcome measures based on brain behavior would overcome limitations with current standardized cognitive measures, which are very blunt in the domains and systems they measure, such as the Clinical Dementia Rating scale or assessments of word-list recall and verbal fluency, which are used as the primary outcome measure. In general, the concept of precision circuit testing to track disease progression behaviorally is still a long way from reality. Below, we discuss how research could be advanced toward this goal.

Alternatively, remote and unsupervised app-based assessments could be made available to the general public to evaluate changes in their navigation ability and prompt them to seek medical advice. This is a nontrivial option that would require a profound transformational change in society. It also raises ethical questions about the feasibility and implications of screening for dementia in individuals. If more drugs for Alzheimer disease become available that have fewer side effect profiles, the need for such a screening approach may be heightened. However, while providing enhanced datasets, unsupervised cognitive assessments can reduce the reliability of the data because testing environments will have unknown effects on performance (Berron et al. 2024).

3.1.2 Real-World Navigation Assessments

Understanding the performance of an individual in a given navigational context without necessarily knowing the underlying computations is relevant to understanding how well an individual is navigating in their everyday lives. Such an understanding has clinical relevance in functional assessments that evaluate whether someone needs to move into assisted living or care because they can no longer find their way around their local community. However, real-world assessments are not limited to clinical applications and can focus on optimizing support for individuals to live independently for longer. This includes designing navigation aids, implementing environmental and building modifications, or developing technology-driven solutions that enhance spatial orientation and wayfinding in everyday settings. For example, a group of vision-science engineers have developed an augmented reality app that can measure how much visual stress a certain design feature would trigger using real-time video footage of a building's features. This technology is targeted at architects and designers interested in improving the human-centeredness of urban design (Hosking 2024). Another example enables individuals in Singapore to actively contribute to the improvement of their community by using the Environment Audit Tool app to identify facilitators and barriers regarding mobility, services, legibility, and safety (Yuen et al. 2022). Despite the far-reaching applications of real-world navigation ability assessments, they do not appear as well-researched or used as navigation systems assessments. This represents a key area for future work.

3.1.2.1 Challenges

Designing real-world assessments that can reliably infer how well someone navigates in everyday life remains an unaddressed gap in the literature. The ecological validity (or degree of transfer to real life relevance) of many established cognitive navigation assessments remains unknown (see Chapter 15). Small-scale studies involving *Sea Hero Quest*, a serious mobile navigation game, have attempted to compare the performance of the mobile game to navigating street networks around London. These studies, however, are not generalizable and findings are mixed in terms of being able to predict the likelihood of getting lost (Goodroe et al. 2025; Puthusseryppady et al. 2022).

One pressing challenge is the replication of the real-world cost and emotional valence of getting lost, which might impact either the navigational systems or context in which individuals operate (Chapter 12). Building large-scale immersive virtual reality replications of environments to assess real-world wayfinding has made some progress, but it is not fully understood how these virtual systems impact strategy selection or neural representation of information.

Even designing self-report questionnaires remains an unresolved problem. Currently, there is no widely accepted questionnaire or agreement on which type of questionnaire should be used. In addition, the majority of studies investigating self-report abilities show that there is male overinflation and female underestimation of navigation abilities, and that these self-reports do not always correspond to objective navigation performance on assessments (Clark and Maguire 2020; van der Ham et al. 2021). These challenges are likely related to an issue around metacognition of navigation abilities, where, generally, people have little awareness of how they navigate in everyday life. This issue is even more pronounced for aging adults, given that existing questionnaires are often validated using data from college students. To our knowledge, there is no validated navigation questionnaire specifically designed to assess navigational abilities in both healthy and pathological aging.

3.1.2.2 Unmet Needs

Clinicians need short, reliable, and valid assessments of how well someone is coping with everyday navigational tasks. Some neurologists and psychiatrists, who perform diagnostic workups for individuals suspected to have early dementia, report that the most useful and sensitive question they ask a patient or their carer is whether or not the patient still drives and, if not, whether it is due to reduced confidence in finding their way around (D. Chan, pers. comm.). Here, one simple intervention could be to introduce a set of extremely well-designed questions that would probe navigational ability into an already commonly used clinical functional questionnaire, such as the Instrumental Activities of Daily Living Scale.

Another under-researched theme related to health care is whether staff at care facilities would find a real-world navigation assessment useful to decide whether certain residents could have greater outdoor access based on their risk of getting lost (Heward et al. 2022). Similarly, it might be useful for the police to have tools to assess the likelihood of certain neighborhoods becoming "hotspots" for older adults to become easily lost based on unique environmental features. For example, street network entropy but not landmark density was demonstrated to be associated with the density of missing person reports for a given area in rural eastern areas of the United Kingdom (Puthusseryppady et al. 2019, 2020).

Using smartphone GPS sensors, or GPS tags, that people carry with them or are installed in their cars, it is possible to collect remote, passive tracking of an individual's personal, naturalistic, day-to-day movements over a defined period. With machine learning analysis of these trajectories, high-level metrics and patterns from the data can be extracted (e.g., how many times an individual leaves a 1 km radius around their house or the number of angles above 90° in their day-to-day movements). In several studies, these metrics have been shown to be predictive of an individual's cognitive status, even if patients might be traveling with the support of a carer (Ghosh et al. 2022; Roe et al. 2019). Intriguingly, these passive, remote GPS-tracking metrics sometimes demonstrate bigger effect sizes in classifying the risk of future Alzheimer disease than active "systems" navigation assessments performed in research settings (Čepukaitytė et al. 2024). How GPS and mobile sensor data could help inform real-world assessment of everyday navigation ability in healthy aging populations is an unaddressed topic in the literature.

3.2 Supporting and Maintaining Ability

The ability to navigate is critical to support an independent life: it enables outdoor mobility fundamental for grocery shopping, attending health appointments, traveling for pleasure, and social interaction with other people. The inability to navigate effectively has serious implications, particularly for people living with Alzheimer disease, as getting lost can be a life-threatening event. Furthermore, difficulties with navigation compounds several other key risk factors for unhealthy aging and age-related diseases, including reduced physical activity and increased social isolation, both of which are key risk factors for dementia (Livingston et al. 2024). Only scant literature addresses how navigation ability and metacognition of one's own navigation ability change with age and disease, and even less is available on how navigation can be reliably supported through interventions.

Based on interview studies with older adults (Heward et al. 2022), we know that people report an increased fear of getting lost, which leads them to alter their behavior. They go out less often and only go when they can hand over navigation responsibilities to someone else. Interestingly, individuals who have sufficient technological skills might use Google Maps and Street View to familiarize themselves with a new environment before traveling, such as when they go on holiday. Others report an aversion to using available SatNav devices, as they fear the technology will distract them while driving or walking, which could lead to accidents or falls.

Returning to the Triadic Navigation Framework, creating interventions (Figure 16.7) for people to continue navigating at their normal level or to feel sufficiently confident to navigate alone can be achieved through following:

1. Support interventions: The external environment or navigation context needs to be made more navigation friendly, and assistive solutions need to be created to help guide people during navigation.
2. Maintenance interventions: Cognitive training paradigms (in isolation or with neurofeedback mechanisms or drugs) should be developed to restore real-world ability or sensory systems, with assistive aids to reduce the noise of environmental and idiothetic cue inputs to people's navigation circuitry.

3.2.1 Support Interventions

Support interventions may include small-scale tools for people to use or large-scale initiatives aimed at building design. An example of the latter can be found in Singapore neighborhoods (Figure 16.8), where large apartment blocks have been painted with murals of different animals, and wayfinding signs were added to walls and floors to assist place recognition and help older people orientate. In another example, hospitals are implementing floor-based wayfinding cues, such as guiding lines, since older adults are more likely to look downward, either to seek geometric cues or because of a stooped gait. Similarly, in dementia care home design, greater emphasis is being placed on the use of lighting and differentiation of different corridors to aid the wayfinding ability of individuals who live there (O'Malley et al. 2018). Navigation-based support interventions can also play a role in public health. For example, there are fewer rates of hip fractures and better long-term postfracture mortality outcomes in more complex rural areas compared to urbanized environments (Solbakken et al. 2019).

More scalable support interventions might include smartphone apps, such as Beeline, that aid navigation. Similar to Google Maps, these apps do not have complex user interfaces but simply display a homing vector arrow to show someone the general direction to their destination. There are also audio versions of wayfinding guidance apps designed for individuals with visual impairments. Such apps are particularly important since aging is often accompanied by a decline in visual acuity as well as conditions (e.g., age-related macular degeneration and glaucoma) that can restrict central or peripheral vision and make traditional visual cues used in GPS systems less effective (see Section 2.2). In future technology, we would expect to see augmented reality glasses that display navigational information or gamified versions of wayfinding apps, similar to the concept of Pokémon Go, which has augmented reality images visible through a smartphone camera that encourage people to go outdoors and perform wayfinding-based tasks for fun.

For all applications, it is important to remember the significance of individual differences in navigation ability, preferences, and practices. As shown in our framework, these can influence how a person's navigation system might work, the type of navigation context they are in, and whether their ability in a particular type of navigation is relevant. If lifelong place-based navigators begin to experience a decline

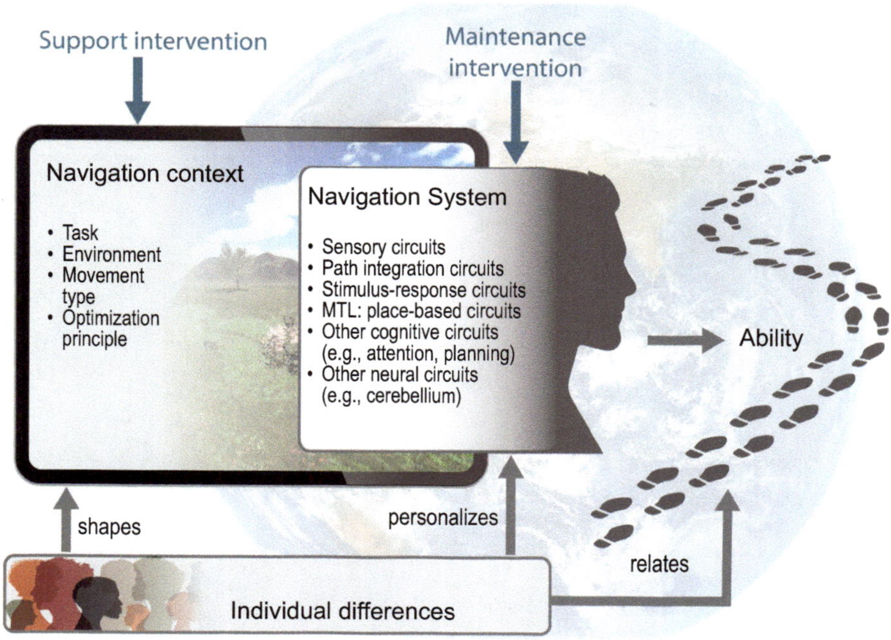

Figure 16.7 Potential points of intervention in the navigation framework that can support, maintain, or enhance people's ability to perform navigational tasks in situations that are meaningful to them.

in navigation ability, an app that offers only egocentric-based guidance will not help them. Similarly, although simple to use, assistive technologies that inform a person where to go, but not where they are, can feel restrictive and paternalistic to the user, limiting choice instead of supporting agency. There is a strong need for navigation researchers to engage in more participatory, qualitative-based research to codesign potential navigation support interventions with their end users.

3.2.2 Maintenance Interventions

Maintenance interventions aim at the functioning of core navigation or supporting sensory brain circuits on which an individual's navigation ability depends. This can involve cognitive rehabilitation, but the field is renowned for having a mixed and poor-quality evidence base, often dominated by commercialized apps claiming evidence of brain training. Generally, the main issue with cognitive training is the issue of transfer—whether the training actually confers real-world functional improvements—and this is similarly problematic in navigation training.

Notwithstanding the literature on London taxi drivers, who demonstrate an enlarged posterior hippocampus (Maguire et al. 2000), only a handful of high-quality studies have shown that navigation training effectively enhances either behavioral ability or hippocampal volume (Lövdén et al. 2012). One study specifically examined the effect of training older adults to be more flexible in switching between allocentric and egocentric strategies, although results revealed an increased ability only in egocentric-based navigation (van der Kuil et al. 2020). Some fMRI studies have

Figure 16.8 Examples of support interventions from the Yio Chu Kang, Khatib, Nee Soon, and Yishun neighborhoods in Singapore, where building and signage design are focused on improving the wayfinding of older adult residents who may become disorientated in more homogeneous environments. (Photos by Victor R. Schinazi)

also explored training with neurofeedback mechanisms during navigational tasks, demonstrating an inability of individuals with biomarkers of Alzheimer disease to downregulate their hippocampal activity with implications for their risk of disease spread (Skouras et al. 2020).

A recent epidemiological study supports the potential of navigation-based cognitive training as a lifelong protective intervention (Patel et al. 2024). Across nine million people in the United States, individuals in navigation-intensive professions (e.g., taxi and ambulance drivers) were less likely to die from Alzheimer disease compared to those in other occupations (for further discussion, see Section 3.1). Interestingly, this protective trend was not observed in other transportation-related jobs that do not require extensive real-time spatial processing (e.g., bus drivers, ship captains, airline pilots), where navigation demands are more commonly offloaded to supporting technologies and systems.

Maintenance interventions can also occur through education. Perhaps the strongest argument for this comes from the navigation game, *Sea Hero Quest*. Coutrot et al. (2018) found that individuals from Scandinavian countries, where orienteering is part of the school curriculum, performed best at the game. It is currently unclear whether educational interventions, even at the school-age level, might have lifelong protective effects on individuals to maintain their navigation ability into older age. Finally, it is important to explore educational interventions in older age that are targeted at developing strategies to overcome difficult navigational scenarios and specific age-related navigation challenges. If successful, such interventions could mitigate age-related spatial anxiety and support independent living, well-being, and quality of life in older age.

3.3 Is Navigation More Than the Sum of Its Parts?

When discussing the utility of applying navigation tools outside of research settings to address unmet societal needs, two questions arise regarding the meaning of navigation: Are the individual component systems, such as path-integration circuity and the mapping to behavior, special, or is it the integration of multiple, different cognitive domains producing an effect that is more than the sum of the individual navigation parts? With respect to maintenance training interventions to support navigation, should one train the whole system or just component parts of navigation to produce effects that transfer to the everyday lives of individuals?

The answer is likely to be both, depending on the particular goal of the navigation assessment. One analogy that may be useful here is that of a professional tennis player. If you were to measure the size of the player's biceps and know how fast they can run, their reaction time, and level of motivation to win, could you reliably infer how good that tennis player is currently or will be in the future? Perhaps it is only possible to infer these things simply by watching them play a match and then looking at the track record of how often they have won matches. Similarly, if you want to make someone better at tennis, do you make them do drills of individual components (serving, rallying, muscle building, mental agility training), or do you make them repeatedly play tennis matches?

Research studies that set out to address these questions would be a useful addition to the field in helping to prioritize the translation of navigation in practical tools for society. Is it the right approach to continue searching for the ultimately controlled and reduced navigation paradigm that is a hypothesized index of a certain navigation circuit, such as in the diagnosis of Alzheimer disease? Alternatively, will more real-world effectiveness be gained from advancing research into passive GPS tracking of the messy, everyday "sum of the parts" navigation?

4 Recommendations for Moving Forward

To explore how the Triadic Navigation Framework can help us conceptualize interactions between aging deficits and spatial navigation deficits, we highlight cases in this section where navigation deficits might be expected to arise. We assess how reliably they can be measured (including transfer) and outline interventions to support or enhance the desired behavior.

4.1 The Emergence of Navigation Deficits with Age

While several experiments demonstrate age-related deficits in spatial navigational tasks (e.g., Barnas et al. 2024; Harris and Wolbers 2012, 2014; Moffat 2009; Wiener et al. 2013), it is currently unknown to which extent a task, environment, and/or movement can affect these deficits (i.e., aspects of navigation context). Moreover, although these features can interact, individual experiments typically consider each of them separately. Older adults show worse path integration than younger adults. These deficits all arise at the level of spatial representations but are nearly always computed within tasks that require path integration during response execution (task); thus, a direct readout of the underlying spatial knowledge is not provided. In addition, experiments nearly always occur in simpler environments at smaller scales, with limited cues (environment). Experiments do vary in movement (i.e., walking, omnidirectional treadmills, joysticks, tablets), but movement is rarely varied within the same experiment (but see Hill et al. 2024). Other forms of spatial deficits (e.g., Head and Isom 2010; Yu et al. 2021) arise in more complex environments, but studies with more complex environments usually employ different tasks (e.g., wayfinding, map building), provide different cues, and use different types of movement that also differ from what is used in simpler navigational tasks like path-integration paradigm.

4.2 Measuring Navigation Deficits Given Individual Differences

Individual variability in strategy selection (e.g., whether an individual follows a familiar route or takes a novel shortcut) has been shown in younger adults and can be driven by factors such as sex or gender (Boone et al. 2019), risk aversion (Aumont et al. 2019), and navigational abilities. These individual difference factors, however, may (a) change the task (i.e., navigation context) the navigator wants to achieve or (b) affect which circuits the navigator might use to achieve that task (i.e., navigation system). Particularly in aging contexts, it could be that the underlying spatial mechanisms are down-weighted in the strategy selection process or that the circuits themselves are impacted by aging (Bates and Wolbers 2014; Harris and Wolbers 2014). Elucidating how individual differences modulate the processing of spatial cues and the adoption of navigational strategies can inform the level of analysis at which behavior is evaluated, be it by examining the transfer across environments, experimentally manipulating cue availability, or assessing the fidelity of grid-like signals in the entorhinal cortex.

4.3 Intervening to Support or Assess Deficits

The needs of the researcher or clinician should dictate which measures are used and whether these relate to functional outcome or underlying circuit. For example, if the researcher's question relates to minimizing older adults becoming lost, measuring which strategy is selected may be less important than measuring success: Did the navigator make it to their goal? If they didn't, why not? If aging impacts spatial attention, then perhaps the ability to encode landmarks at interactions is impaired, causing the navigator to rely on a faulty familiar route strategy. The memory processes that enable the sequence of turns and binding of the landmark to the intersection could be intact, but the attentional mechanisms that direct gaze to the landmark is missing. In this case, independent measures of spatial attention will affect which cues in the navigation context need to be processed by the navigation system to support route-based computations. An intervention could then be developed to enhance spatial attention (e.g., through general cognitive training) or, alternatively, signs could be installed that require fewer spatial attention resources to interpret than landmarks.

4.4 Five Key Recommendations

In formulating the Triadic Navigation Framework and outlining directions for future research, we identified the following key recommendations to advance both basic and applied understanding of navigation in the context of healthy and unhealthy aging:

1. *Psychometric properties*: Despite their growing prominence in research and clinical contexts, navigation assessments have not yet consistently demonstrated strong reliability and construct validity, posing significant challenges to its utility. Reliability refers to the consistency of a measure, such as whether repeated assessments produce stable results under similar conditions. In navigational tasks, factors like varying environmental layouts, differences in test administration and individual participant states (e.g., stress or fatigue) can introduce variability. Construct validity assesses whether a test measures the concept it is intended to assess; in this case, spatial navigation ability. Navigation is a multifaceted skill involving working memory, attention, visual-spatial processing, executive functions, and long-term memory. A poorly designed test might emphasize one domain disproportionately or fail to incorporate key elements, such as sensory integration. Without strong construct validity, a navigation assessment might measure something tangential, such as general cognitive ability, rather than navigation itself. Without established psychometric properties, findings from studies could be misinterpreted, leading to flawed conclusions and even misdiagnosis of cognitive impairments or improper tracking of disease progression. To address these issues, navigation assessments require standardization and rigorous validation studies to establish test-retest reliability and construct validity of navigational tasks across diverse populations and contexts (Coughlan et al. 2020a; Goodroe et al. 2025). This is especially important in longitudinal research and the study of individual differences, where effect sizes may be small and highly reliable measures are needed to detect intraindividual changes or subtle interindividual differences.
2. *Task and data sharing*: "Open science" practices need to be prioritized and fostered through the sharing of tasks and data to ensure transparency, collaboration, and reproducibility. Open repositories, such as *ManySpaces*, provide freely available navigational tasks to researchers worldwide and represent a first step toward addressing barriers. By making tasks accessible, researchers can collaboratively refine assessments, validate their reliability and construct validity across diverse populations, and establish shared benchmarks for performance. Relatedly, the mobile game *Sea Hero Quest*

has provided an enormous normative dataset of navigation performance across the whole lifespan. This data is, however, vastly underutilized, and experimenters could use this and other open datasets as benchmarks for their study populations. Data and task sharing will also facilitate meta-analyses and cross-study comparisons and enable widespread access for researchers, clinicians, and educators worldwide, fostering global collaboration and innovation.

3. *Longitudinal data*: With very few exceptions, the vast majority of studies on age-related decline in spatial navigation have employed cross-sectional designs that compare different age groups at a single time point (Korthauer et al. 2016). This approach can confound age with cohort effects, and it is difficult to ascertain causality of relationships mapped out in the Triadic Navigation Framework. Therefore, it is crucial for the field to start collecting longitudinal datasets that are better suited to isolate true age-related change, to reveal how individuals compensate for cognitive or neural decline over time, and to capture intraindividual changes that reveal fluctuations in navigational function and behavior. Until we have these datasets, it will be impossible to establish how, for example, decline in sensory systems versus decline in entorhinal integrity contribute relatively to the decline in path-integration behaviors.
4. *Bridging insights across species*: A key aspiration for our framework is that it helps researchers working on different species to interact in a common thinking space. Through the shared-species methods of intracortical recordings, MEG/EEG/fMRI, and behavior, understanding can be mapped across the multileveled computational models of the navigation system. By having this holistic framework, we hope to inspire the design of research studies that help address high priority areas relating to construct validity of navigation tools that are on the precipice of achieving impact in addressing unmet clinical and healthy aging society needs.
5. *From the clinic back to basic science*: A core theme in our discussion of addressing unmet needs in healthy and unhealthy aging societies centered on the shift from a "science-push" approach (translating navigation forward given hypothesized relevance in society or the clinic) to a "needs-pull" approach (using more participatory research methods with the public and patients to understand their problems with everyday navigation, and then designing scientific experiments to help address them). We advocate for advancing navigation research toward the "needs-pull" approach, with greater attention given to the needs, barriers, and facilitators that shape navigation in everyday life. More participatory research is needed to support everyday functioning of older people in alignment with their values (e.g., maintaining independence or mobility). To date, this has virtually not been explored in navigation research yet may be crucial to maximizing healthy aging trajectories.

References

Agrawal Y (2008) Prevalence of Hearing Loss and Differences by Demographic Characteristics Among US Adults: Data from the National Health and Nutrition Examination Survey, 1999–2004. Arch Intern Med 168 (14):1522. https://doi.org/10.1001/archinte.168.14.1522

Allen D, Ribeiro L, Arshad Q, and Seemungal BM (2016) Age-Related Vestibular Loss: Current Understanding and Future Research Directions. Front Neurol 7:231. https://doi.org/10.3389/fneur.2016.00231

Allen GL, Kirasic KC, Rashotte MA, and Haun DBM (2004) Aging and Path Integration Skill: Kinesthetic and Vestibular Contributions to Wayfinding. Percept Psychophys 66 (1):170–179. https://doi.org/10.3758/BF03194870

Atchley P, and Andersen GJ (1998) The Effect of Age, Retinal Eccentricity, and Speed on the Detection of Optic Flow Components. Psychol Aging 13 (2):297–308. https://doi.org/10.1037//0882-7974.13.2.297

Aumont É, Blanchette C-A, Bohbot VD, and West GL (2019) Caudate Nucleus-Dependent Navigation Strategies Are Associated with Increased Risk-Taking and Set-Shifting Behavior. Learn Mem 26 (4):101–108. https://doi.org/10.1101/lm.048306.118

Banino A, Barry C, Uria B, et al. (2018) Vector-Based Navigation Using Grid-Like Representations in Artificial Agents. Nature 557:429–433. https://doi.org/10.1038/s41586-018-0102-6

Barhorst-Cates EM, Meneghetti C, Zhao Y, Pazzaglia F, and Creem-Regehr SH (2021) Effects of Home Environment Structure on Navigation Preference and Performance: A Comparison in Veneto, Italy and Utah, USA. J Environ Psychol 74:101580. https://doi.org/10.1016/j.jenvp.2021.101580

Barnas AJ, Kunath J, Perez E, et al. (2024) Spatial Navigation Strategy in Older Adults: Preference or Ability? PsyArXiv Preprint:1–50. https://doi.org/10.31234/osf.io/9mpd7

Bates SL, and Wolbers T (2014) How Cognitive Aging Affects Multisensory Integration of Navigational Cues. Neurobiol Aging 35 (12):2761–2769. https://doi.org/10.1016/j.neurobiolaging.2014.04.003

Beard JR, Officer A, De Carvalho IA, et al. (2016) The World Report on Ageing and Health: A Policy Framework for Healthy Ageing. Lancet 387 (10033):2145–2154. https://doi.org/10.1016/s0140-6736(15)00516-4

Bécu M, Sheynikhovich D, Tatur G, et al. (2020) Age-Related Preference for Geometric Spatial Cues during Real-World Navigation. Nat Hum Behav 4 (1):88–99. https://doi.org/10.1038/s41562-019-0718-z

Berard JR, Fung J, McFadyen BJ, and Lamontagne A (2009) Aging Affects the Ability to Use Optic Flow in the Control of Heading during Locomotion. Exp Brain Res 194 (2):183–190. https://doi.org/10.1007/s00221-008-1685-1

Berron D, Olsson E, Andersson F, et al. (2024) Remote and Unsupervised Digital Memory Assessments Can Reliably Detect Cognitive Impairment in Alzheimer's Disease. Alzheimers Dement (N Y) 20 (7):4775–4791. https://doi.org/10.1002/alz.13919

Bian Z, and Andersen GJ (2013) Aging and the Perception of Egocentric Distance. Psychol Aging 28 (3):813–825. https://doi.org/10.1037/a0030991

Bicanski A, and Burgess N (2018) A Neural-Level Model of Spatial Memory and Imagery. eLife 7 (7052):1–3. https://doi.org/10.7554/elife.33752

Bohbot VD, Lerch J, Thorndycraft B, Iaria G, and Zijdenbos AP (2007) Gray Matter Differences Correlate with Spontaneous Strategies in a Human Virtual Navigation Task. J Neurosci 27 (38):10078–10083. https://doi.org/10.1523/JNEUROSCI.1763-07.2007

Boone AP, Maghen B, and Hegarty M (2019) Instructions Matter: Individual Differences in Navigation Strategy and Ability. Mem Cogn 47 (7):1401–1414. https://doi.org/10.3758/s13421-019-00941-5

Bopp KL, and Verhaeghen P (2005) Aging and Verbal Memory Span: A Meta-Analysis. J Gerontol B Psychol Sci Soc Sci 60 (5):P223–P233. https://doi.org/10.1093/geronb/60.5.p223

Buckley M, McGregor A, Ihssen N, et al. (2024) The Well-Worn Route Revisited: Striatal and Hippocampal System Contributions to Familiar Route Navigation. Hippocampus 34 (7):310–326. https://doi.org/10.1002/hipo.23607

Busche MA, Wegmann S, Dujardin S, et al. (2019) Tau Impairs Neural Circuits, Dominating Amyloid-B Effects, in Alzheimer Models *In Vivo*. Nat Neurosci 22 (1):57–64. https://doi.org/10.1038/s41593-018-0289-8

Byrne P, Becker S, and Burgess N (2007) Remembering the Past and Imagining the Future: A Neural Model of Spatial Memory and Imagery. Psychol Rev 114 (2):340–375. https://doi.org/10.1037/0033-295X.114.2.340

Camponogara I, Turchet L, Carner M, Marchioni D, and Cesari P (2016) To Hear or Not to Hear: Sound Availability Modulates Sensory-Motor Integration. Front Neurosci 10:22. https://doi.org/10.3389/fnins.2016.00022

Castegnaro A, Ji Z, Rudzka K, Chan D, and Burgess N (2023) Overestimation in Angular Path Integration Precedes Alzheimer's Dementia. Curr Biol 33 (21):4650–4661.e4657. https://doi.org/10.1016/j.cub.2023.09.047

Čepukaitytė G, Newton C, and Chan D (2024) Early Detection of Diseases Causing Dementia Using Digital Navigation and Gait Measures: A Systematic Review of Evidence. Alzheimers Dement (N Y) 4:3054–3073. https://doi.org/10.1002/alz.13716

Chen C, Lim J, Koh J, et al. (2024) A Global Analysis of Adaptation to Societal Aging across Low-, Middle- and High-Income Countries Using the Global Aging Society Index. Nat Aging 5 (1):113–121. https://doi.org/10.1038/s43587-024-00772-3

Chrastil ER, and Warren WH (2014) From Cognitive Maps to Cognitive Graphs. PLoS ONE 9 (11):e112544. https://doi.org/10.1371/journal.pone.0112544

Clark IA, and Maguire EA (2020) Do Questionnaires Reflect Their Purported Cognitive Functions? Cognition 195:104114. https://doi.org/10.1016/j.cognition.2019.104114

Coughlan G, Laczó J, Hort J, Minihane A-M, and Hornberger M (2018) Spatial Navigation Deficits—Overlooked Cognitive Marker for Preclinical Alzheimer Disease? Nat Rev Neurol 14 (8):496–506. https://doi.org/10.1038/s41582-018-0031-x

Coughlan G, Puthusseryppady V, Lowry E, et al. (2020a) Test-Retest Reliability of Spatial Navigation in Adults At-Risk of Alzheimer's Disease. PLoS ONE 15:0239077. https://doi.org/10.1371/journal.pone.0239077

Coughlan G, Zhukovsky P, Puthusseryppady V, et al. (2020b) Functional Connectivity between the Entorhinal and Posterior Cingulate Cortices Underpins Navigation Discrepancies in at-Risk Alzheimer's Disease. Neurobiol Aging 90:110–118. https://doi.org/10.1016/j.neurobiolaging.2020.02.007

Coutrot A, Manley E, and Goodroe S (2022) Entropy of City Street Networks Linked to Future Spatial Navigation Ability. 604:104–110. Nature 604:104–110. https://doi.org/10.1038/s41586-022-04486-7

Coutrot A, Silva R, Manley E, et al. (2018) Global Determinants of Navigation Ability. Curr Biol 28 (17):2861–2866.e2864. https://doi.org/10.1016/j.cub.2018.06.009

Craik FIM, and Salthouse TA (2011) The Handbook of Aging and Cognition. Psychology Press, New York

Dahmani L, and Bohbot VD (2015) Dissociable Contributions of the Prefrontal Cortex to Hippocampus- and Caudate Nucleus-Dependent Virtual Navigation Strategies. Neurobiol Learn Mem 117:42–50. https://doi.org/10.1016/j.nlm.2014.07.002

Diamond A (2013) Executive Functions. Annu Rev Psychol 64 (1):135–168. https://doi.org/10.1146/annurev-psych-113011-143750

Edvardsen V, Bicanski A, and Burgess N (2020) Navigating with Grid and Place Cells in Cluttered Environments. Hippocampus 30 (3):220–232. https://doi.org/10.1002/hipo.23147

Etienne AS, and Jeffery KJ (2004) Path Integration in Mammals. Hippocampus 14 (2):180–192. https://doi.org/10.1002/hipo.10173

Gellersen HM, Guell X, and Sami S (2021) Differential Vulnerability of the Cerebellum in Healthy Ageing and Alzheimer's Disease. Neuroimage Clin 30:102605. https://doi.org/10.1016/j.nicl.2021.102605

Ghosh A, Puthusseryppady V, Chan D, Mascolo C, and Hornberger M (2022) Machine Learning Detects Altered Spatial Navigation Features in Outdoor Behaviour of Alzheimer's Disease Patients. Sci Rep 12:3160. https://doi.org/10.1038/s41598-022-06899-w

Goodroe S, Fernandez Velasco P, Gahnstrom CJ, et al. (2025) Predicting Real-World Navigation Performance from a Virtual Navigation Task in Older Adults. PLoS ONE 20 (1):e0317026. https://doi.org/10.1371/journal.pone.0317026

Grove CR, Anson ER, Agrawal Y, Simonsick EM, and Schubert MC (2024) Utricular Dysfunction and Hearing Impairment Affect Spatial Navigation in Community-Dwelling Healthy Adults: Analysis from the Baltimore Longitudinal Study of Aging. Audiol Neurotol 29 (5):357–366. https://doi.org/10.1159/000537769

Harris MA, and Wolbers T (2012) Ageing Effects on Path Integration and Landmark Navigation. Hippocampus 22 (8):1770–1780. https://doi.org/10.1002/hipo.22011

——— (2014) How Age-Related Strategy Switching Deficits Affect Wayfinding in Complex Environments. Neurobiol Aging 35 (5):1095–1102. https://doi.org/10.1016/j.neurobiolaging.2013.10.086

Hartley T, Maguire EA, Spiers HJ, and Burgess N (2003) The Well-Worn Route and the Path Less Traveled: Distinct Neural Bases of Route Following and Wayfinding in Humans. Neuron 37 (5):877–888. https://doi.org/10.1016/s0896-6273(03)00095-3

Head D, and Isom M (2010) Age Effects on Wayfinding and Route Learning Skills. Behav Brain Res 209 (1):49–58. https://doi.org/10.1016/j.bbr.2010.01.012

Hegarty M, Richardson AE, Montello DR, Lovelace K, and Subbiah I (2002) Development of a Self-Report Measure of Environmental Spatial Ability. Intelligence 30 (5):425–447. https://doi.org/10.1016/S0160-2896(02)00116-2

Heward M, Adams A, Hicks B, and Wiener J (2022) 'We Go for a Homely Feel … Not the Clinical Dementia Side': Care Home Managers' Experiences of Supporting Residents with Dementia to Orientate and Navigate Care Environments. Ageing Soc 42 (7):1659–1685. https://doi.org/10.1017/S0144686X20001701

Hill PF, Bermudez S, McAvan AS, et al. (2024) Age Differences in Spatial Memory Are Mitigated during Naturalistic Navigation. Neuropsychol Dev Cogn B Aging Neuropsychol Cogn 31:2023.2001.2023.525279. https://doi.org/10.1080/13825585.2024.2326244

Hilton C, Johnson A, Slattery TJ, Miellet S, and Wiener JM (2021) The Impact of Cognitive Aging on Route Learning Rate and the Acquisition of Landmark Knowledge. Cognition 207:104524. https://doi.org/10.1016/j.cognition.2020.104524

Hosking I (2024) ViStA Tools for Visual Stress Analysis: How Can We Make Public Health More Precise? https://cph.shorthandstories.com/2024showcase/index.html

Howett D, Castegnaro A, Krzywicka K, et al. (2019) Differentiation of Mild Cognitive Impairment Using an Entorhinal Cortex-Based Test of Virtual Reality Navigation. Brain 142 (6):1751–1766. https://doi.org/10.1093/brain/awz116

Hunter LL, Monson BB, Moore DR, et al. (2020) Extended High Frequency Hearing and Speech Perception Implications in Adults and Children. Hear Res 397:107922. https://doi.org/10.1016/j.heares.2020.107922

Hunter SK, Pereira HM, and Keenan KG (2016) The Aging Neuromuscular System and Motor Performance. J Appl Physiol 121 (4):982–995. https://doi.org/10.1152/japplphysiol.00475.2016

Iglói K, Doeller CF, Paradis A-L, et al. (2015) Interaction between Hippocampus and Cerebellum Crus I in Sequence-Based but Not Place-Based Navigation. Cereb Cortex 25 (11):4146–4154. https://doi.org/10.1093/cercor/bhu132

Ishikawa T, and Montello D (2006) Spatial Knowledge Acquisition from Direct Experience in the Environment: Individual Differences in the Development of Metric Knowledge and the Integration of Separately Learned Places. Cogn Psychol 52 (2):93–129. https://doi.org/10.1016/j.cogpsych.2005.08.003

Jacob P-Y, Poucet B, Liberge M, Save E, and Sargolini F (2014) Vestibular Control of Entorhinal Cortex Activity in Spatial Navigation. Front Integr Neurosci 8:38. https://doi.org/10.3389/fnint.2014.00038

Kavcic V, Vaughn W, and Duffy CJ (2011) Distinct Visual Motion Processing Impairments in Aging and Alzheimer's Disease. Vision Res 51 (3):386–395. https://doi.org/10.1016/j.visres.2010.12.004

Kessler F, Frankenstein J, and Rothkopf CA (2024) Human Navigation Strategies and Their Errors Result from Dynamic Interactions of Spatial Uncertainties. Nat Commun 15 (1):5677. https://doi.org/10.1038/s41467-024-49722-y

Korthauer LE, Nowak NT, Moffat SD, et al. (2016) Correlates of Virtual Navigation Performance in Older Adults. Neurobiol Aging 39:118–127. https://doi.org/10.1016/j.neurobiolaging.2015.12.003

Kunz L, Schröder TN, Lee H, et al. (2015) Reduced Grid-Cell-Like Representations in Adults at Genetic Risk for Alzheimer's Disease. Science 350 (6259):430–433. https://doi.org/10.1126/science.aac8128

Lane CA, Hardy J, and Schott JM (2018) Alzheimer's Disease. Eur J Neurol 25 (1):59–70. https://doi.org/10.1111/ene.13439

Lester AW, Moffat SD, Wiener JM, Barnes CA, and Wolbers T (2017) The Aging Navigational System. Neuron 95 (5):1019–1035. https://doi.org/10.1016/j.neuron.2017.06.037

Levine TF, Allison SL, Stojanovic M, et al. (2020) Spatial Navigation Ability Predicts Progression of Dementia Symptomatology. Alzheimers Dement (N Y) 16 (3):491–500. https://doi.org/10.1002/alz.12031

Livingston G, Huntley J, Liu K, et al. (2024) Dementia Prevention, Intervention, and Care: 2024 Report of the Lancet Standing Commission. Lancet 404 (10452):572–628. https://doi.org/10.1016/s0140-6736(24)01296-0

Lövdén M, Schaefer S, Noack H, et al. (2012) Spatial Navigation Training Protects the Hippocampus against Age-Related Changes during Early and Late Adulthood. Neurobiol Aging 33 (3):620.e629–620.e622. https://doi.org/10.1016/j.neurobiolaging.2011.02.013

Maguire EA, Gadian DG, Johnsrude IS, et al. (2000) Navigation-Related Structural Change in the Hippocampi of Taxi Drivers. PNAS 97 (8):4398–4403. https://doi.org/10.1073/pnas.070039597

Mather M, and Carstensen LL (2005) Aging and Motivated Cognition: The Positivity Effect in Attention and Memory. Trends Cogn Sci 9 (10):496–502. https://doi.org/10.1016/j.tics.2005.08.005

Mirino P, Pecchinenda A, Boccia M, et al. (2022) Cerebellum-Cortical Interaction in Spatial Navigation and Its Alteration in Dementias. Brain Sci 12 (5):523. https://doi.org/10.3390/brainsci12050523

Moffat SD (2009) Aging and Spatial Navigation: What Do We Know and Where Do We Go? Neuropsychol Rev 19 (4):478–489. https://doi.org/10.1007/s11065-009-9120-3

Mozolic JL, Hugenschmidt CE, Peiffer AM, and Laurienti PJ (2012) Multisensory Integration and Aging. The Neural Bases of Multisensory Processes. CRC Press, Boca Raton

Nazareth A, Huang X, Voyer D, and Newcombe N (2019) A Meta-Analysis of Sex Differences in Human Navigation Skills. Psychon Bull Rev 26 (5):1503–1528. https://doi.org/10.3758/s13423-019-01633-6

Newton C, Pope M, Rua C, et al. (2024) Entorhinal-Based Path Integration Selectively Predicts Midlife Risk of Alzheimer's Disease. Alzheimers Dement (N Y) 20 (4):2779–2793. https://doi.org/10.1002/alz.13733

Nolte J, and Hanoch Y (2024) Adult Age Differences in Risk Perception and Risk Taking. Curr Opin Psychol 55:101746. https://doi.org/10.1016/j.copsyc.2023.101746

Nyberg L, Lövdén M, Riklund K, Lindenberger U, and Bäckman L (2012) Memory Aging and Brain Maintenance. Trends Cogn Sci 16 (5):292–305. https://doi.org/10.1016/j.tics.2012.04.005

O'Malley M, Innes A, Muir S, and Wiener JM (2018) 'All the Corridors Are the Same': A Qualitative Study of the Orientation Experiences and Design Preferences of UK Older Adults Living in a Communal Retirement Development. Ageing Soc 38 (9):1791–1816. https://doi.org/10.1017/S0144686X17000277

Owsley C, Sekuler R, and Siemsen D (1983) Contrast Sensitivity Throughout Adulthood. Vision Res 23 (7):689–699. https://doi.org/10.1016/0042-6989(83)90210-9

Packard MG, and McGaugh JL (1996) Inactivation of Hippocampus or Caudate Nucleus with Lidocaine Differentially Affects Expression of Place and Response Learning. Neurobiol Learn Mem 65 (1):65–72. https://doi.org/10.1006/nlme.1996.0007

Patel VR, Liu M, Worsham CM, and Jena AB (2024) Alzheimer's Disease Mortality among Taxi and Ambulance Drivers: Population Based Cross Sectional Study. BMJ (Online) 387:e3082194. https://doi.org/10.1136/bmj-2024-082194

Perez E, Kvam P, McKemey R, and Weisberg S (2025) The Role of Risk Tolerance in Navigation Strategy Decisions. J Exp Psychol Learn Mem Cogn 51 (10):1642–1661. https://doi.org/10.1037/xlm0001448

Perez E, and Weisberg SM (2025) Individual Differences in Spatial Navigation. In: Encyclopedia of the Human Brain. Elsevier, Amsterdam, pp 277–293

Puthusseryppady V, Coughlan G, Patel M, and Hornberger M (2019) Geospatial Analysis of Environmental Risk Factors for Missing Dementia Patients. J Alzheimers Dis 71 (3):1005–1013. https://doi.org/10.3233/JAD-190244

Puthusseryppady V, Manley E, Lowry E, Patel M, and Hornberger M (2020) Impact of Road Network Structure on Dementia-Related Missing Incidents—Space Syntax Approach. Sci Rep 10:1–9. https://doi.org/10.1038/s41598-020-74915-y

Puthusseryppady V, Morrissey S, Spiers H, Patel M, and Hornberger M (2022) Predicting Real World Spatial Disorientation in Alzheimer's Disease Patients Using Virtual Reality Navigation Tests. Sci Rep 12:1–12. https://doi.org/10.1038/s41598-022-17634-w

Ramanoël S, Durteste M, Perot V, Habas C, and Arleo A (2022) An Appraisal of the Role of the Neocerebellum for Spatial Navigation in Healthy Aging. Cerebellum 22 (2):235–239. https://doi.org/10.1007/s12311-022-01389-1

Ritchie K, Ropacki M, Albala B, et al. (2017) Recommended Cognitive Outcomes in Preclinical Alzheimer's Disease: Consensus Statement from the European Prevention of Alzheimer's Dementia Project. Alzheimers Dement (NY) 13 (2):186–195. https://doi.org/10.1016/j.jalz.2016.07.154

Robinson JC, Ying J, Hasselmo ME, and Brandon MP (2024) Optogenetic Silencing of Medial Septal GABAergic Neurons Disrupts Grid Cell Spatial and Temporal Coding in the Medial Entorhinal Cortex. Cell Rep 43 (8):114590. https://doi.org/10.1101/2023.11.08.566228

Rodgers MK, Sindone JA, and Moffat SD (2012) Effects of Age on Navigation Strategy. Neurobiol Aging 33 (1):202.e215–202.e222. https://doi.org/10.1016/j.neurobiolaging.2010.07.021

Roe CM, Stout SH, Rajasekar G, et al. (2019) A 2.5-Year Longitudinal Assessment of Naturalistic Driving in Preclinical Alzheimer's Disease. J Alzheimers Dis 68 (4):1625–1633. https://doi.org/10.3233/JAD-181242

Sakurai R, Suzuki H, Ogawa S, Takahashi M, and Fujiwara Y (2021) Hearing Loss and Increased Gait Variability among Older Adults. Gait Posture 87:54–58. https://doi.org/10.1016/j.gaitpost.2021.04.007

Satizabal CL, Beiser AS, Chouraki V, et al. (2016) Incidence of Dementia over Three Decades in the Framingham Heart Study. New England Journal of Medicine 374 (6):523–532. https://doi.org/10.1056/nejmoa1504327

Schinazi VR, and Epstein RA (2010) Neural Correlates of Real-World Route Learning. Neuroimage 53 (2):725–735. https://doi.org/10.1016/j.neuroimage.2010.06.065

Segen V, Kabir R, Streck A, et al. (2025) Path Integration Impairments Reveal Early Cognitive Changes in Subjective Cognitive Decline. Sci Adv 11:eadw6404. https://doi.org/10.1126/sciadv.adw6404

Shayman CS, Earhart GM, and Hullar TE (2017) Improvements in Gait with Hearing Aids and Cochlear Implants. Otol Neurotol 38 (4):484–486. https://doi.org/10.1097/mao.0000000000001360

Skouras S, Torner ÃJ, Andersson P, et al. (2020) Earliest Amyloid and Tau Deposition Modulate the Influence of Limbic Networks during Closed-Loop Hippocampal Downregulation. Brain 143:976–992. https://doi.org/10.1093/brain/awaa011

Solbakken SM, Magnus JH, Meyer HE, et al. (2019) Urban-Rural Differences in Hip Fracture Mortality: A Nationwide NOREPOS Study. JBMR Plus 3 (11):e10236. https://doi.org/10.1002/jbm4.10236

van der Ham IJM, van der Kuil MNA, and Claessen MHG (2021) Quality of Self-Reported Cognition: Effects of Age and Gender on Spatial Navigation Self-Reports. Aging Ment Health 25 (5):873–878. https://doi.org/10.1080/13607863.2020.1742658

van der Kuil MNA, Evers AWM, Visser-Meily JMA, and van der Ham IJM (2020) The Effectiveness of Home-Based Training Software Designed to Influence Strategic Navigation Preferences in Healthy Subjects. Front Hum Neurosci 14 (March):1–13. https://doi.org/10.3389/fnhum.2020.00076

Warren WH, Blackwell AW, and Morris MW (1989) Age Differences in Perceiving the Direction of Self-Motion from Optical Flow. J Gerontol 44 (5):P147–P153. https://doi.org/10.1093/geronj/44.5.p147

Weisberg SM, and Newcombe NS (2016) How Do (Some) People Make a Cognitive Map? Routes, Places, and Working Memory. J Exp Psychol Learn Mem Cogn 42 (5):768–785. https://doi.org/10.1037/xlm0000200

Weisberg SM, Schinazi VR, Newcombe NS, Shipley TF, and Epstein RA (2014) Variations in Cognitive Maps: Understanding Individual Differences in Navigation. J Exp Psychol Learn Mem Cogn 40 (3):669–682. https://doi.org/10.1037/a0035261

WHO (2025) Ageing. https://www.who.int/health-topics/ageing#tab=tab_1

Wiener JM, Büchner SJ, and Hölscher C (2009) Taxonomy of Human Wayfinding Tasks: A Knowledge-Based Approach. Spat Cogn Comput 9 (2):152–165. https://doi.org/10.1080/13875860902906496

Wiener JM, De Condappa O, Harris MA, and Wolbers T (2013) Maladaptive Bias for Extrahippocampal Navigation Strategies in Aging Humans. J Neurosci 33 (14):6012–6017. https://doi.org/10.1523/jneurosci.0717-12.2013

Wolbers T, and Hegarty M (2010) What Determines Our Navigational Abilities? Trends Cogn Sci 1 (3):138–146. https://doi.org/10.1016/j.tics.2010.01.001

Wolters FJ, Chibnik LB, Waziry R, et al. (2020) Twenty-Seven-Year Time Trends in Dementia Incidence in Europe and the United States: The Alzheimer Cohorts Consortium. Neurology 95 (5):519–e531. https://doi.org/10.1212/WNL.0000000000010022

Wood RA, Moodley KK, Lever C, Minati L, and Chan D (2016) Allocentric Spatial Memory Testing Predicts Conversion from Mild Cognitive Impairment to Dementia: An Initial Proof-of-Concept Study. Front Neurol 7:215. https://doi.org/10.3389/fneur.2016.00215

Yao Q, Tang F, Wang Y, et al. (2022) Effect of Cerebellum Stimulation on Cognitive Recovery in Patients with Alzheimer Disease: A Randomized Clinical Trial. Brain Stimulation 15 (4):910–920. https://doi.org/10.1016/j.brs.2022.06.004

Yoder RM, and Taube JS (2014) The Vestibular Contribution to the Head Direction Signal and Navigation. Front Integr Neurosci 8:32. https://doi.org/10.3389/fnint.2014.00032

Yu S, Boone AP, He C, et al. (2021) Age-Related Changes in Spatial Navigation Are Evident by Midlife and Differ by Sex. Psychol Sci 32 (5):692–704. https://doi.org/10.1177/0956797620979185

Yuen B, Bhuyan MDR, Song S, et al. (2022) Age-Friendly Neighbourhood Planning and Design Guidelines: A Singapore Case Study. World Scientific, Singapore

Zhou R, Belge T, and Wolbers T (2023) Reaching the Goal: Superior Navigators in Late Adulthood Provide a Novel Perspective into Successful Cognitive Aging. Top Cogn Sci 15 (1):15–45. https://doi.org/10.1111/tops.12608

Subject Index

A

adaptive behavior 37, 49–52, 56, 57, 82, 263, 280, 300, 305, 341, 433
addiction 263, 301, 303
aging 6, 139, 141, 321–354, 381–386, 431–460
  car driving 385, 386, 391–393
  defined 433, 434
  egocentric strategies 337, 353, 383
  hearing loss 323, 440, 441
  modeling 147, 348–350
  quality of life 7, 379, 392, 433, 434, 453, 456, 460
  sensorimotor changes 343–345
  synaptic changes 323, 325–327
algae 90, 91, 95
allocentric navigation 176, 328, 333, 336–338, 346, 352, 353, 383, 391
allocentric spatial coding 151, 243, 297–299
Alzheimer disease 6, 141, 323, 352, 379, 380, 386, 388, 399, 433, 442, 449
  assessing 390, 393, 400–402, 434
  path-integration performance 174, 389, 390
  pathophysiology 387–389, 393
  risk 232, 390, 406, 407, 418, 446, 456
amoeba (*Dictyostelium*) 67
amygdala 104, 171, 274–276, 279, 282, 299, 302
anterior dorsal thalamic nucleus (ADN) 145, 228, 229, 262, 338, 339
ants 66, 91, 96, 99–101
  celestial compass 69, 81
  desert 76, 77, 81, 108
  magnetic compass 108
anxiety disorders 274, 279, 281, 282, 301, 303, 442
approach-avoidance conflicts 280, 281
archaea 65, 96
arm maze, radial 205, 207, 244, 267
arthropods 81, 140
artificial intelligence (AI) 34, 41, 95, 109, 283
associative memory 172, 206, 211, 405
attention 25, 274, 345, 386, 392, 400, 406, 442, 458, 459
attractor network dynamics 142, 324
  continuous 166, 208–210, 349, 350
autonomous navigation 33, 35–37, 47–49, 52, 79, 103. *See also* robot navigation
  error correction 41, 94
  ethical concerns 55, 56
  execution 44–46

B

bacteria 65, 90, 96, 100
  magnetotactic 75, 108
Balearic shearwaters (*P. mauretanicus*) 21
Barnes maze 205, 207
bar-tailed godwits (*Limosa lapponica*) 16, 19, 65
basal ganglia 98, 152, 342
bats 71, 78, 82, 84, 87, 92, 97, 109
  Egyptian fruit 66, 102
  goal coding 264, 283, 298
  head direction 148
  place cells 171
BaYaka people 83, 93
BB model 7, 151, 438–440, 447
behavioral bridges 221, 223, 237–239, 242
Behnke-Fried electrodes 237
birds 2, 11–32, 65, 66, 77, 87, 95, 97, 227
  clock-and-compass strategy 14, 70, 82
  collective navigation 99
  decision making 92, 101
  directional cues 82
  place cells 171
black-capped chickadees (*Poecile atricapillus*) 17, 227
border cells 212, 215, 230, 240, 294, 330, 381, 389
boundary representations 165, 166, 176–178, 190, 203, 214, 228–232
  environmental geometry 225, 230
boundary vector cells (BVCs) 151, 212, 215, 230, 236, 240, 260, 328, 438
  allocentric 176–178
  egocentric 176–178, 339
Braitenberg's law of uphill analysis and downhill invention 140, 150, 151

C

canonical navigation neurons 6, 143, 145, 146, 291, 293–297. *See also* border, grid, head-direction, and place cells

N. S. Newcombe and K. Cheng (eds.), *Challenges in Navigation Research*, Strüngmann Forum Reports, https://doi.org/10.1007/978-3-032-20563-6

car driving 35, 391, 393, 452, 454
autonomous 35, 55
impact on aging 385, 386, 391–393
caudate nucleus 337, 338, 441
celestial compass 1, 14–17, 65, 67, 69, 70, 75, 77, 81, 82, 105, 307
central module 73, 74, 89–95
cerebellum 98, 322, 339, 442
age-related changes 326–328, 341–343
strategy switching 353
chemokinesis 69, 90
chemosensitivity 76
chemotaxis 67, 69, 95
cilia 69, 76, 81, 90, 91, 95, 96, 100
cingulate cortex 94, 167, 172, 260, 275, 276
anterior (ACC) 263, 334
posterior (PCC) 340, 341
circadian rhythm 14, 16, 25, 90
Clark's nutcrackers (*Nucifraga columbiana*) 15, 17–19, 82, 87
clinical tools, designing 7, 399, 401–403, 416, 417, 434
smartphone apps 454
clock-and-compass strategy 14, 70, 82
cloud-based navigation 40, 43, 53, 55
clustering hypothesis 234, 235
cognitive ecology 11, 12, 23–26
cognitive maps 25, 109, 165, 171, 221, 261–263, 268, 274, 277, 294, 308
bats 87
computational models 348–350
defined 293
egocentric 170, 349
grid cell 230
human 224–233
rodent 87, 92, 224–233, 260
cognitive training 454–456, 459
computational models 150, 151, 176, 181, 221, 234–237, 324, 348–351, 435, 437, 448, 460
ring attractor 147, 295
concept cells 226, 227
conjunctive hypothesis 167–169, 234, 235
continuous attractor network (CAN) 166, 208–210, 349, 350
convergent evolution 84, 96, 144, 145
creativity 261, 268
cross-species comparisons 4, 139–162, 243, 322, 460
behavioral bridge 221, 223, 237–239, 242
methodological bridge 221, 223, 234–237, 242
cue integration 19, 46, 50, 56, 90, 109, 241

D

Dale's principle 147
dark-eyed junco (*Junco hyemalis*) 17
data collection 351, 402, 414, 417, 451, 452, 460
web-based 348, 415
data sharing 242, 404, 459, 460
dead reckoning 39, 49, 83
decision making 3, 5, 25, 43, 63, 89–95, 259–263, 265, 268, 278, 280, 281, 294, 302, 304–306, 333, 442
birds 92, 99–101
real-time 35, 37, 456
rodents 92, 335
deep learning technology 37, 38, 47, 297
deep neural network (DNN) 297
default mode network 259, 260, 340, 341
dementia 6, 390, 391, 399, 433, 445, 452. *See also* Alzheimer disease
frontotemporal 386, 393
global burden 380, 400
risk for 324, 406, 441, 453
vascular 386, 393
dentate gyrus (DG) 146, 275, 325, 329, 331–333, 435
depression 282, 301
desert ants. *See* ants
Desertas petrels (*Pterodroma deserta*) 22
desktop virtual reality (VR) paradigm 345–347, 383, 448
detouring 11, 13, 19–22, 26, 56–59, 224. *See also* obstacle avoidance
development 4, 5, 202–215, 221, 224, 239–242
digital health tools 403, 404, 409, 413
assessing 417, 418, 420, 421
designing 406–415, 419
direction module 73–75, 80–85, 90, 107
disorientation 25, 209, 224, 231, 239, 328, 343, 379, 384, 385, 390–393, 400, 441, 453
dorsal tegmental nucleus (DTN) 145, 146, 151, 228, 338
drones 33, 35–37, 47, 53, 56, 103
Dual Solution Paradigm 447, 448

E

echolocation 71, 75, 78, 92
egocentric spatial coding 202, 264, 297–299
egocentric vector cells 151, 170, 298, 299
Egyptian fruit bats 66, 102
Eleonora's falcons (*Falco eleonorae*) 22
emotions 5, 273–283, 294, 301–303, 442
entorhinal cortex (EC) 104, 163–169, 189, 190, 201, 229, 232, 260–262, 275, 294, 307, 327, 329. *See also* medial entorhinal cortex (MEC)
  boundary representations 176–178
  goal representations 263
  grid cells 88, 166, 264
  lateral 275, 327, 330
  place cells 171, 172, 296
  planning 268
EC–HC–PFC circuit 327–335, 350
environmental variability 2, 11, 13, 15–19, 47, 49–51, 99–104, 237–243, 302
epilepsy patients 3, 4, 141, 179, 180, 223, 351
  boundary representation 177, 236
  grid cells 167, 237
  landmark representation 175
  place cells 171, 237
  recordings from 150, 164, 168–170, 181
episodic memory 151, 226, 227, 259, 307, 380
  deficits 337, 343, 380, 387–389
  patient HM 277
  role of RSC 339, 344
error correction 3, 67, 89–95, 100, 105, 106, 109
  continuous 63, 73, 94
error monitoring 3, 89, 95, 105–107
eukaryotes 65, 90
evolution 2, 12, 79, 140, 143, 144, 153, 154, 307
  convergent 84, 96, 144, 145
executive function 72, 90, 94, 259, 260, 268, 400, 406, 459
  age-related deficits 337, 442
exploration 205, 206, 211, 265, 266, 268, 293, 302, 442

F

familiar area map 14, 25
fear 99–104, 109, 205–207, 263, 274, 278, 302–304, 385, 442, 453
flagella 66, 69, 76, 81, 86, 91, 95, 100
flexible navigation 165, 224, 233, 260, 265, 268, 300, 303, 305, 349
fMRI 3, 4, 163–178, 190, 240, 346
  BOLD response 179–181, 230, 234, 331
  grid cell activity 166–168, 234, 236
  hexadirectional modulation 167, 168, 181
  pediatric 191, 193–195
foraging 13, 24, 68, 93, 104, 223, 233, 293, 334
  bats 71, 92, 97
  birds 11, 19–22, 101
  insects 100, 101
  optimal foraging theory 23
  primates 93, 103, 223
Foster maze 239
frigatebirds (*Fregata minor*) 22, 23
frontostriatal circuits 335–338, 353
functional ultrasound imaging 237
future research 2–7, 107, 175, 177, 238, 241, 242, 244, 301–306, 348, 351–354, 455, 457–460
  aging 340, 379, 386, 391–393, 439–442, 446, 452
  egocentric vector neurons 170, 299
  hippocampal circuitry 277, 281, 282
  magnetosensory pathways 80
  map-based navigation 193, 215
  replay 233
  robotics 56–58, 109
  visually guided navigation 190, 195

G

gain modulation 298
gait 96, 98, 342, 351, 440
  age-related changes 324, 345, 392, 454
gamma oscillations 167, 168, 328, 335
garden warblers (*Sylvia borin*) 17
geometric cues 48, 165, 170, 224, 230, 239, 241, 296, 330, 454
goal 301, 307
  coding 81, 84, 85, 90, 259, 263, 264, 268
  direction cells 84, 85, 107
  pointing-the-goal phase 67–69, 78, 109
  representations 260, 263–265
goal-directed navigation 145, 166, 263, 274, 278, 293, 328, 436, 437
  modeling 324
  mPFC neurons 334
  neuronal mechanisms 232, 233
gorillas 65, 66, 83, 93, 97

GPS 37–39, 54, 79, 85, 106, 390, 443, 453
gravity sensing 76, 79, 86
grid cells 6, 88, 148, 164, 208–210, 214, 224, 229, 232, 240, 277, 294, 330, 349, 381, 439
  computer simulations 234–236
  entorhinal cortex 88, 166–169, 201, 264, 350
  hexadirectional pattern 167, 346
  role in cognitive maps 260–262
  vector computations 166, 221, 238, 239
grid representation 4, 14, 43, 45, 148, 149, 163, 166–169, 178, 181, 268

H
head direction (HD) 140, 142, 144–147, 203, 214, 229, 338–341
  coding 139, 143, 145, 154, 228, 447
  HD cells 84, 164, 221, 240, 262, 294–297, 328, 330, 338, 381
  modeling 147, 151
  predictive coding 153
hearing loss 323, 440, 441
hierarchical look-ahead (HiLAM) model 262, 264
hippocampus (HC) 5, 107, 165, 166, 174, 181, 189–191, 259, 260, 275–277, 281, 283, 302, 323, 329, 441
  age-related changes 325–327, 332–334, 349
  Alzheimer disease 387, 450, 451
  CA1 224, 261, 326, 331–333
  CA3 224, 261, 325, 331–333
  circuit architecture 274–277
  cognitive maps 224, 226
  functional organization 333
  goal representation 104, 263–265, 268
  map-based navigation 191–193
  motivational-emotion systems 273–290
  place coding 227–231
  place representations 163, 166, 171–174, 181
  role in planning 259–272
  theta sequences 265
homeostatic scaling 323
homing 67, 68, 70, 109
  autonomous systems 72, 454
  natal 70
  pigeon (*Columba livia*) 14, 15, 17, 24, 25, 77, 82, 106
  rodent 78, 204, 206, 207
honeybees 87, 100, 101
honey buzzards (*Pernis apivorus*) 22
human-computer interactions 402–404, 419
human intracranial recordings 164, 179–181, 223, 226, 227, 236, 237
humanoids 35, 36, 98
hummingbirds 13, 17, 22, 92

I
imagination 5, 214, 259–268, 438
independent living 7, 379, 385, 392, 433, 434, 453, 456, 460
inertial sensors 39, 40, 50, 79
infralimbic cortex 263, 334
insects 69, 76, 81, 86, 91, 96, 100, 195
  central complex 72, 90, 140
  head direction 84, 140, 144–148
  impact on robotics 35, 38
internal needs 14, 25, 71, 265, 274, 294, 300–303
intracranial techniques 163–181, 221, 223, 226, 227, 236, 237, 351
introspection in robot navigation 54, 57, 94, 103, 106, 109

L
landmark-based navigation 75, 166, 178, 203, 224, 225, 228, 260, 294, 401
  age-related changes 382–385, 441–444
  birds 14–19, 26, 70, 82, 87
  defined 382
  insects 91
  marine animals 70
  neural basis 165, 173–176, 190–192, 209, 298, 339
  primates 71, 83, 88
  robot navigation 49
  selection strategies 194, 384–386
landscape of fear 99–104
large language models 33, 34, 38, 46
lateral mammillary nucleus (LMN) 145, 228, 338
learning 48, 52, 56, 94, 107, 174, 192, 274, 275, 279, 303–305, 405, 449
  birds 15, 17, 22, 25
  fear 205–207, 302–304
  geometry-based 191
  inductive biases 153, 154
  insects 76, 91, 105
  place 239, 241, 389

reinforcement 262, 263, 267, 300
rodents 82, 206–208
route 191, 193, 202, 244, 382, 384
Leiden Navigation Test (LNT) 402
LiDAR 35, 37–40, 47, 55, 79
local field potentials (LFPs) 165, 168–170, 170, 229, 236, 237
intracranial recordings 164, 167, 177, 179, 181
locomotion 24, 73–76, 95–98, 141, 193, 233, 265, 344
locus coeruleus (LC) 276, 278, 279, 352, 381, 387
long-distance navigation 11, 63, 67, 68, 73, 99
bats 78, 82
birds 12, 16, 25, 70, 75
defined 13
insects 69, 77
marine animals 65, 70, 76, 77
primates 93
long-term depression (LTD) 323, 326, 339
long-term potentiation (LTP) 279, 323, 325, 326

M

machine learning 33–37, 303, 453
magnetic cues 3, 12, 14–18, 65, 67, 70, 75–77, 80–82, 86, 108, 109
magnetite 75, 108
magnetosensation 76, 78, 86, 108, 307
Manx shearwaters (*Puffinus puffinus*) 15, 20, 21
map-based navigation 4, 48, 68, 71, 189–193, 202, 206, 260
development of 195, 196, 201, 206, 212, 214
RSComp 189, 190, 193
marine animals 66, 70, 77, 82, 86, 91, 96, 101
Marr's levels of analysis 2, 73, 74, 139, 141–144, 154
medial entorhinal cortex (MEC) 146, 209, 228, 232, 238, 275, 296, 327–330, 439
age-related changes 325, 331
Alzeimer disease 387, 389, 450
border cells 212, 230, 330
HD cells 330, 381
medial temporal lobe (MTL) 4, 174, 179, 189, 190, 196, 224, 227, 237, 260, 337, 387, 388, 438, 448
development in humans 191–193
memory 107, 166, 204–208, 213, 227, 243, 244, 259, 265, 304, 305, 325
age-related changes 337, 343, 347, 380, 387–389
Alzheimer disease 388–390
associative 172, 206, 211, 405
BB model 438–440
birds 21, 25
consolidation 233, 260, 266, 267, 282, 283
deficits 388
episodic 151, 226, 227, 259, 307, 339
formation 95, 171, 180, 283
insects 70, 81
long-term 207, 215, 442, 459
retrieval 100, 171, 180, 215, 260, 267, 382, 438
role of emotions 273–279, 302
working 207, 260, 334, 335, 352, 442
mental navigation 5, 151, 214, 230, 259–261, 442
methodological bridges 221, 223, 234–237, 242
Mind Crowd 348
Mini-Mental State Examination (MMSE) 400
mobile technology 324, 351, 399, 402, 404, 453
mHealth tools 403, 404
modules of navigation 73–98
Monarch butterfly 69, 74
Montreal Cognitive Assessment (MoCA) 400, 406, 407, 417
mood disorders 281, 282
Morris water maze (MWM) 141, 206, 238, 239, 242, 347, 388, 451
virtual 333, 334, 336
motivation 5, 68, 105, 107, 109, 273–283, 294, 300–302, 403, 442, 444
motor cortex 98, 321, 342

N

navigation, defined 202–220, 293, 322
navigational context 74, 90, 99, 101–105, 107, 443
navigation assessment tools 448–453
construct validity 459
navigation toolbox 1, 293, 322, 443, 444
neural dedifferentiation, age-related 343–345

neurodegenerative disease 7, 323, 327, 379, 386–391, 433, 448. *See also* Alzheimer disease
Parkinson 347, 386
Neuroligin-2 knockout mice 282
NeuroSLAM 38, 57
nidopallium caudolaterale (NCL) 72, 74, 90, 94
novel object recognition (NOR) 204, 205, 208, 328
nucleus accumbens 275, 276, 279, 282
nucleus reuniens 146, 264, 276

O

object vector cells 151, 262, 330, 438
obstacle avoidance 35–39, 42, 46, 49, 52, 97, 190, 194. *See also* detouring
occipital place area (OPA) 4, 168, 189, 190, 229–231, 299
development of 193–195
Williams syndrome 244
ocean waves 70, 76, 82
olfaction 14, 17, 20, 69, 75, 77–79, 83, 84, 203
optimal foraging theory 23
optimization principle 436, 443, 446, 448
optogenetics 279
orbitofrontal cortex (OFC) 104, 264, 305, 306, 334
value coding 261, 263
orientation 164–167, 173, 174, 176, 190, 194, 228–232, 330, 338, 441
ospreys (*Pandion haliaetus*) 22

P

pallium 98
parahippocampal cortex (PHC) 163, 166, 173, 176–178
age-related neural dedifferentiation 344
landmarks 173
scene representations 168–171, 178
parahippocampal place area (PPA) 164, 168–170, 176, 193, 229–231
parallel navigation circuits 321–354
parietal cortex 85, 152, 243, 244, 298, 340, 353, 389, 438, 442
lateral 336, 337
posterior 179, 262, 342, 343, 353
Parkinson disease 347, 386
path integration 11, 149, 226, 439, 450
age-related changes 340, 382, 458
Alzheimer disease 389, 393, 449, 450
assessing 344–346, 402, 405
bats 87
birds 21, 26
dlPFC 334
impaired 441
insects 86, 99
neural basis 166, 174, 349, 350
robots 39
rodents 78, 83, 204, 205, 214
SPACE 405, 415, 416
path planning. *See* planning
pedunculopontine tegmental nucleus (PPN) 330, 331
perception
depth 79, 194
robot navigation 38–42, 48, 49
self-motion 83, 382, 440
phase transitions 3, 63, 65, 69–72, 105
pied flycatchers (*Ficedula hypoleuca*) 19
pigeons 14, 15, 17–21, 24, 25, 75, 77, 82, 92, 106
pinpointing the goal 67–69, 78, 109
place-based navigation 342, 353, 442, 446, 449
place cells 88, 89, 107, 164, 171–173, 221, 241, 266, 277, 293–299, 307, 331, 381
cognitive maps 109, 260–262, 268
development 208, 210–212, 240–242
firing 152, 176, 201, 349
nonhuman primates 226–228
omnidirectional tuning 172
remapping 224–226
replay 211, 214, 215
place learning 239, 241, 389
planning 63, 105, 152, 221, 233, 239, 268, 294, 302–307
autonomous systems 38, 39, 43–46, 56
mental 202, 260, 261, 265
neural basis of 259–272
role of insight 261, 265, 268
role of PFC 259–263
role of replay 214, 238, 265, 267
route 180, 334–338, 385, 401, 443
position module 73–75, 85–89, 107
postsubiculum (PoS) 146, 228, 262, 338, 339, 343
post-traumatic stress disorder 282
predation 19, 24, 63, 90, 93, 99–103, 109, 302
predictive map theory 308, 309

prefrontal cortex (PFC) 94, 104, 264, 265, 275–277, 305–307, 324, 327
age-related changes 325, 328, 333–335, 381
cognitive maps 262, 263, 264
role in planning 72, 90, 233, 259–263
–striatal circuit 353
prelimbic cortex (PrL) 263, 334
presubiculum 173, 209, 330, 338
primates 65, 71, 85, 93, 103, 104, 223, 227, 228
gorillas 65, 66, 83, 93, 97
HD system 84, 144, 338
impact of aging on 326, 329, 332, 334, 352
positional cues 88, 89
visual system 78, 79, 83, 148, 330
protein kinase C (PKC) 326, 339
protein phosphatase 2B (PP2B) 326, 339
protists 76, 90

Q

quality of life 7, 379, 385, 392, 433, 434, 453, 456, 460

R

radial arm maze 205, 207, 244, 267
rat (*Rattus norvegicus*) 71, 78, 82, 92, 148, 202–220
RatSLAM 38, 57
Razorbills (*Alco torda*) 11
real-world navigation 104, 302, 352, 383–385, 390, 391
assessing 346, 348, 451–453
robotics 38, 44, 46
reference frame 212, 224, 231, 293, 297–300, 336
allocentric 85, 324, 328
egocentric 260, 262, 298, 324, 328, 342, 349, 438
reinforcement learning 52, 262, 263, 267, 302–304, 324, 349, 350
remapping 51, 89, 211, 212, 224, 225, 277, 296, 331, 332
reorientation task 194, 224, 238, 239, 244
repetition suppression hypothesis 167–169, 234, 235
replay 180, 211, 214, 233, 238, 240, 265–267, 283
retina 12, 77–80, 85, 298
retrosplenial complex (RSComp) 168, 173, 189, 190, 195, 196, 229, 231
age-related neural dedifferentiation 344
development of 193
retrosplenial cortex (RSC) 94, 166, 171, 173–176, 179, 262, 277, 328, 339–341, 343, 438, 441
Alzheimer disease 387
boundary representations 176–178
HD cells 228, 229, 349
reference frames 298, 299
role in episodic memory 339, 344
strategy switching 353
Reynolds number (Re) 95, 96, 105
ring attractor 84, 140, 145–147, 153, 154, 447
computational model 147, 295
one-dimensional 142, 309
robot navigation 2, 33–58, 72, 79, 83, 88, 93, 98, 103
adaptability 49–51
bio-inspired 35, 38, 46, 49, 57, 109
core components 39–46
error correction 94
ethical concerns 55–57
introspection 54, 57, 94, 103, 106, 109
localization 88
multi-robot coordination 53, 54
spatial representations 43, 57
rodents 65, 71, 78, 82, 84, 87, 92, 97, 102, 223, 224, 263
age equivalence of humans 241
grid cell 331
HD cells 140, 144
place cells 171, 227
vector navigation 147–148
wild rat 66
route-based navigation 22, 202, 262, 328, 333, 352, 441, 442
route learning 191, 193, 382, 384
route planning 89–91, 180, 263, 268, 305, 334–338, 385, 401, 443
rufous hummingbirds (*Selasphorus rufus*) 13, 17, 22

S

satellite systems 39, 55, 83, 385, 392
GPS 37–39, 54, 79, 85, 106, 390, 443, 453
SatNav devices 391, 454

scales of navigation 3, 11, 15, 26, 43, 50–53, 65, 68, 74, 237, 242. *See also* phase transitions
long-distance 25, 63, 67, 73, 93
medium-scale 70, 71, 89
scene categorization 193, 194, 231, 244
scene representation 4, 79, 165, 166, 168–171, 178, 180
schizophrenia 281, 282
Sea Hero Quest 348, 352, 392, 402, 452, 456, 459
self-localization 88, 107, 166, 173, 176, 232, 233, 294, 299, 307
self-motion 83, 84, 144, 226, 294, 327, 342, 345, 382, 439, 440, 447
sensor technologies 37–39, 41–43, 47
sensory information processing 3, 73, 273, 278, 306, 327
age-related changes 343–345
auditory cues 75, 78, 79, 439, 440
idiothetic cues 176, 237, 328, 344, 349
sensory module 73–80, 90
sequence-based navigation 278, 342, 353, 442
shearwaters 15, 20, 21, 26
signal alignment 221, 234, 236, 237
single-cell recordings 108, 170, 177–179, 221, 346
Behnke-Fried electrode 237
CA3 neurons 224, 331, 332
grid cells 167, 230, 346
head direction 228
human 150, 167–171, 175
Neuropixels probe 238
SLAM (simultaneous localization and mapping) 37, 43, 50, 56, 88, 89, 260
NeuroSLAM 38, 57
RatSlam 38, 57
sleep 211, 241, 283, 294, 445
replay 233, 266, 267
social interaction 97, 102, 302, 385, 434, 453
sonar 37, 39–41, 79
songbirds 12, 77, 101
night-migratory 70, 74, 80, 106, 108
space-like nonspatial reasoning 308, 309
SPACE (Spatial Performance Assessment for Cognitive Evaluation) 7, 401, 405, 415, 422
construct validity 416, 417
design 406–415
preliminary results 406
sparrows 18
spatial representations 178, 242, 243, 260, 261, 274, 277–280, 322, 323, 343, 344
robotic navigation 43–45
speed cells 330, 331, 339, 349, 439
star compass 16–18, 77
strategy switching 72, 82, 89, 263, 266, 334, 352, 353, 384, 435, 455
age-related differences 352, 381–383
stretchy-bird experiment 307, 308
striatal complex, avian 98
striatum 324, 326, 329, 335–338, 342, 347, 353, 441
age-related differences 336
ventral 274, 302
subiculum 176, 212, 230, 236, 275, 281, 327, 333, 338, 349
successor representations 262, 308, 350
sun compass 12, 14–16, 69, 77, 82, 83
superior parietal lobule (SPL) 189, 190, 196, 231
synaptic plasticity 108, 325, 326, 433
age-related changes 321, 323, 326

T

taxon system 107
thalamus 146, 171, 275, 330, 342, 353
theta sequences 178, 180, 233, 237, 238, 240, 241, 265, 266, 335
timescale representations 265–268
T-maze 205, 207, 211, 215
topological map 42–44, 51, 53, 262
translational navigation research 444, 448–457
Triadic Navigation Framework 7, 431, 435–448
triangulation task 26, 206, 344, 389, 450
Trichoplax 95
turtles 66, 86, 91, 96
sea 65, 70, 76, 77

U

unicellular organisms 63, 67, 76, 90, 96, 100
directional sense 81
positional sense 86
universal principles of navigation 63, 65–69, 72–75, 85, 89, 95

V

vector-based navigation 4, 14, 21, 82, 139, 147–149, 232, 233, 262, 268
egocentric vector cells 151, 170, 298, 299
ventral tegmental area (VTA) 276, 278, 279
vestibular system 40, 78, 79, 83, 98, 203, 209, 228, 343
age-related changes 344, 345, 349, 441, 446
vicarious trial and error (VTE) 92, 233, 261, 265, 266
virtual reality (VR) technology 170, 191–193, 226, 243, 324, 347, 348, 382, 401. *See also* SPACE
Alzheimer disease 388, 389
desktop 345–347, 383, 448
immersive 324, 347, 351, 447, 452
limitations 228, 295, 383–385
vision 75, 78, 79, 84, 148, 189, 190, 294
age-related changes 324, 343, 441, 454
assessing 405, 408
depth perception 79, 194
development in rat 203
retina 12, 77–80, 85, 298
role of OPA 193–195
visual cliff task 194
vista environments 242–244
defined 237

W

water maze 205–208, 244
Morris 141, 238, 239, 242, 388, 451
virtual 333, 334, 336, 347
Williams syndrome 242–244
wind 2, 11, 13, 21–23, 76, 81, 91
W/M mazes 262
working memory 207, 260, 352, 442
age-related changes 334, 335

Strüngmann Forum Reports

Series Editor: Julia R. Lupp
Editorial Assistance: Aimée Gessner, Catherine Stephen
Lektorat: BerlinScienceWorks

Metabolic Neuropsychiatry
Edited by Dost Öngür, Judith M. Ford
Springer Cham
DOI: https://doi.org/10.1007/978-3-032-05630-6
ISBN: 978-3-032-05629-0

Sex and Gender: Toward Transforming Scientific Practice
Edited by L. Zachary DuBois, Anelis Kaiser Trujillo, Margaret M. McCarthy
Springer Cham
DOI: https://doi.org/10.1007/978-3-031-91371-6
ISBN: 978-3-031-91370-9

The Nature and Dynamics of Collaboration
Edited by Paul F. M. J. Verschure
DOI: https://doi.org/10.7551/mitpress/15533.001.0001
ISBN: 9780262548144

The Frontal Cortex: Organization, Networks, and Function
Edited by Marie T. Banich, Suzanne N. Haber and Trevor W. Robbins
DOI: https://doi.org/10.7551/mitpress/15679.001.0001
ISBN: 9780262549530

Digital Ethology: Human Behavior in Geospatial Context
Edited by Tomáš Paus and Hye-Chung Kum
DOI: https://doi.org/10.7551/mitpress/15532.001.0001
ISBN: 9780262548137

Migration Stigma: Understanding Prejudice, Discrimination, and Exclusion
Edited by Lawrence H. Yang, Maureen A. Eger and Bruce G. Link
DOI: https://doi.org/10.7551/mitpress/15529.001.0001
ISBN: 9780262548120

Exploring and Exploiting Genetic Risk for Psychiatric Disorders
Edited by Joshua A. Gordon and Elisabeth Binder
DOI: https://doi.org/10.7551/mitpress/15380.001.0001
ISBN electronic: 9780262377423

Intrusive Thinking: From Molecules to Free Will
Edited by Peter W. Kalivas and Martin P. Paulus
ISBN: 9780262542371

Deliberate Ignorance: Choosing Not to Know
Edited by Ralph Hertwig and Christoph Engel
ISBN 9780262045599

Youth Mental Health: A Paradigm for Prevention and Early Intervention
Edited by Peter J. Uhlhaas and Stephen J. Wood
ISBN: 9780262043977

N. S. Newcombe and K. Cheng (eds.), *Challenges in Navigation Research*,
Strüngmann Forum Reports, https://doi.org/10.1007/978-3-032-20563-6

The Neocortex
Edited by Wolf Singer, Terrence J. Sejnowski and Pasko Rakic
ISBN: 9780262043243

Interactive Task Learning: Humans, Robots, and Agents Acquiring New Tasks through Natural Interactions
Edited by Kevin A. Gluck and John E. Laird
ISBN: 9780262038829

Agrobiodiversity: Integrating Knowledge for a Sustainable Future
Edited by Karl S. Zimmerer and Stef de Haan
ISBN: 9780262038683

Rethinking Environmentalism: Linking Justice, Sustainability, and Diversity
Edited by Sharachchandra Lele, Eduardo S. Brondizio, John Byrne, Georgina M. Mace and Joan Martinez-Alier
ISBN: 9780262038966

Emergent Brain Dynamics: Prebirth to Adolescence
Edited by April A. Benasich and Urs Ribary
ISBN: 9780262038638

The Cultural Nature of Attachment: Contextualizing Relationships and Development
Edited by Heidi Keller and Kim A. Bard
ISBN (Hardcover): 9780262036900 ISBN (ebook): 9780262342865
Winner of the Ursula Gielen Global Psychology Book Award

Investors and Exploiters in Ecology and Economics: Principles and Applications
Edited by Luc-Alain Giraldeau, Philipp Heeb and Michael Kosfeld
ISBN (Hardcover): 9780262036122 ISBN (eBook): 9780262339797

Computational Psychiatry: New Perspectives on Mental Illness
Edited by A. David Redish and Joshua A. Gordon
ISBN: 9780262035422

Complexity and Evolution: Toward a New Synthesis for Economics
Edited by David S. Wilson and Alan Kirman
ISBN: 9780262035385

The Pragmatic Turn: Toward Action-Oriented Views in Cognitive Science
Edited by Andreas K. Engel, Karl J. Friston and Danica Kragic
ISBN: 9780262034326

Translational Neuroscience: Toward New Therapies
Edited by Karoly Nikolich and Steven E. Hyman
ISBN: 9780262029865

Trace Metals and Infectious Diseases
Edited by Jerome O. Nriagu and Eric P. Skaar
ISBN 9780262029193

Pathways to Peace: The Transformative Power of Children and Families
Edited by James F. Leckman, Catherine Panter-Brick and Rima Salah,
ISBN 9780262027984

Rethinking Global Land Use in an Urban Era
Edited by Karen C. Seto and Anette Reenberg
ISBN 9780262026901

Schizophrenia: Evolution and Synthesis
Edited by Steven M. Silverstein, Bita Moghaddam and Til Wykes,
ISBN 9780262019620

Cultural Evolution: Society, Technology, Language, and Religion
Edited by Peter J. Richerson and Morten H. Christiansen,
ISBN 9780262019750

Language, Music, and the Brain: A Mysterious Relationship
Edited by Michael A. Arbib
ISBN 9780262019620

Evolution and the Mechanisms of Decision Making
Edited by Peter Hammerstein and Jeffrey R. Stevens
ISBN 9780262018081

Cognitive Search: Evolution, Algorithms, and the Brain
Edited by Peter M. Todd, Thomas T. Hills and Trevor W. Robbins,
ISBN 9780262018098

Animal Thinking: Contemporary Issues in Comparative Cognition
Edited by Randolf Menzel and Julia Fischer
ISBN 9780262016636

Disease Eradication in the 21st Century: Implications for Global Health
Edited by Stephen L. Cochi and Walter R. Dowdle
ISBN 9780262016735

Better Doctors, Better Patients, Better Decisions: Envisioning Health Care 2020
Edited by Gerd Gigerenzer and J. A. Muir Gray
ISBN 9780262016032

Dynamic Coordination in the Brain: From Neurons to Mind
Edited by Christoph von der Malsburg, William A. Phillips and Wolf Singer,
ISBN 9780262014717

Linkages of Sustainability
Edited by Thomas E. Graedel and Ester van der Voet
ISBN 9780262013581

Biological Foundations and Origin of Syntax
Edited by Derek Bickerton and Eörs Szathmáry
ISBN 9780262013567

Clouds in the Perturbed Climate System: Their Relationship to Energy Balance, Atmospheric Dynamics, and Precipitation
Edited by Jost Heintzenberg and Robert J. Charlson
ISBN 9780262012874
Winner of the Atmospheric Science Librarians International Choice Award

Better Than Conscious? Decision Making, the Human Mind, and Implications For Institutions
Edited by Christoph Engel and Wolf Singer
ISBN 978-0-262-19580-5

MIX
Papier aus verantwortungsvollen Quellen
Paper from responsible sources
FSC® C105338
FSC
www.fsc.org

If you have any concerns about our products,
you can contact us on
ProductSafety@springernature.com

In case Publisher is established outside the EU,
the EU authorized representative is:
Springer Nature Customer Service Center GmbH
Europaplatz 3, 69115 Heidelberg, Germany

Printed by Libri Plureos GmbH
in Hamburg, Germany